Classification Criteria for Persistence and Bioaccumulation.

Water Solubility	
Very soluble	$S>10,000$ ppm
Soluble	$1,000<S<10,000$ ppm
Moderately soluble	$100<S<1,000$ ppm
Slightly soluble	$0.1<S<100$ ppm
Insoluble	$S<0.1$ ppm

Soil sorption	
Very strong sorption	$\mathrm{Log}\ K_{oc}>4.5$
Strong sorption	$4.5>\mathrm{Log}\ K_{oc}>3.5$
Moderate sorption	$3.5>\mathrm{Log}\ K_{oc}>2.5$
Low sorption	$2.5>\mathrm{Log}\ K_{oc}>1.5$
Negligible sorption	$1.5>\mathrm{Log}\ K_{oc}$

Biodegradation	
Rapid	>60% degradation over 1 week
Moderate	>30% degradation over 28 days
Slow	<30% degradation over 28 days
Very slow	<30% degradation over more than 28 days

Volatility (H in atm-m^3/mole)	
Very volatile	$H>10^{-1}$
Volatile	$10^{-1}>H>10^{-3}$
Moderately volatile	$10^{-3}>H>10^{-5}$
Slightly volatile	$10^{-5}>H>10^{-7}$
Nonvolatile	$10^{-7}>H$

Bioaccumulation potential	
High potential	$8.0>\mathrm{Log}\ K_{oc}>4.3$ or BCF>1000
Moderate potential	$4.3>\mathrm{Log}\ K_{ow}>3.5$ or 1000>BCF>250
Low potential	$3.5>\mathrm{Log}\ K_{oc}$ or 250>BCF

GREEN ENGINEERING

Environmentally Conscious Design of Chemical Processes

DAVID T. ALLEN
AND
DAVID R. SHONNARD

Prentice Hall PTR
Upper Saddle River, NJ 07458
www.phptr.com

ISBN 0-13-061908-6

90000

9 780130 619082

Library of Congress Cataloging-in-Publication Data

Allen, David T.
    Green engineering : environmentally conscious design of chemical processes / by David
Allen and David Shonnard.
        p. cm.
    Includes bibliographical references and index.
    ISBN 0-13-061908-6
        1. Environmental chemistry—Industrial applications.   2. Environmental management.
    I. Shonnard, David.   II. Title.

TP155.2.E58 A54 2002
660—dc21

2001034380

Editorial/Production Supervision: *Kerry Reardon*
Production Coordinator: *Anne R. Garcia*
Acquisitions Editor: *Bernard Goodwin*
US EPA Editor: *Sharon L. Austin*
Marketing Manager: *Dan DePasquale*
Manufacturing Manager: *Alexis R. Heydt-Long*
Cover Design Director: *Jerry Votta*
Cover Design: *Anthony Gemmellaro*
Cover Photo: *©2002 www.ArtToday.com*
Art Director: *Gail Cocker-Bogusz*
Editorial Assistant: *Michelle Vincenti*
Composition: *Pine Tree Composition*

© 2002 Prentice Hall PTR
Prentice-Hall, Inc.
Upper Saddle River, NJ 07458

Prentice Hall books are widely used by corporations and government agencies for training, marketing, and resale.

The publisher offers discounts on this book when ordered in bulk quantities.
For more information, contact Corporate Sales Department, phone: 800-382-3419;
fax: 201-236-7141; e-mail: corpsales@prenhall.com
or write: Prentice Hall PTR, Corporate Sales Department, One Lake Street, Upper Saddle River, NJ 07458

Printed in the United States of America on 30% post-consumer recycled fiber.

10 9 8 7 6 5 4 3 2 1

ISBN 0-13-061908-6

Pearson Education LTD.
Pearson Education Australia PTY, Limited
Pearson Education Singapore, Pte. Ltd.
Pearson Education North Asia Ltd.
Pearson Education Canada, Ltd.
Pearson Educacíon de Mexico, S.A. de C.V.
Pearson Education–Japan
Pearson Education Malaysia Pte. Ltd.
Pearson Education, Upper Saddle River, New Jersey

Contents

2   RISK CONCEPTS   35

3   ENVIRONMENTAL LAW AND REGULATIONS: FROM END-OF-PIPE TO POLLUTION PREVENTION   63

6 EVALUATING EXPOSURES   139

7 GREEN CHEMISTRY   177

8 EVALUATING ENVIRONMENTAL PERFORMANCE
DURING PROCESS SYNTHESIS   199

10   FLOWSHEET ANALYSIS FOR POLLUTION PREVENTION

309

11   EVALUATING THE ENVIRONMENTAL PERFORMANCE OF A FLOWSHEET

361

Preface

Chemical processes provide a diverse array of valuable products and materials used in applications ranging from health care to transportation and food processing. Yet these same chemical processes that provide products and materials essential to modern economies also generate substantial quantities of wastes and emissions. Managing these wastes costs tens of billions of dollars each year, and as emission and treatment standards continue to become more stringent, these costs will continue to escalate. In the face of rising costs and increasingly stringent performance standards, traditional end-of-pipe approaches to waste management have become less attractive and a strategy variously known as *environmentally conscious manufacturing, eco-efficient production,* or *pollution prevention* has been gaining prominence. The basic premise of this strategy is that avoiding the generation of wastes or pollutants can often be more cost effective and better for the environment than controlling or disposing of pollutants once they are formed.

The intent of this textbook is to describe environmentally preferable or "green" approaches to the design and development of processes and products. The idea of writing this textbook was conceived in 1997 by the staff of the Chemical Engineering Branch (CEB), Economics, Exposure and Technology Division (EETD), Office of Pollution Prevention and Toxics (OPPT) of the US EPA. In 1997, OPPT staff found that, although there was a growing technical literature describing "green" approaches to chemical product and process design, and a growing number of university courses on the subject, there was no standard textbook on the subject area of green engineering.

So, in early 1998, OPPT initiated the Green Engineering Project with the initial goal of producing a text describing "green" design methods suitable for inclusion in the chemical engineering curriculum.

Years of work, involving extensive interaction between chemical engineering educators and EPA staff, have resulted in this text. The text presents the "green" engineering tools that have been developed for chemical processes and is intended for senior-level chemical engineering students. The text begins (Chapters 1–4) with a basic introduction to environmental issues, risk concepts, and environmental regulations. This background material identifies the types of wastes, emissions, material use, and energy use to determine the environmental performance of chemical processes and products. Once the environmental performance targets have been defined, the design of processes with superior environmental performance can begin. Chapters 5–12 describe tools for assessing and improving the environmental performance of chemical processes. The structure of the chapters revolves around a hierarchy of design, beginning with tools for evaluating environmental hazards of chemicals, continuing through unit operation and flowsheet analysis, and concluding with the economics of environmental improvement projects. The final section of the text (Chapters 13 and 14) describes tools for improving product stewardship and improving the level of integration between chemical processes and other material processing operations.

It is our hope that this text will contribute to the evolving process of environmentally conscious design.

Draft manuscripts of this text have been used in senior-level engineering elective and required courses at the University of Texas at Austin, Michigan Technological University, the University of South Carolina, and West Virginia University. It is suggested, in a typical semester, all of the material in the text is presented. Portions of the textbook have been and can be used in a number of other chemical engineering courses as well as other engineering or environmental policy courses.

Dr. David T. Allen, University of Texas, Austin
Dr. David R. Shonnard, Michigan Technological University, Houghton
Nhan T. Nguyen, U.S. Environmental Protection Agency, Washington D.C.

About the Authors

David T. Allen is the Reese Professor of Chemical Engineering and the Director of the Center for Energy and Environmental Resources at the University of Texas at Austin. His research and teaching interests lie in environmental reaction engineering, particularly issues related to air quality and pollution prevention. He is the author of three books and over 100 papers in these areas. The quality of his research has been recognized by the National Science Foundation (through the Presidential Young Investigator Award), the AT&T Foundation (through an Industrial Ecology Fellowship), and the American Institute of Chemical Engineers (through the Cecil Award for contributions to environmental engineering). Dr. Allen's teaching has been recognized through awards given by both UCLA and the University of Texas. He received his B.S. degree in Chemical Engineering, with distinction, from Cornell University in 1979. His M.S. and Ph.D. degrees in Chemical Engineering were awarded by the California Institute of Technology in 1981 and 1983. He has held visiting faculty appointments at the California Institute of Technology, the Department of Energy, and the University of California, Santa Barbara.

David R. Shonnard is Associate Professor of Chemical Engineering at Michigan Technological University. He has a B.S. in Chemical Engineering from the University of Nevada-Reno, M.S. and PhD. Degrees in Chemical Engineering from the University of California at Davis, postdoctoral experience at Lawrence Livermore National Laboratory, and has been the visiting lecturer at the University of California at Berkley. His research and teaching interests are in the areas of environmental impact and risk assessment, process design and optimization, and environmental biotechnology. Dr. Shonnard is author of over 30 research publications and one edited book titled *Emerging Separation and Separative-Reactor Technologies for*

Process Waste Reduction: Adsorption and Membrane System, published by the American Institute of Chemical Engineers. He has published in engineering education journals on the topic of environmental aspects of Chemical Engineering. Environmental impact assessment software under development in his laboratory has been disseminated widely to faculty at other universities for use in the process design curriculum, and he is a 1998 recipient of a NSF/Lucent Technologies Foundation Industrial Ecology Research Fellowship.

Other Authors and Contributors

Dr. Paul Anastas serves in the National Security and International Activities Division in the White House Office of Science and Technology Policy (OSTP). In addition to bilateral international activities, Dr. Anastas is responsible for furthering international public-private cooperation in areas of Science for Sustainability such as green chemistry. Prior to coming to OSTP in October of 1999, Dr. Anastas had served, since 1989, as the Chief of the Industrial Chemistry Branch of the US Environmental Protection Agency. In 1991, he established the industry-government-university partnership Green Chemistry Program which was expanded to include basic research and the Presidential Green Chemistry Challenge Awards. Dr. Anastas coauthored Chapter 7 on Green Chemistry. Prior to joining the U.S. EPA, he worked as an industrial consultant to the chemical industry in the development of analytical and synthetic chemical methodologies. Dr. Anastas is the author/editor of nine scientific and technical books including *Green Chemistry: Theory and Practice* which has been translated into five languages. Dr. Anastas received his M.A. and Ph.D in Organic Chemistry from Brandeis University and his B.S. in chemistry from the University of Massachusetts at Boston.

Dr. Fred Arnold, PhD., joined EPA in 1994. Previously, he was an engineer at Westinghouse and an Assistant Professor of Engineering at the University of Oklahoma. Dr. Arnold has developed several exposure models during his tenure with CEB. Dr. Arnold coauthored Chapter 2, "Risk Concepts," and authored Chapter 6, "Evaluating Exposures," and contributed to several portions of the book, especially homework problem development. Dr. Arnold is a licensed P.E. in two states, received his PhD in Chemical Engineering from the University of Minnesota, his MBA from College of St. Thomas, and his J.D. from George Mason University.

Mr. John Blouin joined EPA in 1997 after 29 years of experience in the chemical process industry. He is experienced in the design, construction, and operation of pilot plants and new manufacturing processes for production of catalysts, organic, inorganic biochemical and biomedical products. Mr. Blouin holds 10 US patents as a result of his development work. Mr. Blouin contributed to several portions of the book, especially Chapter 2. John received a B.S. in Chemical Engineering from the University of Massachusetts, Lowell.

Ms. Gail Froiman joined EPA in 1989 and recently moved from the Chemical Engineering branch to the TRI program in the Office of Environmental Information (OEI). She started her career as a Research Engineer with Amoco Oil Process Development. Ms. Froiman also worked on advanced control systems in Amoco's Texas City Refinery, and as a process engineer at Vista Chemical. Ms. Froiman primarily authored Chapter 2 on Risk Concepts. Gail holds a B.S. in Chemical Engineering from Princeton University.

Mr. Scott Prothero joined EPA's Chemical Engineering Branch in 1990. He worked for four years as a process engineer in Monsanto Chemical Company's chlorobenzenes and nitroanalines departments before joining EPA. Mr. Prothero coauthored Chapter 8, "Evaluating Environmental Performance During Process Synthesis," providing most of the release data in the text. Mr. Prothero received a B.S. in Chemical Engineering from Washington University in 1985.

Ms. Kirsten Sinclair Rosselot is owner of Process Profiles, a consulting firm that specializes in environmental planning and management tools. She co-authored Chapter 10, "Flowsheet Analysis for Pollution Prevention"; Chapter 12, "Evaluating Environmental Costs and Benefits"; and Chapter 13, "Life-Cycle Concepts, Product Stewardship, and Green Engineering." Ms. Rosselot obtained her B.S. in chemical engineering with honors from the University of California, Los Angeles and is a licensed professional chemical engineer in the state of California.

Acknowledgments

The authors are grateful for the time and effort provided by US EPA staff in the development of this text. The authors especially thank Nhan Nguyen, Chief of the Chemical Engineering Branch, for conceiving the creation of this textbook and for his valuable guidance and comments, and Sharon L. Austin, the Green Engineering Coordinator, for her guidance and support. While the individuals who contributed to the writing of the text are recognized as chapter authors, many more individuals, both within and outside the EPA, were involved in editing and reviewing the text. To these reviewers, too numerous to mention individually, we extend our thanks. And finally, thanks to the many faculties who participated in workshops, in which the material in the text was presented and critiqued. Your input was enormously helpful.

In addition to the five branches and Immediate Office of the Economics, Exposure, and Technology Division (EETD), we want to give special thanks to many staff members from various divisions in the Office of Pollution Prevention and Toxics (OPPT), including the Risk Assessment Division (RAD), the Chemical Control Division (CCD), and the Pollution Prevention Division (PPD) for their valuable contributions. We would also like to thank staff members from other offices within EPA including Office of Research and Development (ORD), Office of Air Quality Planning and Standards (OAQPS), and the Office of General Counsel (OGC) for their input and advice.

PART I

A Chemical Engineer's Guide to Environmental Issues and Regulations

OVERVIEW

This text presents approaches and methodologies for evaluating and improving the environmental performance of chemical processes and chemical products. Prerequisites for understanding this presentation are a basic knowledge of environmental issues and environmental regulations. The group of chapters listed below present this background material at a level suitable for senior to graduate-level chemical engineering students.

1. Chapter 1 presents an introduction to environmental issues. The issues range from global to local, and the emphasis is on the types of wastes and emissions that drive the environmental impacts.
2. Chapter 2 presents the concept of environmental risk. Risk frameworks are commonly used to assess the relative significance of environmental concerns.
3. Chapter 3 describes the regulatory frameworks that have emerged to control environmental risks. The focus is on key statutes that affect chemical engineers and the gradual evolution of regulatory structures from an end-of-pipe focus to a more flexible pollution prevention approach.
4. Chapter 4 summarizes the many contributions that chemical engineers can make in addressing environmental issues, particularly focusing on the role played by chemical process and product design engineers.

More specifically, Chapter 1 provides a general introduction to the data and science underlying environmental issues such as global warming, stratospheric ozone depletion, ecosystem health, atmospheric and aquatic acidification, smog formation, hazardous waste generation, and non-hazardous waste generation. These environmental issues will be considered throughout the remainder of the book in evaluating the environmental performance of chemical processes and products. A basic understanding of the nature of these environmental concerns is important. Concepts of risk will also be used throughout the text, so Chapter 2 presents basic definitions of risk. Chapter 2 also presents a qualitative discussion of the building blocks of risk assessment—emission estimation, environmental fate and transport evaluation, exposure quantification, and dose-response relationships. Chapter 3 provides an overview of the regulatory framework that has been built in the United States to address the environmental issues described in Chapter 1. The main focus is on federal legislation that has a major impact on chemical processes and products. A complete treatment of the topics covered in Chapters 1–3 could fill an entire curriculum but the goal in this volume is to condense this material into a form that can be covered in a few lectures. The treatment of individual topics is therefore brief. References to more complete descriptions are provided.

After reviewing Chapters 1–3, students should have a basic understanding of the environmental issues that a chemical engineer may need to address. The final chapter of this part describes the role that chemical process engineers and chemical product designers can play in solving these environmental problems.

CHAPTER 1

An Introduction to Environmental Issues

by
David R. Shonnard

1.1 INTRODUCTION

Environmental issues gained increasing prominence in the latter half of the 20th century. Global population growth has led to increasing pressure on worldwide natural resources including air and water, arable land, and raw materials, and modern societies have generated an increasing demand for the use of industrial chemicals. The use of these chemicals has resulted in great benefits in raising the standard of living, prolonging human life and improving the environment. But as new chemicals are introduced into the marketplace and existing chemicals continue to be used, the environmental and human health impacts of these chemicals has become a concern. Today, there is a much better understanding of the mechanisms that determine how chemicals are transported and transformed in the environment and what their environmental and human health impacts are, and it is now possible to incorporate environmental objectives into the design of chemical processes and products.

The challenge for future generations of chemical engineers is to develop and master the technical tools and approaches that will integrate environmental objectives into design decisions. The purpose of Chapter 1 is to present a brief introduction to the major environmental problems that are caused by the production and use of chemicals in modern industrial societies. With each environmental problem introduced, the chemicals or classes of chemicals implicated in that problem are identified. Whenever possible, the chemical reactions or other mechanisms responsible for the chemical's impact are explained. Trends in the production, use, or release of those chemicals are shown. Finally, a brief summary of adverse health effects is presented. This chapter's intent is to present the broad range of environmental issues which may be encountered by chemical engineers. Chapter 3 contains a review of selected environmental regulations that may affect chemical engineers. It is hoped that this information will elevate the environmental awareness of chemical engineers and will lead to more informed decisions regarding the design, production, and use of chemicals.

1.2   THE ROLE OF CHEMICAL PROCESSES AND CHEMICAL PRODUCTS

In this text, we cover a number of design methodologies for preventing pollution and reducing risks associated with chemical production. Figure 1.2-1 shows conceptually how chemical processes convert raw materials into useful products with the use of energy. Wastes generated in chemical manufacturing, processing, or use are released to the environment through discharges to streams or rivers, exhausting into the air, or disposal in a landfill. Often, the waste streams are treated prior to discharge.

We may be exposed to waste stream components by three routes: dermal (skin contact), inhalation, and ingestion. The route and magnitude of exposure is influenced by the physical, chemical, and reactivity properties of the waste stream components. In addition, waste components may affect the water quality of streams and rivers, the breathability of ambient air, and the well-being of terrestrial flora and fauna. What information will a chemical engineer need to make informed pollution prevention and risk reduction decisions? A few generalized examples will aid in answering such a question.

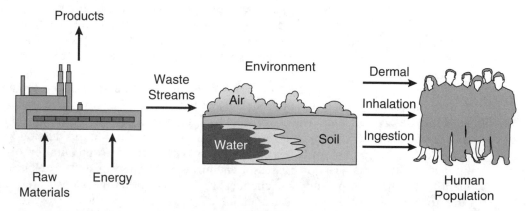

Figure 1.2-1   Generalized scenario for exposure by humans to environmental pollutants released from chemical processes.

Formulation of an Industrial Cleaner

Company A plans to formulate a concentrated, industrial cleaner, and needs to incorporate a solvent within the product to meet customer performance criteria. A number of solvents are identified that will meet cost and performance specification. Further, Company A knows that the cleaning product (with the solvent) will be discharged to water and is concerned about the aquatic toxicity of the solvent. The company conducts a review of the pertinent data to aid in making the choice. In aquatic environments, a chemical will have low risk potential if it has the following characteristics:

a) High Henry's Law constant (substance will volatilize into the air rather than stay in the water)
b) High biodegradation rate (it will dissipate before exerting adverse health effects)
c) Low fish toxicity parameter (a high value of the concentration lethal to a majority of test organisms or LC_{50})
d) Low Bioconcentration Factor, BCF (low tendency for chemicals to partition into the fatty tissue of fish, leading to exposure and adverse health effects upon consumption by humans)

Company A assembles the data and chooses a solvent with the least adverse environmental consequences. Methods are presented in this text to provide estimates of environmental properties. In addition, measured data for some of these properties are tabulated.

Formulation of a Paint Solvent

Company B is formulating a paint for an automobile refinishing. The formulation must contain fast-drying solvents to ensure uniform coating during application. These fast-drying solvents volatilize when the paint is sprayed and are exhausted by a fan. Workers in the booths may be exposed to the solvents during application of the paint and nearby residents may inhale air contaminated by the exhausted solvents.

The company is concerned about the air releases and problems that arise with worker exposure to toxic agents and impact to air quality. A number of solvents having acceptable cost and coating performance characteristics have been identified. A chemical will have low risk potential in the air if it has the following characteristics:

a) Low toxicity properties (high Reference Dose [RfD] for inhalation toxicity to humans or a low cancer potency), and
b) Low reactivity for smog formation (ground level ozone production).

Candidate solvents may be screened for these properties to identify the environmentally optimal candidate.

Choice of Refrigerant for a Low-Temperature Condenser

A chemical engineer is in charge of redesigning a chemical process for expanded capacity. One part of the process involves a vapor stream heat exchanger and a refrigeration cycle. In the redesign, the company decides to use a refrigerant having low potential for stratospheric ozone depletion. In addition, the engineer must also ensure that the refrigerant possesses acceptable performance characteristics such as thermodynamic properties, materials compatibility, and thermal stability. From the list of refrigerants that meet acceptable process performance criteria, the engineer estimates or finds tabulated data for

a) atmospheric reaction-rate constant,
b) global warming potential, and
c) ozone depletion potential.

From an environmental perspective, an ideal refrigerant would have low ozone depletion and global warming potentials while not persisting in the atmosphere.

These three examples illustrate the role the chemical engineer plays by assessing the potential environmental impacts of product and process changes. One important impact the chemical engineer must be aware of is human exposure, which can occur by a number of routes. The magnitude of exposure can be affected by any number of reactive processes occurring in the air, water, and soil compartments in the environment. The severity of the toxic response in humans is determined by the toxicology properties of the emitted chemicals. The chemical engineer must also be aware of the life cycle of a chemical. What if the chemical volatilizes but is an air toxicant? What if the biodegradation products (as, for example, with DDT) are the real concern? For example, terpenes, a class of chemical compounds, were touted as a replacement for chlorinated solvents to avoid stratospheric ozone depletion, but terpenes are highly reactive and volatile and can contribute to photochemical smog formation.

The next sections present a wide range of environmental problems caused by human activities. Trends in the magnitude of these problems are shown in tabular or graphical form, and contributions by industrial sources are mentioned whenever possible. Later chapters develop risk assessment and reduction methods to help answer the questions posed in the previous examples.

1.3 AN OVERVIEW OF MAJOR ENVIRONMENTAL ISSUES

The next several sections present an overview of major environmental issues. These issues are not only of concern to the general public, but are challenging problems for the chemical industry and for chemical engineers. The goal of the following sections is to provide an appreciation of the impacts that human activities can have on the environment. Also, the importance of healthy ecosystems are illustrated as they affect human welfare, the availability of natural resources, and economic sustainability.

When considering the potential impact of any human activity on the environment, it is useful to regard the environment as a system containing interrelated subprocesses. The environment functions as a sink for the wastes released as a result of human activities. The various subsystems of the environment act upon these wastes, generally rendering them less harmful by converting them into chemical forms that can be assimilated into natural systems. It is essential to understand these natural waste conversion processes so that the capacity of these natural systems is not exceeded by the rate of waste generation and release.

The impact of waste releases on the environment can be global, regional, or local in scope. On a global scale, man-made (anthropogenic) greenhouse gases, such as methane and carbon dioxide, are implicated in global warming and climate change. Hydrocarbons released into the air, in combination with nitrogen oxides originating from combustion processes, can lead to air quality degradation over urban areas and extend for hundreds of kilometers. Chemicals disposed of in the soil can leach into underground water and reach groundwater sources, having their primary impact locally, near to the point of release. The timing of pollution releases and rates of natural environmental degradation can affect the degree of impact that these substances have. For example, the build-up of greenhouse gases has occurred over several decades. Consequently, it will require several decades to reverse or stall the build-up that has already occurred. Other releases, such as those that impact urban air quality, can have their primary impact over a period of hours or days.

The environment is also a source of raw materials, energy, food, clean air, water, and soil for useful human purposes. Maintenance of healthy ecosystems is therefore essential if a sustainable flow of these materials is to continue. Depletion of natural resources due to population pressures and/or unwise resource management threatens the availability of these materials for future use.

The following sections of Chapter 1 provide a short review of environmental issues, including global energy consumption patterns, environmental impacts, ecosystem health, and natural resource utilization. Much of the material presented in this section is derived from the review by Phipps (1996) and from US EPA reports (US EPA, 1997).

1.4   GLOBAL ENVIRONMENTAL ISSUES

1.4.1   Global Energy Issues

The availability of adequate energy resources is necessary for most economic activity and makes possible the high standard of living that developed societies enjoy. Although energy resources are widely available, some such as oil and coal are non-renewable, and others, such as solar, although inexhaustible, are not currently cost effective for most applications. An understanding of global energy usage patterns, energy conservation, and the environmental impacts associated with the production and use of energy are therefore very important.

Often, primary energy sources such as fossil fuels must be converted into another form such as heat or electricity. As the Second Law of Thermodynamics dictates, such conversions will be less than 100% efficient. An inefficient user of primary energy is the typical automobile, which converts into motion about 10% of the energy available in crude oil. Some other typical conversion efficiencies are given in Example 1.4-1, below.

Example 1.4-1

Efficiency of Primary and Secondary Energy: Determine the efficiency of primary energy utilization for a pump. Assume the following efficiencies in the energy conversion:

- Crude oil to fuel oil is 90% (.90)
- Fuel oil to electricity is 40% (.40)
- Electricity transmission and distributions is 90% (.90)
- Conversion of electrical energy into mechanical energy of the fluid being pumped is 40% (.40)

Solution: The overall efficiency for the primary energy source is the product of all the individual conversion efficiencies.

$$\text{Overall Efficiency} = (.90)(.40)(.90)(.40) = (.13) \text{ or } 13\%$$

The global use of energy has steadily risen since the dawn of the industrial revolution. More recently, from 1960 to 1990 world energy requirements rose from 3.3 to 5.5 gtoe (gigatonnes oil equivalent) (WEC 1993). Currently, fossil fuels make up roughly 85% of the world's energy consumption (EIA 1998a,b), while renewable sources such as hydroelectric, solar, and wind power account for only about 8% of the power usage. Nuclear power provides roughly 6% of the world energy demand, and its contribution varies from country to country. The United States meets about 20% of its electricity demand, Japan 28%, and Sweden almost 50% from nuclear power

The disparity in global energy use is illustrated by the fact that 65–70% of the energy is used by about 25% of the world's population. Energy consumption per capita is greatest in industrialized regions such as North America, Europe, and Japan. The average citizen in North America consumes almost fifteen times the energy consumed by a resident in sub-Saharan Africa. (However, the per capita income of the U.S. is 33 times greater than that in sub-Saharan Africa.)

Another interesting aspect of energy consumption by industrialized countries and the developing world is the trend in energy efficiency, the energy consumed per unit of economic output. The amount of energy per unit of gross domestic product (GDP) has fallen in industrialized countries and is expected to continue to fall in the future. The U.S. consumption of energy per unit of GDP has fallen

30% from 1980–1995 (Organization for Economic Cooperation and Development (OECD) Environmental Data Compendium). Future chemical engineers will need to recognize the importance of energy efficiency in process design.

World energy consumption is expected to grow by 75% in the year 2020 compared to 1995. The highest growth in energy consumption is predicted to occur in Southeast and East Asia, which contained 54% of the world population in 1997. Energy consumption in the developing countries is expected to overtake that of the industrialized countries by 2020.

Many environmental effects are associated with energy consumption. Fossil fuel combustion releases large quantities of carbon dioxide into the atmosphere. During its long residence time in the atmosphere, CO_2 readily absorbs infrared radiation contributing to global warming. Further, combustion processes release oxides of nitrogen and sulfur oxide into the air where photochemical and/or chemical reactions can convert them into ground level ozone and acid rain. Hydropower energy generation requires widespread land inundation, habitat destruction, alteration in surface and groundwater flows, and decreases the acreage of land available for agricultural use. Nuclear power has environmental problems linked to uranium mining and spent nuclear rod disposal. "Renewable fuels" are not benign either. Traditional energy usage (wood) has caused widespread deforestation in localized regions of developing countries. Solar power panels require energy-intensive use of heavy metals and creation of metal wastes. Satisfying future energy demands must occur with a full understanding of competing environmental and energy needs.

1.4.2   Global Warming

The atmosphere allows solar radiation from the sun to pass through without significant absorption of energy. Some of the solar radiation reaching the surface of the earth is absorbed, heating the land and water. Infrared radiation is emitted from the earth's surface, but certain gases in the atmosphere absorb this infrared radiation, and re-direct a portion back to the surface, thus warming the planet and making life, as we know it, possible. This process is often referred to as the *greenhouse effect.* The surface temperature of the earth will rise until a radiative equilibrium is achieved between the rate of solar radiation absorption and the rate of infrared radiation emission. Human activities, such as fossil fuel combustion, deforestation, agriculture and large-scale chemical production, have measurably altered the composition of gases in the atmosphere. Some believe that these alterations will lead to a warming of the earth-atmosphere system by enhancement of the greenhouse effect. Figure 1.4-1 summarizes the major links in the chain of environmental cause and effect for the emission of greenhouse gases.

Table 1.4-1 is a list of the most important greenhouse gases along with their anthropogenic (man-made) sources, emission rates, concentrations, residence times in the atmosphere, relative radiative forcing efficiencies, and estimated contribution to global warming. The primary greenhouse gases are water vapor, car-

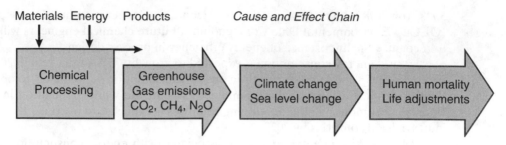

Figure 1.4-1   Greenhouse emission from chemical processes and the major cause and environmental effect chain.

bon dioxide, methane, nitrous oxide, chlorofluorocarbons, and tropospheric ozone. Water vapor is the most abundant greenhouse gas, but is omitted because it is generally not from anthropogenic sources. Carbon dioxide contributes significantly to global warming due to its high emission rate and concentration. The major factors contributing to global warming potential of a chemical are infrared absorptive capacity and residence time in the atmosphere. Gases with very high absorptive capacities and long residence times can cause significant global warming even though their concentrations are extremely low. A good example of this phenomenon is the chlorofluorocarbons, which are, on a pound-for-pound basis, more than 1000 times more effective as greenhouse gases than carbon dioxide.

For the past four decades, measurements of the accumulation of carbon dioxide in the atmosphere have been taken at the Mauna Loa Observatory in Hawaii, a location far removed from most human activity that might generate carbon dioxide. Based on the current level of CO_2 of 360 parts-per-million (ppm), levels of CO_2 are increasing at the rate of 0.5%/year (from about 320 ppm in 1960). Atmospheric concentrations of other greenhouse gases have also risen. Methane has increased from about 700 ppb in pre-industrial times to 1721 ppb in 1994, while N_2O rose from 275 to 311 ppb over the same period. While it is clear that atmospheric concentrations of carbon dioxide, and other global warming gases are increasing, there is significant uncertainty regarding the magnitude of the effect on climate that these concentration changes might induce (interested readers should consult the reports of the Intergovernmental Panel on Climate Change (IPCC), see references at the end of the chapter).

1.4.3 Ozone Depletion in the Stratosphere

There is a distinction between "good" and "bad" ozone (O_3) in the atmosphere. Tropospheric ozone, created by photochemical reactions involving nitrogen oxides and hydrocarbons at the earth's surface, is an important component of smog. A po-

Table 1.4-1 Greenhouse Gases and Global Warming Contribution. M stands for million. Phipps (1996), IPCC (1996).

Gas	Source (Natural and Anthropogenic)	Estimated Anthropogenic Emission Rate	Pre-Industrial Global Concentration	Approximate Current Concentration	Estimated Residence Time in the Atmosphere	Radiative Forcing Efficiency (absorptivity capacity) ($CO_2 = 1$)	Estimated Contribution to Global Warming
Carbon Dioxide (CO_2)	Fossil fuel combustion; deforestation	6,000 M tons/yr	280 ppm	355 ppm	50–200 yrs	1	50 %
Methane (CH_4)	Anaerobic decay (wetlands, landfills, rice paddies), ruminants, termites, natural gas, coal mining, biomass burning	300– 400 M tons/yr	0.8 ppm	1.7 ppm	10 yrs	58	12–19 %
Nitrous Oxide (N_2O)	Estuaries and tropical forests; agricultural practices, deforestation, land clearing, low-temperature fuel combustion	4–6M tons/yr	0.285 ppm	0.31 ppm	140–190 yrs	206	4–6 %
Chlorofluoro-carbons (CFC-11 & CFC-12)	Refrigerants, air conditioners, foam-blowing agents, aerosol cans, solvents	1 M tons/yr	0	.0004– .001 ppm	65–110 yrs	4,860	17–21 %
Tropospheric Ozone (O_3)	Photochemical reactions involving VOCs and NOx from transportation and industrial sources	not emitted directly	NA	.022 ppm	hours– days	2,000	8 %

tent oxidant, ozone irritates the breathing passages and can lead to serious lung damage. Ozone is also harmful to crops and trees. Stratospheric ozone, found in the upper atmosphere, performs a vital and beneficial function for all life on earth by absorbing harmful ultraviolet radiation. The potential destruction of this stratospheric ozone layer is therefore of great concern.

The stratospheric ozone layer is a region in the atmosphere between 12 and 30 miles (20–50 km) above ground level in which the ozone concentration is elevated compared to all other regions of the atmosphere. In this low-pressure region, the concentration of O_3 can be as high as 10 ppm (about 1 out of every 100,000 molecules). Ozone is formed at altitudes between 25 and 35 km in the tropical regions near the equator where solar radiation is consistently strong throughout the year. Because of atmospheric motion, ozone migrates to the polar regions and its highest concentration is found there at about 15 km in altitude. Stratospheric ozone concentrations have steadily declined over the past 20 years.

Ozone equilibrates in the stratosphere as a result of a series of natural formation and destruction reactions that are initiated by solar energy. The natural cycle of stratospheric ozone creation and destruction has been altered by the introduction of man-made chemicals. Two chemists, Mario Molina and Sherwood Rowland of the University of California, Irvine, received the 1995 Nobel Prize for Chemistry for their discovery that chlorofluorocarbons (CFCs) take part in the destruction of atmospheric ozone. CFCs are highly stable chemical structures composed of carbon, chlorine, and fluorine. One important example is trichlorofluoromethane, CCl_3F, or CFC-11.

CFCs reach the stratosphere due to their chemical properties; high volatility, low water solubility, and persistence (non-reactivity) in the lower atmosphere. In the stratosphere, they are photo-dissociated to produce chlorine atoms, which then catalyze the destruction of ozone (Molina and Rowland, 1974):

$$Cl + O_3 \rightarrow ClO + O_2$$
$$ClO + O \rightarrow O_2 + Cl$$

$$\overline{}$$

$$O_3 + O \rightarrow O_2 + O_2$$

The chlorine atom is not destroyed in the reaction and can cause the destruction of up to 10,000 molecules of ozone before forming HCl by reacting with hydrocarbons. The HCl eventually precipitates from the atmosphere. A similar mechanism as outlined above for chlorine also applies to bromine, except that bromine is an even more potent ozone destroying compound. Interestingly, fluorine does not appear to be reactive with ozone. Figure 1.4-2 summarizes the major steps in the environmental cause and effect chain for ozone-depleting substances.

CFC's were first introduced in the 1930's for use as refrigerants and solvents. By the 1950's significant quantities were released into the atmosphere. Releases

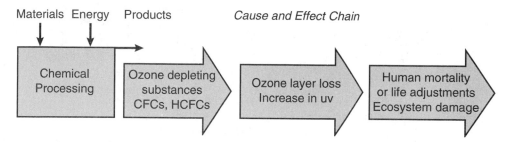

Figure 1.4-2 Ozone-depleting chemical emissions and the major steps in the environmental cause and effect chain.

reached a peak in the mid-eighties (CFC-11 and CFC-12 combined were about 700 million kg). Releases have been decreasing since about 1990 (1995 data: 300 million kg, same level as 1966). The Montreal Protocol, which instituted a phase-out of ozone-depleting chemicals, is the primary reason for the declining trend. Figures 1.4-3 and 1.4-4 show recent trends in the production of several CFCs and in the resulting remote tropospheric concentrations from releases. The growth in accumulation of CFCs in the environment has been halted as a result of the Montreal Protocol.

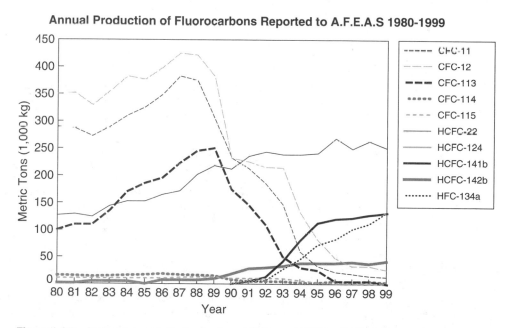

Figure 1.4-3   Recent trends in the production of CFCs and HCFCs. (AFEAS 2000)

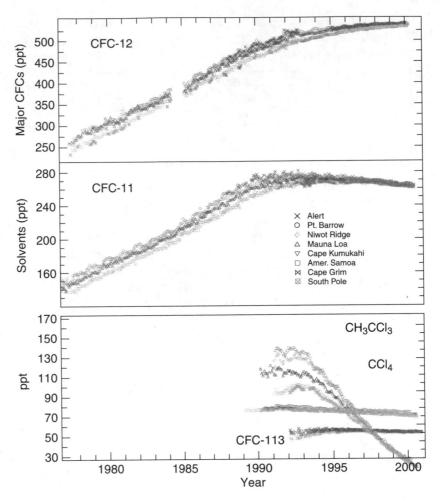

Figure 1.4-4 Trends of controlled ozone-depleting substances under the 1987 Montreal Protocol and its subsequent amendments from the NOAA/CMDL Network. Mixing ratios (dry mole fraction) of CFC-12 are shown in the top panel from flasks except for the 1993-1994 where in situ values are used. Mixing ratios of CFC-11 (collected from flasks) in the middle panel follow the pattern of total equivalent chlorine in the troposphere. Mixing ratios of CH_3CCl_3 and CCl_4 from in situ measurements and CFC-113 from flask measurements are shown in the bottom panel (updated data from Elkins et al., 1993 and Montzka et al., 1999).

1.5 AIR QUALITY ISSUES

Air pollution arises from a number of sources, including stationary, mobile, and area sources. Stationary sources include factories and other manufacturing processes. Mobile sources are automobiles, other transportation vehicles, and recreational vehicles such as snowmobiles and watercraft. Area sources are emissions associated with human activities that are not considered mobile or stationary. Examples of area sources include emissions from lawn and garden equipment, and residential heating. Pollutants can be classified as primary, those emitted directly to the atmosphere, or secondary, those formed in the atmosphere after emission of precursor compounds. Photochemical smog (the term originated as a contraction of smoke and fog) is an example of secondary pollution that is formed from the emission of volatile organic compounds (VOCs) and nitrogen oxides (NOx), the primary pollutants. Air quality problems are closely associated with combustion processes occurring in the industrial and transportation sectors of the economy. Smog formation and acid rain are also closely tied to these processes. In addition, hazardous air pollutants, including chlorinated organic compounds and heavy metals, are emitted in sufficient quantities to be of concern. Figure 1.5-1 shows the primary environmental cause and effect chain leading to the formation of smog.

1.5.1   Criteria Air Pollutants

Congress in 1970 passed the Clean Air Act which charged the Environmental Protection Agency (EPA) with identifying those air pollutants which are most deleterious to public health and welfare, and empowered EPA to set maximum allowable ambient air concentrations for these criteria air pollutants. EPA identified six substances as criteria air pollutants (Table 1.5-1) and promulgated primary and secondary standards that make up the National Ambient Air Quality Standards (NAAQS). Primary standards are intended to protect the public health with an adequate margin of safety. Secondary standards are meant to protect public welfare, such as damage to crops, vegetation, and ecosystems or reductions in visibility.

Criteria pollutants are a set of individual chemical species that are considered to have potential for serious adverse health impacts, especially in susceptible populations. These pollutants have established health-based standards and were among the first airborne pollutants to be regulated, starting in the early 1970's.

Since the establishment of the NAAQS, overall emissions of criteria pollutants have decreased 31% despite significant growth in the U.S. population and economy. Even with such improvements, more than a quarter of the U. S. population lives in locations with ambient concentrations of criteria air pollutants above the NAAQS (National Air Quality Emission Trends Report, *www.epa.gov/oar/aqtrnd97/*). These criteria pollutants and their health effects will be discussed next.

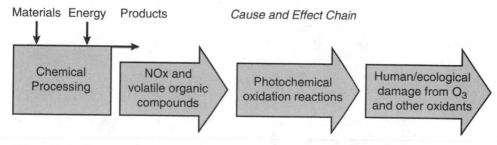

Figure 1.5-1   Environmental cause and effect chain for photochemical smog formation.

TABLE 1.5-1   Criteria Pollutants and the National Ambient Air Quality Standards.

	Primary Standard (Human Health Related)		Secondary (Welfare Related)	
Pollutant	*Type of Average*	*Concentration[a]*	*Type of Average*	*Concentration*
CO				
[-38%][h]	8-hour[b]	9 ppm (10 mg/m^3)	No Secondary Standard	
{-25%}[i]	1-hour[b]	35 ppm (40 mg/m^3)	No Secondary Standard	
Pb				
[-67%]	Maximum Quarterly	1.5 mg/m^3	Same as Primary Standard	
{-44%}	Average			
NO$_2$				
[-14%]	Annual Arithmetic Mean	0.053 ppm	Same as Primary Standard	
{-1%}		(100 µg/m^3)		
O$_3$				
[-19%]	1-hour[c]	0.12 ppm	Same as Primary Standard	
		(235 µg/m^3)		
	8-hour[d]	0.08 ppm	Same as Primary Standard	
		(157 µg/m^3)		
PM$_{10}$				
[-26%]	Annual Arithmetic Mean	50 µg/m^3	Same as Primary Standard	
{-12%}	24-hour[e]	150 µg/m^3	Same as Primary Standard	
PM$_{2.5}$	Annual Arithmetic Mean[f]	15 µg/m^3	Same as Primary Standard	
	24-hour[g]	65 µg/m^3	Same as Primary Standard	
SO$_2$				
[-39%]	Annual Arithmetic Mean	0.03 ppm	3-hour[b]	0.50 ppm
{-12%}		(80 µg/m^3)		(1,300 µg/m^3)
	24-hour[b]	0.14 ppm		
		(365 µg/m^3)		

[a]Parenthetical value is an equivalent mass concentration.
[b]Not to be exceeded more than once per year.
[c]Not to be exceeded more than once per year on average.
[d]3-year average of annual 4th highest concentration.
[e]The pre-existing form is exceedance-based. The revised form is the 99th percentile.
[f]Spatially averaged over designated monitors.
[g]The form is the 98th percentile.
[h]Air quality concentration, % change 1988–1997.
[i]Emissions, % change 1988–1997.

Source: 40 Code of Federal Register (CFR) Part 50, revised standards issued July 18, 1997.
Adapted from U.S. EPA (1998).

1.5.1.1   NOx, Hydrocarbons, and VOCs—Ground-Level Ozone

Ground-level ozone is one of the most pervasive and intractable air pollution problems in the United States. We should again differentiate between this "bad" ozone created at or near ground level (tropospheric) from the "good" or stratospheric ozone that protects us from UV radiation.

Ground-level ozone, a component of photochemical smog, is actually a secondary pollutant in that certain precursor contaminants are required to create it. The precursor contaminants are nitrogen oxides (NOx, primarily NO and NO_2) and hydrocarbons. The oxides of nitrogen along with sunlight cause ozone formation, but the role of hydrocarbons is to accelerate and enhance the accumulation of ozone.

Oxides of nitrogen (NOx) are formed in high-temperature industrial and transportation combustion processes. In 1997, transportation sources accounted for 49.2% and non-transportation fuel combustion contributed 45.4% of total NOx emissions. Health effects associated with short-term exposure to NO_2 (less than three hours at high concentrations) are increases in respiratory illness in children and impaired respiratory function in individuals with pre-existing respiratory problems. Figure 1.5-2 shows the NOx emission trends from 1988 to 1997 for major source categories. Industry makes a significant contribution to the "fuel combustion" category from the energy requirements of industrial processes.

Major sources of hydrocarbon emissions are the chemical and oil refining industries, and motor vehicles. In 1997, industrial processes accounted for 51.2% while the transportation sector contributed 39.9% of the total of man-made (non-biogenic) hydrocarbon sources. Solvents comprise 66% of the industrial emissions and 34% of total VOC emissions. It should be noted that there are natural (biogenic) sources of HCs/VOCs, such as isoprene and monoterpenes that can contribute significantly to regional hydrocarbon emissions and low-level ozone levels. Figure 1.5-3 summarized recent trends in VOC emissions.

Ground-level ozone concentrations are exacerbated by certain physical and atmospheric factors. High-intensity solar radiation, low prevailing wind speed (dilution), atmospheric inversions, and proximity to mountain ranges or coastlines (stagnant air masses) all contribute to photochemical smog formation.

Human exposure to ozone can result in both acute (short-term) and chronic (long-term) health effects. The high reactivity of ozone makes it a strong lung irritant, even at low concentrations. Formaldehyde, peroxyacetylnitrate (PAN), and other smog-related oxygenated organics are eye irritants. Ground-level ozone also affects crops and vegetation adversely when it enters the stomata of leaves and destroys chlorophyll, thus disrupting photosynthesis. Finally, since ozone is an oxidant, it causes materials with which it reacts to deteriorate, such as rubber and latex painted surfaces.

1.5.1.2   Carbon Monoxide (CO)

CO is a colorless, odorless gas formed primarily as a by-product of incomplete combustion. The major health hazard posed by CO is its capacity to bind with hemoglobin in the blood stream and thereby reduce the oxygen-carrying ability of

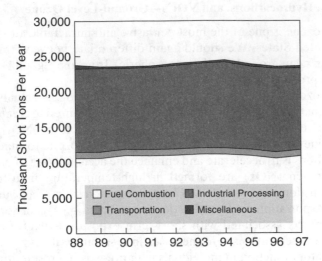

Figure 1.5-2 Emission trends for major categories of NOx emission sources (US EPA 1998).

the blood. Transportation sources account for the bulk (76.6%) of total national CO emissions. As noted in Table 1.5-1, ambient CO concentrations have decreased significantly in the past two decades, primarily due to improved control technologies for vehicles. Areas with high traffic congestion generally will have high ambient CO concentrations. High localized and indoor CO levels can come from cigarettes (second-hand smoke), wood-burning fireplaces, and kerosene space heaters.

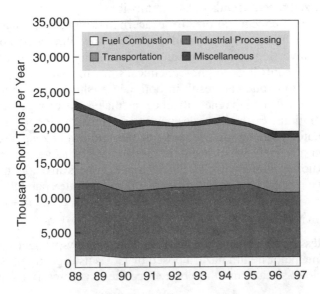

Figure 1.5-3 Emission trends for major categories of VOC emission sources (US EPA 1998).

1.5.1.3 Lead

Lead in the atmosphere is primarily found in fine particulates, up to 10 microns in diameter, which can remain suspended in the atmosphere for significant periods of time. Tetraethyl lead $((CH_3CH_2)_4-Pb)$ was used as an octane booster and antiknock compound for many years before its full toxicological effects were understood. The Clean Air Act of 1970 banned all lead additives and the dramatic decline in lead concentrations and emissions has been one of the most important yet unheralded environmental improvements of the past twenty-five years. (Table 1.5-1). In 1997, industrial processes accounted for 74.2% of remaining lead emissions, with 13.3% resulting from transportation, and 12.6% from non-transportation fuel combustion (US EPA 1998).

Lead also enters waterways in urban runoff and industrial effluents, and adheres to sediment particles in the receiving water body. Uptake by aquatic species can result in malformations, death, and aquatic ecosystem instability. There is a further concern that increased levels of lead can occur locally due to acid precipitation that increases lead's solubility in water and thus its bioavailability. Lead persists in the environment and is accumulated by aquatic organisms.

Lead enters the body by inhalation and ingestion of food (contaminated fish), water, soil, and airborne dust. It subsequently deposits in target organs and tissue, especially the brain. The primary human health effect of lead in the environment is its effect on brain development, especially in children. There is a direct correlation between elevated levels of lead in the blood and decreased IQ, especially in the urban areas of developing countries that have yet to ban lead as a gasoline additive.

1.5.1.4 Particulate Matter

Particulate matter (PM) is the general term for microscopic solid or liquid phase (aerosol) particles suspended in air. PM exists in a variety of sizes ranging from a few Angstroms to several hundred micrometers. Particles are either emitted directly from primary sources or are formed in the atmosphere by gas-phase reactions (secondary aerosols).

Since particle size determines how deep into the lung a particle is inhaled, there are two NAAQS for PM, $PM_{2.5}$, and PM_{10}. Particles smaller than 2.5 μm are called "fine," are composed largely of inorganic salts (primarily ammonium sulfate and nitrate), organic species, and trace metals. Fine PM can deposit deep in the lung where removal is difficult. Particles larger than 2.5 μm are called "coarse" particles, and are composed largely of suspended dust. Coarse PM tends to deposit in the upper respiratory tract, where removal is more easily accomplished. In 1997, industrial processes accounted for 42.0% of the emission rate for traditionally inventoried PM_{10}. Non-transportation fuel combustion and transportation sources accounted for 34.9% and 23.0%, respectively. As with the other criteria pollutants, PM_{10} concentrations and emission rates have decreased modestly due to pollution control efforts (Table 1.5-1).

Coarse particle inhalation frequently causes or exacerbates upper respiratory difficulties, including asthma. Fine particle inhalation can decrease lung functions and cause chronic bronchitis. Inhalation of specific toxic substances such as asbestos, coal mine dust, or textile fibers are now known to cause specific associated cancers (asbestosis, black lung cancer, and brown lung cancer, respectively).

An environmental effect of PM is limited visibility in many parts of the United States including some National Parks. In addition, nitrogen and sulfur containing particles deposited on land increase soil acidity and alter nutrient balances. When deposited in water bodies, the acidic particles alter the pH of the water and lead to death of aquatic organisms. PM deposition also causes soiling and corrosion of cultural monuments and buildings, especially those that are made of limestone.

1.5.1.5 SO_2, NOx, and Acid Deposition

Sulfur dioxide (SO_2) is the most commonly encountered of the sulfur oxide (SOx) gases, and is formed upon combustion of sulfur-containing solid and liquid fuels (primarily coal and oil). SOx are generated by electric utilities, metal smelting, and other industrial processes. Nitrogen oxides (NOx) are also produced in combustion reactions; however, the origin of most NOx is the oxidation of nitrogen in the combustion air. After being emitted, SOx and NOx can be transported over long distances and are transformed in the atmosphere by gas phase and aqueous phase reactions to acid components (H_2SO_4 and HNO_3). The gas phase reactions produce microscopic aerosols of acid-containing components, while aqueous phase reactions occur inside existing particles. The acid is deposited to the earth's surface as either *dry deposition* of aerosols during periods of no precipitation or *wet deposition* of acid-containing rain or other precipitation. There are also natural emission sources for both sulfur and nitrogen-containing compounds that contribute to acid deposition. Water in equilibrium with CO_2 in the atmosphere at a concentration of 330 ppm has a pH of 5.6. When natural sources of sulfur and nitrogen acid rain precursors are considered, the "natural" background pH of rain is expected to be about 5.0. As a result of these considerations, "acid rain" is defined as having a pH less than 5.0. Figure 1.5-4 shows the major environmental cause and effect steps for acidification of surface water by acid rain.

Major sources of SO_2 emissions are non-transportation fuel combustion (84.7%), industrial processes (8.4%), transportation (6.8%), and miscellaneous (0.1%) (US EPA 1998). As shown in Table 1.5-1, SO_2 concentrations and emissions have decreased significantly from 1988 to 1997. Emissions are expected to continue to decrease as a result of implementing the Acid Rain Program established by EPA under Title IV of the Clean Air Act. The goal of this program is to decrease acid deposition significantly by controlling SO_2 and other emissions from utilities, smelters, and sulfuric acid manufacturing plants, and by reducing the average sulfur content of fuels for industrial, commercial, and residential boilers.

There are a number of health and environmental effects of SO_2, NOx, and acid deposition. SO_2 is absorbed readily into the moist tissue lining the upper respi-

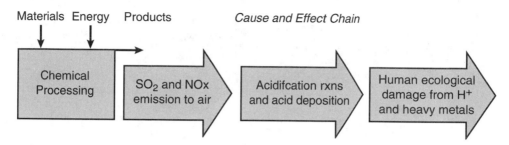

Materials Energy   Products   *Cause and Effect Chain*

Chemical Processing → SO₂ and NOx emission to air → Acidifcation rxns and acid deposition → Human ecological damage from H⁺ and heavy metals

Figure 1.5-4   Environmental cause and effect for acid rain.

ratory system, leading to irritation and swelling of this tissue and airway constriction. Long-term exposure to high concentrations can lead to lung disease and aggravate cardiovascular disease. Acid deposition causes acidification of surface water, especially in regions of high SO_2 concentrations and low buffering and ion exchange capacity of soil and surface water. Acidification of water can harm fish populations, by exposure to heavy metals, such as aluminum which is leached from soil. Excessive exposure of plants to SO_2 decreases plant growth and yield and has been shown to decrease the number and variety of plant species in a region (USEPA 1998). Figures 1.5-5 shows recent trends in the emission and concentrations of SO_2.

1.5.2   Air Toxics

Hazardous air pollutants (HAPs), or air toxics, are airborne pollutants that are known to have adverse human health effects, such as cancer. Currently, there are over 180 chemicals identified on the Clean Air Act list of HAPs (US EPA 1998). Examples of air toxics include the heavy metals mercury and chromium, and organic chemicals such as benzene, hexane, perchloroethylene (perc), 1,3-butadiene, dioxins, and polycyclic aromatic hydrocarbons (PAHs).

The Clean Air Act defined a major source of HAPs as a stationary source that has the potential to emit 10 tons per year of any one HAP on the list or 25 tons per year of any combination of HAPs. Examples of major sources include chemical complexes and oil refineries. The Clean Air Act prescribes a very high level of pollution control technology for HAPs called MACT (Maximum Achievable Control Technology). Small area sources, such as dry cleaners, emit lower HAP tonnages but taken together are a significant source of HAPs. Emission reductions can be achieved by changes in work practices such as material substitution and other pollution prevention strategies.

HAPs affect human health via the typical inhalation or ingestion routes. HAPs can accumulate in the tissue of fish, and the concentration of the contaminant increases up the food chain to humans. Many of these persistent and bioaccumulative chemicals are known or suspected carcinogens.

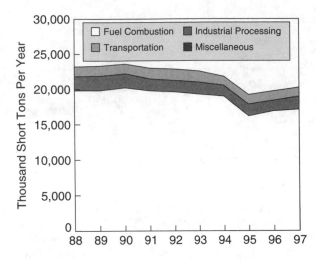

Figure 1.5-5 Emission trends for SO_2 from 1988–1997 for different source categories.

1.6 WATER QUALITY ISSUES

The availability of freshwater in sufficient quantity and purity is vitally important in meeting human domestic and industrial needs. Though 70% of the earth's surface is covered with water, the vast majority exists in oceans and is too saline to meet the needs of domestic, agricultural, or other uses. Of the total 1.36 billion cubic kilometers of water on earth, 97% is ocean water, 2% is locked in glaciers, 0.31% is stored in deep groundwater reserves, and 0.32% is readily accessible freshwater (4.2 million cubic kilometers). Freshwater is continually replenished by the action of the hydrologic cycle. Ocean water evaporates to form clouds, precipitation returns water to the earth's surface, recharging the groundwater by infiltration through the soil, and rivers return water to the ocean to complete the cycle. In the United States, freshwater use is divided among several sectors; agricultural irrigation 42%, electricity generation 38%, public supply 11%, industry 7%, and rural uses 2% (Solley et al. 1993). Groundwater resources meet about 20% of U.S. water requirements, with the remainder coming from surface water sources.

Contamination of surface and groundwater originates from two categories of pollution sources. Point sources are entities that release relatively large quantities of wastewater at a specific location, such as industrial discharges and sewer outfalls. Non-point sources include all remaining discharges, such as agricultural and urban runoff, septic tank leachate, and mine drainage. Another contributor to water pollution is leaking underground storage tanks. Leaks result in the release of pollution into the subsurface where dissolution in groundwater can lead to the extensive destruction of drinking water resources.

Besides the industrial and municipal sources we typically think of in regard to water pollution, other significant sources of surface and groundwater contamination include agriculture and forestry. Contaminants originating from agricultural activities include pesticides, inorganic nutrients such as ammonium, nitrate, and phosphate, and leachate from animal waste. Forestry practices involve widespread disruption of the soil surface from road building and the movement of heavy machinery on the forest floor. This activity increases erosion of topsoil, especially on steep forest slopes. The resulting additional suspended sediment in streams and rivers can lead to light blockage, reduced primary production in streams, destruction of spawning grounds, and habitat disruption of fisheries.

Transportation sources also contribute to water pollution, especially in coastal regions where shipping is most active. The 1989 Exxon Valdez oil spill in Prince William Sound in the state of Alaska is a recent well-known case that coated the shoreline with crude oil over a vast area. Routine discharges of petroleum from oil tanker operations is on the order of 22 million barrels per year (UNEP 1991), an amount 87 times the size of the Exxon Valdez spill. Transportation activities can also be a source of non-point pollution as precipitation runoff from roads carries oil, heavy metals, and salt into nearby streams.

1.7   ECOLOGY

Ecology is the study of material flows and energy utilization patterns in communities of living organisms in the environment, termed ecosystems. This area of science is very important in pollution prevention because of the possibility that pollutants entering sensitive ecosystems might disrupt the cycling of essential nutrients and elements for life, with potentially unforeseen negative consequences. Ecosystems, whether aquatic or terrestrial, share a common set of characteristics. They extract energy from the sun and store this energy in the form of reduced carbon-based compounds (biomass) in a process termed *photosynthesis*. Another very important function of ecosystems is to cycle elements and molecules through the environment, alternating between organic and inorganic forms of carbon, nitrogen, phosphorus, and sulfur.

Organisms that capture solar energy are *primary producers* which inhabit the *first trophic level* of the food chain in ecosystems. Examples of primary producers are plants in terrestrial ecosystems. For aquatic systems, members include aquatic plants, algae, and phytoplankton. The *second trophic level* is inhabited by the primary consumers, such as grazing animals on land and zooplankton and insects in aquatic environments, which prey upon the primary producers. The *third trophic level* is occupied by the secondary consumers, which prey upon the primary consumers. Examples are birds of prey, mammalian carnivores, fish, and many others. Additional trophic levels are possible depending upon the particular ecosystem.

Carnivores at the highest trophic levels in ecosystem food chains can encounter increased exposure to certain classes of anthropogenic pollutants. Chemicals that are hydrophobic (water-hating, non-polar organic compounds of high molecular weight), persistent (do not biodegrade or react biologically in ecosystems), and toxic are of particular concern because these chemicals bioaccumulate in animal fat tissue and are transferred from lower to higher trophic levels in the food chain. High levels of polychlorinated biphenyls (PCBs), certain pesticides, and mercury compounds have been detected in fish of the Great Lakes. The use of the pesticide DDT in the 1950s and 1960s caused dramatic reductions in birth rates of certain birds of prey that were consuming contaminated fish and other contaminated animals. Such examples demonstrate the need to understand the workings of ecosystems so that one can mitigate the harm that chemicals released into the environment can cause to ecosystems.

1.8 NATURAL RESOURCES

The production of industrial materials and products begins with the extraction of natural resources from the environment. The availability of these resources is vital for the sustained functioning of both industrialized and developing societies. Examples of natural resources include water, minerals, energy resources like fossil fuels, solar radiation, wind, and lumber. Renewable resources have the capacity to be replenished, while non-renewable resources are only available in finite quantities. The management of natural resources is intended to assure an adequate supply of these materials for anticipated future uses, also known as sustainable use. Non-renewable resources are of particular importance because of their inherently finite supply. For example, most energy requirements of today and of the foreseeable future will be met using non-renewable fossil fuels, such as oil, coal, and natural gas. As the availability of resources is diminished, the costs and energy consumption for producing these materials are likely to increase. Resource management techniques like conservation, recycling of materials, and improved technologies can be used to ensure the availability of these materials for the future. In some cases, materials already in use can be continuously recycled into new products (for instance, lead from batteries, steel from scrap cars, aluminum from beverage cans).

1.9 WASTE FLOWS IN THE UNITED STATES

There is no single source of national industrial waste data in the United States. Instead, the national industrial waste generation, treatment, and release picture is a composite derived from several sources of data. A major source of industrial waste data is the United States Environmental Protection Agency, which compiles various national inventories in response to legislative statutes. A sampling of the many laws requiring EPA to collect environmental data include the Clean Air Act, Resource Conservation and Recovery Act (RCRA), Superfund Amendments and Reauthorization Act (SARA), and the Emergency Planning and Community Right-to-Know Act (EPCRA). In addition to these federal government sources of data, there is also information collected by industry consortia such as the American Chemistry Council (formerly the Chemical Manufacturer's Association) and the American Petroleum Institute. Table 1.9-1 lists a number of national industrial waste databases. Due to the many inventories and the fact that the data sources might contain inconsistent data, the assembly of the national waste picture is difficult. However, from these data sources one is able to identify the major industrial sectors involved and the magnitude of their contributions.

Non-hazardous industrial waste represents the largest contribution to the national industrial waste picture. From 1986 data, almost 12 billion tons of non-hazardous waste was generated and disposed of by U.S. industry (Allen and

Table 1.9-1   Sources of National Industrial Waste Trends Data. See Appendix F for additional information.

Non-Hazardous Solid Waste

Report to Congress: Solid Waste Disposal in the United States, Volumes I and II, US Environmental Protection Agency, EPA/530-SW-88-011 and EPA/530-SW-88-011B, 1988.

Criteria Air Pollutants

Aerometric Information Retrieval System (AIRS); US EPA Office of Air Quality Planning and Standards, Research Triangle Park, NC.

National Air Pollutant Emission Estimates; US EPA Office of Air Quality Planning and Standards, Research Triangle Park, NC.

Hazardous Waste (Air Releases, Wastewater, and Solids)

Biennial Report System (BRS); available through TRK NET, Washington, DC.

National Biennial Report of Hazardous Waste Treatment, Storage, and Disposal Facilities Regulated under RCRA; US EPA Office of Solid Waste, Washington, DC.

National Survey of Hazardous Waste Generators and Treatment, Storage, Disposal and Recycling Facilities in 1986; available through National Technical Information Service (NTIS) as PB92-123025.

Generation and Management of Residual Materials; Petroleum Refining Performance (replaces *The Generation and Management of Wastes and Secondary Materials* series); American Petroleum Institute, Washington, DC.

Preventing Pollution in the Chemical Industry: Five Years of Progress (replaces the *CMA Hazardous Waste Survey* series); Chemical Manufacturers Association (CMA), Washington, DC.

Report to Congress on Special Wastes from Mineral Processing; US EPA Office of Solid Waste, Washington, DC.

Report to Congress: Management of Wastes from the Exploration, Development, and Production of Crude Oil, Natural Gas, and Geothermal Energy, Vol. 1, *Oil and Gas*; US EPA Office of Solid Waste, Washington, DC.

Toxic Chemical Release Inventory (TRI); available through National Library of Medicine, Bethesda, Maryland and RTK NET, Washington, DC.

Toxic Release Inventory: Public Data Release (replaces *Toxics in the Community: National and Local Perspectives*); EPCRA hotline (800)-535-0202. www.epa.gov/TRI

Permit Compliance System; US EPA Office of Water Enforcement and Permits, Washington, DC.

Economic Aspects of Pollution Abatement

Manufacturers' Pollution Abatement Capital Expenditures and Operating Costs; Department of Commerce, Bureau of the Census, Washington, DC.

Minerals Yearbook, Volume 1 Metals and Minerals; Department of the Interior, Bureau of Mines, Washington, DC. Census Series: *Agriculture, Construction Industries, Manufacturers-Industry, Mineral Industries*; Department of Commerce, Bureau of the Census, Washington, DC.

Source: US Department of Energy (DOE), "Characterization of Major Waste Data Sources," DOE/CE-40762T-H2, 1991.

Rosselot, 1997; US EPA 1988a and 1988b). That amount is about 240 pounds of industrial waste per person each day using today's population numbers. This amount is about 60 times higher than the rate of waste generation by households in the United States (municipal solid waste). The largest industrial contributors to non-hazardous waste are the manufacturing industry (7,600 million tons/yr), oil and gas industry (2,095–3,609 million tons/yr), and the mining industry (>1,400 million tons/yr). Lesser amounts are contributed by electricity generators (fly ash and flue-gas desulfurization waste), construction waste, hospital infectious waste, and waste tires.

Hazardous waste is defined under the provisions of the Resource Conservation and Recovery Act (RCRA) as residual materials which are ignitable, reactive, corrosive, or toxic. Once designated as hazardous, the costs of managing, treating, storing, and disposing of this material increase dramatically. The rate of industrial hazardous waste generation in the United States is approximately 750 million tons/yr (1986 data, Baker and Warren, 1992; Allen and Rosselot, 1997). This rate is 1/16th the rate at which non-hazardous solid waste is generated by industry. Furthermore, hazardous waste contains over 90% by weight of water, having only a relatively minor fraction of hazardous components. Therefore, the rate of generation of hazardous components in waste by industry is estimated at 10–100 million tons/yr, though there is significant uncertainty in the exact amount due to differing definitions of hazardous waste. As shown in Figure 1.9-1, the chemical and allied products industries generate about 51% by weight of the hazardous wastes produced in the United States each year (about 380 million tons/yr on a wet basis). Electronics, petroleum refining and related products, primary metals, and transportation equipment manufacturers each contribute from 50 to 70 million tons/yr.

Releases and waste generation rates for more than 600 chemicals and chemical categories are currently reported to the US EPA in the Toxic Release Inventory (TRI). Manufacturing operations (those with Standard Industrial Classification (SIC) Codes between 20 and 39) and certain federal facilities are required to report their releases of listed chemicals. Facilities must report releases of toxic chemicals to the air, water, and soil, as well as transfers to off-site recycling or treatment, storage, and disposal facilities. The release rate estimates only include the toxic chemicals of any waste stream, thus water or other inerts are not included, in contrast with the industrial hazardous waste reporting system. The total releases and transfers reported to the TRI in 1994 (TRI 1994) was three million tons and the distribution of this amount among several manufacturing categories is shown in Figure 1.9-2. Again, as in hazardous waste, a relatively few industrial sectors release the majority of the toxic pollutants. More recent versions of the TRI have included more industrial sectors (such as electricity generation and mining) in the reporting, resulting in somewhat different distributions. Nevertheless, a few industrial sectors continue to dominate the releases.

What happens to all of the hazardous waste generated by the United Stated industry each year? Table 1.9-2 shows several management methods, the quantity of hazardous waste managed, and the number of facilities involved. Note that the

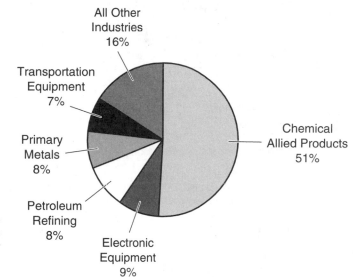

Figure 1.9-1 Industrial hazardous waste generation in the United States by industry sector (1986 data, Baker and Warren, 1992).

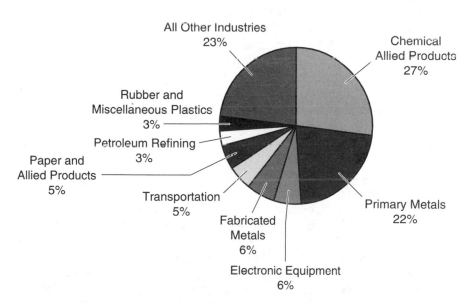

Figure 1.9-2 Adapted from Allen and Rosselot, 1997. Toxic chemical releases from United States industry as a percentage of the total release (3 million tons per year). *Source:* Allen and Rosselot, *Pollution Prevention for Chemical Processes* © 1997. This material is used by permission of John Wiley & Sons, Inc.

Table 1.9-2   Hazardous Waste Managed for Each Management Technology (1986 data).

Management Method	Quantity Managed in 1986 (million tons)	Number of Facilities
Metal Recovery	1.4	330
Solvent Recovery	1.2	1,500
Other Recycling	0.96	240
Fuel Blending	0.75	180
Reuse as Fuel	1.4	300
Incineration	1.1	200
Solidification	0.77	120
Land Treatment	0.38	58
Wastewater Treatment	730	4,400
Disposal Impoundment	4.6	70
Surface Impoundment	230	300
Landfill	3.2	120
Waste Pile	0.68	71
Underground Injection	29	63
Storage (RCRA permitted)	190	1,800
Other Treatment	2.0	130

quantities managed in Table 1.9-2 add up to more than 750 million tons/year because the same waste may be counted in more than one management method. For example, some wastewater may be stored or may be temporarily placed in surface impoundments before treatment. It is also interesting to note that 96% of hazardous wastes are managed on-site at the facilities that generated them in the first place. Most hazardous waste is managed using wastewater treatment. This is not surprising because over 90% of hazardous waste is water. Also, very little recycling and recovery of hazardous waste components occurs.

SUMMARY

In this chapter a wide array of environmental issues were introduced, and their impacts were related to chemical production and use. The pertinent chemicals and the environmental reactions of those chemicals were discussed. For many environmental problems, the chemicals causing the adverse environmental or health impacts were not the same chemical originally emitted from the production process or from the use of a chemical. Thus, the environment is a complex system with a large number of transport and transformation processes occurring simultaneously. Fortunately for the chemical engineer, it is not necessary to understand these processes in great detail in order to gain the insights needed to design chemical processes to be more efficient and less polluting. A focal point for improving process designs is to understand that the properties of chemicals can have an important influence on their ultimate fate in the environment and on their potential impact on the environment and human health. The influences of chemical properties on how chemicals may behave in the environment will be discussed in detail in Chapters 5 and 6. With a basic understanding of environmental issues, the chemical engineer will be able to spot environmental problems earlier and will contribute to the solution of those problems by improving the environmental performance of chemical processes and products.

REFERENCES

AFEAS, *Alternative Fluorocarbons Environmental Acceptability Study,* 1333 H Street NW, Washington, DC 20005 USA, *http://www.afeas.org/.* Sept. 2000.

Allen, D.T. and Rosselot, K.S., *Pollution Prevention for Chemical Processes,* John Wiley and Sons, New York, NY, 1997.

Baker, R.D. and Warren, J.L., "Generation of hazardous waste in the United States," *Hazardous Waste & Hazardous Materials, 9(1),* 19–35, Winter 1992.

EIA, Energy Information Agency, *International Energy Outlook* 1998, US Department of Energy, DOE/EIA-0484(98), April 1998a.

EIA, Energy Information Agency, *Annual Energy Review,* U.S. Department of Energy, DOE/EIA-0484(98), July 1998b.

Elkins, J.W., Thompson, T.M., Swanson, T.H., Butler, J.H., Hall, B.D., Cummings, S.O., Fisher, D.A., and Raffo, A.G., Decrease in the growth rates of atmospheric chlorofluorocarbons 11 and 12, *Nature, 364,* 780–783, 1993.

IPCC, Intergovernment Panel on Climate Change, *Climate Change 1995: The Science of Climate Change,* ed. Houghton, J.T., Milho, L.G.M., Callander, B.A., Harris, H. Kattenberg, A., and Maskell, K., Cambridge University Press, Cambridge, UK, 1996.

Molina, M.J. and Rowland, R.S. "Stratospheric sink for chlorofluoromethanes: Chlorine atom-catalyzed destruction of ozone," *Nature,* V. 249: 810–812 (1974).

Montzka, S.A., Butler, J.H., Elkins, J.W., Thompson, T.M., Clarke, A.D., and Lock, L.T., Present and future trends in the atmospheric burden of ozone-depleting halogens, *Nature, 398,* 690–694, 1999.

NOAA, National Oceanic and Atmospheric Administration, Climate Monitoring and Diagnostics Laboratory, Boulder, CO, *http://www.cmdl.noaa.gov/,* September 2000.

Phipps, E., *Overview of Environmental Problems,* National Pollution Prevention Center for Higher Education, University of Michigan, Ann Arbor, MI, 1996, http://www.css.snre.umich.edu.

Solley, W.B., Pierce, R.R., and Perlman, II.A., *Estimated Use of Water in the United States* 1990, US Geological Survey Circular 1081 (Washington: Government Printing Office, 1993).

TRI94, *Toxic Chemical Release Inventory for 1994,* Bethesda, MD, National Library of Medicine, July 1996.

UNEP, United Nations Environment Programme, *The State of The World Environment: 1991, 37,* May 1991.

US EPA, United States Environmental Protection Agency, "Report to Congress: Solid Waste Disposal in the United States, Volume 1," EPA/530-SW-88-011, 1988a.

US EPA, United States Environmental Protection Agency, "Report to Congress: Solid Waste Disposal in the United States, Volume 1," EPA/530-SW-88-011B, 1988b.

US EPA, United States Environmental Protection Agency, 1997 National Air Quality and Emissions Trends Report, Office of Air Quality Planning and Standards, Research Triangle Park, NC 27711, EPA 454/R-98-016, December 1998, *http://www.epa.gov/oar/aqtrnd97/*.

Wallace, J.M. and Hobbs, P.V., *Atmospheric Science: An Introductory Survey,* Academic Press, New York, NY, 1977.

WEC, World Energy Council, *Energy for Tomorrow's World,* St. Martin's Press, New York, NY, pg. 111, 1993.

PROBLEMS

1. **Electric Vehicles: Effects on Industrial Production of Fuels.** Replacing automobiles having internal combustion engines with vehicles having electric motors is seen by some as one solution to urban smog and tropospheric ozone. Write a short report (1–2 pages double spaced) on the likely effects of this transition on industrial production of fuels. Assume for this analysis that the amount of energy required per mile traveled is roughly the same for each kind of vehicle. Consider the environmental impacts of using different kinds of fuel for the electricity generation to satisfy the demand from electric vehicles. This analysis does not include the loss of power over the lines/grid. Background reading for this problem is found in *Industrial Ecology and the Automobile* by Thomas Graedel and Braden Allenby, Prentice Hall, 1998.

2. **Global Energy Balance: No Atmosphere (adapted from Wallace and Hobbs, 1977).** The figure below is a schematic diagram of the earth in radiative equilibrium with its surroundings assuming no atmosphere. Radiative equilibrium requires that the rate of radiant (solar) energy absorbed by the surface must equal the rate of radiant energy emitted (infrared). Let S be the incident solar irradiance (1,360 Watts/meter^2), E the infrared planetary irradiance (Watts/meter^2), R_E the radius of the earth (meters), and A the planetary albedo (0.3). The albedo is the fraction of total incident solar radiation reflected back into space without being absorbed.

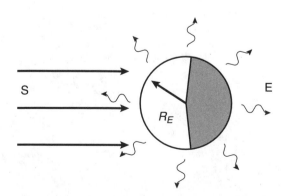

(a) Write the steady-state energy balance equation assuming radiative equilibrium as stated above. Solve for the infrared irradiance, E, and show that its value is 241 W/ meter^2.

(b) Solve for the global average surface temperature (K) assuming that the surface emits infrared radiation as a black body. In this case, the Stefan-Boltzman Law for a blackbody is $E = \sigma\, T^4$, σ is the Stefan-Boltzman Constant (5.67×10^{-8} Watts/(m$^2 \bullet °$K^4)), and T is absolute temperature ($°$K). Compare this temperature with the observed global average surface temperature of 280 K. Discuss possible reasons for the difference.

3. **Global Energy Balance: with a Greenhouse Gas Atmosphere (adapted from Wallace and Hobbs, 1977).** Refer to the schematic diagram below for energy balance calculations on the atmosphere and surface of the earth. Assume that the atmosphere can be regarded as a thin layer with an absorbtivity of 0.1 for solar radiation and 0.8 for infrared radiation. Assume that the earth surface radiates as a black body (absorbtivity = emissivity = 1.0).

Let x equal the irradiance (W/m^2) of the earth surface and y the irradiance (both upward and downward) of the atmosphere. E is the irradiance entering the earth-atmosphere system from space averaged over the globe ($E = 241$ W/m^2 from problem 2). At the earth's surface, a radiation balance requires that

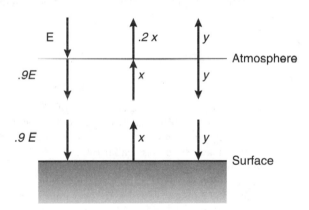

$$0.9E + y = x$$
(irradiance in = irradiance out)

while for the atmosphere layer, the radiation balance is

$$E + x = 0.9E + 2y + .2x$$

(a) Solve these equations simultaneously for y and x.

(b) Use the Stefan-Boltzman Law (see problem 2) to calculate the temperatures of both the surface and the atmosphere. Show that the surface temperature is higher than when no atmosphere is present (problem 2).

(c) The emission into the atmosphere of infrared absorbing chemicals is a concern for global warming. Determine by how much the absorbtivity of the atmosphere for infrared radiation must increase in order to cause a rise in the global average temperature by 1°C above the value calculated in part b.

4. **Global Carbon Dioxide Mass Balance.** Recent estimates of carbon dioxide emission rates to and removal rates from the atmosphere result in the following schematic diagram (EIA, 1998a)

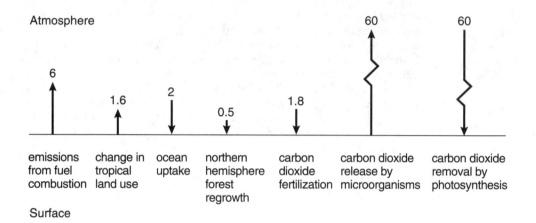

The numbers in the diagram have units of 10^9 metric tons of *carbon* per year, where a metric ton is equal to 1000 kg. To calculate the emission and removal rates for *carbon dioxide,* multiply each number by the ratio of molecular weights (44 g CO_2/12 g C).

(a) Write a steady state mass balance for carbon dioxide in the atmosphere and calculate the rate of accumulation of CO_2 in the atmosphere in units of kg/yr. Is the accumulation rate positive or negative?

(b) Change the emission rate due to fossil fuel combustion by +10% and recalculate the rate of accumulation of CO_2 in the atmosphere in units of kg/yr. Compare this to the change in the rate of accumulation of CO_2 in the atmosphere due to a +1% change in carbon dioxide release by micro-organisms.

(c) Calculate the rate of change in CO_2 concentration in units of ppm per year, and compare this number with the observed rate of change stated in section 1.4.2. Recall the definition of parts per million (ppm), which for CO_2, is the mole fraction of CO_2 in the air. Assume that we are only considering the first 10 km in height of the atmosphere and that its gases are well mixed. Take for this calculation that the total moles of gas in the first 10 km of the atmosphere is approximately 1.5×10^{20} moles.

(Note: $\text{ppm}_v = \dfrac{C_{CO_2}}{C} \times 10^6$, where C_{CO_2} is the number of moles of CO_2 and C is the total moles of air.)

(d) Describe how the rate of accumulation of CO_2 in the atmosphere, calculated in parts b and c, would change if processes such as carbon dioxide fertilization and forest growth increase as CO_2 concentrations increase. What processes releasing CO_2 might increase as atmospheric concentrations increase? (Hint: assume that temperature will rise as CO_2 concentrations rise).

5. **Ozone Depletion Potential of Substitute Refrigerants.** A chemist is trying to develop new alternative refrigerants as substitutes for chlorofluorocarbons. The chemist decides that either bromine or fluorine will substitute for the chlorines on existing compounds. Which element, bromine or fluorine, would be more effective in reducing the ozone depletion potential for the substitute refrigerants? Explain your answer based on the information contained in this chapter.

CHAPTER 2

Risk Concepts

by
Fred Arnold & Gail Froiman
with John Blouin

Risk: the probability that a substance or situation will produce harm under specific conditions. Risk is a combination of two factors: the probability that an adverse event will occur and the consequences of the adverse event.

The Presidential/Congressional Commission
on Risk Assessment & Risk Management, Vol. 1, 1997

2.1 INTRODUCTION

Risk is a concept used in the chemical industry and by practicing chemical engineers. The term risk is multifaceted and is used in many disciplines such as: finance (rate of return for a new plant or capital project, process improvement, etc.), raw materials supply (single source, back integration), plant design and process change (new design, impact on bottom line), and site selection (foreign, political stability). Though the term risk used in these disciplines can be discussed either qualitatively or quantitatively, it should be obvious that these qualitative or quantitative analyses are not the same in all fields (financial risk ≠ process change risk). This chapter will focus on the basic concept of environmental risk and risk assessment as applied to a chemical's manufacturing, processing, or use, and the impact of exposure to these chemicals on human health or the environment.

Risk assessment is a systematic, analytical method used to determine the probability of adverse effects. A common application of risk assessment methods is to evaluate human health and ecological impacts of chemical releases to the environment. Information collected from environmental monitoring or modeling is incorporated into models of human or worker activity and exposure, and conclusions on the likelihood of adverse effects are formulated. As such, risk assessment is an important tool for making decisions with environmental consequences. Almost always, when the results from environmental risk assessment are used, they are incorporated into the decision-making process along with economic, societal, technological, and political consequences of a proposed action.

Section 2.2 provides a general description of risk, risk categories, and a conceptual expression of chemical risk. The value of risk assessment in design and the pertinent environmental regulations to the engineering profession are described in Sections 2.3 and 2.4. The rest of the chapter covers risk assessment and its four major components: hazard assessment, dose-response, exposure assessment, and risk characterization. (Sections 2.5 through 2.9). Later chapters further expand on these risk concepts and their applications.

2.2 DESCRIPTION OF RISK

Risks can be grouped into three general categories:

- **Voluntary risk:** A result of actions taken by choice or out of necessity. Examples include firefighting, driving, bungee cord jumping, and lifestyle choices such as diet and smoking.
- **Natural disasters:** These include floods, hurricanes, earthquakes, and other disasters that are beyond human control. However, the risk to natural disasters can be exacerbated by such voluntary actions as living in a known flood plain or on an active earthquake fault.
- **Involuntary risk:** Risk resulting from uncontrollable actions of others. Examples include pesticide residues or pathogens in food, occupational exposure to industrial chemicals or being murdered. These risks tend to have more uncertainty and are not as well known.

Quantitatively, in the above categories, risk in the first two groupings voluntary and natural) is frequently determined by actuarial-based statistics (e.g., fatalities are correlated with activity, location, and other parameters). In involuntary exposure, such as those to chemicals, risk, for the most part, is based on inferred data (animal tests, analogs, extrapolation). People are more familiar with expressions of risk associated with various activities than they are with risks associated with chemical exposure. Table 2.2-1 lists one assessor's evaluation of various risk factors, where being the unmarried male causes the greatest loss of life expectancy.

Risks from toxic chemicals, depending on the context, may be defined, described, and calculated in different ways. Risk is normally defined as the probability for an individual to suffer an adverse effect from an event. What is the probability that certain types of cancer will develop in people exposed to aflatoxin in peanut products or benzene from gasoline? What is the likelihood that workers exposed to lead will develop nervous system disorders? In the context of this text, a

Table 2.1-1   Loss of Life Expectancy from Various Societal Activities and Phenomena.

Risk Factor	Loss of life expectancy (days)
Cancer risks associated with environmental pollutants	
Indoor radon	30
Worker chemical exposure	30
Pesticide residues in food	12
Indoor air pollution	10
Consumer products use	10
Stratospheric ozone depletion	22
Inactive hazardous waste sites	2.5
Carcinogens in air pollution	4
Drinking water contaminants	1.3
Noncancer risks associated with environmental pollutants	
Lead	20
Carbon monoxide	20
Sulfur dioxide	20
Radon	0.2
Air pollutants (e.g., carbon tetrachloride, chlorine)	0.2
Drinking water contaminants (e.g., lead, pathogens, nitrates, chlorine disinfectants)	0.2
Industrial discharge into surface water	Few minutes
Sewage treatment plant sludge	Few minutes
Mining wastes	Few minutes
Lifestyle/demographic status risks	
Being an unmarried male	3500
Smoking cigarettes and being male	2250
Being an unmarried female	1600
Being 30% overweight	1300
Being 20% overweight	900
Having less than an 8th-grade education	850
Smoking cigarettes and being female	800
Being poor	700
Smoking cigars	330
Having a dangerous job	300
Driving a motor vehicle	207
Drinking alcohol	130
Having accidents in the home	95
Suicide	95
Being murdered	90
Misusing legal drugs	90

Source: Fan and Chang (1996), pg. 247.

chemical release is an example of an event. As with any relationship expressing or using probability, there is no defined way of expressing (mathematically or with scientific rigor) a single deterministic value of a phenomenon that is probabilistic. A fairly simple conceptual way of expressing chemical risk is shown below.

$$\text{Risk} = f(\text{Hazard, Exposure})$$

Hazard is the potential for a substance or situation to cause harm or to create adverse impacts on persons or the environment. The magnitude of the hazard reflects the potential adverse consequences, including mortality, shortened life-span, impairment of bodily function, sensitization to chemicals in the environment, or diminished ability to reproduce. Exposure denotes the magnitude and the length of time the organism is in contact with an environmental contaminant, including chemical, radiation, or biological contaminants.

When risk is in terms of probability, it is expressed as a fraction, without units. It has values from 0.0 (absolute certainty that there is no risk) to 1.0 (absolute certainty that an adverse outcome will occur).

For chemicals the term hazard is typically associated with the toxic properties of a chemical specific to the type of exposure. Similar chemicals would have similar innate hazards. However, one must examine the exposure to that hazard to determine the risk. For example, let us say you have three pumps that are all transporting the same chemical (same hazard), but one pump has a seal leak. Which pump poses the greatest risk to the worker? The pump with the seal leak has the greatest potential for exposure, while the hazards are equal (same chemical), so the seal leak pump poses the greatest risk. To expand, let's say we have three pumps that are transporting different chemicals; which one poses the greatest risk to the worker? In this case the engineer would need to examine the hazard—or innate inherent toxicity—of each of the chemicals, as well as the operation of the pumps to determine which poses the greatest risk. Assessment of the inherent toxicity of various chemicals (hazards) is covered in greater detail in later chapters.

If a chemical is known to present dermal hazard, the exposure would be expressed as surface area of potentially exposed skin multiplied by the mass of the chemical per unit of surface area of skin that it contacts. In this text, the exposure term in the above equation, unless otherwise stated, will be for human exposure (ingestion, inhalation, and dermal). A detailed discussion of the pathways for worker and general population exposure can be found in Chapter 6.

The concept of exposure and hazard equating to risk may be applied in different ways, depending on the information available. In addition, risks may be described across pathways or routes, or as a comparison between, say, using one chemical versus another. In the future, research is likely to reveal completely new sources of chemical risks which were previously unknown. For example, stratospheric ozone depletion and endocrine disrupters were emerging concepts when the authors of this text were engineering students. Today, the level of risk can be much more accurately characterized. The risk assessment framework presented in this chapter is sufficiently flexible to apply even to new sources of risk from chemical releases as they are recognized.

Example 2.2-1: Interaction of Toxic Agents.

Smoking may act synergistically with toxic agents found in the workplace to cause more severe health damage than that anticipated from adding the separate influences of the occupational hazard and smoking. In a study of 370 asbestos workers, 24 of 283 cigarette smokers died of bronchogenic carcinoma during the four year period of the study, while not one of the 87 non-smokers died of this cancer (Selikoff, 1968). This study suggested that asbestos workers who smoke have eight times the risk of lung cancer as compared to all other smokers and 92 times the risk of nonsmokers not exposed to asbestos. This same group of insulation workers was restudied five years later, at which time 41 of the 283 smokers had died of bronchogenic cancer. Only 1 of the 87 noncigarette smokers, a cigar smoker, died of lung cancer (Hammond, 1973).

Other chemicals and occupational exposures which appear to act synergistically with tobacco smoke include radon daughters, gold mine exposures, and exposures in the rubber industry. (Lednar, 1977)

2.3   VALUE OF RISK ASSESSMENT IN THE ENGINEERING PROFESSION

Risk assessment may be conceptualized as simply a means of organizing and analyzing all available scientific information that addresses the question, what are the risks associated with a chemical manufacturing process or use of a chemical product? If an engineer is asked to conduct a comprehensive assessment, such as developing an Environmental Impact Statement for a proposed new facility, a major study of this magnitude would necessitate the formation of a team of appropriate professionals (engineer, toxicologist, ecologist, chemist, industrial hygienist, medical and legal staff, etc.). It is critical that the resulting assessment focus not only on the quantitative aspects of risk but also the qualitative character of risk. This need for qualitative assessment is often driven by serious data gaps in health and eco-toxic effects, which preclude precise quantification of all impacts.

From an engineering perspective, it may be useful to think of risk as safety issues extrapolated from the present to the long term. That is, safety may be thought of as the likelihood of immediate adverse consequences, and risk as the likelihood of long-term adverse consequences. Engineers can elevate risk concerns from chronic exposures to toxic chemicals to the same level of concern as safety issues. As with safety issues, the potential for chemical risks is only one of many factors that influences decisions. Financial considerations will always be paramount in business decisions. However, significant chemical risks may be a vital consideration in some instances and may carry financial consequences as well. The task before the engineer is to understand, quantify, and communicate risk issues as comprehensively as possible.

One important distinction between long-term risk and conventional safety issues is that while the consequences of chemical accidents are readily linked to their cause, chronic exposures from chemicals often are not. For chemical accidents, injuries and property damage can be anticipated via some level of process hazard analysis (PHA) such as fault tree analysis, or Hazards and Operability Studies (HAZOP). A simple case study utilizing one of these methods is shown in Example 2.3-1. If a facility experiences a chemical explosion that shatters windows and injures nearby residents, everyone in the community and the facility management knows the source of the injury or damage. The results are immediate. In contrast, it is often extremely difficult to link a local epidemic of cancers to a chemical exposure that may have occurred decades before. The uncertainties associated with long-term risks render them difficult for managers to grapple with effectively. These distinctions between chronic risks and traditional safety issues create an important barrier to elevating chronic emission and release issues to the same level as safety concerns. (See Chapter 4, The Roles & Responsibilities of the Chemical Engineer, for further coverage of this topic.)

It is worth stressing that any risk assessment should be carefully and fully documented, including specific references for data used and calculations used to reach a conclusion. One of the more significant differences among many standard engineering tasks and a risk assessment is the "volatility" of the input data. It is for this reason that careful, complete documentation with narrative is a requirement for a comprehensive risk assessment. Risk assessment, when incorporated into process design, can have a positive impact on the environment *as well as* positive economic benefits. (See Chapter 12, Environmental Cost Accounting, for further coverage of this topic.)

Whether the risk assessment is broad or narrow in scope, the concepts presented here remain the same. As documented throughout this text, mathematical and database-derived computer models can assist in providing estimates of hazards and exposures. Methods presented in the subsequent chapters will enable the engineer to make design decisions based on chemical risk, even when some of the data gaps have not been filled.

Example 2.3-1: Fault Tree Analysis.

Underground gas pipelines can fail when an operator of construction equipment punctures the pipeline. The pipeline can also fail due to corrosion when the coating separating the pipeline from the soil is damaged and the sacrificial cathode fails to inhibit rusting of the pipeline. Damage to the coating may be due to abrasion by human activity or degradation in the environment. Based on this statement, draw a fault tree for the possible failure of a gas pipeline (Cooke, 1998).

Solution:

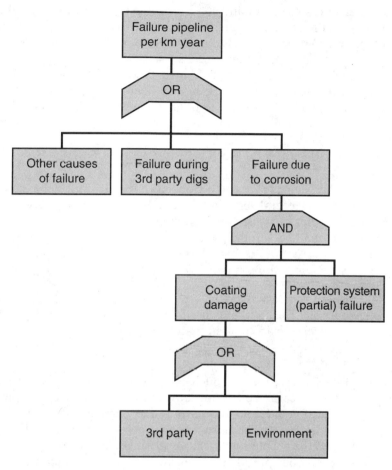

Figure 2.3-1   Fault tree analysis of gas pipelines demonstrating possible modes of failure.

2.4   RISK-BASED ENVIRONMENTAL LAW

Many environmental statutes (laws) incorporate risk management as a goal of the legislation. Some environmental laws consider economic impacts of risk management as well. For example, the provisions of the Clean Air Act pertaining to National Ambient Air Quality Standards call for standards that "protect the public health allowing an adequate margin of safety." That is, these standards mandate protection of public health based only on risk, without regard to technology or cost factors. In contrast, the Clean Water Act requires industries to install specific treatment technologies. These have descriptions like "best practicable control technology," and "best available technology economically achievable." Pesticides are licensed if they don't cause "any unreasonable risks to man or the environment taking into account the

economic, social, and environmental costs and benefits of the use of any pesticide."
In other words, economic and other factors may or may not be combined with risk issues as regulations are developed. These details become important if the engineer is
required to understand and follow the regulations or even requested to comment regarding proposed regulations (see Patton, 1993).

Table 2.4-1 lists selected United States safety, health, and environmental
statutes that require or suggest human health risk assessment before regulations
are promulgated. The list is enormous, and will probably grow with time. Chapter 3
describes some of these statutes in greater detail.

2.5   GENERAL OVERVIEW OF RISK ASSESSMENT CONCEPTS

In 1983, the National Research Council (NRC, 1983) developed a risk assessment
framework for federal regulatory agencies that is still in place today. That framework states that a risk assessment should estimate adverse impacts to health or the
environment and determine whether these impacts pose a serious threat. Risk assessment consists of four major components: *hazard assessment, dose-response, exposure assessment, and risk characterization.*

1. **Hazard Assessment.** What are the adverse health effects of the chemical(s) in
   question? Under what conditions? For example, does it cause a certain kind
   of cancer? Toxicologists usually perform this analysis. Since this information
   is pertinent to use of a chemical, sometimes hazard information can be obtained from reference data.

2. **Dose-Response.** How much of the chemical causes a particular adverse effect? There may be multiple adverse health effects, or responses, for the same
   chemical at different concentrations. Each adverse effect has a unique dose-
   response curve. The dose-response curve is non-linear because some members of the population are more sensitive than others.

   For our purposes, dose is defined as the quantity of a chemical that
   crosses a boundary to get into a human body or organ system. The term applies regardless of whether the substance is inhaled, ingested, or absorbed
   through the skin. Dose-response, then, is a mathematical relationship between the magnitude of a dose and the extent of a certain negative response
   in the exposed population.

3. **Exposure Assessment.** Who is exposed to this chemical? How much of the
   chemical reaches the boundary of a person, and how much enters the person's
   body? Exposure may be measured, estimated from models, or even back-cal-
   culated from measurements called biomarkers taken from exposed people.

4. **Risk Characterization.** How great is the potential for adverse impact from
   this chemical? What are the uncertainties in the analyses? How conclusive
   are the results of these analyses?

This general risk assessment framework has been tailored to human health risk
assessment from exposure to chemicals. A risk assessment team may decide that spe-

Table 2.4-1   United States Safety, Health, and Environmental Statutes
That Imply Risk Assessment.

United States Environmental Protection Agency	
Atomic Energy Act (also NRC)	42.U.S.C.2011
Comprehensive Environmental Response, Compensation and Liability Act (CERCLA, or Superfund)	42.U.S.C.9601
Clean Air Act	42.U.S.C.7401
Clean Water Act	33.U.S.C.1251
Emergency Planning and Community Right to Know Act	42.U.S.C.11001
Federal Food and Drug, and Cosmetics Act (also HHS)	21.U.S.C.301
Federal Insecticide, Fungicide, and Rodenticide Act	7.U.S.C.136
Lead Contamination Control Act of 1988	42.U.S.C.300j-21
Marine Protection, Research, and Sanctuaries Act (also DA)	16.U.S.C.1431
Nuclear Waste Policy Act	42.U.S.C.10101
Resource Conservation and Recovery Act	42.U.S.C.6901
Safe Drinking Water Act	42.U.S.C.300f
Toxic Substances Control Act	7.U.S.C.136
Food Quality Protection Act of 1996	7.U.S.C.6

Consumer Product Safety Commission	
Consumer Product Safety Act	15.U.S.C.2051
Federal Hazardous Substance Act	15.U.S.C.1261
Lead-Based Paint Poisoning Act (also HHS and HUD)	42.U.S.C.4801
Lead Contamination Control Act of 1988	42.U.S.C.300j-21
Poison Prevention Packaging Act	15.U.S.C.1471

Department of Agriculture	
Eggs Products Inspection Act	21.U.S.C.1031
Federal Meat Inspection Act	21.U.S.C.601
Poultry Products Inspection Act	21.U.S.C.451

Department of Labor	
Federal Mine Safety and Health Act	30.U.S.C.801
Occupational Safety and Health Act	29.U.S.C.651

Department of Transportation	
Hazardous Liquid Pipeline Safety Act	49.U.S.C.1671
Hazardous Materials Transportation Act	49.U.S.C.1801
Motor Carrier Safety Act	49.U.S.C.2501
National Traffic and Motor Vehicle Safety Act	15.U.S.C.1381
National Gas Pipeline Safety Act	49.U.S.C.2001

Source: Federal Focus, 1991; Roberts and Abernathy, 1997; and Fort, 1996

cific aspects of the eco-assessment require attention. This level of activity is critical for new plant siting (grass-roots), which must include a thorough examination of the eco-systems in-place as well as unique areas (wetlands, forests, endangered species habitat).

The risk assessment process can be iterative. That is, if a cursory or screening risk assessment identifies concerns, a more rigorous process may be called for. This process may in turn illustrate that there are important data gaps that need to be filled to render the process sufficiently conclusive for risk management. The data gaps may be filled with recommendations for special studies with varying cost and time requirements, such as:

- proceeding with testing for health effects;
- evaluating the effectiveness of engineering controls and personnel protective equipment (PPE) to limit exposures;
- defining the kinetics and decomposition products of a waste stream and the impact of the chemical waste and its degradation products on local flora and fauna.

If it is reasonably clear from the risk assessment that a risk exists, the next step is risk management.

> Risk management is the process of identifying, evaluating, selecting, and implementing actions to reduce risk to human health and to ecosystems. The goal of risk management is scientifically sound, cost effective, integrated actions that reduce or prevent risks while taking into account social, cultural, ethical, political, and legal considerations. (Presidential Commission, 1997)

Risk managers must clearly answer many questions, some of which are:

- What level of exposure to a chemical risk agent is an unacceptable risk?
- How great are the uncertainties and are there any mitigating circumstances?
- Are there any trade-offs between risk reduction, benefits, and additional cost?
- What are the chances of risk shifting, that is, transferring risk to other populations?
- Are some of the risks worse than others?

The answers to these questions often depend on the culture and values of the organization that commissioned the risk assessment. Minimizing risk through improved engineering design and proactive process development should be core values of the engineer.

2.6 HAZARD ASSESSMENT

In the context of this text, a hazard is an adverse health effect related to chemical exposure. This section begins with a discussion of hazard assessment. It continues with a discussion about structural activity relationships, which are tools used to screen hazards in the absence of chemical-specific laboratory data, based on the known hazards of materials with similar chemical structures. A brief description of readily available references concerning known hazards may be found at the end of this Hazard Assessment section.

As noted above, a chemical exposure hazard assessment answers the question: What are the adverse effects of a chemical? The most common adverse effects, or endpoints, studied are various kinds of cancer, but other types of adverse health effects such as endocrine disruption or reproductive toxicity are also currently being studied. Effects immediately dangerous to life or health may result from a high but brief or acute exposure, while long-term effects may result from chronic exposures to low levels of a toxin that are insufficient to cause any acute effects. Health effect studies are usually performed on rodents; these studies are called subchronic effects studies, and provide the basis for estimating a particular hazard or hazards.

Example 2.6-1: Endocrine Disruptors.

There is evidence that domestic animals and wildlife have suffered adverse consequences from exposure to environmental chemicals that interact with the endocrine system. These problems have been identified primarily in species exposed to relatively high concentrations of organochlorine pesticides, PCBs, dioxins, and synthetic and plant-derived estrogens. Whether similar effects are occurring in the general human or wildlife populations from exposures to ambient environmental concentrations is unknown. For example, while there have been reports of declines in the quantity and quality of sperm production in humans over the last four decades, other studies show no decrease. Reported increases in incidences of certain cancers (breast, testes, prostate) may also be related to endocrine disruption. Because the endocrine system plays a critical role in normal growth, development, and reproduction, even small disturbances in endocrine function may have profound and lasting effects. This is especially true during highly sensitive prenatal periods, such that small changes in endrocrine status may have delayed consequences that are evident much later in adult life or in a subsequent generation. Furthermore, the potential for synergistic effects from multiple contaminants exists.

2.6.1   Cancer and Other Toxic Effects

Cancers of various organs or systems are among the most thoroughly studied toxic effects. Other examples of toxic effects which are known to be caused by chemical substances are decreased pulmonary capacity caused by inhalation of asbestos, and damage to the nervous system and internal organs resulting from ingestion of lead, mercury, and other metals. Chemical exposures may also induce neurotoxicity, reproductive toxicity, or developmental toxicity. A relatively new discipline, developmental toxicity, refers to birth defects and other toxic effects which become apparent after birth, and which may be rooted in the prenatal period.

From the perspective of hazard assessment, cancer can be caused by two different types of chemical substances—genotoxic carcinogens and nongenotoxic carcinogens. Genotoxic chemicals are believed to have no threshold amount below which they will NOT cause cancer. Theoretically, one molecule of a genotoxic carcinogen could alter DNA and cause a mutation. In most cases, such an exposure would not cause cancer because of natural mechanisms which can repair internal damage caused by exposures at this level. Unfortunately, only (expensive) mechanistic studies can distinguish whether a carcinogen is genotoxic or not. In the absence of these studies, genotoxicity is generally assumed (Velazquez et al. 1997).

In contrast, nongenotoxic carcinogens are believed to have a safe threshold quantity. This becomes clearer from dose-response assessment, which is discussed below. For the purpose of risk assessment, nongenotoxic substances are analyzed much like chemicals with endpoints other than cancer by using the concept of a Hazard Quotient, which is also discussed below. There are other concepts for addressing quantitative estimates of risk, but they are beyond the scope of this chapter.

2.6.2 Hazard Assessment for Cancer

US EPA has developed guidelines for hazard assessment of chemical carcinogens. There are three types of information used to make the hazard determination: human data, animal data, and supporting data. The data are first evaluated as to their conclusiveness. Then, the substance is classified, usually as part of one of the following groups:

Group A:	Carcinogenic to humans (There are currently only about 20 of these chemicals.)
Group B1:	Probably carcinogenic to humans based on limited human evidence of carcinogenicity
Group B2:	Probably carcinogenic to humans based on sufficient animal evidence, but inadequate human evidence
Group C:	Possibly carcinogenic to humans
Group D:	Not classifiable for human carcinogenicity
Group E:	Evidence of noncarcinogenicity for humans

Organizations other than US EPA have developed alternative classification schemes for toxic chemicals. For example, Table 2.6-1 lists thirteen chemical substances, by name and CAS number, regulated by OSHA as human carcinogens. The previous use of the substance is also listed. Due to the ecotoxic concerns of these chemicals, many are no longer in commerce and/or have been replaced with less hazardous alternative chemistries. Other organizations such as the American Conference of Governmental Industrial Hygienists (ACGIH) also classify chemicals according to the evaluated risk for human carcinogenicity.

Once this determination of Carcinogen Group is made using US EPA's or other guidelines, supporting information, such as mechanistic data, is considered. This information may be used to shift the above classification. The determination of classification requires careful professional judgment and peer review.

Example 2.6-2: Cancer Slope Factor.

A study of the potential of acrylonitrile to produce brain tumors in Fischer 344 rats was conducted by administering the carcinogen in drinking water for twenty-four

Table 2.6-1   Thirteen OSHA-Regulated Carcinogens (29CFR 1910.1003).

CAS Number	Chemical Name	Previous Use
53-96-3	2-Acetylaminofluorene	hazardous air pollutant—no use
92-67-1	4-Aminodiphenyl	antifungal agent
92-87-5	Benzidine	manufacture of azo dyes
542-88-1	Bis-chloromethyl Ether	manufacturing ion exchange resins
91-94-1	3,3'-Dichlorobenzidine	manufacture of azo dyes, yellow pigments
60-11-7	4-Dimethylaminoazo-benzene	pH indicator
151-56-4	Ethyleneimine	treatment (etherification) of cotton
107-30-2	Methyl Chloromethyl Ether	manufacturing ion exchange resins
134-32-7	Alpha-Naphthylamine	manufacturing dyes
91-59-8	Beta-Naphthylamine	manufacturing dyes
92-93-3	4-Nitrobiphenyl	manufacturing p-biphenylamine
62-75-9	N-Nitrosodimethylamine	antioxidant in lubricants, polymer softener
57-57-8	Beta-Propiolactone	disinfectant

months. The results of the study for female rats are tabulated below. Use a linear model of the relationship between the administered dose and the incidence of tumors to calculate the slope factor for acrylonitrile (Monsanto, 1980).

Tumor data for acrylonitrile-induced brain tumors in the Fischer 344 rats

Dose (mg/kg-day)	Brain tumor incidence
0	1/179
0.12	1/90
0.36	2/91
1.25	4/85
3.65	6/90
10.89	23/88

Solution: Convert the data to decimal equivalents and subtract the deaths at zero concentration to obtain the excess risk at each dose level. Fit the data with a linear equation, excess deaths=m * dose rate (mg/kg-day), where m is the slope factor. Finally compare the deaths predicted with the regression data with the observed frequencies.

Dose (mg/kg-day)	Brain Tumor Incidence	Excess Risk	Linear Estimate of Excess Risk
0	0.0056		
0.12	0.0111	0.0055	0.0028
0.36	0.0220	0.0164	0.0084
1.25	0.0471	0.0415	0.0292
3.65	0.0667	0.0611	0.0853
10.89	0.2614	0.2558	0.2545

$m = \Sigma$ (excess risk)$/\Sigma$ (dose, mg/kg-day) = 0.3802/16.27 mg/kg-day=0.0234/(mg/kg-day)

2.6.3 Hazard Assessment for Non-Cancer Endpoints

Adverse effects other than cancer and gene mutations are generally assumed to have a dose or exposure threshold. As a result, a different approach is used to evaluate potential risk for these *non-cancer* effects, which include liver toxicity, neurotoxicity, and kidney toxicity. The first step in this approach requires the identification of a critical effect for which the magnitude of the response can be assessed. The Reference Dose (RfD) or Reference Concentration (RfC) approach is used to evaluate such chronic effects. The RfD is defined as "an estimate (with uncertainty spanning perhaps an order of magnitude) of a daily exposure to the human population that is likely to be without appreciable risk of deleterious effects during a lifetime" (US EPA, 2000) and is expressed as a mg pollutant/kg body weight/day. The RfC is expressed as a concentration, or mg/m^3. Roughly speaking, it is the baseline "safe" dose or concentration to which a real exposure may be compared (US EPA, 2000).

The RfD or RfC is usually based on the most sensitive known effect—i.e., the effect that occurs at the lowest dose. The basic approach for deriving an RfD or RfC involves determining a "no-observed-adverse-effect level (NOAEL)" or "lowest-observed-adverse-effect level (LOAEL)" from an appropriate animal study or human epidemiologic study, and applying various uncertainty and modifying factors to arrive at the RfD/RfC. Each factor represents a specific area of uncertainty. The uncertainty surrounding these terms spans about an order of magnitude. For example, an RfD based on an NOAEL from a long-term animal study might incorporate a factor of 10 to account for the uncertainty in extrapolating from the test species to humans, and another factor of 10 to account for the variation in sensitivity within the human population. Another common uncertainty factor may be used to extrapolate from subchronic test exposures to potentially chronic human exposures. An RfD based on an LOAEL typically contains an additional factor of 10 to account for the extrapolation from LOAEL to NOAEL. Finally, another modifying factor (between 1 and 10) is sometimes applied to account for uncertainties in data quality (Roberts and Abernathy, 1997; Cicmanec et al., 1997).

The combination of these uncertainty factors can result in highly conservative interpretations. One can conclude from an RfD that a chemical is quite toxic when the reality is that little is known about the chemical's toxicity. The engineer must be sure to use appropriate caveats when presenting data, and should understand the reason or reasons for a specific RfD value.

The NOAEL described above is based on a single study, or data point from a more complete data set. However, it is well known that when drawing conclusions, it is preferable to use all of the available data rather than a single point. To this end, the US EPA is moving to a method that entails developing the RfD or RfC from a dose-response relationship derived from all the data. This new approach, the Benchmark Dose concept, has a goal of improving the quality of the RfD and RfC estimates, and reducing the number of uncertainty factors used (US EPA 2000).

Example 2.6-3: Reference Dose.

Reference doses are used to evaluate noncarcinogenic effects resulting from exposure to chemical substances. The reference dose (RfD) is the threshold of exposure below which protective mechanisms are believed to guard an organism from adverse effects resulting from exposure over a substantial period of time. When valid human toxicological data are available, it forms the basis for the reference dose. When human exposure data are not available, the animal species believed to be most sensitive to the chemical of concern is used to determine the lowest level at which an adverse effect is detected, often called the LOAEL. Similarly the NOAEL is the greatest test-dose level at which no adverse effect is noted. When animal data are used the reference dose for human populations is adjusted by extrapolation factors to convert the NOAEL or LOAEL into a human subthreshold or reference dose.

$$RfD = \frac{NOAEL}{F_A F_H F_S F_L F_D}$$

where F_A is an adjustment factor to extrapolate from animal to human populations;

F_H is an adjustment factor for differences in human susceptibility;

F_S is an adjustment factor used when data are obtained from subchronic studies;

F_L is an adjustment factor applied when the LOAEL is used instead of the NOAEL; and,

F_D is an adjustment factor applied when the data set is dubious or incomplete.

Each adjustment factor should account for the systematic difference between the two measures bridged by the extrapolation and incorporate a margin of safety in accordance with the uncertainty associated with the extrapolation. For example, in a three-month subchronic study in mice, the NOAEL for tris (1,3-dichloro-2-propyl) phosphate was 15.3 mg/kg body weight per day; the LOAEL was 62 mg/kg at which dose abnormal liver effects were noted. (Kamata, 1989) If each of the adjustment factors is equal to 10, the reference dose for this chemical is:

Using the NOAEL:

$$RfD = \frac{NOAEL}{F_A F_H F_S} = \frac{15.3\text{mg/kg–day}}{10x10x10} = 0.015\text{mg/kg–day}$$

Using the LOAEL:

$$RfD = \frac{LOAEL}{F_A F_H F_L F_S} = \frac{62\text{mg/kg–day}}{10x10x10x10} = 0.0062\text{mg/kg–day}$$

The lesser of the two values, 0.0062 mg/kg-day, would be selected as the reference dose for humans in this instance.

2.6.4   Structure Activity Relationships (SAR)

Structure Activity Relationships (SAR) are an effective technique for estimating the hazard, as well as other properties (see Chapter 5), of a chemical. The US EPA often use SARs when estimating hazard and other elements of risk. SARs estimate hazards by drawing analogies with chemically similar substances whose hazard has been studied. The similar substance is called a structural analog. This technique requires the expert judgment of toxicologists. Although an engineer should not be expected to perform this kind of analysis (for assessing hazard), an engineer might request that such an analysis be conducted. Therefore, a brief explanation is provided here.

The definition of structural activity, as it applies here, is the relationship between the structural property of a molecule and its biological activity. Health effects which can be evaluated are many, and include absorption into the body; metabolism by the body, oncogenicity (capability to produce tumors); mutagenicity (capability to induce DNA mutations); and acute, chronic, and subchronic toxicity, neurotoxicity, developmental, and reproductive effects (adverse effects on fertility). Some examples of chemical classes of concern that are amenable to SAR review are: acrylamides, vinyl sulfones, dianilines, sulfoniums, epoxides, benzothiazoliums, hindered amines, acrylates, and dichlorobenzene pigments.

The basis for choosing an appropriate structural analog may be structure, substructure, or physical/chemical properties. For example, an unsaturated ketone may be a good analog for an unsaturated ester. In some instances, the toxicologist predicts metabolites (biotransformation products) of the chemical of interest and assesses the hazards of the metabolites.

Some information about environmental fate is required to complete an SAR assessment. Environmental fate is determined by physical-chemical properties of the chemical. Examples are octanol/water partition coefficient and water solubility. Models for estimating these parameters are frequently used by engineers and are described in Chapter 5.

The intrinsic problem with using Structural Activity Relationships to estimate toxicity is the uncertainty associated with extrapolating information from one chemical to another. This uncertainty limits the accuracy of toxicity estimates made using SAR techniques, although they can be helpful when no other data are available. However, direct, accurate data on hazard should be used rather than SAR estimates whenever the data are available.

2.6.5   Readily Available Hazard References

Listed below are references commonly used to inform hazard assessment. The list is intended as a starting point for the engineer charged with hazard assessment and is not comprehensive. A comprehensive list for the entire text is located in Appendix F.

1. MSDS. The Material Safety Data Sheet is a document developed by chemical manufacturers. The MSDS contains safety and hazard information, physical and chemical characteristics, and precautions on safe handling and use. It may also include hazards to animals, especially aquatic species. The manufacturer is required to keep it up to date. Any employer that purchases a chemical is required by law to make the MSDS available to employees. Development of an MSDS is required under OSHA's Hazard Communication Standard.

2. NIOSH Pocket Guide to Chemical Hazards. NIOSH is the National Institute for Occupational Safety and Health; this is the organization that performs research for OSHA, the Occupational Safety and Health Administration. The Pocket Guide may be found on-line at:

www.cdc.gov/niosh/npg/pgdstart.html

It includes safety information, some chemical properties, and OSHA Permissible Exposure Limit concentrations, or PELs. The lower the permissible concentration, the greater the hazard to human health. PELs are human exposure concentration limits that have been set by OSHA for commonly used chemicals. The lower the permissible concentration, the greater the hazard to human health. By law, these concentration values cannot be exceeded in the workplace. Concentration measurements of these chemicals must be taken so that exposure levels are known and any documented overexposures can be addressed.

3. IRIS. IRIS is a database maintained by the United States Environmental Protection Agency. IRIS stands for Integrated Risk Information System. It is available through:

http://www.epa.gov/ngispgm3/iris/index.html

IRIS is a database of human health effects that may result from exposure to various substances found in the environment.

4. The National Library of Medicine has a Hazardous Substances Data Bank. The web address is:

http://chem.sis.nlm.nih.gov/hsdb/

The Hazardous Substances Data Bank (HSDB®) is a toxicology data file that focuses on the toxicology of potentially hazardous chemicals. It is enhanced with information on human exposure, industrial hygiene, emergency handling procedures, environmental fate, regulatory requirements, and related areas.

5. Also available from the National Library of Medicine is Toxnet. The web address is:

http://sis.nlm.nih.gov/sis1/

Toxnet is a cluster of databases on toxicology, hazardous chemicals, and related areas. Both IRIS and the HSDB are available through Toxnet.

6. Casarett and Doull's text *Toxicology, the Basic Science of Poisons* (Casarett and Doull, 1995). This is the classic text in the field for interested readers. It is published by Macmillan and is updated every few years.

7. Patty's *Industrial Hygiene and Toxicology.* This set of volumes is a starting point for readers who want more information than exposure limits, but who are not experts in toxicology. It is published by John Wiley & Sons.

8. The American Conference of Governmental Industrial Hygienists (ACGIH). This organization publishes workplace chemical exposure concentration limits which are voluntary, unlike the legally enforceable OSHA PELs. These limits are known at Threshold Limit Values (TLVs). The documentation for the TLVs contains detailed information on the relevant toxicity and exposure concerns related to each chemical with an established TLV.

2.7 DOSE-RESPONSE

A dose response curve (Figure 2.7-1) is a graph of the quantitative relationship between exposure and toxic effect. This analysis enables risk assessors to estimate a "safe" dose. Actual dose is compared to safe dose to estimate risk. Dose-response answers the question: How large a dose causes what magnitude of effect? Larger doses cause greater and more serious effects. For a given chemical, there is a separate curve for each adverse health effect.

The basic shape of the dose-response curve is determined by the biological mechanism of action. On a subtler level, the curve illustrates the sensitivity of different members of the population. It is a plot of dose in mg chemical per kg of body weight, versus percent of the population affected by that dose. For example, an LD50, or lethal dose 50%, is a statistic frequently tabulated for some chemicals. It is the dose, in mg/kg, at which 50% of the rats or other tested species die. This statistic emerges from a dose-response assessment. Rats, mice and rabbits are frequently tested species. They are like humans in that they are mammals, but they are also small, and breed and mature quickly, which can make the testing process more manageable. Nonetheless, these species may react very differently from humans to exposure to a particular chemical. Significant research efforts have been under way for some time to find reliable substitutes for animal testing of chemical hazards.

The curvature of the dose-response curve illustrates the varying sensitivity of different members of the exposed population. That is, if sensitivity to the chemical were constant, dose-response would be a straight line. The curvature illustrates that some people (or, more likely, rodents) are especially vulnerable, while others are more resistant. Among humans, common examples of sensitive subpopulations are children, the elderly, and the immunosuppressed.

Section 8.2 of Chapter 8 uses Threshold Limit Values (TLVs) and Permissible Exposure Limits (PELs) to generate dose-response curves, and lists TLVs and PELs for several compounds (Table 8.2-4).

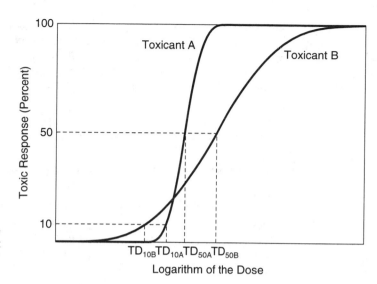

Figure 2.7-1 Dose-response curves for two compounds that have different relative Threshold Limit Values.

Example 2.7-1

The toxic response of two chemicals, A and B, as a function of dose is shown in Figure 2.7-1. Chemical A has a higher threshold concentration, at which no toxic effects are observed, than chemical B. Once the threshold dose is exceeded, however, chemical A has a greater response to increasing dose than chemical B. If the TLV were based on the dose at which 10% of the population experienced health effects, then chemical B would have a lower TLV than chemical A. In contrast, if the TLV were based on the dose at which 50% of the population experienced a health impact, chemical A would have the lower TLV. So, which chemical is more toxic?

Solution: The answer depends on the precise definition of toxicity and the specifics of the dose-response relationship. This conceptual example is designed to illustrate the dangers of using simple indices as precise, quantitative indicators of environmental impacts. There is value, however, in using these simple indicators in rough, qualitative evaluations of potential environmental impacts.

Developing the data to support a dose-response curve is expensive, time-consuming, and rigorous. It is generally not performed until some screening has suggested that it could be useful. When this testing is performed, it often begins with a rangefinder study. The purpose of this preliminary study is to determine what order of magnitude of dose generates adverse effects. This improves the quality of the dose-response testing.

The outcome of the overall dose-response effort helps tell the assessor what the toxicological endpoint of concern is. Are we concerned about neurotoxicity in young children, whose nervous system is still developing? Are we studying cancer in a particular organ? The dose-response study also provides the NOAEL and the benchmark dose (BMD). These quantities can provide a basis for risk assessment.

Since dose-response testing is so resource-intensive, risk assessors sometimes use structural-activity relationships to estimate a NOAEL or BMD, generally incorporating a coefficient to account for uncertainty. That is, we find a chemical whose NOAEL or BMD is known and has similar (chemical) functional groups to the substance of interest. The structural analog is then used to estimate a NOAEL or BMD for the substance with no dose-response curve available (Auer et al., 1990).

For cancer, dose-response analysis is appropriate for Group A and B substances. Fewer than 10% of the 80,000 or so chemicals in commerce currently have dose-response curves.

There are several important concerns associated with dose-response analysis.

1. Different species may have different responses. We don't know if humans are more or less sensitive than the most sensitive species of rodent. In the absence of data, risk assessors use a safety factor of 10 to account for this uncertainty. With data, a scaling factor of body weight to the 3/4 power is used to convert from rodents to humans. Similar scaling factors are available for a large number of laboratory animals.

2. Very high doses, to the point of acute poisoning of the test animal, are sometimes necessary to generate a statistically significant effect. The shape of the curve below the lowest dose tested is truly unknown, and often very relevant. Actual exposures are often well below the lowest tested dose. Models have been developed to approximate this portion of the dose-response curve.

3. Since it may take a long time for cancers to be detected in laboratory animals, some otherwise well-designed experiments may have been too brief. Furthermore, the time-to-tumor may be a function of dose, which further complicates the entire analysis.

4. The route of exposure can also effect the outcomes of an analysis. For example, Chromium (VI) is hazardous when inhaled; however, laboratory experiments have not shown evidence that exposure through ingestion causes any adverse effects. Therefore, it is extremely important to be cognizant of the route of exposure when assessing risk.

2.8 EXPOSURE ASSESSMENT

The amount of a substance that comes into contact with the external boundaries of a person is called exposure. The quantity that crosses the external boundary is defined as dose and the amount absorbed is the internal dose. The ratio of the internal dose to exposure is called the bioavailability of the substance. While some organizations and older sources of information use slightly different definitions, these have been adopted by the US EPA.

Two common routes of exposure to chemicals are through the skin (dermal) and the lungs (inhalation). Because exposure to chemical in the workplace can occur through inhalation and skin absorption, the engineer must be aware of potential pathways into the body. In Section 6.2.2, the exposure pathway model highlights potential pathways leading from the process to the worker and provides a framework for evaluating pathways for exposure to chemical in the workplace. Most dermal exposures result from hand contact and may occur while performing common worker activities such as sampling, drumming, filter changing, and maintenance. Since skin is a protective barrier for many kinds of chemicals, the bioavailability of these substances is often low, perhaps 5%. Inhalation exposure may be in the form of vapors, aerosols, or solid particulates. Exposure to vapors, for example, may occur due to vapors generated during activities such as drumming and sampling, or from fugitive emissions from small process leaks. Unlike dermal exposure, the bioavailability of these inhaled vapors can be quite high, often close to 100%. Section 8.3 describes US EPA's AP-42, a document on emissions issues and estimation methods.

A third route of exposure is ingestion, through either eating or drinking. Exposure through ingestion is not usually of interest in scenarios likely to be assessed in occupational settings. However, the engineer should note that ingestion can be a major source of exposure to workers who may eat, drink, or smoke on the job without adequate time or facilities for washing up, or where clean rooms are not available in situations where surfaces are contaminated with chemicals and eating, drinking, or smoking is allowed. A fourth route of exposure is percutaneous exposure, or injection through the skin. However, this type of exposure is rarely seen in the workplace. A notable example is the potential for needlesticks in healthcare settings.

The preferred approach for assessing exposure is to use personal monitoring data for the chemical of interest at the site. If not available, monitoring data for the chemical at sites with similar operations is the next choice. If there are no data available on the chemical of interest, exposure can be assessed using data for a surrogate chemical. A surrogate chemical is one whose physical and chemical properties are as similar as possible, and is used in similar operations. Finally, in the absence of any relevant data, exposure can be assessed using models. For example, a mass balance model can be used to estimate inhalation exposure to vapors (see Chapter 6).

A different approach to addressing exposure is to measure some appropriate biomarker. This applies to people who have already been exposed. A biomarker is a measurable substance whose presence in the body is a direct result of exposure to a specific chemical. Exposure may be estimated from models and based upon the biomarker measurements. Unfortunately, there are few substances that pose an exposure concern for which a biomarker has already been identified and measured. Some substances have metabolites which can be detected in blood or urine; these are common testing approaches for biomarkers.

As the engineer proceeds down this hierarchy of methods for assessing exposure, the degree of uncertainty increases. Information about this uncertainty must be communicated before risk management decisions are made. On the other hand, a high degree of uncertainty may be acceptable for some decisions. In addition to workplace exposure, exposures in the ambient environment resulting from plant emissions may also be part of the exposure assessment (Ott and Roberts, 1998). Using emissions as a starting point, exposure can be estimated from a variety of models which consider environmental fate and transport.

Techniques for modeling workplace exposure are described in more detail in Chapter 6. A discussion of exposure assessment for the ambient environment is also presented.

Example 2.8-1

A facility produces a liquid waste which contains several hazardous chemicals. There are several improper methods of disposal which, if used, would potentially expose the general population in the area surrounding the plant to these chemicals.

1. Set open-top drums containing the hazardous waste behind the plant and allow the waste to evaporate.
2. Pour the liquid waste into a ditch leading to a creek behind the plant.
3. Bury drums containing the liquid waste behind the plant.

For each of these improper disposal options, identify the pathways by which the population in the area surrounding the plant might be exposed to the hazardous chemicals.

Solution:

1. Allowing volatile wastes to evaporate will expose persons near or downwind from the open drums. The primary route of exposure will be by inhalation of air containing the volatile chemicals. If the chemical substances are corrosive, they may also irritate the skin and eyes.
2. Liquids discharged into drainage ways, either intentionally or by accident spillage, will flow quickly into nearby rivers or streams. The flow may be enhanced by precipitation runoff which washes the liquid discharge into the waterway. Persons whose drinking water comes from downstream surface waters could be exposed to the chemical substance if it is not removed during treatment by the local water supply authority. In arid areas, the chemical may seep into the ground and be carried downward by subsequent precipitation. Persons in these areas using ground water as a source of drinking water would be exposed to the chemical by ingestion of the chemical.
3. Improperly buried drums may leak their contents into surrounding soils. The leakage will flow downward by gravity until it reaches the water table. If soluble, the chemical will dissolve in the ground water and may be ingested by persons using local ground water as a source of water supply. Also, liquid leaking from buried drums can be carried by shallow ground water to nearby residences; the chemical could vaporize and accumulate in basements when air flow is stagnant. Persons in these residences would be exposed to the chemical vapors when inhaling the basement air.

2.9   RISK CHARACTERIZATION

Risk characterization is the amalgamation of available hazard and exposure information—i.e., risk, as well as all major issues developed during the assessments, including the uncertainty of all aspects of the analysis. It embodies the effects of potential concern, the route and the magnitude of the expected exposure and the numbers of the populations estimated to be exposed. As stated above, the primary human health concern is carcinogenicity. Generally, the potential for carcinogenicity is assessed using pharmaco-kinetics, chronic toxicity data from analogs, and mechanistic information (when these data are available).

2.9.1   Risk Characterization of Cancer Endpoints

The classical treatment of cancer risk defines risk as the probability of developing cancer from a particular chemical if a sub-population is exposed to that chemical over a lifetime. A person can contract cancer from many sources besides exposure to a particular chemical. This concept is called the background cancer level, and must be separated from the probability of developing cancer from a particular chemical exposure. Thus, risk is defined in this particular context as the cancer probability in excess of the background cancer level. Our basic equation of risk is:

$$\text{Risk} = f(\text{Hazard, Exposure})$$

The basis for cancer risk assessment is the dose-response curve (risk of incidence of cancer vs. dose of an agent). Unfortunately, for cancer, bioassays are usually run at only two doses to describe the carcinogenic response of the test species. The relationship is typically non-linear. Since it is assumed that carcinogens do not have thresholds, the "cancer" model generates a non-linear curve. There is never enough data to provide a complete dose-response curve. To deal with this reality, the risk assessor is left with the option of applying one of a number of mathematical models to the limited data set so as to describe the relationship. For a new chemical, with limited dose-response data, one methodology is to use the slope of the dose-response curve or (percent response per mg pollutant per kg of body weight per day) as a measure of hazard. Exposure is the quantity that arrives at the surface of a person's body, in mg of pollutant per kg body weight per day. This simple application of the basic risk equation often provides the risk manager with sufficient information to make risk management decisions.

2.9.2   Risk Characterization of Non-Cancer Endpoints

Non-cancer risk also has a dose-response curve. The model relationship in this case is linear. Therefore, simplifying assumptions allow us to characterize the risk of adverse health effects as a simple ratio or Hazard Quotient. The Hazard Quotient is the ratio of the estimated chronic dose or exposure level to the RfD or RfC.

Hazard Quotient values below unity imply that adverse effects are very unlikely. The more the Hazard Quotient exceeds unity, the greater the level of concern. However, the Hazard Quotient is not a probabilistic statement of risk.

2.9.3 Adding Risks

The discussion above presumes risk occurs from one chemical at one source. In fact, there are multiple chemicals, multiple pathways, and multiple exposure routes. It is necessary either to estimate what the most important risks are, or to calculate all sources and pathways. Aggregate and Cumulative Risk are fairly recent terms in the lexicon. Aggregate means adding risks together from multiple exposure routes: dermal, inhalation and ingestion.

The use of the term endpoint becomes important in the emerging area of Cumulative Risk assessment. Sometimes, the risks from one chemical may be too low to generate concern. However, several different chemicals may have the same toxicological endpoint. That is, they affect an organ or system adversely in the same way. Exposures from these chemicals need to be combined to determine whether the adverse effect may occur as a result of a combination of chemical exposures.

SUMMARY

Risk is a quantitative assessment of the probability of an adverse outcome. Risk may result from voluntary exposure to hazardous conditions in one's occupation, involuntary exposure to radiation, chemicals, pathogens, or the reckless behavior of others, or natural disasters.

There are four components of risk assessment: hazardous assessment; dose-response; exposure assessment; and risk characterization. The engineer should work with chemists, toxicologists, and others when a risk assessment is needed. Although there may be uncertainties in performing risk assessments, it can assist in choosing between process options.

The risk concepts presented will be expanded on in later chapters throughout the text, and their direct application in assessing risk in the manufacturing and use of chemical processes and products will be shown.

REFERENCES

Auer, C.M., Nabholz, J.V., and Baetcke, K.P., Mode of Action and the Assessment of Chemical Hazards in the Presence of Limited Data: Use of Structure-Activity Relationships (SAR) under TSCA, Section 5, *Environmental Health Perspectives.* Vol. 87, pp. 183–197, 1990.

Casarett, L.J. and Doull, J., *Toxicology, the Basic Science of Poisons* Fifth Edition, Macmillan Publishing Co., Inc. New York, 1995.

Cicmanec, J.L., Dourson, M.L., and Hertzberg, R.C., *Noncancer Risk Assessment: Present and Emerging Issues in Toxicology and Risk Assessment,* Chapter 17, Fan, A.M., Academic Press, 1997.

Cooke, R. and Jager, E., A probabilistic model for failure frequency of underground gas pipelines, Risk Analysis, 18(4), 511–523, 1998.

Fan, A.M. and Chang, L.W., *Toxicology and Risk Assessment: Principles, Methods, and Applications,* Marcel Dekker, Inc. New York, NY, 1996, p. 247.

Federal Focus, Inc. *Towards Common Measures: Recommendations for a Presidential Executive Order on Environmental Risk Assessment and Risk Management Policy,* Federal Focus Inc. and The Institute for Regulatory Policy, Washington, D.C., 1991.

Fort, D.D., "Environmental Laws and Risk Assessment," in *Toxicology and Risk Assessment: Principles, Methods, and Applications,* Fan, A.M. and Chang, L.W. ed., Marcel Dekker, Inc. New York, NY, p. 653–677, 1996.

Hammond, E.C., and Selikoff, I.J.: Relation of cigarette smoking to risk of death of asbestos-associated disease among insulation workers in the United States, in Biological Effects of Asbestos, P. Bogovski, J.C. Gilson, V. Timbrell, J.C. Wagner, and W. Davis, editors, *Int. Agency Res. Cancer,* Scientific Publication No. 8, Lyon, France, p. 312–317, 1973.

Kamata E. et al., Acute and subacute toxicity studies of tris(1,3-dichloro-2-propyl) phosphate on mice. *Bull Natl Inst Hyg Sci,* 107:36–43 (1989).

Lednar, W.M., Tyroler, H.A., McMichael, A.J., and Shy, C.M., The occupational determinants of chronic disabling pulmonary disease in rubber workers, *J. Occup. Med.* 19(4), p. 263–268, 1977.

Monsanto Company, St. Louis, MO, 1980.

NRC, National Research Council, "Risk Assessment in the Federal Government: Managing the Process," Committee on Institutional Means for Assessment of Risks to Public Health, National Academy Press, Washington, D.C., 1983.

Ott, W.R. and Roberts, J.W. "Everyday Exposure to Toxic Pollutants," *Scientific American,* February 1998.

Patton, D.E., "The ABCs of Risk Assessment," *EPA Journal,* Volume 19, Number 1. January 1993. Document Control Number EPA 175-N-93-014

Presidential/Congressional Commission on Risk Assessment and Risk Management; Final Report, Vol 1, 1997; p.1.

Roberts, W.C. and Abernathy, C.O., "Risk Assessment: Principles and Methodologies," in *Toxicology and Risk Assessment: Principles, Methods, and Applications,* Fan, A.M. and Chang, L.W. ed., Marcel Dekker, Inc. New York, NY, p. 245–270, 1996.

Roberts, W.C. and Abernathy, C.O., "Risk Assessment: Principles and Methodologies," in *Toxicology and Risk Assessment,* Chapter 15, Fan, A.M. ed., Academic Press, 1997.

Selikoff, I.J., Hammond, E.C., and Churg, J., Asbestos Exposure, Smoking, and Neoplasia, *J. Am. Med. Assoc.* 204(2): p. 106–112, 1968.

US EPA, United States Environmental Protection Agency, *Guidelines for Exposure Assessment,* FR Vol. 57, No. 104, May 29, 1992.

US EPA, United States Environmental Protection Agency, Terminology Reference System (TRS 2.0), September 11, 2000, *http://oaspub.epa.gov/trs/prc_qry.keyword.htm.*

US EPA, United States Environmental Protection Agency, National Center for Environmental Assessment, Benchmark Dose Software, *http://www.epa.gov/ncea/bmds.htm.*

Velazquez et al. "Cancer Risk Assessment: Historical Perspectives, Current Issues, and Future Directions," Chapter 14 of *Toxicology and Risk Assessment,* Fan, A.M., Academic Press, 1997

PROBLEMS

1. Each year, approximately 45,000 persons lose their lives in automobile accidents in the United States (population 281 million according to the 2000 census). How many fatalities would be expected over a three-day weekend in the Minneapolis-St. Paul, Minnesota metropolitan area (population 2 million)?

2. A collection sump has two control systems to prevent overflowing of the sump (see figure below). The first is a level sensor connected to an alarm which alerts the operator to the high liquid level. A second level sensor is connected to a solenoid valve which opens a drain to lower the liquid level in the sump. Draw the fault tree diagram for this system.

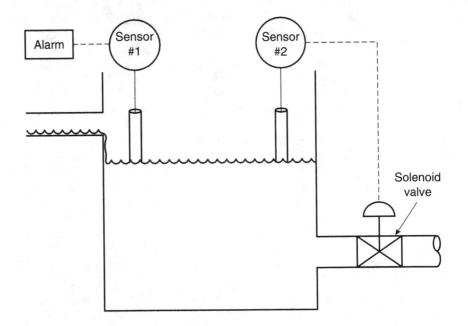

3. Toxicological testing is performed on non-human species: mice, rats, rabbits, dogs, and others. If a no observed adverse effect level (NOAEL) is determined in a non-human species, what safety factor should be applied to this NOAEL to set an exposure level that is acceptable for humans?

4. Repeat the reference dose calculation of Example 2.6-3 for tris(2-chloroethyl) phosphate for which the NOAEL was determined to be 22 mg/kg-day and the LOAEL was found to be 44 mg/kg-day for increased weights of liver and kidneys in rats.

5. A colleague has requested your advice on selection of a safe solvent for a photoresist. A photoresist consists of an acrylate monomer, polymeric binder, and photoinitiator applied to the surface of a copper-clad laminate or silicon wafer. After the solvent evaporates, the photoresist is exposed to ultraviolet light through a mask containing the pattern to be etched on the circuit board or silicon wafer. When exposed, the resist polymerizes and becomes insoluble to the developer. The circuit board or silicon wafer is subsequently washed with the developer solution to remove unpolymerized photoresist, exposing the pattern to be etched with acid into the copper metal or the silicon wafer. Your colleague has identified the following solvents as suitable for formulation of the photoresist.

Solvent	CAS Number	Vapor Pressure kPa at 25 C	OSHA Permissible Exposure Limit, parts per million
furfuryl alcohol	98-00-0	0.1	50
diethylamine	109-89-7	30.1	25
ethyl acetate	141-78-6	12.6	400
monomethyl ether	109-86-4	1.3	25
methyl ethyl ketone	79-93-3	12.1	200
n butyl acetate	123-86-4	1.3	150

(a) Using the OSHA Permissible Exposure Limit as a surrogate for relative hazard, a higher OSHA PEL connoting a lower hazard, rank these solvents from highest hazard to lowest.

(b) Using the vapor pressure as a surrogate for the magnitude of worker exposure to the solvent vapors, rank these solvents from highest exposure potential to lowest.

(c) Considering both hazard and exposure potential, which of these solvents would you recommend to your colleague for the photoresist solution?

(d) What alternatives can be used to reduce the risk associated with solvents even further?

6. **Carcinogenic Risk Assessment Near a Petroleum Refinery:** A petroleum refinery on the east coast has initiated a voluntary program to evaluate sources of environmental emissions. As one important step in this evaluation, the company wishes to perform a quantitative risk assessment on the atmospheric releases of volatile organic compounds, some of which are toxic, from the facility. As a test case, perform a risk assessment on benzene released to the air from the facility and its impact on human health (carcinogenic impact, inhalation only) in a hypothetical residential area located 1 km from the center of the facility (assume that the center is the emission source). The dose-response carcinogenic slope factor (SF) for benzene inhalation is 2.9×10^{-2} [mg benzene/(kg body weight $\bullet$ day)]$^{-1}$). The maximum average annual concentration of benzene in the *outside* air (C_A) within a residential area downwind from the facility is 82 $\mu g/m^3$.

Using the following exposure properties,

Exposure Properties:

BW-	Average Adult Body Weight (kg)	70
CR-	Air breathing rate (m^3/day)	19.92
RR-	Retention rate, inhaled air	1.0
ABS-	Absorption rate, inhaled air	1.0
EF-	Exposure frequency (exposure days/yr)	365
ED-	Exposure duration (yr)	70
AT-	Averaging time (days)	25,550

where RR is the efficiency of the lungs to retain benzene and ABS is the efficiency of the lung tissue to absorb the retained chemical. These values were to set to default values (1.0) for this problem and may actually be much lower.

(a) Calculate the inhalation dose of benzene to a typical resident using the following equation

$$\text{Inhalation Dose (mg benzene/(kg body weight} \cdot \text{day))} =$$
$$\frac{C_A \times CR \times EF \times ED \times RR \times ABS}{BW \times AT}$$

(watch your units!)

(b) Calculate the inhalation carcinogenic risk for this scenario using the following equation.

$$\text{Inhalation Carcinogenic Risk (dimensionless)} = \text{Inhalation Dose} \times SF$$

(c) Is the risk greater than the recommended range of $< 10^{-4}$ to 10^{-6} for carcinogenic risk?

(d) Discuss possible reasons that this methodology might over-predict the actual risk based on the discussions in this chapter and on the information given in the problem statement.

CHAPTER 3

Environmental Law and Regulations: From End-of-Pipe to Pollution Prevention

by
David R. Shonnard

3.1 INTRODUCTION

Chemical engineers practice a profession and must obey rules governing their professional conduct. One important set of rules that all chemical engineers should be aware of is environmental statutes, which are laws enacted by Congress. Regulations are promulgated by administrative agencies based on authority conferred by the statute. The environmental statutes are designed to protect human health and the environment by placing limits on the quantity and chemical make-up of waste streams that are released from manufacturing processes. For example, one statute places restrictions on how hazardous waste from industry is stored, transported, and treated. Another statute places strict liability on the generators of hazardous waste, requiring responsible parties to clean up disposal sites that fail to protect the environment. For manufacturers of new chemicals, there are regulatory requirements that require filing of a premanufacture notice (PMN) before introducing a new chemical into the marketplace. While many companies have Health, Safety, and Environment (HS&E) staff that can help the engineer interpret and implement environmental requirements, it is nevertheless important that chemical engineers be aware of prominent federal environmental laws, and adhere to the requirements of these statutes.

The purpose of this chapter is to provide an overview of environmental regulation. Much of the material on regulations in this chapter has been adapted from the excellent review of environmental law by Lynch (1995). More comprehensive sources on this topic include the United States Code (U.S.C.) and the Code of Federal Regulations (C.F.R.), which are sets of environmental statutes and regulations, respectively; they are available online at the site maintained by the federal government printing office. The Environmental Law Handbook (Sullivan and Adams, 1997) and West's Environmental Law Statutes (West Publishing Co.) are compendia of existing statutes. Most of these sources can be found online.

There are approximately 20 major federal statutes, hundreds of state and local ordinances, thousands of federal and state regulations, and even more federal and state court cases and administrative adjudications, etc., that deal with environmental issues. Taken together, they make up the field of environmental law, which has seen explosive growth in the last 30 years, as shown in Figure 3.1-1. Chemical engineers should be familiar with environmental laws and regulations because they affect the operation of chemical processes and the professional responsibilities of chemical engineers. *Environmental regulations* and the *common law system* of environmental law require actions by affected entities. For example, the Clean Water Act (an environmental statute) requires facilities which discharge pollutants from a point source into navigable waters of the United States to apply for a National Pollutant Discharge Elimination System (NPDES) permit. In many firms, chemical engineers are responsible for applying for and obtaining these permits. The common law created by judicial decision also encourages chemical engineers to act responsibly when performing their professional duties because environmental laws and regulations do not cover every conceivable environmental wrong. Chemical engineers need to be aware of potential legal liability resulting from violation of environmental laws and regulations to protect their company and themselves from legal and administrative actions.

The sources of environmental law and regulations are legislatures, administrative agencies, and the courts. When drafting environmental laws, federal and state legislatures often use broad language to describe the objectives, regulatory programs, and enforcement provisions of the statute. Often, legislators do not have the time or resources needed to implement the statute and therefore leave the detailed development of regulations to administrative agencies. Administrative agencies, such as the Environmental Protection Agency, give meaning to statutory provisions through a procedure known as rule making. Federal rule making consists of giving notice of proposed new regulations by publication in the *Federal Register*, providing an opportunity for public comment, altering the proposed rule, where appropriate, to incorporate the comments received, and publishing final regulations in the Federal Register. Final rules have the force of law. As such, administrative agencies fulfill a legislative function delegated to them by Congress.

Administrative agencies can be created by the executive or legislative branches of government. In 1970, President Nixon established the United States Environmental Protection Agency by executive order to consolidate federal programs for regulating air and water pollution, radiation, pesticides, and solid waste disposal. However, administrative agencies are most often established by statute (for example, the Occupational Safety and Health Act established the Occupational Safety and Health Administration), and in these cases, the agency powers are derived from their enabling legislation. Administrative agencies also have the authority to resolve disputes that arise from the exercise of their administrative

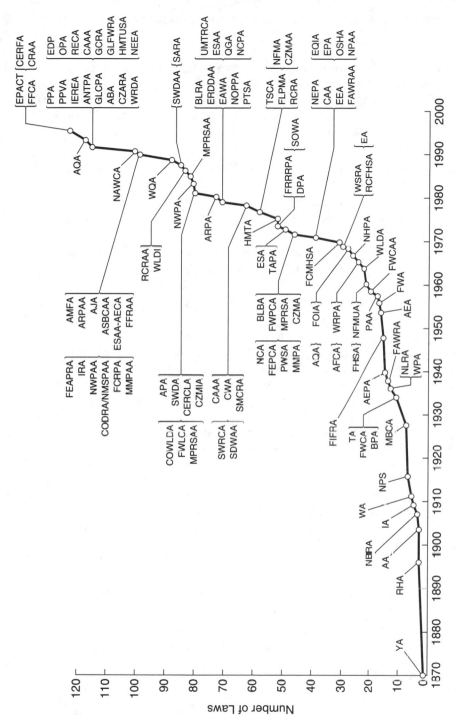

Figure 3.1-1 Cumulative growth in federal environmental laws and amendments.

powers. Regulated entities have the right to appeal decisions made by administrative agencies to an administrative law judge, who is appointed by the agency. Thus administrative agencies have a judicial function in addition to a legislative function.

Courts are a third government actor that defines the field of environmental law. The role of the courts in environmental law is:

1. To determine the coverage of environmental statutes (which entities are covered by regulations);
2. To review administrative rules and decisions (ensuring that regulations are promulgated following proper procedures and within the limits of statutorily delegated authority); and,
3. To develop the common law (a record of individual court cases and decisions that set a precedent for future judicial decisions).

Section 3.2 provides a brief description of the most important features of nine federal environmental statutes that most significantly affect chemical engineers and the chemical industry. This brief survey is meant to be representative, not comprehensive, and the focus will be on federal laws because they have national scope and often serve as models for state environmental statutes. We begin with three statutes that regulate the creation, use, and manufacture of chemical substances. Next, we cover the key provisions of three statutes that seek to control the discharge of pollutants to specific environmental media—air, water, and soil. Next, a statute that initiated a clean-up program for the many sites of soil and groundwater contamination is discussed. The final two statutes involve the reporting of toxic substance releases and a voluntary program for preventing pollution generation and release at industrial facilities. Section 3.3 describes the evolution in environmental regulation from end-of-pipe pollution control to more proactive pollution prevention approaches. Section 3.4 presents the key features of pollution prevention, including its position in the hierarchy of environmental management alternatives, a short review of terminology, and examples of pollution prevention strategies and applications.

3.2 NINE PROMINENT FEDERAL ENVIRONMENTAL STATUTES

This section provides the key provisions of nine federal environmental statutes that every chemical engineer should know. Taken together, these laws regulate chemicals throughout their life cycle, from creation and production to use and disposal. The nine laws are:

a) The Toxic Substances Control Act (TSCA), 1976 (regulating testing and necessary use restrictions on chemical substances).
b) The Federal Insecticide, Fungicide, and Rodenticide Act (FIFRA), 1972 (the manufacture and use of pesticides).

c) The Occupational Safety and Health Act (OSHA), 1970 (to protect health and safety in the workplace).

d) The Clean Air Act (CAA), 1970 (to protect and enhance the quality of the Nation's air resources).

e) The Clean Water Act (CWA), 1972 (to restore and maintain the chemical, physical, and biological integrity of the Nation's water resources).

f) The Resource Conservation and Recovery Act (RCRA), 1976 (the regulation of hazardous and non-hazardous waste treatment, storage, and disposal).

g) The Comprehensive Environmental Response, Compensation, and Liability Act (CERCLA), 1980 (the cleanup of abandoned and inactive hazardous waste sites).

h) The Emergency Planning and Community Right-to-Know Act (EPCRA), 1986 (responding to chemical emergencies and reporting of toxic chemical usage).

i) The Pollution Prevention Act (PPA), 1990 (a proactive approach to reducing environmental impact).

A summary of these prominent federal environmental statutes is provided in Table 3.2–1. The most important regulatory provisions for each statute are stated along with a listing of some key requirements for chemical processing facilities. A more complete description of these federal statutes is included in Appendix A.

3.3   EVOLUTION OF REGULATORY AND VOLUNTARY PROGRAMS: FROM END-OF-PIPE TO POLLUTION PREVENTION

Many of the environmental laws listed in the previous section were enacted to ensure the protection of a single environmental medium. For example, the Clean Air Act instituted a strategy for pollution control on atmospheric emissions. Similarly, the Clean Water Act and the Resource Conservation and Recovery Act provided systems for the protection of the water and the soil environments, respectively. Although these legislative actions have been extremely effective in restoring and maintaining environmental media, they have not ensured that the total amount of hazardous materials entering the environment will eventually decrease. In fact, despite more than twenty years of regulation the volumes and hazards of toxic chemical releases into the environment continued to grow through the 1970s and 1980s (Johnson, 1992).

Beginning in the mid to late 1980s, however, the absolute amounts of toxic releases to the environment in many categories began to decrease. If one uses the Toxics Release Inventory (TRI) as a gauge, the amount of "toxics" released decreased from 3.4 billion pounds in 1986 to less than 2.0 billion pounds in 1998 (USEPA, 2000). The amount released decreased every year from 1988 through 1996 (releases in 1997 were slightly up from 1996 as a result of a booming

Table 3.2-1 Summary Table for U.S. Environmental Laws

Environmental Statute	Background	Key Provisions
Regulation of Chemical Manufacturing		
The Toxic Substances Control Act (TSCA) 1976	Highly toxic substances, such as polychlorinated biphenyls (PCBs), began appearing in the environment and in food supplies. This prompted the federal government to create a program to assess the risks of chemicals before they are introduced into commerce and to test existing chemical substances.	Chemical manufacturers, importers, or processors must submit a report detailing chemical and processing information for each chemical. Extensive testing by companies may be required for chemicals of concern. For newly created chemicals, a Premanufacturing Notice must be submitted.
The Federal Insecticide, Fungicide, and Rodenticide Act (FIFRA) Enacted, 1947 Amended, 1972	Because all pesticides are toxic to some plants and animals, they may pose an unacceptable risk to human health and the environment. FIFRA is a federal statute whose purpose is to assess the risks of pesticides and to control their usage so that any exposure that may result poses an acceptable level of risk.	Before any pesticide can be distributed or sold in the US, it must be registered with the EPA. The registration data are difficult and expensive to develop and must prove that the chemical is effective and safe to humans and the environment. Labels must be placed on pesticide products that indicate approved uses and restrictions.
The Occupational Safety and Health Act (OSH Act) 1970	The agency that oversees the implementation of the OSH Act is the Occupational Safety and Health Administration (OSHA). All private facilities having more than 10 employees must comply with the OSH Act requirements.	Companies must adhere to all OSHA health standards (exposure limits to chemicals) and safety standards (physical hazards from equipment). The OSH Act's Hazard Communication Standard requires companies to develop hazard assessment data (material safety data sheets (MSDS)), label chemical substances, and inform and train employees in the safe use of chemicals.
Regulation of Discharges to the Air, Water, and Soil		
Clean Air Act (CAA) Enacted 1970, Amended 1990	The CAA is intended to control the discharge of air pollution by establishing uniform ambient air quality standards that are in some instances health-based and in others, technology-based. The CAA also addresses specific air pollution problems such as hazardous air pollutants, stratospheric ozone depletion, and acid rain.	The CAA established the National Ambient Air Quality Standards (NAAQS) for maximum concentrations in ambient air of CO, Pb, NO_2, O_3, particulate matter, and SO_2. States must develop source-specific emission limits to achieve the NAAQS. States issue air emission permits to facilities. Stricter requirements are often established for hazardous air pollutants (HAPs) and for new sources.

Clean Water Act (CWA)
Enacted, 1972

The Clean Water Act (CWA) is the first comprehensive federal program designed to reduce pollutant discharges into the nation's waterways ("zero discharge" goal). Another goal of the CWA is to make water bodies safe for swimming, fishing, and other forms of recreation ("swimmable" goal). This act has resulted in significant improvements in the quality of the nation's waterways since its enactment.

The CWA established the National Pollutant Discharge Elimination System (NPDES) permit program that requires any point source of pollution to obtain a permit. Permits contain either effluent limits or require the installation of specific pollutant treatment. Permit holders must monitor discharges, collect data, and keep records of the pollutant levels of their effluents. Industrial sources that discharge into sewers must comply with EPA pretreatment standards by applying the best available control technology (BACT).

Resource Conservation and Recovery Act (RCRA)
Enacted 1976

The Resource Conservation and Recovery Act was enacted to regulate the "cradle-to-grave" generation, transport, and disposal of both non-hazardous and hazardous wastes to land; encourage recycling; and promote the development of a tentative energy sources based on solid waste materials.

Generators must maintain records of the quantity of hazardous waste generated and where the waste was sent for treatment, storage, or disposal, and file this data in biennial reports to the EPA. Transporters and disposal facilities must adhere to similar requirements for record keeping as well as for monitoring the environment.

Clean-Up, Emergency Planning, and Pollution Prevention

The Comprehensive Environmental Response, Compensation, and Liability Act (CERCLA) 1980

CERCLA began a process of identifying and remediating uncontrolled hazardous waste at abandoned sites, industrial complexes, and federal facilities. EPA is responsible for creating a list of sites ranked by level of risk, which is termed the National Priority List (NPL). CERCLA was amended by the Superfund Amendments and Reauthorization Act (SARA) of 1986.

After a site is listed in the NPL, EPA identifies potentially responsible parties (PRPs) and notifies them of their potential CERCLA liability, which is strict, joint and several, and retroactive. PRPs are 1) present or 2) past owners of hazardous waste disposal facilities, 3) generators of hazardous waste, and 4) transporters of hazardous waste.

The Emergency Planning and Community Right to Know Act (EPCRA) 1986

Title III of SARA contains a separate piece of legislation called EPCRA. There are two main goals of EPCRA: 1) to have states create local emergency units that must develop plans to respond to chemical release emergencies, and 2) to require EPA to compile an inventory of toxic chemical releases to the air, water, and soil from manufacturing facilities.

Facilities must work with state and local entities to develop emergency response plans in case of an accidental release. Affected facilities must report annually to EPA data on the maximum amount of the toxic substance on-site in the previous year, the treatment and disposal methods used, and the amounts released to the environment or transferred off-site for treatment and/or disposal.

Pollution Prevention Act (PPA) 1990

The Pollution Prevention Act established pollution prevention as the nation's primary pollution management strategy with emphasis on source reduction and established a Pollution Prevention Information Clearinghouse whose goal is to compile source reduction information and make it available to the public.

The PPA requires owners and operators of facilities that are required to file a Form R under the SARA Title III to report to the EPA information regarding the source reduction and recycling efforts that the facility has undertaken during the previous year.

economy). In addition, concentrations of many categories of pollutants in the environment are going down over time. This is true for ozone, lead, volatile organic compounds (VOCs), and carbon monoxide (CO). Other environmental indicators are also showing improvement. For example, the amount of energy used per dollar of Gross National Product has decreased from about 15,000 to 11,000 Btu/1990 dollar (EIA, 1998) over the last 10 years.

As additional reductions in emissions to individual environmental media are sought, it is important to guard against moving pollutants from one environmental medium into another. For example, traditional air pollution control devices such as scrubbers transfer pollutants from a gaseous stream to a liquid stream. The liquid stream would require further treatment to either remove or destroy the original contaminant. Conversely, some wastewater streams containing volatile organic compounds are contacted with an air stream, transferring the pollutants from the water to air. A more subtle form of media shifting can occur when pollutants are destroyed or transformed into less harmful forms by reaction during waste treatment. These processes can be very energy intensive, and energy use can result in the formation of pollutants.

It is clear from the trends just discussed that a complementary strategy is needed to reduce the amounts and the hazardous characteristics of industrial wastes released into all media of the environment. This strategy should also decrease the amounts of contaminants entering traditional waste treatment processes. In the next section of this chapter, we will review the environmental management hierarchy as outlined in the Pollution Prevention Act of 1990 and define important terms, such as pollution prevention, source reduction, and others. These definitions will provide a proper context and categorization for much of the pollution prevention design activities discussed in the remainder of the text.

3.4 POLLUTION PREVENTION CONCEPTS AND TERMINOLOGY

A logical starting point for understanding pollution prevention concepts is the waste management hierarchy established in the Pollution Prevention Act of 1990. The waste management hierarchy is defined as follows (U.S.C. §§13101–13109):

> The Congress hereby declares it to be the national policy of the United States that pollution should be prevented or reduced at the source whenever feasible; pollution that cannot be prevented should be recycled in an environmentally safe manner, whenever feasible; pollution that cannot be prevented or recycled should be treated in an environmentally safe manner whenever feasible; and disposal or other release into the environment should be employed only as a last resort and should be conducted in an environmentally safe manner.

Based on this definition and distinctions between recycle options, we can place the waste management hierarchy in the following descending order, from the most to the least preferable:

1. Source reduction
2. In-process recycle
3. On-site recycle
4. Off-site recycle
5. Waste treatment
6. Secure disposal
7. Direct release to the environment

The distinction between these seven elements of the waste management hierarchy are shown in Figure 3.4-1, using a simple reactor/separator sequence of units in chemical processes (adapted from Allen and Rosselot, 1997).

1. Source reduction—the reactor is modified so that less waste is generated or so that the waste is less hazardous.
2. In-process recycle—unreacted feed is separated and recycled back to the reactor.
3. On-site recycle—waste from the reactor is converted to a commercial product by a second reactor within the facility.
4. Off-site recycle—waste from the reactor is separated and then transferred off-site where it is converted to a commercial product within another facility.
5. Waste treatment—waste from the reactor is separated and then treated to render it less hazardous.
6. Secure disposal—waste from the reactor is separated and sent to a secure disposal facility (landfill).
7. Direct release to the environment—waste is separated from product and released to the environment.

The waste management hierarchy introduces a number of terms that require definition if the scope of pollution prevention activities is to be understood. In the federal Pollution Prevention Act of 1990, source reduction is defined as:

A. The term "source reduction" means any practice that
   1. Reduces the amount of any hazardous substance, pollutant, or contaminant entering any waste stream or otherwise released into the environment (including fugitive emissions) prior to recycling, treatment, or disposal.
   2. Reduces the hazards to public health and the environment associated with the release of such substances, pollutants, or contaminants.
   The term includes equipment or technology modifications, process or procedure modifications, reformation or redesign of products, substitution of raw materials, and improvements in housekeeping, maintenance, training, or inventory control.

Following are a few examples of source reduction (Hunt, 1995). *Inventory control* aims to reduce waste generation resulting from "out-of-date" or "off-spec" raw materials or final products. Effective techniques for inventory control might include ordering only the amount of raw material needed for one production run or reviewing purchasing procedures to eliminate hazardous chemicals and substitute environmentally-friendly alternatives. Other techniques for inventory control might be more challenging, such as adopting just-in-time manufacturing techniques. Modifying production procedures can lead to waste reduction and increased profits. A joint DuPont/EPA pollution prevention study showed that a cleaning solvent waste from a specialty chemical multiple batch process could be completely eliminated (US EPA, 1993). A source reduction project installed drains at low points in the process to recover chemicals from the prior campaign, yielding a Net Present Value of $2,212,000. More examples of source reduction methods for all industries are available in several references (US EPA, 1992 and 1993; Hunt, 1995). We will present several examples of unit operation-specific pollution prevention methods in Chapter 9.

The federal legislation continues to define what source reduction is *not*.

B. The term "source reduction" does not include any practice which alters the physical, chemical, or biological characteristics or the volume of a hazardous substance, pollutant, or contaminant through a process or activity which itself is not integral to and necessary for the production of a product or the providing of a service.

The federal definition of source reduction is controversial because it seems to exclude activities that may reduce the amounts of hazardous substances entering waste streams by processes that may not be "integral to and necessary for the production of a product or the providing of a service." These potentially beneficial, although excluded, processes would typically fall into the categories of on-site and off-site recycle according to the federal definition. In addition to the federal definition of source reduction, there are many state legislatures and other pertinent bodies having similar definitions that are either more or less exclusive in terms of allowable activities (Foecke, 1992).

In order to help clarify which activities constitute pollution prevention and which do not, the Pollution Prevention Act of 1990 provides a definition (Habitch, 1992).

Pollution prevention means "source reduction," as defined under the Pollution Prevention Act, and other practices that reduce or eliminate the creation of pollutants through
- Increased efficiency in the use of raw materials, energy, water, or other resources, or
- Protection of natural resources by conservation.

The act (Habitch, 1992) goes further to state what recycling activities are included within pollution prevention activities.

Drawing an absolute line between prevention and recycling can be difficult. "Prevention" includes what is commonly called "in-process recycling," but not "out-of-process recycling." Recycling conducted in an environmentally sound manner shares many of the advantages of prevention, such as energy and resource conservation, and reducing the need for end-of-pipe treatment or waste containment. . . . Some practices commonly described as "in-process recycling" may qualify as pollution prevention.

Thus, the EPA considers the first two elements of the waste management hierarchy as pollution prevention: source reduction and in-process recycling. However, many on-site and off-site recycling activities are consistent with the intent of pollution prevention because of the resulting increased efficiency in the use of raw materials, energy, and other resources. At the state level, there is some consensus in the definition of pollution prevention. For example, 17 states exclude off-site recycling and 14 exclude treatment or incineration (Foecke, 1992). Thus, many state legislatures consider pollution prevention to include the first two and perhaps three elements of the waste management hierarchy. The Pollution Prevention Task force of the American Petroleum Institute, an industry group, provides a more expansive definition of pollution prevention, to include environmentally sound recycling and multi-media reductions in discharges to air, water, and soil (API, 1993). Other definitions exist, some more restrictive and others more expansive (California EPA, 1991).

One of the key concepts that is useful in determining whether a process change is considered pollution prevention or not is defining what is and what is not a process. Is a process only a simple sequence of a reactor and a separator as shown in Figure 3.4-1 or can we consider a process to be comprised of a set of integrated subprocesses? For example, consider the process change shown in Figure 3.4-2 where two reactors are shown (adapted from Allen and Rosselot, 1997). The first reactor converts feeds A and B to product C and byproduct D and a second reactor converts feeds D and E to product F. This modification would be considered in-process recycling and thus pollution prevention by the federal definition if the two reactions are considered to comprise a single integrated process. If each reactor is considered to be a separate process, then this modification would be considered on-site (out-of-process) recycle and not pollution prevention according to the federal definition.

Another important consideration is what consitiutes a waste that would need to be treated to render the stream less hazardous and what consitiutes an intermediate stream composed of byproducts that can be transformed into commercial products. Resolving this issue can help categorize process modifications as waste treatment or intermediate recycling. Consider the process flow diagram on Figure 3.4-3 (adapted from Allen and Rosselot, 1997). The top process diagram features in-process recycling of components A and B to the reactor for further reaction and separation of product C from waste D, which is disposed into the environment. After the process modification, the stream containing D is reacted and separated further on-site to render a recycle stream containing A only and one containing E, which is transferred to an off-site recycle operation. The key issue is whether we consider stream D as a waste stream or another intermediate stream which is processed further into salable products. Thus it is difficult to know whether this process modification is waste treatment or recycling. Furthermore, the Pollution

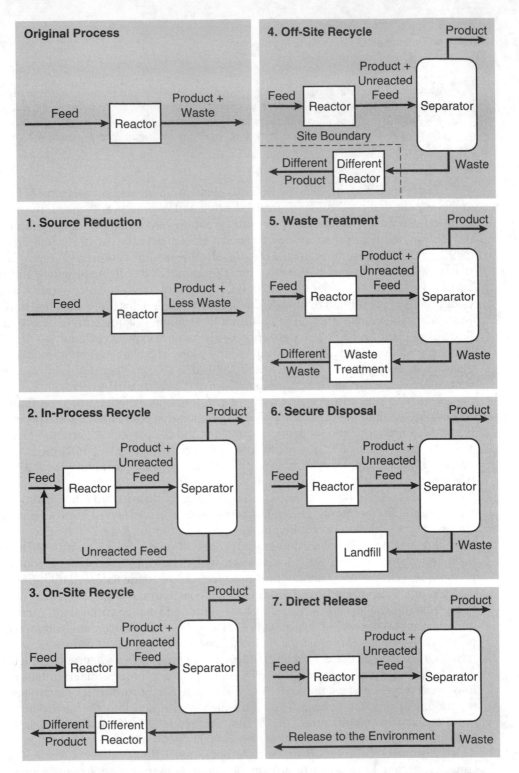

Figure 3.4-1 Waste management modifications for a simple reactor/separator process classified according to the waste management hierarchy. *Source:* Allen and Rosselot, *Pollution Prevention for Chemical Processes* © 1997. This material is used by permission of John Wiley & Sons, Inc.

Prevention Act of 1990 provides no guidance for this and many other situations that must be considered on a case-by-case basis.

Because of the ambiguities in the definition of pollution prevention and distinguishing between key elements of the waste management hierarchy, we will adopt a more expansive definition of pollution prevention in this text. Process design modification for pollution prevention will constitute the first four elements of the waste management hierarchy: source reduction, in-process recycle, on-site (out of process) recycle, and off-site recycle. The justification for this expanded definition is the many cases where recycle modifications accomplish the primary goals of pollution prevention: improving the efficiency of raw materials conversion and reducing the consumption of energy, water, and other resources.

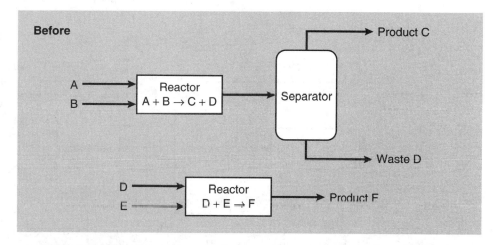

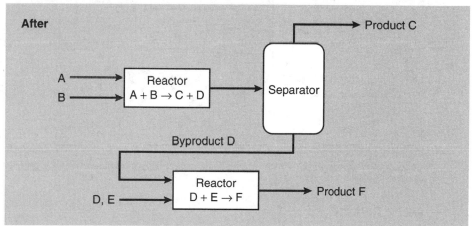

Figure 3.4-2 A process modification involving two reactors that are part of the same industrial facility. It is in-process recycling or on-site recycling? *Source:* Allen and Rosselot, *Pollution Prevention for Chemical Processes* © 1997. This material is used by permission of John Wiley & Sons, Inc.

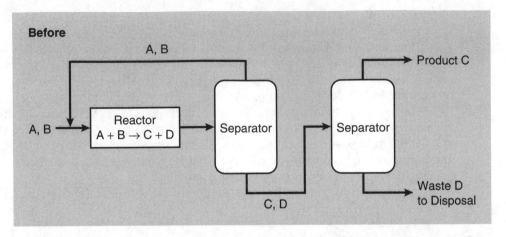

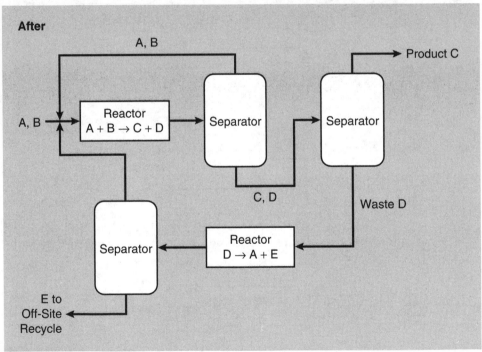

Figure 3.4-3 A process modification. Is it waste treatment or recycling? *Source:* Allen and Rosselot, *Pollution Prevention for Chemical Processes* © 1997. This material is used by permission of John Wiley & Sons, Inc.

REFERENCES

Allen, D.T. and Rosselot, K.S., *Pollution Prevention for Chemical Processes,* John Wiley & Sons, New York, 1997.

API, American Petroleum Institute, *Environmental Design Considerations for Petroleum Refining Crude Processing Units,* Publication 311, Feb. 1993.

California EPA (California Environmental Protection Agency), "Report of the 90 Day External Program Review of California's Toxic Substances Control Program," 1991.

EIA, Energy Information Administration, US Department of Energy, "International Energy Outlook—1998", DOE/EIA-0484(98), 1998.

Foecke, T. "Defining pollution prevention and related terms," Pollution Prevention Review, 2(1), 103–112, Winter 1991/1992.

Habitch, F.H., Memo to all EPA personnel, May 28, 1992.

Hunt, G.E., "Overview of waste reduction techniques leading to pollution prevention," in *Industrial Pollution Prevention Handbook,* (ed.) Freeman, H.M., McGraw-Hill, pg 9–26, 1995.

Johnson, S. "From reaction to proaction: The 1990 Pollution Prevention Act," 17 Columbia Journal of Env. Law, 153, 156 (1992).

Lynch, H., "A Chemical Engineer's Guide to Environmental Law and Regulation," National Pollution Prevention Center for Higher Education, University of Michigan, Ann Arbor, MI, 1995. http://css.snre.umich.edu.

Sullivan, T.F.P. and Adams, T.L., *Environmental Law Handbook,* Government Institutes, Rockville, MD, 1997.

US EPA, "Pollution Prevention Case Studies Compendium," United States Environmental Protection Agency, Office of Research and Development, EPA/600/R-92/046, April 1992.

US EPA, "DuPont Chambers Works Waste Minimization Project," United States Environmental Protection Agency, Office of Research and Development, EPA/600/R-93/203, pg. 86–91, November 1993.

US EPA, "33/50 Program: The Final Record," United States Environmental Protection Agency, Office of Pollution Prevention and Toxics, IPA-745-R-99-004, March 1999.

US EPA, United States Environmental Protection Agency, 1997 National Air Quality and Emissions Trends Report, Office of Air Quality Planning and Standards, Research Triangle Park, NC 27711, EPA 454/R-98-016, December 1998, http://www.epa.gov/oar/aqtrnd97/.

US EPA, "1998 Toxics Release Inventory Public Data Release," United States Environmental Protection Agency, Office of Information Analysis and Access, EPA 745-R-00-007, September 2000.

PROBLEMS

1. Provide definitions for the following terms
   (a) pollution prevention
   (b) source reduction
   (c) in-process versus on-site versus off-site recycling
   (d) waste treatment
   (e) disposal
   (f) direct release

2. Categorize the following solvent recovery operation in terms of the waste management hierarchy. Discuss the pollution prevention features of this process. Assess

whether this process is pollution prevention, using both the federal definition and also the expanded definition adopted in this text.

Process Description: The automotive industry uses robots to paint automobile bodies before attaching them to the chassis, and installing other components such as the drive train, lights, trim, and upholstery. In order to accommodate different colors, the paint lines must be flushed with a solvent and then re-charged with the new color paint. In the past, this solvent and paint residue was disposed of as hazardous waste or incinerated. The current process of spray painting automobiles uses a closed-loop solvent recovery process as outlined in the diagram below (Gage Products, Ferndale, MI).

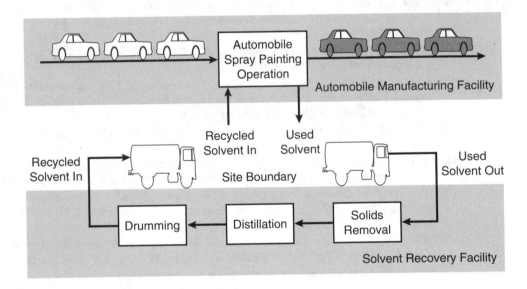

3. Choose one of the nine federal environmental statutes listed in Table 3.2-1 and then analyze the regulatory provisions for the potential to impact a chemical production facility's capital and operating costs. What are the key provisions requiring action? What is the nature of those actions? What are the cost implications of those actions? The information contained in Appendix A will be helpful in answering these questions.

4. Categorize the following chemical process source reduction case studies using *one or more* of the following source reduction categories:
   - equipment or technology modifications,
   - process or procedure modifications,
   - reformation or redesign of products,
   - substitution of raw materials,
   - improvements in housekeeping,
   - maintenance,
   - training, or
   - inventory control.
   (a) A specialty aromatic compound (SAC) process includes a reaction and a distillation train. This process relies on the quality of feed from a separate raw materials process.

Due to poor "acid" control from the raw materials process, acid tars are generated within the SAC process, together with thermal tars, at the rate of 0.07 pounds of incinerable tars per pound of SAC product. A relatively large fraction of the tar mass is entrained in the SAC product. Installation of on-line instrumentation for pH control on the raw materials process allowed operators to maintain low acidity levels in the product leaving this processing step. Due to this effort and a lower reactor temperature in the SAC process, SAC waste was reduced by 60% (to 0.03 lb waste/lb SAC product) and had a Net Present Value (12%) of almost $1,000,000. (This case study demonstrates that waste generation can and often does result from complex interactions between separate processing steps in a chemical production facility.)

(b) The crude product from a specialty alcohol process required two washing steps to remove corrosive chlorinated compounds and residual acidity. The wash steps were conducted using two vessels; a wash kettle and an accumulation drum, with the wash solution being composed of water and isopropyl alcohol. The wash solution was sent to an on-site wastewater treatment plant. The washing operations were a severe bottleneck step for the entire process. Over time, the reaction steps leading to the crude specialty alcohol were improved, resulting in a nearly impurity-free crude product with only residual acidity. Because of this realization, the wash steps were completely unnecessary and were replaced with a neutralization step, resulting in elimination of the wastewater stream. The capital cost for this was $40,000 and the project had a NPV (12%) of $272,000.

CHAPTER 4

The Roles
and Responsibilities
of Chemical Engineers

by
David R. Shonnard

4.1 INTRODUCTION

Many chemical engineers design and operate large-scale and complex chemical production facilities supplying diverse chemical products to society. In performing these functions, a chemical engineer will likely assume a number of roles during a career. The engineer may become involved in raw materials extraction, intermediate materials processing, or production of pure chemical substances; in each activity, the minimization and management of waste streams will have important economic and environmental consequences. Chemical engineers are involved in the production of bulk and specialty chemicals, petrochemicals, integrated circuits, pulp and paper, consumer products, minerals, and pharmaceuticals. Chemical engineers also find employment in research, consulting organizations, and educational institutions. The engineer may perform functions such as process and production engineering, process design, process control, technical sales and marketing, community relations, and management.

As engineers assume such diverse roles, it is increasingly important that they be aware of their responsibilities to the general public, colleagues and employers, the environment, and also to their profession. One of the central roles of chemical engineers is to design and operate chemical processes yielding chemical products that meet customer specifications and that are profitable. Another important role is to maintain safe conditions for operating personnel and for residents in the immediate vicinity of a production facility. Finally, chemical process designs need to be protective of the environment and of human health. Environmental issues must be considered not only within the context of chemical production but also during other stages of a chemical's life cycle, such as transportation, use by customers, recycling activities, and ultimate disposal.

This chapter introduces approaches to designing safe chemical processes (Section 4.2). The point of briefly introducing this important topic is to demonstrate that the evolution of the methods used to design safe processes mirrors the evolution of methods described in this text, which are used to design processes that minimize environmental impacts. Section 4.3 reviews, in slightly more detail, the types of procedures that will be used in designing processes that minimize environmental impacts, and the responsibilities of chemical engineers to reduce pollution generation within chemical processes. Section 4.4 briefly notes some of the other professional responsibilities of chemical engineers, i.e., issues dealing with engineering ethics.

4.2 RESPONSIBILITIES FOR CHEMICAL PROCESS SAFETY

A major objective for chemical process design is the inclusion of safeguards that minimize the number and severity of accidental releases of toxic chemicals and the incidence of fires and explosions. A number of chemical plant accidents have occurred in the relatively recent past illustrating the importance of integrating safety into process designs. These accidents resulted in loss of life, permanent disability, and the destruction of chemical plant, process equipment and neighboring residences. The most famous accidents occurred in Flixborough, England (1974) and Bhopal, India (1984).

Flixborough

The Flixborough Works of Nypro Limited was designed to produce 70,000 tons per year of caprolactam, a raw material for the production of nylon. The process used cyclohexane as a raw material and oxidized it to cyclohexanol in the presence of air within a series of six catalytic reactors. Under process conditions, cyclohexane vaporizes immediately upon depressurization, forming a cloud of flammable cyclohexane vapor mixed with air. Reactor 5 was found to have a small crack in the stainless steel structure and was removed. The number 4 reactor was connected to the last reactor in the series using a 20" pipe, even though the reactors are normally connected using 28" pipe. The temporary section of piping was not properly supported and it ruptured upon pressurization, releasing an estimated 30 tons of cyclohexane in a large cloud. An unknown ignition source caused the cloud to explode, leveling the entire plant facility. A total of 28 people died, another 36 were injured, and damage extended to nearby homes, shops, and factories. The resulting fire in the plant burned for over 10 days. The accident could have been prevented by following proper safety design and operating procedures, including reducing the inventory of flammable liquids on site.

Bhopal

Bhopal is located in a central state of India and on December 3, 1984, an accidental release of methyl isocyanate (MIC) occurred, killing 2,000 nearby residents and injuring over 20,000. The plant, which was partially owned by Union Carbide and partially owned by local investors, manufactured pesticides. One of the intermediates was MIC. MIC is a liquid at ambient conditions, it boils at 39.1°C, its vapor is heavier than air, and it is very toxic even at low concentrations. The maximum allowable exposure concentration of MIC for workers during an eight-hour period is only 0.02 parts per million (ppm). Death at large dose is due to respiratory damage. MIC reacts with water exothermically, but slowly, and the heat released can cause MIC to boil if cooling is not provided. On the day of the accident, the unit using MIC was not operating due to a labor dispute. The storage tank holding the MIC was contaminated with water from an unknown source. A reaction between MIC and water occurred in the tank causing the temperature to rise above the boiling point of MIC. The vapors generated escaped the pressure relief valve on the tank and were diverted into a scrubber and flare system designed to control MIC releases. Unfortunately, the release control system was not operating on this day and an estimated 25 tons of MIC vapor was released into the surrounding community with catastrophic effects. The accident could have been prevented by any number of steps, including the use of proper safety review procedures, by redesigning the process to accommodate a lower inventory of MIC, or by using alternative reaction chemistries that eliminate MIC.

In incidents such as this, loss of life and injuries are tragic, and economic consequences are severe. Engineers have a special role to play in preventing such incidents. Part of an engineer's professional responsibility is to design processes and products that are as safe as possible. Traditionally, this has meant identifying hazards, evaluating their severity and then applying several layers of protection as a means of mitigating the risk of an accident. Figure 4.2-1 shows the layer of protection concept and includes examples of layers that might be found in a typical chemical plant. This approach can be very effective and has resulted in significant improvement of the safety performance of chemical processes. However, the layer of protection approach has disadvantages that place limitations on its effectiveness: (1) the layers are expensive to build and maintain, and (2) the hazard remains and there is always a finite risk that an accident will happen despite the layers of protection.

Inherently safer design is a fundamentally different approach to chemical process safety. Instead of working with existing hazards in a chemical process and adding layers of protection, the engineer is challenged to reconsider the design and eliminate or reduce the source of the hazard within the process. Approaches to the design of inherently safer processes have been grouped into the four categories listed below. This list contains a short checklist of questions related to inherently safer processes. A more extensive checklist can be found in the Center for Chemical Process Safety (CCPS) publications (CCPS, 1993a; Crowl, 1996).

Minimize   Use smaller quantities of hazardous substances.
             • Have all in-process inventories of hazardous materials in
               storage tanks been minimized?

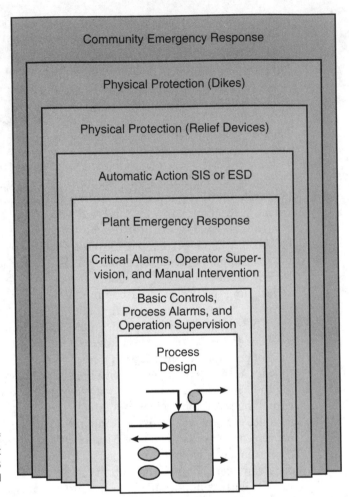

Figure 4.2-1 Typical layers of protection for a chemical plant (CCPS 1993b, Crowl 1996). SIS is safety interlock system and ESD is emergency shutdown.

- Are all of the proposed in-process storage tanks really needed?
- Can other types of unit operations or equipment reduce material inventories (for example, continuous in-line mixers in place of mixing vessels)?

Substitute   Use a less hazardous material in place of a more hazardous substance.

- Is it possible to completely eliminate hazardous raw materials, process intermediates, or byproducts by using an alternative process or chemistry?

	• Is it possible to substitute less hazardous raw materials or to substitute noncombustible for flammable solvents?

Moderate   Use less hazardous conditions or facilities which minimize the impacts of a release of a hazardous material or energy.

• Can the supply pressure of raw materials be limited to less than the working pressure of the vessels they are delivered to?

• Can reaction conditions (temperature, pressure) be made less severe by using a catalyst, or by using a better catalyst?

Simplify   Design facilities which eliminate unnecessary complexity and make operating errors less likely, and which are forgiving of errors that are made.

• Can equipment be sufficiently designed to totally contain the maximum pressure generated, even if the "worst credible event" occurs?

Textbooks (Crowl and Louvar, 1990), case studies, and other materials (Crowl, 1996) document procedures for improving the safety of chemical processes and that material is not duplicated in this text. Instead, the focus in this text is the prevention of chronic (slow, continuous) as opposed to acute (fast, rare and intermittent) releases, and the role that chemical process and product design can play in minimizing these releases. As these tools for minimizing environmental impacts are described in this text, however, it is useful to recognize analogies between chemical process safety and the design of processes that minimize environmental impacts. As noted in this section, traditional approaches to chemical process safety rely on designing layers of protection around process hazards. Similarly, traditional approaches to environmental management have focused on designing processes to treat wastes. A new generation of inherently safer processes relies on designs that reduce hazards, rather than providing protection from hazards. Similarly, new generations of processes that minimize environmental impact do not rely on treating wastes, but instead are designed so that they do not generate wastes.

4.3 RESPONSIBILITIES FOR ENVIRONMENTAL PROTECTION

When the method for managing environmental performance is to treat wastes, the process is designed, wastes are generated, and treatment technologies are deployed. The design method for meeting environmental objectives is sequential. In contrast, if the primary design, rather than the design of peripheral waste treatment units, is to be modified to meet environmental objectives, a key question to answer is "At what stage in the design should environmental considerations be considered?"

Designs for new processes and retrofitting of existing procedures are multistep procedures (Seider et al., 1999). The first step is the definition of a primitive problem, such as identifying the chemical to be produced and the annual quantity.

Table 4.3-1   CMA Pollution Prevention Code of Management Practices (Now the American Chemistry Council)

This Code is designed to achieve ongoing reductions in the amount of all contaminants and pollutants released to the air, water, and land from member company facilities. The Code is also designed to achieve ongoing reductions in the amount of wastes generated at facilities. These reductions are intended to help relieve the burden on industry and society of managing such wastes in future years.

Management Practices

Each member company shall have a pollution prevention program that shall include:

1. A clear commitment by senior management through policy, communications, and resources, to ongoing reductions at each of the company's facilities, in releases to the air, water, and land and in the generation of wastes.
2. A quantitative inventory at each facility of wastes generated and releases to the air, water, and land, measured or estimated at the point of generation or release. **(Chapter 8)**
3. Evaluation, sufficient to assist in establishing reduction priorities, of the potential impact of releases on the environment and the health and safety of employees and the public. **(Chapters 1, 2, 5, 8, and 11)**
4. Education of, and dialogue with, employees and members of the public about the inventory, impact evaluation, and risks to the community.
5. Establishment of priorities, goals and plans for waste and release reduction, taking into account both community concerns and the potential health, safety, and environmental impacts as determined under items 3 and 4.
6. Ongoing reduction of wastes and releases, giving preference first to source reduction, second to recycle/reuse, and third to treatment. These techniques may be used separately or in combination with one another. **(Chapters 7, 9, and 10)**
7. Measurement of progress at each facility in reducing the generation of wastes and in reducing releases to the air, water, and land, by updating the quantitative inventory at least annually. **(Chapter 8)**
8. Ongoing dialogue with employees and members of the public regarding waste and release information, progress in achieving reductions, and future plans. This dialogue should be at a personal, face-to-face level, where possible, and should emphasize listening to others and discussing their concerns and ideas.
9. Inclusion of waste and release prevention objectives in research and in design of new or modified facilities, processes, and products.
10. An ongoing program for promotion and support of waste and release reduction by others, which may, for example, include:
    a. Sharing of technical information and experience with customers and suppliers;
    b. Support of efforts to develop improved waste and release reduction techniques;
    c. Assisting in establishment of regional air monitoring networks;
    d. Participation in efforts to develop consensus approaches to the evaluation of environmental, health, and safety impacts of releases;
    e. Providing educational workshops and training materials;
    f. Assisting local governments and others in establishment of waste reduction programs benefiting the general public.
11. Periodic evaluation of waste management practices associated with operations and equipment at each member company facility, taking into account community concerns and health, safety, and environmental impacts and implementation of ongoing improvements.
12. Implementation of a process for selecting, retaining, and reviewing contractors and toll manufacturers taking into account sound waste management practices that protect the environment and the health and safety of employees and the public.
13. Implementation of engineering and operating controls at each member company facility to improve prevention of and early detection of releases that may contaminate groundwater.
14. Implementation of an ongoing program for addressing past operating and waste management practices and for working with others to resolve identified problems at each active or inactive facility owned by a member company taking into account community concerns and health, safety, and environmental impacts.

This is followed by a process creation step that includes choosing reaction chemistry, the use of design heuristics to identify process equipment and operating conditions, development of a base case flowsheet, and process simulation. The third step is a more detailed process synthesis of separation trains and a heat/power integration analysis. What follows is a detailed design and simulation of the flowsheet, profitability analysis, and optimization. The final steps include a plantwide controllability assessment, startup assessment, and reliability and safety analysis. In Part II of this text, systematic methods are presented for incorporating environmental considerations into *all* of these steps of chemical process design.

As part of their professional responsibilities, engineers should, through their designs, continuously improve the environmental performance of chemical processes. Recently the Chemical Manufacturers Association (CMA, now the American Chemistry Council) has adopted the Pollution Prevention Code of Management Practice, which outlines tangible steps along a path to continuous reductions in the amounts of all contaminants released to air, water, and soil. Table 4.3-1 shows the set of management practices and specific chapters in this textbook that will aid engineers and other decision makers in achieving pollution prevention objectives. These practices demonstrate a clear commitment by senior management, a path to quantify waste generation and prioritize waste reduction, a preference for source reduction and reuse/recycle rather than pollution control, and a plan to measure and report on progress in achieving reduction goals.

4.4 FURTHER READING IN ENGINEERING ETHICS

Process safety and environmental protection are not the only responsibilities of professional engineers. Engineers also have responsibilities to clients, to colleagues, and to the profession. The American Institute of Chemical Engineers has assembled a Code of Ethics that highlights the main issues in the area of professional conduct. This code can be found at AIChE website *(http://www.aiche.org/membership/ethics.htm)*. Case studies in engineering ethics are available in the journal *Chemical Engineering* (March 2, 1987). Nine ethical dilemmas pertinent to chemical engineers are presented and reader responses are reported in a subsequent issue (Sept. 28, 1987 issue). Some of the responses dealt with putting health, safety, and environmental issues ahead of profits; placing self-respect as professionals above loyalty to companies; working within organizations versus whistleblowing to promote ethical behavior; and taking career risks in order to get a company to do the right thing. Further discussion of engineering ethics is provided by Mitcham and Shannon Duval (2000).

REFERENCES

CCPS, Center for Chemical Process Safety, *Guidelines for Engineering Design of Process Safety*, New York, NY, American Institute of Chemical Engineers, 1993a.

CCPS, Center for Chemical Process Safety, *Guidelines for Safe Automation of Chemical Processes*, New York, NY, American Institute of Chemical Engineers, 1993b.

CMA, Pollution Prevention Codes of Management Practices, Chemical Manufacturers Association (renamed American Chemistry Council), http://www.cmahq.com/.

Crowl, D.A. and Louvar, J.F., *Chemical Process Safety: Fundamentals with Applications*, Prentice Hall PTR, Englewood Cliffs, NJ, 1990.

Crowl, D.A. ed., *Inherently Safer Chemical Processes: A Life Cycle Approach*, Center for Chemical Process Safety, American Institute of Chemical Engineers, 1996

Mitcham, C. and Shannon Duval, R., *Engineering Ethics*, Prentice Hall, Upper Saddle River, NJ (2000).

Seider, W.D., Seader, J.D., and Lewin, D.R., *Process Design Principles: Synthesis, Analysis, and Evaluation*, John Wiley & Sons, New York, 1999.

PROBLEMS

1. Compare and contrast the Inherently Safer Design concepts presented in this chapter with the Pollution Prevention concepts from Chapter 3. Note in particular that design methods for improving process safety are focused on preventing catastrophic releases, while pollution prevention design methods are primarily concerned with reducing chronic emissions.

2. What chemical properties will be most important in evaluating the potential for catastrophic releases? What chemical properties would be most important in developing methods to prevent chronic releases? Draw upon the material presented in this chapter and in Chapter 1.

3. You are the chemical engineer responsible for all new processes for your facility. Your facility operates processes that extract valuable natural products from various botanicals. Presently, the extraction process uses hot water to extract the requisite material, followed by concentration, crystallization, separation, drying and packaging.

   You have been asked to evaluate a new process for a different botanical yielding a new "natural" product. That process could use the same unit operations as the present process. The new process would however, use either n-hexane or USP grade ethanol (anhydrous) as the extractant. There are no storage tanks on your site for use as a solvent storage tank. The extraction, with either solvent, is done by recirculating the solvent throught a packed bed of botanical operating at 40°C. N-hexane performs the extraction in half the time (12 hrs) when compared to ethanol.

   You have been requested to analyze the proposed process (each solvent) and define its impact (regulatory, permit, safety) on this facility. The product from this new process will be new and is not on the TSCA inventory. Solvent recovery is available on-site. The only cooling media available on-site is a source of 85°F water operating in a closed cycle.

   The questions that need to be addressed are:
   - What are the physical properties, including toxicological data on the proposed solvents?
   - Are utilities available to support the process?
   - What are the local, state and federal permits issues?

- Is annual reporting required?
- Is there any additional medical monitoring required?
- Are there other unit operations you could use to do the same thing?

(a) What is your recommendation? Justify with a careful analysis.

(b) If the plant may have to shut down if it does not get a new process, eliminating your job, could this impact your decision?

(c) If your facility cannot (or will not) implement the process, it has been suggested that the process could be done in an overseas facility in Asia or other locations (other states) in the US. What is you recommendation? Justify.

PART II

Evaluating and Improving Environmental Performance of Chemical Processes

OVERVIEW

Evaluating the environmental impacts of chemical processes and improving the environmental performance of chemical process designs are complex tasks involving a wide variety of analysis tools. To systematically present these tools, the group of chapters that follows is organized into a framework commonly employed in the design of chemical processes. The key steps in this framework are listed below.

1. Specify the product to be manufactured and evaluate potential environmental fate (Chapter 5), releases, and exposures (Chapter 6).
2. Establish the input/ouput structure of the chemical process, including chemical synthesis pathways and potential byproducts (Chapter 7).
3. Evaluate potential emissions and environmental impacts associated with the conceptual process (Chapter 8).
4. Specify the unit operations and process flows and identify pollution prevention opportunities (Chapter 9).
5. Systematically examine the flowsheet to identify opportunities for environmental improvements and identify opportunities for energy and mass integration (Chapter 10).
6. Evaluate the environmental performance of the detailed process flowsheet (Chapter 11).
7. Evaluate the environmental costs associated with the process (Chapter 12).

The first step in the analysis is to identify and evaluate potential environmental impacts of the chemicals to be manufactured. Evaluating environmental impacts requires knowledge of emission rates, the environmental fate of those emissions, and the potential human health and ecosystem impacts associated with the chemical's environmental fate. Estimates of emissions, fate, exposures and impacts, in turn, rely on a host of other chemical and physical properties and data for these properties may or may not be available. Thus, the task of evaluating potential environmental impacts of new chemicals is formidable. Nevertheless, such evaluations are prudent because they can identify key environmental issues at the earliest stages of the design process. Chapters 5 and 6 describe qualitative and quantitative tools that have been developed to identify and evaluate potential environmental fates, exposures, and impacts.

The next step in the design process involves selection of raw materials and chemical synthesis pathways. For many chemicals, a variety of synthesis pathways are possible, and each pathway will require slightly different raw materials and reaction conditions and may generate different byproducts. Chapter 7 describes some of the emerging tools available for identifying and evaluating alternative chemical synthesis pathways. These tools can be used in conjunction with the environmental assessment tools described in Chapters 5 and 6 to select preferred raw materials for chemical processes.

Once the basic input/output structure of the flowsheet has been established, it is prudent to perform a preliminary evaluation of environmental impacts. Chapter 8 presents methods for assessing the environmental performance of a process when only limited, conceptual information on the process design is available.

The next major opportunity for risk reduction occurs when the process flowsheet has been established. At this stage, pollution prevention opportunities should be considered for each of the unit operations in the process. Chapter 9 presents pollution prevention methods for common unit operations.

While Chapter 9 focuses on qualitative and semi-quantitative flowsheet evaluation tools, Chapter 10 examines more quantitative approaches. The goals of these quantitative analyses will be to improve the energy and mass efficiencies of the flowsheets. Pinch analysis methods will be presented and demonstrated through case studies.

Once the structure of the flowsheet has been established, a comprehensive evaluation of potential environmental impacts can be performed. Detailed estimates of process emissions can be performed. Environmental fates of the emissions and wastes can be estimated. Potential environmental impacts can be assessed. Chapter 11 presents the analysis tools that are available for evaluating the environmental performance of a chemical process. These tools are analogous to those presented in Chapters 5 and 6, but incorporate much more detail in estimating releases and assessing environmental fates.

Finally, Chapter 12 examines the integration of information about the environmental performance of a chemical process into more traditional economic evaluations. Hidden environmental costs, potential liability costs, and less tangible costs and benefits are described. Case studies are used to illustrate the environmental accounting principles.

CHAPTER 5

Evaluating Environmental Fate: Approaches Based on Chemical Structure

by
David T. Allen

5.1 INTRODUCTION

A new chemical is to be manufactured. Will its manufacture or use pose significant environmental or human health risks? If there are risks, what are the exposure pathways? Will the chemical degrade if it is released into the environment or will it persist? If the chemical degrades, will the degradation products pose a risk to the environment?

The challenges involved in answering these questions are formidable. Over 9,000 chemicals are produced commercially and every year, a thousand or more new chemicals are developed. For any chemical in use, there are a number of potential risks to human health and the environment. In general, it will not be possible to rigorously and precisely evaluate all possible environmental impacts. Nevertheless, a preliminary screening of the potential environmental impacts of chemicals is necessary and is possible. Preliminary risk screenings allow businesses, government agencies, and the public to identify problem chemicals and to identify potential risk reduction opportunities. The challenge is to perform these preliminary risk screenings with a limited amount of information.

This chapter presents qualitative and quantitative methods for estimating environmental risks when the only information available is a chemical structure. Many of these methods have been developed by the US Environmental Protection Agency (US EPA) and its contractors. The methods are routinely used in evaluating premanufacture notices submitted under the Toxic Substances Control Act (TSCA). Under the provisions of TSCA, before a new chemical can be manufactured in the

United States, a premanufacture notice (PMN) must be submitted to the US EPA. The PMN specifies the chemical to be manufactured, the quantity to be manufactured, and any known environmental impacts including potential releases from the manufacturing site. Based on these limited data, the US EPA must assess whether the manufacture or use of the proposed chemical may pose an unreasonable risk to human or ecological health. To accomplish that assessment, a set of tools has been developed that relate chemical structure to potential environmental risks.

Figure 5.1-1 provides a qualitative summary of the processes that determine environmental risks. Table 5.1-1 identifies the chemical and physical properties that will influence each of the processes that determine environmental exposure

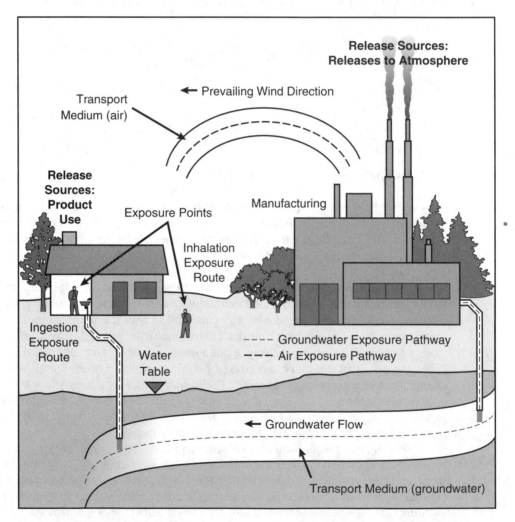

Figure 5.1-1   The chemical and physical properties that will influence each of the processes that determine environmental exposure and hazard.

Table 5.1-1   Chemical Properties Needed to Perform Environmental Risk Screenings.

Environmental Process	Relevant Properties
Dispersion and fate	Volatility, density, melting point, water solubility, effectiveness of waste water treatment
Persistence in the environment	Atmospheric oxidation rate, aqueous hydrolysis rate, photolysis rate, rate of microbial degradation, and adsorption
Uptake by organisms	Volatility, lipophilicity, molecular size, degradation rate in organism
Human uptake	Transport across dermal layers, transport rates across lung membrane, degradation rates within the human body
Toxicity and other health effects	Dose-response relationships

and hazard. The table makes clear that a wide range of properties need to be estimated to perform a screening level assessment of environmental risks.

The first group of properties that must be estimated in an assessment of environmental risk are the basic physical and chemical properties that describe a chemical's partitioning between solid, liquid, and gas phases. These include melting point, boiling point, vapor pressure, and water solubility. Additional molecular properties, related to phase partitioning, that are frequently used in assessing the environmental fate of chemicals include octanol-water partition coefficient, soil sorption coefficients, Henry's Law constants and bioconcentration factors. (Each of these properties is defined in Section 5.2). Once the basic physical and chemical properties are defined, a series of properties that influence the fate of chemicals in the environment are estimated. These include estimates of the rates at which chemicals will react in the atmosphere, the rates of reaction in aqueous environments and the rate at which the compounds will be metabolized by organisms. If environmental concentrations can be estimated based on release rates and environmental fate properties, then human exposures to the chemicals can be estimated. Finally, if exposures and hazards are known, then risks to humans and the environment can be estimated.

The remainder of this chapter describes estimation tools for the properties outlined above. Section 5.2 describes estimation tools for physical and chemical properties. Section 5.3 describes how properties that influence environmental fate are estimated. Methods for estimating hazards to ecosystems are discussed in Section 5.4, and Section 5.5 presents simple models that can be used to characterize the environmental partitioning of chemicals. Finally, Section 5.6 describes how chemical property data can be used to classify the risks associated with chemicals.

5.2   CHEMICAL AND PHYSICAL PROPERTY ESTIMATION

Although many chemical and physical properties can influence the way in which a chemical partitions in the environment, most screening-level evaluations focus on only a small number of properties. These properties describe the partitioning of chemicals between solid, liquid and gaseous phases and include melting point, boil-

Table 5.2-1   Properties that Influence Environmental Phase Partitioning.

Property	Definition	Significance in estimating environmental fate and risks
Melting point (T_m)	Temperature at which solid and liquid coexist at equilibrium	Sometimes used as a correlating parameter in estimating other properties for compounds that are solids at ambient or near-ambient conditions
Boiling point (T_b)	Temperature at which the vapor pressure of a compound equals atmospheric pressure; normal boiling points (temperature at which pressure equals one atmosphere) will be used in this text	Characterizes the partitioning between gas and liquid phases; frequently used as a correlating variable in estimating other properties
Vapor pressure (P_{vp})	Partial pressure exerted by a vapor when the vapor is in equilibrium with its liquid	Characterizes the partitioning between gas and liquid phases
Henry's Law constant (H)	Equilibrium ratio of the concentration of a compound in the gas phase to the concentration of the compound in a dilute aqueous solution (sometimes reported as atm-m^3/mol; dimensionless form will be used in this text)	Characterizes the partitioning between gas and aqueous phases
Octanol-water partition coefficient (K_{ow})	Equilibrium ratio of the concentration of a compound in octanol to the concentration of the compound in water	Characterizes the partitioning between hydrophilic and hydrophobic phases in the environment and the human body; frequently used as a correlating variable in estimating other properties
Water solubility (S)	Equilibrium solubility in mol/L	Characterizes the maximum concentration in the aqueous phase
Soil sorption coefficient (K_{oc})	Equilibrium ratio of the mass of a compound adsorbed per unit weight of organic carbon in a soil (in μg/g organic carbon) to the concentration of the compound in a liquid phase (in μg/ml)	Characterizes the partitioning between solid and liquid phases in soil which in turn determines mobility in soils; frequently estimated based on octanol-water partition coefficient, and water solubility
Bioconcentration factor (BCF)	Ratio of a chemical's concentration in the tissue of an aquatic organism to its concentration in water (reported as L/kg)	Characterizes the magnification of concentrations through the food chain

ing point, and vapor pressure. Additional molecular properties, related to phase partitioning, that are frequently used in assessing the environmental fate of chemicals include Henry's law constants, octanol-water partition coefficient, water solubility, soil sorption coefficients and bioconcentration factors. Table 5.2-1 defines each of these properties and describes the significance of the property in estimating environmental fate.

This section describes how each of these properties can be estimated based on the structure of the chemical. The review of estimation methods will not be comprehensive. Rather, the focus is on presenting commonly used methods that can produce property estimates based only on the chemical structure of the target compound. More complete presentations are available in texts on environmental property estimation (e.g., Lyman, et al., 1990). More complete compilations of data are available from Howard (1997), Mackay, et al. (1992), Reinhard and Drefahl (1999), and the sources listed in Appendix F.

The methods described in this section generally assume that a molecule is composed of a collection of functional groups or molecular fragments and that each fragment contributes in a well-defined manner to the properties of the molecule. These methods are generally described as group contribution methods, structure activity relationships (SARs) or quantitative structure activity relationships (QSARs).

5.2.1 Boiling Point and Melting Point

As a first example of a structure activity relationship, consider the estimation of boiling point (at one atmosphere pressure). Boiling point is influenced by molecular weight and intermolecular attractions. It can be estimated using a relatively simple group contribution method, developed by Joback and Reid (1987) and modified by Stein and Brown (1994), that relates the boiling point to the number and type of functional groups present in the molecule.

$$T_b \, (K) = 198.2 + \Sigma \, n_i \, g_i \qquad \text{(Eq. 5-1)}$$

where T_b is the normal boiling point (at one atmosphere pressure) in degrees Kelvin, n_i is the number of groups of type i in the molecule, g_i is the contribution of each group to the boiling point, and the summation is taken over all groups. The boiling point predicted by Equation 5-1 is corrected using one of the following equations:

$$T_b \, (\text{corrected}) = T_b - 94.84 + 0.5577 T_b - 0.0007705 (T_b)^2 \quad [T_b \leq 700K] \quad \text{(Eq. 5-2)}$$

$$T_b \, (\text{corrected}) = T_b + 282.7 - 0.5209 T_b \qquad\qquad [T_b > 700K] \, \text{(Eq. 5-3)}$$

Structural groups and group contributions (g_i) for boiling point estimation are listed in Table 5.2-2. When tested against a set of more than 4000 organic compounds, this method yielded an average error of 3.2% (Stein and Brown, 1994).

Table 5.2-2 Structural Groups and Group Contributions for Boiling Point Estimation (Stein and Brown, 1994).

Structural group	Contribution (g_i) to normal boiling point	Structural group	Contribution (g_i) to normal boiling point
Carbon groups		**Nitrogen groups**	
$-CH_3$	21.98	$-NH_2$	61.98
$>CH_2$	24.22	Aromatic-NH_2	86.63
$>C_{ring}H_2$	26.44	$>NH$	45.28
$>CH-$	11.86	$>N_{ring}H$	65.50
$>C_{ring}H-$	21.66	$>N-$	25.78
$>C<$	4.50	$>N_{ring}-$	32.77
$>C_{ring}<$	11.12	$>NOH$	104.87
$=CH_2$	16.44	$>NNO$	184.68
$=CH-$	27.95	anN	39.88
$=C_{ring}H-$	28.03	$=NH$	73.40
$=C<$	23.58	$=N-$	31.32
$=C_{ring}<$	28.19	$=N_{ring}-$	43.54
aaCH*	28.53	$=N_{ring}N_{ring}H-$	179.43
aaC$-$	30.76	$-N_{ring}=C_{ring}N_{ring}H-$	284.16
aaaC	45.46	$-N=NNH-$	257.29
$\equiv CH$	21.71	$-N=N-$	90.87
$\equiv C-$	32.99	$-NO$	30.91
		$-NO_2$	113.99
Oxygen groups		$-CN$	119.16
$-OH$	106.27	Aromatic-CN	95.43
Primary $-OH$	88.46		
Secondary $-OH$	80.63	**Halogen groups**	
Tertiary $-OH$	69.32	$-F$	0.13
Aromatic $-OH$	70.48	Aromatic-F	-7.81
$-O-$	25.16	$-Cl$	34.08
$-O_{ring}-$	32.98	Primary-Cl	62.63
$-OOH$	72.92	Secondary-Cl	49.41
		Tertiary-Cl	36.23
Carboxyl groups		Aromatic-Cl	36.79
$-CHO$	83.38	$-Br$	76.28
$>CO$	71.53	Aromatic-Br	61.85
$>C_{ring}O$	94.76	$-I$	111.67
$-C(O)O-$	78.85	Aromatic-I	99.93
$-C_{ring}(O)O_{ring}-$	172.49		
$-C(O)OH$	169.83	**Sulfur groups**	
$-C(O)NH_2$	230.39	$-SH$	81.71
$-C(O)NH-$	225.09	Aromatic-SH	77.49
$-C_{ring}(O)N_{ring}H-$	246.13	$-S-$	69.42
$-C(O)N<$	142.77	$-S_{ring}-$	69.00
$-C_{ring}(O)N_{ring}<$	180.22	$>SO$	154.50
		$>SO_2$	171.58
		$>CS$	106.20
		$>C_{ring}S$	179.26

*The symbol a denotes an aromatic bond.

Although Equations 5-1 to 5-3 are not the only method or even the most accurate method for estimating boiling point (for a more complete discussion, see Reid, et al., 1987), the approach does illustrate the basic principles of a group contribution method. Each functional group in a molecule is assumed to make a well-defined contribution (in this case, g_i) to the property. The group contributions may be simply added together, as in Equation 5-1, or a more complex mathematical form may be used. The application of the method is illustrated in Example 5.2-1.

Example 5.2-1

Estimate the normal boiling point for ethanol, toluene, and acetaldehyde.

Solution: *Ethanol* has the molecular structure CH_3-CH_2-OH. Referring to the groups in Table 5.2-2, this structure can be represented by one $-CH_3$ group, one $-CH_2$ group and one $-OH$ group. The uncorrected normal boiling point, from Equation 5-1, is given by:

$$T_b \, (K) = 198.2 + 21.98 + 24.22 + 88.46 = 332.9 \text{ K}$$

The corrected value is:

$$T_b \, (\text{corrected}) = T_b - 94.84 + 0.5577T_b - 0.0007705(T_b)^2 = 338.3 \text{ K}$$

The actual boiling point is 351 K, so the predicted value is in error by -3.6 %

Toluene has the molecular structure $CH_3-C_6H_5$. Referring to the groups in Table 5.2-2, this structure can be represented by one $-CH_3$ group, one $-aaC-$ group (a substituted carbon bound to two aromatic carbons) and five $-aaCH$ groups. The uncorrected normal boiling point, from Equation 5-1, is given by:

$$T_b \, (K) = 198.2 + 21.98 + 30.76 + 5(28.53) = 393.6 \text{ K}$$

The corrected value is:

$$T_b \, (\text{corrected}) = T_b - 94.84 + 0.5577T_b - 0.0007705(T_b)^2 = 398.9 \text{ K}$$

The actual boiling point is 384 K, so the predicted value is in error by $+3.9\%$

Acetaldehyde has the molecular structure $CH_3-CH=O$. Referring to the groups in Table 5.2-2, this structure can be represented by one $-CH_3$ group and one $-CHO$ group. The uncorrected normal boiling point, from Equation 5-1, is given by:

$$T_b \, (K) = 198.2 + 21.98 + 83.38 = 303.6 \text{ K}$$

The corrected value is:

$$T_b \, (\text{corrected}) = T_b - 94.84 + 0.5577T_b - 0.0007705(T_b)^2 = 307.0 \text{ K}$$

The actual boiling point is 294 K, so the predicted value is in error by $+4.2\%$

While group contribution methods can produce accurate estimates of chemical and physical properties, it is important to recognize their limitations. Group contribution equations are empirical. They are designed to accurately reflect a particular set of property data. If a group contribution method is used to estimate the properties of molecules that have structures significantly different from those used in the original data set, substantial errors can result. Consider, for example, what might happen if a group contribution method, originally developed with data on alcohols, was used to estimate the properties of glycols. Since glycols have two hydroxyl groups per molecule, they can form chains of molecules (n-mers) held together by hydrogen bonding forces. In contrast, alcohols, with only one hydroxyl group per molecule, can only form dimers in solution. A group contribution method for boiling point, developed with data on alcohols, would likely underpredict the boiling point of glycols (see Example 5.2-2).

Example 5.2-2

Estimate the normal boiling point for ethylene glycol.

Solution: *Ethylene glycol* has the molecular structure $HO-CH_2-CH_2-OH$. Referring to the groups in Table 5.2-2, this structure can be represented by two $-CH_2$ groups and two $-OH$ groups. The uncorrected normal boiling point, from Equation 5-1, is given by:

$$T_b \text{ (K)} = 198.2 + 2(24.22) + 2(88.46) = 424 \text{ K}$$

The corrected value is:

$$T_b \text{ (corrected)} = T_b - 94.84 + 0.5577T_b - 0.0007705(T_b)^2 = 427 \text{ K}$$

The actual boiling point is 470 K, so the predicted value is in error by -9%.

As shown in Example 5.2-1, the estimation of boiling point, using Equations 5-1 to 5-3, is relatively straightforward. Boiling points, in turn, can be used to estimate a variety of other properties. One property, which is occasionally used in estimating the phase partitioning of solids, is melting point. Melting point is sometimes expressed as a simple fraction of boiling point (Lyman, 1985):

$$T_m \text{ (K)} = 0.5839 \, T_b \text{ (K)} \tag{Eq. 5-4}$$

5.2.2 Vapor Pressure

The vapor pressure of a chemical plays a significant role in its environmental partitioning. High vapor pressure materials will generally have higher atmospheric concentrations than lower vapor pressure materials, and therefore, have the potential to be transported over long distances as gases or inhaled as gases. The temperature dependence of vapor pressure also plays a role in environmental transport and partitioning. If a chemical's vapor pressure varies significantly between daytime and nighttime conditions, strong daily cycling of the chemical between environmental media can be expected, assuming no degradation or soil adsorption. Finally, vapor pressures are used in a variety of ways in estimations of exposure and environmental risk. Therefore, reliable estimates of vapor pressure, over a range of temperatures, will be important in screening chemicals for environmental risk.

A number of approaches are available for estimating vapor pressures. Some approaches are based on critical temperatures and pressures; others rely on heats of vaporization (Lyman, et al., 1990). Still other methods use estimates of vapor pressure at a reference temperature (such as the boiling point) to estimate vapor pressure. The methods based on boiling point and heat of vaporization will be the focus of this section—not because they are necessarily more accurate than the other methods, but rather, because they are conveniently estimated from chemical structure.

One method for estimating vapor pressure from boiling point and heat of vaporization uses the mathematical form associated with the Antoine equation:

$$\ln P_{vp} = A + B/(T - C) \qquad \text{(Eq. 5-5)}$$

Where P_{vp} is the vapor pressure, A and C are empirical constants, B is a parameter that is related to the heat of vaporization and T is absolute temperature. A derivation of this equation based on thermodynamic concepts is available in Lyman, et al. (1990) and in most thermodynamic textbooks.

Note that if we apply Equation 5-5 at the boiling point and define the units of vapor pressure as atmospheres, then:

$$\ln (1 \text{ atm}) = 0 = A + B/(T_b - C) \qquad \text{(Eq. 5-6)}$$

Equation 5-6 can be used to express the parameter B in terms of A, C and T_b. Lyman, et al. (1990) provide a derivation of the following equation,

$$\ln P_{vp} \text{ (atm)} = \frac{[A(T_b - C)^2]}{[0.97 \, R \, T_b]} \times [1/(T_b - C) - 1/(T - C)] \qquad \text{(Eq. 5-7)}$$

where R is the gas constant (1.987 l-atm $°K^{-1}mol^{-1}$). Empirical correlations are available for estimating the parameters A and C from boiling point:

$$C = -18 + 0.19 \, T_b \qquad \text{(Eq. 5-8)}$$

$$A = K_F (8.75 + R \ln T_b) \qquad \text{(Eq. 5-9)}$$

Equations 5-7 through 5-9 allow vapor pressure to be estimated, as a function of temperature, based only on the boiling point and the parameter K_F. Values of K_F are given in Tables 5.2-3 and 5.2-4. For any compound not given in the tables, assume $K_F = 1.06$.

Equations 5-7 through 5-9 work well in estimating vapor pressures that range from 10^{-2} to one atmosphere, yielding average errors of 2.7%. The performance deteriorates at lower pressures, with average errors of 86% for vapor pressures ranging from 10^{-6} to 10^{-2} atmosphere (Lyman, et al., 1990).

For solids, a slightly different form is generally used:

$$\ln P = -(4.4 + \ln T_b)[1.803 \, (T_b/T - 1) - 0.803 \ln(T_b/T)] - 6.8 \, (T_m/T - 1) \qquad \text{(Eq. 5-10)}$$

where P is the vapor pressure in atmospheres, T_b is the normal boiling point (K), T is the temperature at which the vapor pressure is to be evaluated (K), and T_m is the melting point (K).

Care must be taken in defining the units for Equations 5-7 through 5-10. Example 5.2-3 illustrates the proper use of units.

Table 5.2-3 Factors (K_F) Used in Estimating Boiling Points (Lyman, et al., 1990).

Compound type	Number of carbon atoms in compound											
	1	2	3	4	5	6	7	8	9	10	11	12-20
Hydrocarbons (consider a phenyl group as a single carbon atom)												
n-alkanes	0.97	1.00	1.00	1.00	1.00	1.00	1.00	1.00	1.00	1.00	1.00	1.00
Alkane isomers				0.99	0.99	0.99	0.99	0.99	0.99	0.99	0.99	0.99
Mono- and diolefins and isomers		1.01	1.01	1.01	1.01	1.01	1.01	1.01	1.01	1.01	1.01	1.01
Cyclic saturated hydrocarbons			1.00	1.00	1.00	1.00	1.00	1.00	1.00	1.00	1.00	1.00
Alkyl derivatives of cyclic saturated hydrocarbons				0.99	0.99	0.99	0.99	0.99	0.99	0.99	0.99	0.99
Halides												
Monochlorides	1.05	1.04	1.03	1.03	1.03	1.03	1.03	1.03	1.02	1.02	1.02	1.01
Monobromides	1.04	1.03	1.03	1.03	1.03	1.03	1.02	1.02	1.02	1.01	1.01	1.01
Monoiodides	1.03	1.02	1.02	1.02	1.02	1.02	1.01	1.01	1.01	1.01	1.01	1.01
Polyhalides (not entirely halogenated)	1.05	1.05	1.05	1.04	1.04	1.04	1.03	1.03	1.03	1.02	1.02	1.01
Mixed halides (completely halogenated)	1.01	1.01	1.01	1.01	1.01	1.01	1.01	1.01	1.01	1.01	1.01	1.01
Perfluorocarbons	1.00	1.00	1.00	1.00	1.00	1.00	1.00	1.00	1.00	1.00	1.00	1.00
Compounds containing a keto group												
Esters		1.14	1.09	1.08	1.07	1.06	1.05	1.04	1.04	1.03	1.02	1.01
Ketones			1.08	1.07	1.06	1.06	1.05	1.04	1.04	1.03	1.02	1.01
Aldehydes		1.09	1.08	1.08	1.07	1.06	1.05	1.04	1.04	1.03	1.02	1.01
Nitrogen compounds												
Primary amines	1.16	1.13	1.12	1.11	1.10	1.10	1.09	1.09	1.08	1.07	1.06	1.05
Secondary amines		1.09	1.08	1.08	1.07	1.07	1.06	1.05	1.05	1.04	1.04	1.03
Tertiary amines			1.01	1.01	1.01	1.01	1.01	1.01	1.01	1.01	1.01	1.01
Nitriles		1.05	1.07	1.06	1.06	1.05	1.05	1.04	1.04	1.03	1.02	1.01
Nitro compounds	1.07	1.07	1.07	1.06	1.06	1.05	1.05	1.04	1.04	1.03	1.02	1.01
Sulfur compounds												
Mercaptans	1.05	1.03	1.02	1.01	1.01	1.01	1.01	1.01	1.01	1.01	1.01	1.01
Sulfides		1.03	1.02	1.01	1.01	1.01	1.01	1.01	1.01	1.01	1.01	1.01
Alcohols and miscellaneous compounds												
Alcohols (single -OH group)	1.22	1.31	1.31	1.31	1.31	1.30	1.29	1.28	1.27	1.26	1.24	1.24
Diols		1.33	1.33	1.33	1.33	1.33	1.33	1.33				
Triols			1.38	1.38	1.38							
Cyclohexanol, cyclohexyl methyl alcohol, etc.						1.20	1.20	1.21	1.24	1.26		
Aliphatic esters		1.03	1.03	1.02	1.02	1.02	1.01	1.01	1.01	1.01	1.01	1.01
Oxides (cyclic ethers)		1.08	1.07	1.06	1.05	1.05	1.04	1.03	1.02	1.01	1.01	1.01

Table 5.2-4   Factors (K_F) Used in Estimating Boiling Points for Aromatics (Lyman, et al., 1990).

Compound type	K_F
Phenols (single $-OH$)	1.15
Phenols (more than one $-OH$)	1.23
Anilines (single $-NH_2$)	1.09
Anilines (more than one $-NH_2$)	1.14
N-substituted anilines (C_6H_5NHR)	1.06
Naphthols (single $-OH$)	1.09
Naphthylamines (single $-NH_2$)	1.06
N-substituted naphthylamines	1.03

Example 5.2-3

Estimate the vapor pressure at 298 K for toluene (a liquid) and naphthalene (a solid).

Solution: *Toluene* has the molecular structure $CH_3-C_6H_5$ and in Example 5.2-1, its boiling point was estimated to be 399 K. The experimental value for the boiling point is 384 K. We will estimate the vapor pressure using both the predicted and the experimental value for boiling point. Using the predicted value of 399 K:

$$C = -18 + 0.19\, T_b = 57.8$$

$$A = K_F (8.75 + R \ln T_b) = 1.0(8.75 + 1.987 \times \ln(399)) = 20.6$$

$$\text{Ln } P_{vp} = \frac{[A(T_b - C)^2]}{[0.97\, R\, T_b]}[1/(T_b - C) - 1/(T - C)] = \frac{[20.6 \times (399 - 57.8)^2]}{[0.97 \times 1.987 \times 399]}[1/341 - 1/240]$$

$$\text{Ln } P_{vp} = -3.83; \quad P_{vp} = 0.021 \text{ atm} = 16 \text{ mm Hg}$$

Repeating the calculation for the experimental boiling point leads to a vapor pressure estimate of 19 mm Hg.

   Naphthalene has the formula $C_{10}H_8$ and is a solid with a melting point of 81°C. The boiling point can be estimated from the methods described earlier in this section. The uncorrected group contribution estimate is:

$$T_b = 198.2 + 2(45.46) + 8(28.53) = 517 \text{ K}$$

The corrected value is: $T_b = 505$ K
   Applying Equation 5-10:

$$\ln P = -(4.4 + \ln T_b)[1.803\,(T_b/T - 1) - 0.803 \ln(T_b/T)] - 6.8\,(T_m/T - 1)$$

$$\ln P = -(4.4 + \ln 505)[1.803\,(505/298 - 1) - 0.803 \ln(505/298)] - 6.8\,(354/298 - 1)$$

$$P = 4.4 \times 10^{-5} \text{ atm} = 0.03 \text{ mm Hg}$$

5.2.3   Octanol-Water Partition Coefficient

While melting points, boiling points, and vapor pressures are familiar properties used in many applications, properties such as the octanol-water partition coefficient are more specialized parameters used in environmental fate modeling. The octanol-water partition coefficient is used to characterize the partitioning of a molecule between largely aqueous phases, such as rivers and lakes, and largely hydrophobic phases, such as the organic fraction of sediments suspended in water bodies. Because the octanol-water partition coefficient (K_{ow}) characterizes partitioning between aqueous and organic, lipid-like phases, it is used to estimate a variety of toxicological, and environmental fate parameters. Therefore, accurate estimates of K_{ow} are critical to successful estimates of other environmental properties.

One specific and simple use of the octanol-water partition coefficient is as a gauge for the potential for bioaccumulation. If a chemical tends to partition into the organic phase (is lipophilic), then the chemical can be stored in fatty tissue of fish and will bioaccumulate in animals that consume the fish. Table 5.2-5 describes the approximate relationship between K_{ow} and bioaccumulation.

Group contribution methods (structure activity relationships) have been developed for octanol-water partition coefficients (Meylan and Howard, 1995), and they have a form very similar to the form used for boiling point.

$$\log K_{ow} = 0.229 + \Sigma\, n_i\, f_i \qquad \text{(Eq. 5-11)}$$

where $\log K_{ow}$ is the base 10 logarithm of the ratio of the chemical's concentration in octanol to the chemical's concentration in water, n_i is the number of groups of type i in the molecule, f_i is the contribution of each group to the partition coefficient, and the summation is taken over all groups. Structural groups and group contributions (f_i) for estimating octanol-water partition coefficients are listed in Table 5.2-6.

Just as was done for boiling point, corrections are introduced to the preliminary estimate. In this case, corrections account for the unusual behavior of selected functional groups. The equation for estimating the corrected value of K_{ow} is:

$$\log K_{ow} = 0.229 + \Sigma\, n_i\, f_i + \Sigma\, n_j\, c_j \qquad \text{(Eq. 5-12)}$$

where n_j is the number of groups of type j in the molecule, c_j is the correction factor for each group, and the summation is taken over all groups that have correction factors. Structural groups and correction factors (c_j) are listed in Table 5.2-7. The method yields a mean error of 0.31 log units (Meylan and Howard, 1995).

On first inspection, the correction factors listed in Table 5.2-7 may seem a bit baffling and arbitrary; however, more careful analysis reveals the rationale behind the corrections. For example, many of the corrections account for electronic interactions between multiple substituents on aromatic rings (e.g., all of the ortho-corrections in Table 5.2-7). Recall from organic chemistry that substituents on

Table 5.2-5   Classification Criteria for Bioaccumulation.

Bioaccumulation potential	
High Potential	$8.0 > \log K_{ow} > 4.3$
Moderate Potential	$4.3 > \log K_{ow} > 3.5$
Low Potential	$3.5 > \log K_{ow}$

Table 5.2-6 Structural Groups and Group Contributions for Estimating Octanol-Water Partition Coefficients (Meylan and Howard, 1995).

Aromatic atoms	Contribution (f_i) to octanol-water partition coefficient	Aliphatic nitrogen groups	Contribution (f_i) to octanol-water partition coefficient
Carbon	0.2940	$-NO_2$ (aliphatic attach.)	-0.8132
Oxygen	-0.0423	$-NO_2$ (aromatic attach.)	-0.1823
Sulfur	0.4082	$-N=C=S$ (aliph. attach.)	0.5236
		$-N=C=S$ (arom. attach.)	1.3369
Aromatic nitrogen		$-NP$	-0.4367
Nitrogen at a fused ring	-0.0001	$-N$ (2 aromatic attach.)	-0.4657
N in a 5 member ring	-0.5262	$-N$ (1 aromatic attach.)	-0.9170
N in a 6 member ring	-0.7324	$-N=C$ (aliph. attach.)	-0.0010
		$-NH_2$ (aliphatic attach.)	-1.4148
Aliphatic Carbon		$-NH$ (aliphatic attach.)	-1.4962
$-CH_3$	0.5473	$-N<$ (aliphatic attach.)	-1.8323
$-CH_2-$	0.4911	$-N(O)$ (nitroso)	-0.1299
$-CH<$	0.3614	$-N=N-$ (azo)	0.3541
$>C<$	0.2676		
Other C, no H attached	0.9723	*Aliphatic oxygen*	
		$-OH$ (nitrogen attach.)	-0.0427
Olefinic/acetylenic C		$-OH$ (P attachment)	0.4750
$=C<$ (2 aromatic bonds)	-0.4186	$-OH$ (olefinic attach.)	-0.8855
$=CH_2$	0.5184	$-OH$ (carbonyl attach.)	0.0
$=CH-$ or $=C<$	0.3836	$-OH$ (aliphatic attach.)	-1.4086
$\equiv CH$ or $\equiv C-$	0.1334	$-OH$ (aromatic attach.)	-0.4802
		$=O$	0.0
Carbonyls		$-O-$ (carbonyl attach.)	0.0
$-CHO$ (aliphatic attach.)	-0.9422	O (aliphatic attach.)	1.2566
$-CHO$ (aromatic attach.)	-0.2828	$-O-$ (1 aromatic attach.)	-0.4664
$-C(O)OH$ (aliph. attach)	-0.6895	$-O-$ (2 aromatic attach.)	0.2923
$-C(O)OH$ (arom. attach.)	-0.1186		
$-NC(O)N-$ (urea type)	1.0453	*Aliphatic sulfur*	
$NC(O)O$ (carbamate)	0.1283	$-SO_2N$ (aliph. attach)	-0.4351
$NC(O)S$ (thiocarbamate)	0.5240	SO_2N (arom. attach)	-0.2079
$-C(O)O-$ (aliph. attach.)	-0.9505	$-S-$ (aliphatic attach.)	-0.4045
$-C(O)O-$ (arom. attach.)	-0.7121	$-S-S-$ (disulfide)	0.5497
$-C(O)N$ (aliph. attach)	-0.5236	$-SO_2OH$ (sulfonic acid)	-3.1580
$-C(O)N$ (arom. attach)	0.1599		
$-C(O)S-$ (aliph. attach)	-1.100	*Halogen groups*	
$-C(O)-$ (aliph. attach)	-1.5586	$-F$ (aliph. attach)	-0.0031
$-C(O)-$ (1 arom. attach)	-0.8666	$-F$ (arom. attach)	0.2004
$-C(O)-$ (cyclic, 2 arom. attach)	-0.2063	$-Cl$ (aliph. attach)	0.3102
		$-Cl$ (arom. attach)	0.6445
$-C(O)-$ (olefinic attach)	-1.2700	$-Cl$ (olefinic attach)	0.4923
$-C(O)-$ (cyclic, arom., olefinic attach.)	-0.5497	$-Br$ (aliph. attach)	0.3997
		$-Br$ (arom. attach)	0.8900

Table 5.2-7 Correction Factors for Estimating Octanol-Water Partition Coefficients (Meylan and Howard, 1995).

Structural group	Correction factor
Correction factors involving ortho substituents on aromatic rings	
$-COOH/-OH$	1.1930
$-OH$/ester	1.2556
Amino (at 2 position) on pyridine	0.6421
Alkyloxy (or alkylthio) ortho to 1 aromatic nitrogen	0.4549
Alkyloxy ortho to two aromatic nitrogens (or pyrazine)	0.8955
Alkylthio ortho to two aromatic nitrogens (or pyrazine)	0.5415
Carboxamide ($-C(O)N$) ortho to an aromatic nitrogen	0.6427
Any group other than hydrogen ortho to $-NHC(O)C$ (e.g., 2 methylacetanilide)	-0.5634
Any two groups other than hydrogen ortho to $-NHC(O)C$ (e.g., 2,6 dimethylacetanilide)	-1.1239
Any group other than hydrogen ortho to $-C(O)NH$ (e.g., 2 methylbenzamide)	-0.7352
Any two groups other than hydrogen ortho to $-C(O)NH$ (e.g., 2,6 dimethylbenzamide)	-1.1284
Correction factors involving non-ortho substituents on aromatic rings	
$-N</-OH$ (e.g., 4-aminophenol)	-0.3510
$-N<$/ester (e.g., 4 aminobenzoic acid methyl ester)	0.3953
$-OH$/ester	0.6487
Correction factors involving ortho or non-ortho substituents on aromatic rings	
$-NO_2$ with $-OH$, $-N<$, or $-N=N-$	0.5770
$-C\equiv N$ with $-OH$ or $-N$ (e.g., cyanophenols)	0.5504
Amino group on triazine, pyrimidine, or pyrazine	0.8566
NC(O)N S on triazine or pyrimidine (2-position)	-0.7500
Additional (non-aromatic) correction factors	
Carbonyl correction factors	
More than one aliphatic $-C(O)OH$	-0.5865
Cyclic ester (non-olefinic)	-1.0577
Cyclic ester (olefinic)	-0.2969
$-C(O)-C-C(O)N$	0.9734
Ring correction factors	
Triazine ring	0.8856
Pyridine ring (non-fused)	-0.1621
Fused aliphatic ring	-0.3421
Alcohol, ether and nitrogen corrections	
More than one aliphatic $-OH$	0.4064
$-NC(C-OH)C-OH$	0.6365
$-NCOC$	0.5494
$HO-CHCOCH-OH$	1.0649
$HO-CHC(OH)CH-OH$	0.5944
$-NH-NH-$	1.1330
$>N-N<$	0.7306

aromatic rings can be electron donating or electron withdrawing and that these electronic effects are different at ortho-, meta- and para- positions. Thus, if there are two or more substituents on an aromatic ring, the substituents will interact with one another through their electronic effects on the ring. The corrections in the table account for this effect. Other corrections in Table 5.2-7 account for other types of interactions between groups and the presence of ring structures. A final type of correction in the table accounts for the presence of multiple hydrogen bonding groups in a molecule. Molecules with one hydrogen bonding group can form dimers, while molecules with more than one hydrogen bonding group can form n-mers. The potential formation of polymer-like chains held together by hydrogen bonds can dramatically influence chemical and physical properties, necessitating correction factors for molecules containing multiple hydrogen bonding groups.

While a list of correction factors could potentially be endless, in practice, only a few types of corrections are normally accounted for. Ring correction factors, factors accounting for multiple hydrogen bonding groups, and corrections for substitution positions are among the most common. Example 5.2-4 illustrates the use of the group contribution method and the application of correction factors.

Example 5.2-4

Estimate the octanol-water partition coefficient for 1,1 dichloroethylene and the structure shown below (a herbicide).

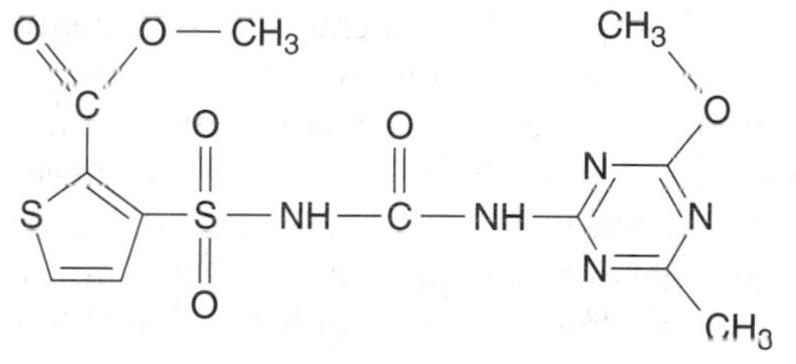

Figure 5.2-1 Structure of herbicide.

Solution: *1,1-Dichloroethylene* has the molecular structure $CH_2{=}CCl_2$. Referring to the groups in Table 5.2-6, this structure can be represented by one $=CH_2$ group, one $=CH-$ or $=C<$ group and two $-Cl$ (olefinic attachment) groups. The uncorrected value of K_{ow} from Equation 5-4 is given by:

$$\log K_{ow} = 0.229 + 0.5184 + 0.3836 + 2(0.4923) = 2.11$$

$$K_{ow} = 130$$

Dichloroethylene does not contain any groups that have correction terms. The experimental value for $\log K_{ow}$ is 2.13, so the predicted value of K_{ow} is in error by 3.3%.

The herbicide can be represented by three $-CH_3$ groups, one $-NH-$ (aliphatic attachment), 7 aromatic carbons, 3 aromatic nitrogens, one $-O-$ (one aromatic attachment) group, one $-N$ (one aromatic attachment) group, one aromatic sulfur group, one $-C(=O)O$ (ester, aromatic attachment) group, one $-SO_2N$ (aromatic attachment) group and one $-NC(=O)N-$ (urea type carbonyl) group. Note that the $-NC(=O)N-$ is listed as a carbonyl group in Table 5.2-7 and accounts only for the carbonyl $(C=O)$, not the nitrogens. The uncorrected value of K_{ow} from Equation 5-4 is given by:

$$\log K_{ow} = 0.229 + 3(0.5473) - 1.4962 + 7(0.2940) - 3(0.7324) - 0.4664 - 0.9170 + 0.4082 - 0.7121 - 0.2079 + 1.0453 = -0.614$$

The herbicide contains several groups that require correction factors. There is one triazine ring correction (0.8856), one correction for an amino-type triazine (0.8566), one correction for an alkoxy ortho to two aromatic nitrogens (0.8955) and one correction for a $-NC(=O)NS$ on a triazine (-0.7500). The total of these correction facors is 1.887, leading to

$$\log K_{ow} = 1.273$$

The octanol-water partition coefficient for this compound is strongly pH-dependent, but this estimation method leads to reasonable estimates for slightly basic solutions.

5.2.4 Bioconcentration Factor

One of the primary reasons for estimating the octanol-water partition coefficient is to assess the partitioning of a chemical between aqueous and lipid phases in living organisms. This partitioning is normally expressed as a bioconcentration factor (BCF). The BCF is defined as the ratio of a chemical's concentration in the tissue of an aquatic organism to its concentration in water (in L/kg). This parameter is called a bioconcentration factor because high values of BCF indicate that a living organism will tend to extract a material from an aqueous phase, such as ingested water or blood, and concentrate it in lipid tissues (e.g., fats). Thus, high values of BCF can be cause for concern. For example, a compound with a high bioconcentration factor may tend to accumulate in fish, resulting in a health hazard if the fish is eaten.

As shown in Table 5.2-8, BCF values can be used to gauge bioaccumulation potential, just as octanol-water partition coefficients were (Table 5.2-5).

Veith and Kosian (1983) propose this correlation between octanol-water partition coefficients and BCF:

Table 5.2-8 Classification Criteria for Bioaccumulation.

Bioaccumulation potential	
High Potential	BCF>1000
Moderate Potential	1000>BCF>250
Low Potential	250>BCF

$$\log BCF = 0.79 \, (\log K_{ow}) - 0.40 \qquad \text{(Eq. 5-13)}$$

More recently, correction factors have been introduced into this correlation, in a manner analogous to the estimation methods for K_{ow}. For non-ionic compounds, Meylan, et al. (1997), propose:

$$\log BCF = 0.77 \, (\log K_{ow}) - 0.70 + \Sigma j_j \qquad \text{(Eq. 5-14)}$$

where j_j is the correction factor for each group, and the summation is taken over all groups that have correction factors. The correction factors are listed in Table 5.2-9. Mean errors of approximately 0.5 log units can be expected with this method.

Note that there are fewer correction factors in Table 5.2-9 than in Table 5.2-7. This is not because BCF is more straightforward to estimate than K_{ow}. If anything, BCF is more difficult to reliably estimate than K_{ow} because of the variability in lipid tissues. The reason why Table 5.2-9 is relatively sparse is because experimental values, on which the correction factors are based, are considerably scarcer for BCF than for K_{ow}. Therefore, estimates of BCF for structurally complex compounds that typically require correction factors may have considerable uncertainty.

Example 5.2-5

Estimate the bioconcentration factor for 2,2,4 trimethyl-1,3 pentanediol, and 2,4',5 trichlorobiphenyl.

Solution: 2,2,4 trimethyl-1,3 pentanediol has the structure $HO-CH_2-(C)(CH_3)_2-CH(OH)-(CH)(CH_3)-CH_3$. Before estimating BCF, it is first necessary to estimate K_{ow}. Referring to the groups in Table 5.2-6, the structure can be represented by four $-CH_3$ groups, one $-CH_2-$ group, one $>C<$ group, two $-CH<$ groups and two $-OH$ groups (aliphatic attachment). The uncorrected value of K_{ow} from Equation 5-4 is given by:

$$\log K_{ow} = 0.229 + 4(0.5473) + 0.4911 + 0.2676 + 2(0.3614) + 2(-1.4086) = 1.08$$

$$K_{ow} = 12.1$$

2,2,4 trimethyl-1,3 pentanediol requires a correction for molecules containing two or more aliphatic $-OH$ (0.4064). The corrected value for $\log K_{ow}$ is 1.49. The experimental value for $\log K_{ow}$ is 1.24. Using the corrected calculated value of $\log K_{ow}$ and Equation 5-13 (Equation 5-14 does not apply because none of the correction factors in Table 5.2-9 are appropriate), proceed to calculate BCF for 2,2,4 trimethyl-1,3 pentanediol.

$$\log BCF = 0.79(1.49) - 0.40 = 0.7771$$

$$BCF = 5.99$$

Referring to Table 5.2-8, we see that because the BCF of 2,2,4 trimethyl-1,3 pentanediol is less than 250, it has low potential for bioaccumulation.

2,4',5 trichlorobiphenyl can be represented by 12 aromatic carbons and 3 $-Cl$ (aromatic attachment) groups. The uncorrected value of K_{ow} from Equation 5-4 is given by:

$$\log K_{ow} = 0.229 + 12(0.2940) + 3(.6445) = 5.69$$

Table 5.2-9 Correction Factors for BCF of Non-Ionic Compounds (Meylan and Howard, 1997).

Structural group	Correction factor
Ketone (with one or more aromatic connections)	−0.84
Phosphate ester, O=P(O-R)(O-R)(O-R) where at least two of the R groups are carbon	−0.78
Multihalogenated biphenyls and polyaromatics	0.62
Compounds containing an aromatic ring and an aliphatic alcohol in the form of −CH-OH (e.g., benzyl alcohol)	−0.65
Compounds containing an aromatic alcohol (e.g., phenol) with two or more halogens attached to the aromatic ring	−0.40
Compounds containing an aromatic triazine ring	−0.32
Compounds containing an aromatic ring with a tert-butyl group in an ortho position to a hydroxyl group	−0.45
Compounds containing a phenanthrene ring	0.48
Compounds containing a cyclopropyl ester	−1.65
Compounds with an alkyl chain containing 8 or more −CH$_2$− groups (4< log K$_{ow}$<6)	−1.00
Compounds with an alkyl chain containing 8 or more −CH$_2$− groups (6< log K$_{ow}$<10)	−1.50
Azo compounds	Log BCF = 1

No corrections are required. The experimental value for log K_{ow} is 5.81. Using this uncorrected calculated value of log K_{ow} and Equation 5-14, proceed to calculate BCF for 2,4',5 trichlorobiphenyl. With the correction factor from Table 5.2-9 for multihalogenated biphenyls and polyaromatics (0.62), Equation 5-14 becomes:

$$\log BCF = 0.77(5.69) - 0.70 + 0.62 = 4.30$$

$$BCF = 20000$$

Referring to Table 5.2-9, it is evident that 2,4',5 trichlorobiphenyl has a very high potential for bioaccumulation.

5.2.5 Water Solubility

In assessing environmental transport and partitioning, it is often necessary to predict maximum, or saturation, concentrations. In the gas phase, this is done by estimating vapor pressure. In aqueous phases, saturation concentrations are estimated using water solubilities.

Water solubility can be estimated in many ways. Activity coefficients, solubility parameters, and other chemical and structural properties can be used as a basis for estimating water solubility. For environmental applications, however, water solubility is most often estimated based on octanol-water partition coefficients. This is not because K_{ow} is the most accurate or reliable parameter for estimating water solubility. Rather, it is a matter of convenience. K_{ow} is used to estimate a wide variety of parameters in evaluating environmental fate and risk. Therefore, K_{ow} is generally available in environmental assessments, while properties such as activity coefficients are not frequently calculated in environmental screening studies. Table 5.2-10 classifies the numerical values for solubility (S) into general solubility categories.

Table 5.2-10 Classification Criteria for Water Solubility.

Water Solubility	
Very Soluble	S>10,000 ppm
Soluble	1,000<S<10,000 ppm
Moderately Soluble	100<S<1,000 ppm
Slightly Soluble	0.1<S<100 ppm
Insoluble	S<0.1 ppm

Meylan, et al. (1996) have used K_{ow}, along with correction factors, to estimate water solubilities. Their correlations are:

$$\log S = 0.342 - 1.0374 \log K_{ow} - 0.0108(T_m - 25) + \Sigma\, h_j \quad \text{(Eq. 5-15)}$$

$$\log S = 0.796 - 0.854 \log K_{ow} - 0.00728(MW) + \Sigma\, h_j \quad \text{(Eq. 5-16)}$$

$$\log S = 0.693 - 0.96 \log K_{ow} - 0.0092(T_m - 25) - 0.00314(MW) + \Sigma\, h_j \quad \text{(Eq. 5-17)}$$

where S is the water solubility in mol/L; K_{ow} is the octanol-water partition coefficient, T_m is the melting point in °C, MW is the molecular weight, and h_j is the correction factor for each group, and the summation is taken over all groups that have correction factors. Note that the correction factors are different for each equation. They are listed in Table 5.2-11. Mean errors are in the range of 0.3 to 0.4 log units.

Any of the three equations can be used, but generally, if more information is available for the correlation (Equations 5-16 or 5-17), the estimate is more accurate.

Example 5.2-6

Estimate the water solubility of 2-hexanol and diphenyl ether.

Solution: 2-hexanol has the molecular structure $CH_3-(CH-OH)-C_4H_9$. Before estimating water solubility, we must estimate K_{ow}. Referring to the groups in Table 5.2-6, this structure can be represented by two $-CH_3$ groups, three $-CH_2-$ groups, one $-CH-$ group and one $-OH$ (aliphatic attachment) group. The uncorrected value of K_{ow} from Equation 5-11 is given by:

$$\log K_{ow} = 0.229 + 2(0.5473) + 3(0.4911) + 0.3614 - 1.4086 = 1.75$$

2-hexanol does not contain any groups that have correction terms. The experimental value for $\log K_{ow}$ is 1.76, so the predicted value of K_{ow} is in error by −0.6 %.

The water solubility can be estimated from Equation 5-16 with a correction term for one aliphatic $-OH$ group.

$$\text{Log } S = 0.796 - 0.854\,(\log K_{ow}) - 0.00728(MW) + \Sigma\, h_j$$

$$\text{Log } S = 0.796 - 0.854(1.75) - 0.00728(102.2) + 0.510 = -0.932$$

$$S = 0.12 \text{ mol/L}$$

Diphenyl ether has the molecular structure $C_6H_5-O-C_6H_5$. Before estimating water solubility, we must estimate K_{ow}. Referring to the groups in Table 5.2-6, this structure can be represented by twelve aromatic carbons and one $-O-$ group (two

Table 5.2-11   Correction Factors for Estimating Water Solubility (Meylan, et al., 1996).

Structural group	Correction factor (Eq. 5-15)	Correction factor (Eq. 5-16)	Correction factor (Eq. 5-17)
Aliphatic alcohols with one $-OH$ attached to aliphatic carbon, except acetamide, amino, azo or $-S=O$ compounds	0.466	0.510	0.424
Aliphatic acids with acid attached to aliphatic group, except amino acids and compounds with $C(O)-N-C-COOH$	0.689	0.395	0.650
Primary, secondary, and tertiary aliphatic, liquid amines	0.883	1.008	0.838
Aromatic acids except amino-substituted compounds	1.104	–	0.898
Phenols, except amino-phenols	1.092	0.580	0.961
Alkylpyridines	1.293	1.300	1.243
Azo compounds $(-C-N=N-C-)$	−0.638	−0.432	−0.341
Nitrile compounds except $(N-C-CN)$	−0.381	−0.265	−0.362
Hydrocarbons (aliphatics containing only carbon and hydrogen)	−0.112	−0.537	−0.441
Aliphatic and aromatic nitro compounds, except aromatic compounds with $-OH$ or amino substitutions	−0.555	−0.390	−0.505
Aromatic sulfonamide and aliphatic compounds with $S-(O)-C-C(O)-C$	−1.187	−1.051	−0.865
Alkanes with two or more fluorines	−0.832	−0.742	−0.945
Polyaromatichydrocarbons	–	−1.110	–
Compounds with two or more aliphatic N, one attached to C(O), S(O) or C(=S); compounds with 4 or more aromatic N, compounds with 2 or more aromatic N and one or more aliphatic N attached to C(O), S(O) or C(=S); except N in nitrile, nitro, azo, barbituate and metal compounds	–	−1.310	–
Amino acids	–	−2.070	–

aromatic attachments) group. The uncorrected value of K_{ow} from Equation 5-11 is given by:

$$\log K_{ow} = 0.229 + 12(0.2940) + 0.2923 = 4.05$$

Diphenyl ether does not contain any groups that have correction terms. The experimental value for $\log K_{ow}$ is 4.21, so the predicted value of K_{ow} is in error by −3.8 %.

The water solubility can be estimated from Equation 5-16. There are no correction terms that apply.

$$\text{Log } S = 0.796 - 0.854 (\log K_{ow}) - 0.00728(MW) + \Sigma\, h_j$$

$$\text{Log } S = 0.796 - 0.854 (4.05) - 0.00728(170.2) + 0.0 = -3.90$$

$$S = 1.2 \times 10^{-4}\ \text{mol/L}$$

5.2.6   Henry's Law Constant

The Henry's Law constant, as commonly used in describing environmental fate, is the ratio of a compound's concentration in air to its concentration in water, at equilibrium. In other words, it shows a compound's affinity for air over water.

Therefore, compounds with high values of Henry's Law constant (H) tend to partition into the air, while compounds with low values of H tend to partition into the water. H is generally expressed either as a dimensionless ratio, or in units of atm-m^3/mole. Table 5.2-12 classifies Henry's Law coefficients into a set of volatility categories.

A group contribution method can also be used to estimate the value of the Henry's Law constant. In this case, the group contribution method is structured differently from the previous methods. The structural elements are bonds rather than functional groups. Consider 1-propanol as an example of how a bond approach differs from a functional group approach to structural characterization.

$$
\begin{array}{ccccccc}
\text{H} & & \text{H} & & \text{H} & & \\
| & & | & & | & & \\
\text{H} - \text{C} - & \text{C} - & \text{C} - & \text{O} - & \text{H} \\
| & & | & & | & & \\
\text{H} & & \text{H} & & \text{H} & &
\end{array}
$$

Using the groups from Table 5.2-2, 1-propanol would consist of one $-CH_3$ group, two $>CH_2$ groups, and one primary $-OH$ group. Expressed as a collection of bonds, 1-propanol consists of 7 C-H bonds, 2 C-C bonds, one C-O bond and one O-H bond.

A preliminary estimate of the Henry's Law constant is obtained by summing each of the bond contributions. This preliminary estimate is then adjusted by correction factors for selected functional groups (Meylan and Howard, 1991).

$$- \log H = \log \text{(air-water partition coefficient)} = \Sigma\, n_i\, h_i + \Sigma\, n_j\, c_j \quad \text{(Eq. 5-18)}$$

where H is the dimensionless Henry's Law constant, n_i is the number of bonds of type i in the molecule, h_i is the bond contribution to the air-water partition coefficient, n_j is the number of groups of type j in the molecule, c_j is the correction factor for each group, and the summations are taken over all bonds and all groups that have correction factors. Bond contributions (h_i) and correction factors (c_j) are listed in Tables 5.2-13 and 5.2-14. Mean errors in log units range from 0.06 for alkanes and alkylbenzenes to 0.4 for haloalkenes (Meylan and Howard, 1991).

Table 5.2-12   Classification Criteria for Volatility.

Volatility (H in atm-m^3/mole)	
Very Volatile	$H > 10^{-1}$
Volatile	$10^{-1} > H > 10^{-3}$
Moderately Volatile	$10^{-3} > H > 10^{-5}$
Slightly Volatile	$10^{-5} > H > 10^{-7}$
Nonvolatile	$10^{-7} > H$

Table 5.2-13   Structural Groups and Group Contributions for Estimating Henry's Law Constants (Meylan and Howard, 1991).

Bond type	Contribution (h_i) to Henry's Law constant	Bond type	Contribution (h_i) to Henry's Law constant
C-H	−0.1197	$C_{aromatic}$−OH	0.5967
C-C	0.1163	$C_{aromatic}$−O	0.3473
C-$C_{aromatic}$	0.1619	$C_{aromatic}$−$N_{aromatic}$	1.6282
C-$C_{olefinic}$	0.0635	$C_{aromatic}$−$S_{aromatic}$	0.3739
C-$C_{acetylenic}$	0.5375	$C_{aromatic}$−$O_{aromatic}$	0.2419
C-CO	1.7057	$C_{aromatic}$−S	0.6345
C-N	1.3001	$C_{aromatic}$−N	0.7304
C-O	1.0855	$C_{aromatic}$−I	0.4806
C-S	1.1056	$C_{aromatic}$−F	−0.2214
C-Cl	0.3335	$C_{aromatic}$−$C_{olefinic}$	0.4391
C-Br	0.8187	$C_{aromatic}$−CN	1.8606
C-F	−0.4184	$C_{aromatic}$−CO	1.2387
C-I	1.0074	$C_{aromatic}$−Br	0.2454
C-NO2	3.1231	$C_{aromatic}$−NO_2	2.2496
C-CN	3.2624	CO-H	1.2102
C-P	0.7786	CO-O	0.0714
C=S	−0.0460	CO-N	2.4261
$C_{olefinic}$−H	−0.1005	CO-CO	2.4000
$C_{olefinic}$=$C_{olefinic}$	0.0000	O-H	3.2318
$C_{olefinic}$−$C_{olefinic}$	0.0997	O-P	0.3930
$C_{olefinic}$−CO	1.9260	O-O	−0.4036
$C_{olefinic}$−Cl	0.0426	O=P	1.6334
$C_{olefinic}$−CN	2.5514	N-H	1.2835
$C_{olefinic}$−O	0.2051	N-N	1.0956
$C_{olefinic}$−F	−0.3824	N=O	1.0956
$C_{acetylenic}$−H	0.0040	N=N	0.1374
$C_{acetylenic}$≡$C_{acetylenic}$	0.0000	S-H	0.2247
$C_{aromatic}$−H	−0.1543	S-S	−0.1891
$C_{aromatic}$−$C_{aromatic}$ (fused)	0.2638	S-P	0.6334
$C_{aromatic}$−$C_{aromatic}$ (ext.)	0.1490	S=P	−1.0317
$C_{aromatic}$−Cl	−0.0241		

Example 5.2-7

Estimate the Henry's Law constant for 1-propanol.

Solution: 1-propanol consists of 7 C-H bonds, 2 C-C bonds, one C-O bond and one O-H bond.

The uncorrected value of log (air to water partition constant) is given by:

$$-\log H = \log \text{(air-water partition coefficient)} =$$
$$7(-0.1197) + 2(0.1163) + 1.0855 + 3.2318 = 3.7112$$

The correction is for linear or branched alcohols (−0.20) giving a net value of 3.5112 for log H^{-1}. The experimental value is 3.55, an error of −1.1% in the logarithm of the

Table 5.2-14   Correction Factors for Henry's Law Constants (Meylan and Howard, 1991).

Structural group	Correction factor
Linear or branched alkane	−0.75
Cyclic alkane	−0.28
Monoolefin	−0.20
Cyclic monoolefin	0.25
Linear or branched aliphatic alcohol	−0.20
Adjacent aliphatic ethers (−C−O−C−O−C−)	−0.70
Cyclic monoether	0.90
Epoxide	0.50
Each additional aliphatic −OH above one	−3.00
Each additional aromatic nitrogen within a single ring above one	−2.50
A fluoroalkane with only one fluorine	0.95
A chloroalkane with only one chlorine	0.50
A fully chlorinated chloroalkane	−1.35
A fully fluorinated fluoroalkane	−0.60
A fully halogenated haloalkane	−0.90

air to water partition constant. Note that this is a dimensionless value (mol/m^3 divided by mol/m^3). To convert to units of atmospheres-m^3/mol, the dimensionless value should be adjusted using the ideal gas law, the gas constant, and the temperature.

5.2.7   Soil Sorption Coefficients

Soil-water partitioning is generally described using soil sorption coefficients. The coefficient (K_{oc}) is defined as the ratio of the mass of a compound adsorbed per unit weight of organic carbon in a soil (in μg/g organic carbon) to the concentration of the compound in a liquid phase (in μg/ml). Values of K_{oc} are categorized in Table 5.2-15.

The property estimation methods just described for water solubility, bioconcentration factor, and Henry's Law constant used the octanol-water partition coefficient as the primary correlating variable. This was possible because both the properties of interest and the correlating variable were bulk properties.

Table 5.2-15   Classification Criteria for Soil Sorption.

Soil Sorption	
Very Strong Sorption	Log K_{oc}>4.5
Strong Sorption	4.5> Log K_{oc}>3.5
Moderate Sorption	3.5> Log K_{oc}>2.5
Low Sorption	2.5> Log K_{oc}>1.5
Negligible Sorption	1.5> Log K_{oc}

Correlations for soil sorption coefficients, based on octanol-water partition coefficients and water solubility, are also available.

Lyman, et al. (1990) have given the following equations for soil sorption coefficient estimation.

$$\log K_{oc} = 0.544 \log K_{ow} + 1.377 \qquad \text{(Eq. 5-19)}$$

$$\log K_{oc} = -0.55 \log S + 3.64 \qquad \text{(Eq. 5-20)}$$

These equations, however, are restricted to quite specific classes of compounds. They are limited in their applicability because of the nature of soil sorption. The soil sorption coefficient describes the physical adsorption and chemical absorption of a compound onto a surface. The coefficient therefore depends not only on bulk properties, but also on steric properties that influence the interaction of a molecule with a surface. Meylan, et al. (1992) have proposed a relatively simple correlation for estimating soil sorption coefficients that incorporates both bulk and steric effects through a structural parameter called the molecular connectivity.

$$\log K_{oc} = 0.53\,{}^1\chi + 0.62 + \Sigma\, n_j P_j \qquad \text{(Eq. 5-21)}$$

where K_{oc} is the soil sorption coefficient expressed as the ratio of the mass of a compound adsorbed per unit weight of organic carbon in a soil (in $\mu g/g$ organic carbon) to the concentration of the compound in a liquid phase (in $\mu g/ml$); ${}^1\chi$ is the first order molecular connectivity index, as described in an appendix; n_j is the number of groups of type j in the molecule; P_j is the correction factor for each group, and the summation is taken over all groups that have correction factors. The correction factors are listed in Table 5.2-16. Mean errors of approximately 0.6 log units can be expected.

Example 5.2-8

Estimate the soil sorption coefficient of 2-hexanol.

Solution: As noted in Example 5.2-6, *2-hexanol* has the molecular structure $CH_3-(CH-OH)-C_4H_9$. Log K_{ow} was estimated to be 1.75 and log S was estimated to be -0.932. Estimating soil sorption coefficients using Equation 5-19 and Equation 5-20:

$$\text{Log } K_{oc} = 0.544 \log K_{ow} + 1.377 = 2.329$$

$$\text{Log } K_{oc} = -0.55 \log S + 3.64 = 4.15$$

Both of these estimates are substantially different from the experimental value of 1.01. Using instead a correlation based on molecular connectivity (Equation 5-21):

$$\text{Log } K_{oc} = 0.53\,{}^1\chi + 0.62 + \Sigma\, n_j P_j$$

Where the value of ${}^1\chi$ is 3.27 gives an uncorrected value of 2.35. Adding in the correction term for an aliphatic alcohol (-1.519) yields an estimate of 0.83.

Table 5.2-16   Correction Factors for Soil Sorption Coefficients (Meylan and Howard, 1992).

Structural group	Correction factor
N containing groups	
Azo	−1.028
N, C containing groups	
Nitrile/cyanide	−0.722
Nitrogen bound to noncyclic aliphatic C	−0.124
Nitrogen bound to cycloalkane	−0.822
Nitrogen bound to non-fused aromatic ring	−0.777
Pyridine ring with no other fragments	−0.700
Aromatic ring with 2 nitrogens	−0.965
Triazine ring	−0.752
N, O containing groups	
Nitro	−0.632
N, C, O containing groups	
Urea group (N−CO−N)	−0.922
Acetamide (N−CO−C)	−0.811
Uracil (−N−CO−N−CO−C=C− ring)	−1.806
N−CO−O−N−	−1.920
Carbamate (N−CO−O-phenyl)	−2.002
N-phenyl carbamate	−1.025
C, O containing groups	
Aromatic ether	−0.643
Aliphatic ether	−1.264
Ketone	−1.248
Ester	−1.309
Aliphatic alcohol	−1.519
Carboxylic acid	−1.751
Carbonyl	−1.200
P, O containing groups	
Aliphatic organophosphorus, P=O	−1.698
Aromatic organophosphorus, P=O	−2.878
P, S containing groups	
P=S	−1.263
C, S containing groups	
Thiocarbonyl	−1.100
S, O containing groups	
Sulfone	−0.995

Summary

This section has examined methods for estimating chemical and physical properties that influence phase partitioning in the environment. These methods will serve as the basis for estimation of a broad range of parameters that describe environmental persistence and environmental impacts. Therefore, any errors or uncertainties associated with the estimates described in this section are likely to propagate through the entire environmental assessment.

SECTION 5.2   QUESTIONS FOR DISCUSSION

1. How would you estimate properties for molecules that contain groups that are not explicitly represented in the group contribution methods (for example, could you estimate the Henry's Law constant for the herbicide listed in Example 5.2-3?).

2. The methodologies presented in this chapter are only a small selection of the group contribution methods available for these properties. How would you select the most accurate estimation methods?

3. Do the functional forms of the group contribution methods seem appropriate? For example, is it reasonable to assume that a boiling point estimation method should be a simple linear function? Would this approach work equally well for carboxylic acids and dicarboxylic acids? Would it work equally well for alcohols and glycols?

4. Can you rationalize the values of the group contributions? For example, does it make sense that the $-OH$ group has a large positive group contribution for boiling point?

5.3 ESTIMATING ENVIRONMENTAL PERSISTENCE

Section 5.2 described estimation tools for properties that influence the phase partitioning of chemicals in the environment. This section will examine methods for estimating the persistence of chemicals in the atmosphere and in aqueous and sediment environments. These methods are, by necessity, extremely simplified attempts to characterize the complex chemistries that occur in ambient environments. Thus, they should not be viewed as precise tools for estimating environmental lifetimes of chemicals. Rather, they should be viewed as semi-quantitative screening tools for ranking relative persistence.

5.3.1 Estimating Atmospheric Lifetimes

Chemicals emitted to the atmosphere undergo oxidation through a wide range of processes. One of the critical steps in these oxidations, particularly for organic compounds, is the rate of reaction with the hydroxyl radical. Hydroxyl radicals are extremely reactive species and can abstract hydrogen from saturated organics, add to double bonds or add to aromatic rings. Some of these reactions are shown below.

Hydrogen abstraction from propane

$$C_3H_8 + OH\cdot \Rightarrow \overset{\cdot}{C}H_3-CH-CH_3 + H_2O$$

$$\Downarrow$$

oxidized products

Hydroxyl radical addition to propene

$$C_3H_6 + OH\cdot \Rightarrow \overset{\cdot}{C}H_3-CH-CH_2\,OH$$

$$\Downarrow$$

oxidized products

Hydroxyl radical addition to an aromatic ring

$$C_6H_6 + OH\cdot \Rightarrow \overset{\cdot}{C}_6H_6-OH$$

$$\Downarrow$$

oxidized products

These reactions with hydroxyl radicals are often the first step in a series of re-actions that lead to the oxidation of organics in the atmosphere. We do not exam-ine the details of these pathways (the interested reader is referred to Seinfeld and Pandis, 1998); however, the relative rate at which a hydroxyl radical reacts with a compound is a semi-quantitative indicator of how long the compound will persist in the atmosphere. For example, for the three reactions listed above (hydrogen ab-straction from propane, addition to propene, and addition to benzene), the rates of reaction are 1.2×10^{12}, 26.0×10^{12}, and 2.0×10^{12} cm^3/molecule-sec., respectively. This indicates that if reaction with a hydroxyl radical is the dominant reaction path-way leading to oxidation in the atmosphere, then the rates of disappearance should be in the ratio 1.2:26:2. As shown in Example 5.3-1, this implies a ratio of atmos-pheric lifetimes of 106 hours:5 hours:64 hours.

So, one method of assessing atmospheric lifetimes is to estimate rate of reac-tion with hydroxyl radical. Once again, group contribution methods are a viable ap-proach. The mechanics of the method are similar to those discussed in Section 5.2. A molecule is divided into a collection of functional groups and each group makes a defined contribution to the overall rate of reaction. The method is slightly differ-ent from the methods discussed in Section 5.2, however, in that a single compound might have multiple rate parameters. Consider, for example, the reactions of propene. Hydroxyl radical can add to the double bond of propene. To estimate that rate constant, we would note that the olefinic group in propene has the structure ($CH_2=CH-$), and based on the data in Table 5.3-2, the rate constant for hydroxyl radical addition would be 26.3×10^{12} cm^3/molecule-sec. But hydroxyl radical can also abstract hydrogen from the terminal methyl group. This reaction, however, occurs much more slowly than the addition reaction. The group contribution for abstraction from a terminal methyl group is only 0.136×10^{12} cm^3/molecule-sec (Table 5.3-1). Thus, although propene can react via two pathways, only one is significant.

Identifying and estimating the rates of hydroxyl radical reactions with all of the functional groups in a molecule requires extensive experience. In this chapter, we will limit our estimations to addition reactions for olefins, and abstraction reactions. The group contributions for estimating these rates are listed in Tables 5.3-1 to 5.3-3. As with the property estimations described in Section 5.2, there are correction factors that can be applied to the estimations. These correction factors account for the electron donating and withdrawing characteristics of substituent groups, ring strain energy, and other parameters. A detailed discussion of these correction factors is beyond the scope of this chapter. Examples 5.3-1 through 5.3-3 illustrate how the basic estimations of hydroxyl radical reaction rates and atmospheric half-life are performed and provide simple illustrations of how the correction factors are applied.

Example 5.3-1

Using the rate of reaction of propene with the hydroxyl radical, estimate the atmospheric half-life of propylene.

Solution: The rate of reaction implies a rate of disappearance of propene:

$$(d[C_{propene}]/dt) = k\,[OH\cdot]\,[C_{propene}]$$

where $[OH\cdot]$ is the concentration of the hydroxyl radical and $[C_{propene}]$ is the concentration of propene.

Assuming that the concentration of hydroxyl radical is at steady state—the pseudosteady-state assumption (see, for example, Fogler, 1995)—leads to the following expression for the concentration of propene:

$$\ln\left([C_{propene}]/[C_{0\text{-}propene}]\right) = -(k\,[OH\cdot])t$$

where $[C_{0\text{-}propene}]$ is the initial concentration of propene, $(k\,[OH\cdot])$ is the rate constant multiplied by the steady state concentration of hydroxyl radicals and t is the time of reaction.

Since $([C_{propene}]/[C_{0\text{-}propene}]) = \frac{1}{2}$ when the concentration has reached one half of its original value, the half life is given by:

$$t_{1/2} = \ln(2) / (k\,[OH\cdot])$$

Assuming a value of 1.5×10^6 molecules/cm^3 for the concentration of the hydroxyl radical (while 1.5×10^6 molecules/cm^3 is a typical value, summertime concentrations in urban areas can reach 10^7 molecules/cm^3) and a value of 26×10^{-12} cm^3/molecule-sec for k:

$$t_{1/2} = \ln(2) / (39 \times 10^{-6}\,\text{sec}^{-1})$$

So, the half life for propene in the atmosphere is:

$$t_{1/2} = 5.0\ \text{hr}$$

Repeating this calculation for propane and benzene, with reaction rates of 1.2×10^{-12} and 2.0×10^{-12} cm^3/molecule-sec, leads to atmospheric half lives of 106 and 64 hours, respectively.

Example 5.3-2

Estimate the rate of reaction of octane with the hydroxyl radical.

Solution: Octane has the molecular structure $CH_3-(CH_2)_6-CH_3$. Since there are no aromatic, olefinic, or acetyl groups, the primary reaction pathway will be hydrogen atom abstraction. Referring to the groups in Table 5.3-1, this structure can be represented by two $-CH_3$ groups and six $-CH_2$ groups.

Both of the $-CH_3$ groups are bound to $-CH_2$ groups, so the abstraction rate from the $-CH_3$ groups is the group contribution for $-CH_3$ multiplied by the substituent factor for the $-CH_2$ group:

$$K(-CH_3) \, F(-CH_2) = 0.136 \, (1.23)$$

Two of the $-CH_2$ groups are bound to one $-CH_2$ group and one $-CH_3$ group, so the abstraction rate from these $-CH_2$ groups is the group contribution for $-CH_2$ multiplied by the substituent factors for the $-CH_2$ group and the $-CH_3$ group:

$$K(-CH_2) \, F(-CH_3) \, F(-CH_2) = 0.934 \, (1.00) \, (1.23)$$

Four of the $-CH_2$ groups are bound to one $-CH_2$ group and one $-CH_2$ group, so the abstraction rate from these $-CH_2$ groups is the group contribution for $-CH_2$ multiplied twice by the substituent factor for the CH_2 group:

$$K(-CH_2) \, F(-CH_2) \, F(-CH_2) = 0.934 \, (1.23) \, (1.23)$$

The sum of the contributions from each of these groups:

$$k - [2(0.136)(1.23) + 2(.934)(1.00)(1.23)$$
$$+ \ 4(0.934)(1.23)(1.23)] \times 10^{-12} \ cm^3/molecule\text{-}sec$$
$$k = 8.28 \times 10^{-12} \ cm^3/molecule\text{-}sec$$

The experimental value is $8.68 \times 10^{-12} \ cm^3/molecule\text{-}sec$.

Example 5.3-3

Estimate the rate of reaction of cis-2-butene with the hydroxyl radical.

Solution: Cis-2-butene has the molecular structure $CH_3-(CH=CH)-CH_3$. Since there are no aromatic groups, the primary reaction pathways will be hydrogen atom abstraction and hydrogen addition to the double bond. The rate of addition is given simply by the rate constant for addition to the cis $(-CH=CH-)$ structure: 56.4×10^{-12} $cm^3/molecule\text{-}sec$. The substituent factors are both 1.00.

The rate of abstraction is the rate due to abstraction from the two $-CH_3$ groups. Both of the $-CH_3$ groups are bound to $=CH$ groups, so the abstraction rate from the $-CH_3$ groups would be the group contribution for $-CH_3$ multiplied by the substituent factor for the $=CH-$ group, if it were available. Since it is not available, a value of 1.0 will be assumed:

$$K(-CH_3) \, F(-CH_2) = 0.136 \, (1.0)$$

The sum of the contributions from each of these routes:

$$k = [56.4 + 2(0.136)(1.0)] \times 10^{-12} \ cm^3/molecule\text{-}sec$$
$$k = 56.7 \times 10^{-12} \ cm^3/molecule\text{-}sec$$

Table 5.3-1 Group Contributions and Substituent Factors for Hydrogen Abstraction Rate Constants (Kwok and Atkinson, 1995).

Structural group	Group rate constant 10^{-12} cm^3/molecule-sec
Group contributions	
K($-$CH$_3$)	0.136
K($-$CH$_2-$)	0.934
K($>$CH$-$)	1.94
K($>$C$<$)	0
K($-$OH)	0.14
K($-$NH$_2$) (aliphatic)	21
K($-$NH$-$) (aliphatic)	63
K($>$N$-$) (aliphatic)	66
K($-$SH) (aliphatic)	32.5
K($-$S$-$)	1.7
K($-$S$-$S$-$)	225
K($>$N$-$NO)	0
K($>$N$-$NO$_2$)	1.3
K(P($=$O))	0
K(P($=$S))	53
Substituent factors	F(X) at 298 K
F($-$CH$_3$)	1.00
F($-$CH$_2-$)	1.23
F($>$CH$-$)	1.23
F($>$C$<$)	1.23
F($-$OH)	3.5
F($-$F)	0.094
F($-$Cl)	0.38
F($-$Br)	0.28
F($-$C(O)OH)	0.74

5.3.2 Estimating Lifetimes in Aqueous Environments

Chemicals emitted to aqueous environments undergo a wide range of reactions. One of the most significant reaction pathways is hydrolysis, which can be catalyzed by acids and bases; hydrolysis can also occur in neutral waters. Calculating the rate at which a compound reacts in water helps in estimating the concentration of that compound in the surface waters of the environment.

Hydrolysis rates can be estimated for a limited number of compound types using correlations based on structure-activity relationships (Mill, et al., 1987). The structure-activity relationships are generally based on linear free energy relationships. A linear free energy relationship assumes that the ratio of a rate constant to some reference rate is linearly proportional to a structural parameter that in some way characterizes the free energy of the transition state for the reaction. So, for

Table 5.3-2 Group Contributions to Rate Constants for Hydroxyl Radical Additions to Olefins and Acetylenes (Kwok and Atkinson, 1995).

Structural group	Group rate constant 10^{-12} cm^3/molecule-sec
CH$_2$=CH−	26.3
CH$_2$=C<	51.4
−CH=CH− (cis−)	56.4
−CH=CH− (trans−)	64.0
−CH=C<	86.9
>C=C<	110.0
−CH=CH− (cyclic)	56.4
CH≡C−	7.0
−C≡C−	27.0
Substituent factors	F(X) at 298 K
F(−CH$_3$)	1.00
F(−CH$_2$−)	1.00
F(>CH−)	1.00
F(>C<)	1.00
F(−F)	0.21
F(−Cl)	0.21
F(−Br)	0.26
F(−Phenyl)	1.00

hydrolysis reactions, the rate of hydrolysis can be correlated using an equation like 5-22.

$$\log\,(\text{hydrolysis rate}) = \log\,(\text{hydrolysis rate of a reference compound}) + \text{Constant} \times \sigma$$

$$\log\,(\text{hydrolysis rate}) = A + B\sigma \qquad \text{(Eq. 5-22)}$$

where σ is a structural parameter commonly used in linear free energy relationships, the Hammet constant. The Hammet constant characterizes the electron donating or electron withdrawing properties of a functional group. The details of

Table 5.3-3 Group Contributions for Rate Constants for Hydroxyl Radical Additions to (−C=C=C−) (Kwok and Atkinson, 1995).

Structural group	Group rate constant 10^{-12} cm^3/molecule-sec
(CH$_2$=C=CH−)	31.0
(−CH=C=CH−)	57.0
(CH$_2$=C=C<)	57.0
(−CH=C=C<)	85.0
(>C=C=C<)	110.0

estimating the Hammet constant and other parameters used in hydrolysis rate estimations are beyond the scope of this chapter, and it should be noted that empirical values for the constants A and B in Equation 5-22 must be determined for individual classes of reactants (e.g., the values for esters would be different than the values for epoxides). The parameter A is reaction and compound class specific because it depends on the reference reaction chosen. The parameter B is reaction and compound class specific because the dependence of rate on structural features depends on the type of reaction being considered.

An added complexity is that rates of reactions, such as hydrolysis, depend not just on the structure of the reactant, but also on the characteristics (e.g., pH) of the receiving waters. Thus, an estimate of hydrolysis rates requires both good rate estimation methods—which are scarce—and a detailed understanding of local environmental conditions.

5.3.3   Estimating Overall Biodegradability

In addition to all of the reactions that may occur with other chemicals in the atmosphere and in aqueous environments, we must also be concerned with the rate at which compounds are metabolized by living organisms. Developing an overall assessment of biodegradation will be difficult. Nevertheless, semi-quantitative assessments are possible. An ideal framework for estimating biodegradation would distinguish between the initial structural change of the compound (primary biodegradation) and the complete conversion to stable reaction products such as CO_2 and H_2O (ultimate biodegradation). It would also distinguish between aerobic (oxygen present) and anaerobic degradation.

Unfortunately, primary and ultimate, aerobic and anaerobic biodegradation rates are available for only a small number of compounds. Therefore, the approach described in previous sections—statistical regression of measured environmental data to yield group contribution parameters—will not work because there are not enough biodegradation data. Nevertheless, it is extremely important to have a qualitative sense of the persistence of compounds in the environment, and biodegradation is one of the most significant removal pathways for compounds in ambient environments. One pragmatic response to this problem has been to rely on estimations of biodegradation by expert panels. As described by Howard, et al. (1992) and Boethling, et al. (1994), expert panels can provide estimates of whether biodegradation occurs over hours, days, weeks, months or longer. These expert assessments can then be used as the basis for a group contribution method for biodegradation.

One such method (Boethling, et al., 1994) involves calculating an index that characterizes aerobic biodegradation rate in ambient environments.

$$I = 3.199 + a_1f_1 + a_2f_2 + \ldots + a_nf_n + a_mMW \qquad \text{(Eq. 5-23)}$$

where I is an indicator of the aerobic biodegradation rate. A value of 5 indicates that the compound is expected to degrade over hours; a value of 4 corresponds to a lifetime of days; 3, 2 and 1 correspond to weeks, months, and longer, respectively. These values of I should not be viewed as an accurate quantitative predictor of biodegradation rate. Rather, they should be viewed as a relative ranking of the probability that a material will biodegrade. The parameter f_n is the number of groups of type n in the molecule, and a_n is the contribution of group n to degradation rate. Group contribution parameters are listed in Table 5.3-4 and sample calculations are given in Example 5.3-4.

Example 5.3-4

Estimate the biodegradation index for 1-propanol and diphenyl ether.

Solution: 1-propanol has a molecular weight of 60 and contains an aliphatic $-OH$. Its biodegradation index is:

$$I = 3.199 + 0.160 - 0.00221\,(60) = 3.22$$

This implies a lifetime of weeks.

Diphenyl ether has a molecular weight of 170 and contains an aromatic ether and two mono-aromatic rings. Its biodegradation index is:

$$I = 3.199 + 2(0.022) - 0.058 - 0.00221\,(170) = 2.81$$

This implies a lifetime of weeks; literature data indicate a lifetime of months.

Table 5.3-4   Group Contributions to Ultimate Aerobic Biodegradation Index (Boethling, et al., 1994).

Structural group	Group contribution (a_n)
Molecular weight	−0.00221
Functional groups	
Unsubstituted mono-, di-, or tri-aromatic ring	−0.586
Unsubstituted phenyl group	0.022
Aromatic acid ($-COOH$)	0.088
Linear 4 carbon terminal chain ($-CH2-CH2-CH2-CH3$)	0.298
Aliphatic acid ($-COOH$)	0.365
Alkyl substituent on a ring	−0.075
Aromatic F	−0.407
Aromatic I	−0.045
Tetra aromatic or larger ring	−0.799
Aromatic amine	−0.135
Aliphatic amine	0.024
Aliphatic Cl	−0.173
Aromatic Cl	−0.207
Aromatic $-OH$	0.056
Aliphatic $-OH$	0.160
Aliphatic ether	−0.0087
Aromatic ether	−0.058

Summary

This section has provided a limited introduction to methods for estimating environmental persistence. The methods are generally specific to a particular environmental medium (air, water, or sediment/soil) and to particular reaction pathways (e.g., reaction with hydroxyl radical in the atmosphere or hydrolysis in aqueous environments). Often the methods will depend on local characteristics, such as the acidity or alkalinity of a water body and the concentration of oxidizing species in the atmosphere. With all of these restrictions, the appropriate use of these methods in performing screening assessments is simply for relative rankings of environmental persistence.

SECTION 5.3   QUESTIONS FOR DISCUSSION

1. When we examine atmospheric oxidation, we monitor only the disappearance of the chemical of interest. Should we be concerned about the reaction products that are formed?
2. The methodologies presented in this chapter represent only a small fraction of possible environmental degradation pathways. How would you use these limited data to perform an overall assessment of environmental persistence?

5.4   ESTIMATING ECOSYSTEM RISKS

Structure activity relationships may also be used to assess ecosystem and human health impacts. The range and variety of such relationships are enormous. Therefore, this section will present only a few, simple relationships that are used to assess ecosystem risk. For a more comprehensive review, the interested reader is referred to extensive literature on structure activity relationships (e.g., Hansch, et al., 1995a,b), as well as extensive literature of experimental data (see, for example, on-line databases of the US EPA available at the EPA website: *http://www.epa.gov*, or the databases cited in Appendix F).

In assessing ecosystem hazard, the standard practice is to estimate toxicity for a variety of species. For example, mortality for daphnids, fish, and guppies are frequently used in assessing the ecosystem hazard of chemicals described in premanufacture notices submitted to the US EPA under the Toxic Substances Control Act. The mortality for guppies can be correlated with the octanol-water partition coefficient using Equation 5-24

$$\log (1/LC_{50}) = 0.871 \log K_{ow} - 4.87 \qquad \text{(Eq. 5-24)}$$

where LC_{50} is the concentration that is lethal to 50% of the population over a 14-day exposure (expressed in μmol/L). This equation was developed using data from a variety of different compounds, including chlorobenzenes, chlorotoluenes, chloroalkanes, diethyl ether, and acetone (Konemann, 1981).

Other equations used to estimate ecosystem hazard are specific to certain compound classes. For example, toxicities for daphnids and fish can be estimated for more than 50 different compound classes. The correlations for acrylates are given below:

$$\log LC_{50} = 0.00886 - 0.51136 \log K_{ow} \qquad \text{(Eq. 5-25)}$$
$$\text{(Daphnids, mortality after 48 hr exposure)}$$

$$\log LC_{50} = -1.46 - 0.18 \log K_{ow} \qquad \text{(Eq. 5-26)}$$
$$\text{(Fish, mortality after 96 hr exposure)}$$

where LC_{50} is expressed in units of millimoles/L.

Example 5.4-1

Compare the fish, guppy and daphnid mortailities for an acrylate with log K_{ow} =1.28 (e.g. methyl methacylate).

Solution: The concentrations yielding 50% mortality are:

Guppies (14 day):	5690 μmol/L
Daphnids (48 hour):	0.226 millimoles/L = 226 μmol/L
Fish (96 hour):	0.020 millimoles/L = 20 μmol/L

SECTION 5.4   QUESTIONS FOR DISCUSSION

1. Why are ecotoxicities evaluated for immature amphibians and similar biota?
2. Why are the lethal concentrations negatively correlated with the octanol-water partition coefficient for these species?

5.5 USING PROPERTY ESTIMATES TO ESTIMATE ENVIRONMENTAL FATE AND EXPOSURE

The previous sections have described methods that can be used to estimate the properties that will govern a chemical's environmental partitioning and fate. This section will illustrate, through a few simple examples, how those properties can be employed to estimate partitioning and fate. These properties will also be used in Chapter 6 to estimate exposures.

Consider, for example, the problem of estimating exposure to a chemical via inhalation. To calculate inhalation exposure, it is necessary to know atmospheric concentrations. Breathing rates are multiplied by atmospheric chemical concentrations to determine inhalation exposures. The atmospheric chemical concentration depends on emission rate, mixing rate, and atmospheric lifetime. A simple case study is given in Example 5.5-1.

Example 5.5-1

Propylene is emitted at a rate of 10 metric tons per year into an airshed that has a volume of 10^4 cubic kilometers. Assume that the airshed has a residence time of one day and is well mixed. Calculate the steady state concentration of propylene, accounting for chemical reaction. Calculate an inhalation exposure for an adult, assuming an inhalation rate of 20 l/min.

Solution: Perform a mass balance to calculate the steady state concentration of propylene:

$$\text{In} - \text{out} - \text{disappearance due to reaction} = 0$$

$$\text{In} = 10^4\,\text{kilogram/yr} = 7.5 \times 10^{-3}\,\text{gram moles/sec}$$
$$\text{(based on a molecular weight of 42)}$$

$$\text{Out} = \text{flow rate} \times \text{steady state concentration of propylene}$$
$$= 10^4\,\text{cubic kilometers/day} \times C_{\text{propylene, ss}} = 1.16 \times 10^{14}\,\text{cm}^3/\text{sec} \times C_{\text{propylene, ss}}$$

$$\text{Disappearance due to reaction} = \text{Volume} \times \text{rate}$$
$$\text{(note that the rate of reaction for propylene was discussed in Section 5.3)}$$
$$= 10^4\,\text{cubic kilometers} \times 26 \times 10^{-12}\,\text{cm}^3/\text{molecule-sec} \times 1.5 \times 10^6\,\text{molecule/cm}^3 \times C_{\text{propylene, ss}}$$
$$= 10^{19}\,\text{cm}^3 \times 39 \times 10^{-6}/\text{sec} \times C_{\text{propylene, ss}}$$

$$C_{\text{propylene, ss}} = 1.5 \times 10^{-17}\,\text{moles/cm}^3$$

Assuming one mole of air occupies 22,400 cm^3 at ambient conditions,

$$C_{\text{propylene, ss}} = 3.3 \times 10^{-13}\,\text{moles propylene/mole air} = 0.3\,\text{ppt}$$

The exposure, assuming an inhalation rate of 20 L/min is:

$$20000 \times 1.5 \times 10^{-17}\,\text{moles/cm}^3 = 7.5 \times 10^{-14}\,\text{moles/min} = 6.4 \times 10^{-6}\,\text{g/yr}$$

Far more sophisticated models than the well-mixed box model, used in Example 5.5-1, can be used to estimate atmospheric concentrations and inhalation rates. Many such models are available and calculating atmospheric concentrations, in order to estimate inhalation rates, is done relatively routinely. The problems associated with estimating environmental exposures via other routes become far more complex. Consider the relatively simple example of calculating exposure through drinking contaminated surface water. Assume that a chemical is released to a river upstream of the intake to a public drinking water treatment plant. To evaluate the exposure we would need to determine:

- What fraction of the chemical was adsorbed by river sediments?
- What fraction of the chemical was volatilized to the atmosphere?
- What fraction of the chemical was taken up by living organisms?
- What fraction of the chemical was biodegraded or was lost through other reactions?
- What fraction of the chemical was removed by the treatment processes in the public water system?

Thus, exposure estimates will require information on the soil sorption coefficient, vapor pressure, water solubility, bioconcentration factor, and biodegradability of the compound, as well as river flow rates, surface area, sediment concentration and other parameters. A simple, yet typical, set of calculations is shown in Examples 5.5-2 through 5.5-4.

Example 5.5-2

Assume that a chemical, with a molecular weight of 150, is released at a rate of 300 kg/day to a river, 100 km upstream of the intake to a public water system. Estimate the initial partitioning of the chemical in the water, sediment, and biota.

Data
Water solubility: 100 ppm
Soil sorption coefficient: 10,000
Organic solids concentration in suspended solids: 15 ppm
River flow rate: 500 million liters per day
Bioconcentration factor: 100,000
Biota loading: 100 g per 100 cubic meter

Solution: The ratio of concentrations in water, sediment and biota will be approximately:

$$1 : 10,000 : 100,000$$

Based on the river flow rate, the total flow rates of water, sediment, and biota are:

Water: (500 million liter/day $\times$ 1 kg/liter) = 500 million kg/day

Sediment: 500 million kg/day $\times$ 15 kg sediment/million kg water =
7500 kg sediment/day

Biota: 500 million kg/day $\times$ 0.1 kg biota/million kg water = 50 kg biota/day

Performing a mass balance:

$$300 \text{ kg/day} = 500 \text{ million kg water/day } (C_{water})$$
$$+ 7500 \text{ kg sediment/day } (10,000 \ C_{water}) +$$
$$50 \text{ kg biota/day } (100,000 \ C_{water})$$

where (C_{water}) is the chemical concentration in the water phase:

$$(C_{water}) = 0.5 \times 10^{-6} \text{ kg chemical/kg water} = 0.5 \text{ ppm}$$

This is well below the solubility of 100 ppm. The ratio of the mass in water, sediment, and biota is:

$$500,000,000 : 75,000,000 : 5,000,000$$

$$84 : 13 : 1$$

Thus, although the concentrations are much higher in the biota and the sediment, more than 80% of the mass remains in the water phase.

Example 5.5-3

For the discharge described in Example 5.5-2, calculate the equilibrium vapor pressure above the river at the discharge point. Is volatilization from the river likely to be significant?

> Data
> Vapor pressure: 10^{-1} mm Hg
> River flow rate: 500 million liters per day
> River velocity: 0.5 m/sec
> River width: 30 m

Solution: Assuming ideal behavior and the concentration determined in Example 5.5-2, the equilibrium vapor pressure should be:

$$0.5 \times 10^{-6} \text{ g chemical/g water} \times 1 \text{ mole chemical/150 g} \times$$
$$18 \text{g/mole water} \times 10^{-1} \text{ mm Hg} = 0.6 \times 10^{-9} \text{ mmHg} = 8.0 \times 10^{-12} \text{ atm}$$

To determine if the loss rate is significant, assume that a volume 10 m above the river reached this concentration for the length of the river to the public water system inlet (a total volume of $100,000 \times 10 \times 30$ m^3). Noting that 1 gram-mole of air at standard conditions occupies 22.4 liters:

$$30 \times 10^6 \text{ m}^3 \times (1 \text{ mole air/0.0224 m}^3) \times 8.0 \times 10^{-12} \text{ moles chemical/mole air} \times$$
$$150 \text{ g/mole} = 1.6 \text{ g}$$

This is the mass required to saturate the atmosphere to a height of 10 m above the river for the 100 km length of the river. Compare this to the total discharge rate of 300 kg/day, and it is clear that volatilization will be negligible.

Example 5.5-4

For the discharge described in Examples 5.5-2 and 5.5-3, estimate what fraction of the initial discharge might still be in the water at the public water intake. If the treatment efficiency of this chemical in the water treatment plant is 95%, what would be the concentration in drinking water?

> Data
>
> Biodegradation half life: 300 hours

Solution: Based on a river velocity of 0.5 m/sec and a travel distance of 100 km, the transit time is 2.3 days. If the half life is 300 hours, the disappearance rate constant is (see Example 5.3-1):

$$t_{1/2} = 300 \text{ hours} = \ln(2)/(k)$$

This can be used to calculate the ratio of final to initial concentration:

$$\ln([C]/[C_0]) = -(\ln(2)/300 \text{ hours})t = -(\ln(2)/300 \text{ hours}) \, 55 \text{ hours}$$

$$([C]/[C_0]) = 0.88$$

The concentration entering the treatment plant is 0.88×0.5 ppm.
The concentration in the drinking water is $0.05 \times 0.88 \times 0.5$ ppm = 20 ppb.

Summary

The purpose of this section has been to illustrate how the properties evaluated in Sections 5.2 and 5.3 can be used to estimate environmental partitioning. Again, the models presented have been simple, demonstrating basic concepts of environmental partitioning, fate, and exposure. More complex and accurate models are available, but are beyond the scope of these simple screening methods.

SECTION 5.5   QUESTIONS FOR DISCUSSION

1. Why is most of the mass of the chemical considered in Example 5.5-2 in the water phase, while the concentrations in the sediment and biota phases are so high?
2. For Example 5.5-3, what vapor pressure would result in significant volatilization rates?
3. How would you develop an accurate estimate for volatilization rate in Example 5.5-3 if the losses were significant?

5.6   CLASSIFYING ENVIRONMENTAL RISKS BASED ON CHEMICAL STRUCTURE

The previous sections have described procedures for estimating the chemical and physical properties that are needed to assess potential environmental risks for chemicals. Our goals in this section are to put these property values in perspective and to introduce the tools that will be needed to perform an overall assessment of environmental hazards.

Three types of criteria are typically considered in risk-based evaluations—persistence, bioaccumulation and toxicity. For any one of these criteria, it may be necessary to consider a number of properties in performing an evaluation. For example, in evaluating persistence, it may be necessary to consider atmospheric half-lives and biodegradation half-lives. In evaluating toxicity, it may be necessary to consider a variety of eco-toxicity measures and human toxicity measures. Because there is such a wide variety of criteria that can be used in evaluating environmental risks—ranging from human carcinogenicity to biodiversity—and because opinions vary widely on the relative importance of the evaluation criteria, *there is no single evaluation methodology that is universally accepted for evaluating the environmental hazards of chemicals.*

Therefore, our approach in this text will be to present approximate classifications that can be used to categorize chemicals according to their persistence, bioaccumulation potential, and toxicity. The classifications will group chemicals into categories of high, moderate, and low concern, using values established by the US EPA in evaluating chemicals under the Toxic Substances Control Act. For example, Table 5.6-1 is a summary of the categories used to classify the persistence and bioaccumulation of chemicals and Figure 5.6-1 shows distributions of one measure of ecotoxicity.

These qualitative screenings can be useful in assigning areas of concern. Ranking risks, however, is more problematic, and approximate methods for ranking chemical risks will be described in Chapter 8. For now, Table 5.6-1 and Figure 5.6-1 can be used to provide perspective on the values for properties generated using the methods of Sections 5-2 and 5-3.

Table 5.6-1 Classification Criteria for Persistence and Bioaccumulation.

Water Solubility	
Very soluble	S>10,000 ppm
Soluble	1,000<S<10,000 ppm
Moderately Soluble	100<S<1,000 ppm
Slightly Soluble	0.1<S<100 ppm
Insoluble	S<0.1 ppm
Soil sorption	
Very Strong Sorption	Log K_{oc}>4.5
Strong Sorption	4.5> Log K_{oc}>3.5
Moderate Sorption	3.5> Log K_{oc}>2.5
Low Sorption	2.5> Log K_{oc}>1.5
Negligible Sorption	1.5> Log K_{oc}
Biodegradation	
Rapid	>60% degradation over 1 week
Moderate	>30% degradation over 28 days
Slow	<30% degradation over 28 days
Very Slow	<30% degradation over more than 28 days
Volatility (H in atm-m^3/mole)	
Very Volatile	H>10^{-1}
Volatile	10^{-1}>H>10^{-3}
Moderately Volatile	10^{-3}>H>10^{-5}
Slightly Volatile	10^{-5}>H>10^{-7}
Nonvolatile	10^{-7}>H
Bioaccumulation potential	
High Potential	8.0>Log K_{oc}>4.3 or BCF>1000
Moderate Potential	4.3>Log K_{ow}>3.5 or 1000>BCF>250
Low Potential	3.5>Log K_{oc} or 250>BCF

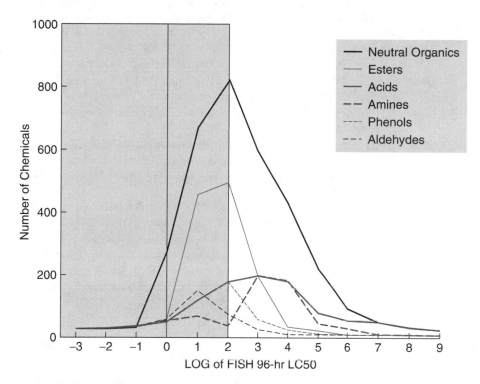

Figure 5.6-1 Distributions of measures of eco-toxicity for several thousand compounds. These distributions can be used to classify compounds into categories of concern. (Note that a high effect concentration implies that a large amount of the material can be released before an effect is observed.) (Zeeman, et al., 1993)

REFERENCES

Atkinson, R. and Carter, W.P.L. 1984, "Kinetics and mechanisms of the gas-phase reactions of ozone with organic compounds under atmospheric conditions," *Chem. Rev.* 84: 437–470.

Atkinson, R. 1985, "Kinetics and mechanisms of the gas-phase reactions of the hydroxyl radical with organic compounds under atmospheric conditions," *Chem. Rev.* 85: 69–201.

Atkinson, R. 1986, "Estimations of OH radical rate constants from H-atom abstraction from C-H and O-H bonds over the temperature range 250-1000 K," *Intern. J. Chem. Kinet.* 18: 555–568.

Atkinson, R. 1988, "Estimation of gas-phase hydroxyl radical rate constants for organic chemicals," *Environ. Toxicol. Chem.* 7: 435–442.

Atkinson, R. 1987, "A structure-activity relationship for the estimation of rate constants for the gas-phase reactions of OH radicals with organic compounds," *Intern. J. Chem. Kinet.* 19: 799–828.

Atkinson, R. 1989. "Kinetics and mechanisms of the gas-phase reactions of the hydroxyl radical with organic compounds," *J. Phys. Chem. Ref. Data Monograph No. 1*, NY: Amer. Inst. Physics & Amer. Chem Soc.

Hansch, C., Leo, A. and Hoekman, eds., "Exploring QSAR: Volume 1, Fundamentals and Applications in Chemistry and Biology," American Chemical Society, Washington, D.C., 1995a.

Hansch, C., Leo, A. and Hoekman, eds., "Exploring QSAR: Volume 2, Hydrophobic, Electronic and Steric Constants," American Chemical Society, Washington, D.C., 1995b.

Howard, P.H., *Handbook of Environmental Fate and Exposure Data for Organic Chemicals,* Lewis Publishers, Chelsea, Mich., 1997.

Howard, P.H., Boethling, R.S., Stiteler, W.M., Meylan, W.M., Hueber, A.E., Beauman, J.A., and Larosche, M.E., "Predictive model for aerobic biodegradability developed from a file of evaluated biodegradation data," *Environmental Toxicology and Chemistry,* 11, 593–603 (1992).

Joback, K.G. and Reid, R.C., "Estimation of Pure-Component Properties from Group Contributions," *Chemical Engineering Communications,* 57, 233–243 (1987).

Koneman, H. 1981, "Fish toxicity tests with mixtures of more than two chemicals: a proposal for a quantitative approach and experimental results," *Toxicology* 19: 229–238.

Kwok, E.S.C. and Atkinson, R. 1995, "Estimation of Hydroxyl Radical Reaction Rate Constants for Gas-Phase Organic Compounds Using a Structure-Reactivity Relationship: An Update," *Atmospheric Environment* (29: 1685-95). [from Final Report to CMA Contract No. ARC-8.0-OR, Statewide Air Pollution Research Center, Univ. of CA, Riverside, CA 92521].

Lyman, W.J., "Estimation of Physical Properties," *Environmental Exposure from Chemicals Volume 1,* Neely, W.B. and Blau, G.E., eds., CRC Press, Boca Raton, FL 38-44 (1985).

Lyman, W.J., Reehl, W.F. and Rosenblatt, D.H., eds, *Handbook of Chemical and Physical Property Estimation Methods: Environmental Behavior of Organic Compounds,* American Chemical Society, Washington, D.C., 1990.

Mackay, D., Shiu, W.Y., and Ma, K.C., *Illustrated Handbook of physical-chemical properties and environmental fate for organic chemicals,* Lewis Publishers, Boca Raton, 1992.

Meylan, W.M. and Howard, P.H., "Bond Contribution Method for Estimating Henry's Law Constants," *Environmental Toxicology and Chemistry,* 10, 1283–1293 (1991).

Meylan, W.M., Howard, P.H., and Boethling, R.S., "Molecular Topology/Fragment Contribution Method for Predicting Soil Sorption Coefficients," *Environ. Sci. Technol.,* 26, 1560–1567 (1992).

Meylan, W.M. and Howard, P.H., "Atom/Fragment Contribution Method for Estimating Octanol-Water Partition Coefficients," *Journal of Pharmaceutical Sciences,* 84, 83–92 (1995).

Meylan, W.M., Howard, P.H., and Boethling, R.S., "Improved Method for Estimating Water Solubility from Octanol/Water Partition Coefficient," *Environmental Toxicology and Chemistry,* 15, 100–106 (1996).

Mill, T., Haag, W., Penwell, P., Pettit, T., and Johnson, H. "Environmental Fate and Exposure Studies Development of a PC-SAR for Hydrolysis: Esters, Alkyl Halides and Epoxides," EPA Contract 68-02-4254. Menlo Park, Ca.: SRI International (1987).

Reid, R.C., Prausnitz, J.M., and Sherwood, T.K., *The Properties of Gases and Liquids, 4th ed.,* McGraw Hill, 1987.

Reinhard, M. and Drefahl, A., *Handbook for Estimating Physicochemical Properties of Organic Compounds,* Wiley, New York, 1999.

Seinfeld, J.H. and Pandis, S. N., *Atmospheric Chemistry and Physics,* Wiley Interscience, New York, 1998.

Stein, S.E. and Brown, R.L., "Estimation of Normal Boiling Points from Group Contributions," *J. Chem. Inf. Comput. Sci.,* 34, 581–587 (1994).

Zeeman, M., Clements, R.G., Nabholz, J.V., Johnson, D. and Kim, A., "SAR/QSAR Ecological Assessment at EPA/OPPT: Ecotoxicity Screening of the TSCA Inventory," Society of Environmental Toxicology and Chemistry Annual Meeting, Houston, November, 1993.

PROBLEMS

Use the methods described in the chapter. In addition, available software to estimate these properties may be used. (http://www.epa.gov/oppt/exposure/docs/episuite.htm)

1. Estimate the properties listed in the table given below.

Property	Nitrobenzene
Boiling point (T_b)	
Vapor pressure (P_{vp})	
Henry's Law constant (H)	
Octanol-water partition coefficient (K_{ow})	
Water solubility (S)	
Soil sorption coefficient (K_{oc})	
Atmospheric half life	
Biodegradability	

2. Estimate the properties listed in the table given below. *If group contributions are not available for the necessary groups, use reasonable judgment in estimating parameters.*

Property	2-Chloroaniline
Boiling point (T_b)	
Vapor pressure (P_{vp}) at 300 K	
Henry's Law constant (H)	
Octanol-water partition coefficient (K_{ow})	
Bioconcentration factor (BCF)	
Water solubility (S)	
Soil sorption coefficient (K_{oc})	
Atmospheric half life	
Biodegradability	

3. Estimate the properties listed in the table given below.

Property	Ethanol	1-propanol	1-hexanol	n-propane	n-hexane
Boiling point (T_b)					
Vapor pressure (P_{vp})					
Henry's Law constant (H)					
Octanol-water partition coefficient (K_{ow})					

Property	Ethanol	1-propanol	1-hexanol	n-propane	n-hexane
Water solubility (S)					
Soil sorption coefficient (K_{oc})					
Atmospheric half life					
Biodegradability					

For each of the properties, comment on whether molecular weight or the presence of a hydrogen bonding group has a more pronounced effect on chemical properties.

4. Benzene in the wastewaters from a manufacturing facility is sent, at a rate of 2000 kg/day, to a publicly owned wastewater treatment works (POTW). The POTW treats the benzene in the wastewater and removes 85% of the organic before discharging to a local river. One hundred kilometers downriver of the discharge point is the intake to a public water system.

> Data
> River flow rate: 1250 million liter per day
> River velocity: 0.5 m/sec
> River width: 50 m
> Organic solids concentration in suspended sediment: 15 ppm
> Biota concentration: 100 g per 100 cubic meter

(a) Estimate the fraction of benzene in water, sediment, and biotic phases at the discharge point.

(b) Determine whether volatilization of benzene from the river is likely to be significant.

(c) Estimate the fraction of the benzene that biodegrades before the effluent reaches the water intake.

(d) Estimate the potential toxicity of the releases to aquatic life.

5. During pesticide application, 1 kg of hexachlorobenzene is accidentally applied to a 10^8 liter pond. Estimate the amount of hexachlorobenzene that would be ingested if a person were to eat a 0.5 kg fish from the pond. Assume that the pond is well mixed and that the organic sediment content is 10 ppm and the total fish loading is 100 g per 100 cubic meter.

6. The Great Lakes Basin is one of the largest freshwater ecosystems in the world. Recently there has been some concern that persistent, bioaccumulative and toxic compounds have been accumulating in the basin, possibly compromising this valuable natural resource. Of particular concern are the chlorinated organics. In 1993 (the most recent year for which data are available) the chlorinated organic released in greatest quantity in the Great Lakes Basin was tetrachloroethylene. The emission rates to air, land and water for the basin were 1.8×10^7 pounds per year, 2.6×10^6 pounds per year and 8.4×10^2 pounds per year, respectively.

(a) Calculate the equilibrium partitioning of tetrachloroethylene in the air, water, soil and sediment of the Great Lakes Basin. Use one year of emissions as your basis. Assume no degradation, initial concentrations are zero, and that the Great Lakes Basin has the properties listed below.

Great Lakes Basin

Property	air	water	soil	sediment
Volume (m^3)	7.6×10^{14}	2.3×10^{13}	2.6×10^{10}	4.8×10^8
Area (m^2)	7.6×10^{11}	2.4×10^{11}	5.2×10^{11}	2.4×10^{11}
Organic fraction			0.02	0.04
Density (kg/m^3)	1.2	1000	1500	1280
Residence time (hr)	130	272,000	550	1700

(b) Will the system that you modeled in part (a) ever reach a steady state? Explain your reasoning.

(c) Estimate the atmospheric residence time and the biodegradability of tetrachloroethylene. Based on these values, estimate the steady state concentration of tetrachloroethylene in each environmental compartment.

(***Hint:*** for steady state to be reached, the total mass input to the systems must equal the total mass lost due to reaction. Assume that biodegradation occurs in water, sediment and soils and that degradation occurs in the atmosphere. Set up a mass balance where you have only one concentration as an independent variable and solve for that concentration.)

(d) In parts a–c you assumed that the environmental compartments were closed (e.g., you effectively assumed that the atmosphere was not ventilated by winds from other regions). Now assume that you want to account for advection in your calculation of steady state concentrations. Describe qualitatively how you would include the atmospheric, water, soil and sediment residence time information provided in the Table for Part (a) into your analysis. Use equations in your explanation if you wish, *but do not attempt a quantitative analysis.*

7. Design a solvent molecule that has a vapor pressure greater than 1 mm Hg, a molecular weight between 75 and 150, and will biodegrade in less than one month.

8. Design a solvent molecule that has a vapor pressure less than 1 mm Hg (at 300 K), a molecular weight between 75 and 150, and will biodegrade in less than one month.

9. The group contribution equation for estimating boiling point is:

$$T_b = 198.2 + \Sigma n_i \, g_i$$

Without consulting the tables in Chapter 5, estimate the relative magnitude of the group contributions for the following three functional groups: $-OH$, $-CH_3$, $-Cl$ (aliphatic). Report your answer as x<y<z, where x, y, and z are the three functional groups. For example if you believed that the values of the three group contributions for $-OH$, $-CH_3$, $-Cl$ were $-1, 0$ and 1, respectively, your answer would be $-OH< -CH_3<-Cl$. *Explain your reasoning.*

10. Without consulting the tables in Chapter 5, estimate the relative magnitude of the group contributions for octanol-water partition coefficient for the following three functional groups: $-OH$, $-CH_3$, $-Cl$ (aliphatic). Report your answer as x<y<z, where x, y, and z are the three functional groups. For example if you believed that the values of the three group contributions for $-OH$, $-CH_3$, $-Cl$ were $-1, 0$ and 1, respectively, your answer would be $-OH<-CH_3<-Cl$. *Explain your reasoning.*

CHAPTER 6

Evaluating Exposures

by
Fred Arnold

6.1 INTRODUCTION

The human health risk associated with a chemical is dependent on the rate at which the chemical is released, the fate of the chemical in the environment, human exposure to the chemical, and human health response resulting from exposure to the chemical. In simpler terms, as described in Chapter 2, risk is a function of hazard (or toxicity) and exposure. Chapter 5 discusses methods of predicting physical-chemical properties from chemical structure to infer the fate of a chemical in the environment. Chapter 7 discusses green chemistry techniques to select chemicals that are less toxic. Chapters 5 and 7 are useful in designing chemical structures with low hazard, one of the two components of the risk equation. This chapter, Chapter 6, addresses the *exposure* component of the risk equation. Ideally, exposure is quantified by monitoring the work area or environmental setting where a chemical will be used or released; however, when monitoring data are not available to measure exposures, exposures can be estimated using methods described in this chapter.

The methods for estimating exposure will be separated into two sections—occupational and community. Occupational exposure occurs in the workplace. Workers in chemical production facilities may be exposed to toxins used or produced in the chemical process. Exposure to chemicals may occur from the inhalation of workplace air, ingestion of dust or contaminated food, or from contact of the chemical substance with the skin or eyes. In addition, chemical engineers must be aware of community exposures resulting from releases into the air and water, and from solid and hazardous waste disposal. Chemical releases to rivers, lakes, and streams may accumulate in fish and other marine life, which are subsequently used as a source of food, or may be ingested by persons using the downstream reaches of rivers as a supply of potable water. Persons living downwind of a chemical manufacturing facility may be exposed to fugitive and point source releases of chemical toxins to the atmosphere.

Disposal of solid and hazardous wastes on the land, either in repositories such as landfills or into subterranean strata by injection into wells may result in contamination of potable groundwater if the waste is not isolated from the water supplies.

The intent of Chapter 6 is to introduce students to some methods for predicting potential exposure, in particular, occupational exposure and community exposure. During process design, it may be useful to predict potential exposures to workers from chemical emissions (i.e., "occupational exposure"), or potential exposures to nearby residents from chemical emissions or releases from the plant (i.e., "community or general population exposure"). There are other exposure areas, such as consumer exposure, which are not discussed in this chapter. The chemical engineer, in addition to selecting chemicals with low toxicity, also needs to select solvent chemicals and design unit operations to minimize potential exposure as well.

There are many good references on exposure assessment. Interested students are encouraged to consult references on other types of exposure not covered in this chapter. EPA has a website specifically for exposure which contains computerized tools for all exposure areas (http://www.epa.gov/oppt/exposure). This information can be useful in selecting and designing unit operations. Many of these references are listed in Appendix F.

6.2 OCCUPATIONAL EXPOSURES: RECOGNITION, EVALUATION, AND CONTROL

The basic components of assessing occupational exposure are to recognize all sources of exposure to chemicals, evaluate the exposure, determine if the exposure is within permissible limits, and at the minimum, control those exposures that exceed permissible limits.

Recognizing exposures involves developing a list of all sources of chemical exposure in the work environment. Workers may be exposed to chemical substances during the performance of tasks making or utilizing chemicals, in sampling reaction vessels, or in transfer of chemicals from the reactor to storage or transportation containers. As mentioned before, contact with the chemicals may occur through inhalation of vapors or by dermal contact as the chemicals are sampled or transferred. Although the highest exposures usually result from tasks performed directly by the worker, significant exposures may occur from nearby tasks performed by other workers or from incidental contact with background contamination in the workplace.

To evaluate the significance of an occupational exposure to a chemical substance, both the level and the duration of exposure must be known. Exposures to chemicals that have no cumulative or persistent effects may be tolerated at low levels in the workplace over long periods of time. However, short-term exposures to higher concentrations may result in acute toxicity to the worker. For other chemicals, exposures to low levels of the chemicals over long periods of time may result in chronic effects even though no acute effects are seen in short-term exposures.

Limitations on occupational exposures to chemicals are set by the Occupational Safety and Health Administration (OSHA), a division of the U.S. Depart-

Table 6.2-1   OSHA Permissible Exposure Limits for Air Contaminants.

Chemical Substance	CAS No.*	ppm by volume**	mg/m³***
Acetic acid	64-19-7	10	25
Acetone	67-64-1	1000	2400
Acrolein	107-02-8	0.1	0.25
Ammonia	7664-41-7	50	35
Bromine	7726-95-6	0.1	0.7
2-Butanone(Methyl ethyl ketone)	78-93-3	200	590
Ethyl benzene	100-41-4	100	435
Hydrogen cyanide	74-90-8	10	11
Nitric oxide	10102-43-9	25	30
Dichlorodifluoro-methane (CFC 12)	75-71-8	1000	4950

*The CAS Number is a unique number assigned as a means of identification to distinct chemical substances by the Chemical Abstracts Service.

**Parts of vapor or gas per million parts of contaminated air by volume at 25° C and 760 torr.

***Milligrams of chemical substance per cubic meter of air.

ment of Labor. The limitations, often called OSHA Permissible Exposure Limits or OSHA PELs, are listed in Title 29, Part 1910.1000 of the Code of Federal Regulations. Listed in Table 6.2-1 are limitations for air contaminants set by OSHA for representative chemical substances. The relative toxicity of a chemical substance can be gauged by comparing the OSHA PEL for a chemical substance with that for a known poison, hydrogen cyanide, or an irritating but generally nontoxic gas, ammonia. All limitations given in Table 6.2-1 are expressed as time-weighted averages for the chemical substance in any 8-hour work shift of a forty-hour work week. For PELs the action level is not the actual PEL but one-half the PEL, meaning action must be taken at this level to reduce the emissions. An overexposure is observed when monitoring demonstrates an average concentration of a chemical in the workplace greater than the occupational exposure limit over the appropriate time period.

Control and elimination of unacceptable exposures require information on the source, pathway, and worker exposed to the chemical substance. Control measures can be applied at any step; e.g., process changes can reduce the amount of emissions from various sources. Adjustments in ventilation systems can intercept chemical contaminants and eliminate the pathway for exposure. Finally, personal protective equipment can provide additional protection when other measures are inadequate.

6.2.1   Characterization of the Workplace

The first step in an occupational exposure assessment is to characterize the workplace. Description of the workplace begins with a schematic or written description of the chemical manufacturing process and identification of unit operations where exposure to chemicals may occur. The schematic diagram is used to highlight unit operations and activities where exposure to chemicals may occur, provide a description of production activities and process chemistry, and identify ventilation and other mechanisms that reduce worker exposures. Written descriptions should also include releases and exposures that do not take place in the chemical manufacturing facility, such as transportation and disposal of empty shipping containers.

From an occupational exposure viewpoint, the key elements of a process flow diagram are the sources of potential exposure. A source of potential exposure is a unit operation or worker task that brings the worker into potential contact with the chemical substance. For this reason, sampling points and transfer operations must be highlighted in the schematic diagram or description. Likewise, transfers of materials entering or leaving the process should be described (bagging, drumming, tank truck filling, etc.) since this highlights handling problems that could result in exposure to chemical substances. Waste streams leaving the process should be identified to indicate possible sources of exposure and to provide a resource for environmental studies. The completed flow diagram should highlight possible sources of exposure and minimize the possibility that potential hazards will be overlooked.

The written description should explain the activities occurring in the work area and should emphasize locations where potential exposure to chemicals may occur. It should also include important details such as component stream concentrations, operating temperatures, and pressures. Other factors that affect the potential for exposure (ventilation systems, open-top or closed vessels, use of protective equipment) should be noted in the description. If respiratory or dermal protection is used to limit exposures, the appropriate protection factor provided by the protective equipment should be listed.

Knowledge of the process and its component operations is needed to assess the likelihood and magnitude of exposure of workers to chemical substances. The frequency and duration of sampling events, the duration of batch processes, the type and frequency of transfer operations, and the number of workers involved in each operation are needed to make quantitative estimates of exposures to chemical substances. For convenience, the workers may be separated into groups performing similar operations and thus having similar exposures. A detailed description of the time engaged in each work task (sampling, monitoring unit operations, transferring raw materials and products) and in each work area should be developed where the potential for significant exposure exists.

The schematic and written descriptions can be used to prepare a relatively complete inventory of the chemicals that may be encountered in the work environment and the rates of use or generation of each chemical. For each chemical of concern, the engineer can assemble physical property data (boiling point, vapor pressure, particle size distribution, etc.) which will be of assistance in assessing the potential for exposure. For solid particles, knowledge of the particle size distribution will enable the engineer to evaluate the fraction of airborne particles that are potentially respirable. An excellent source of information on the health effects (nuisance, irritant, toxicity, carcinogenicity, potential for birth defects, etc.) of chemical substances is the Material Safety Data Sheet (MSDS) prepared for each chemical substance by the manufacturer. The MSDS will also provide occupational exposure guidelines established by regulatory or consensus organizations. These include the OSHA PELs, the American Conference of Governmental Industrial Hygienists' Threshold Limit Values (TLVs), and the American Industrial Hygiene Association's Workplace Environmental Exposure Level (WEEL) guides. OSHA PELs and TLVs are discussed in more detail in Chapter 8 of this text.

6.2.2   Exposure Pathways

Because exposure to chemicals in the work environment can occur through inhalation, skin absorption, or ingestion, the engineer must be aware of these potential pathways into the body. The exposure pathway model in Figure 6.2-1 highlights potential pathways leading from process to worker and provides a framework for evaluating pathways for exposure to chemicals in the workplace. Used in conjunction with the schematic diagram, process description, and physical properties of the chemical substance, the important exposure pathways and controls to minimize exposure can be identified.

Inhalation exposure is often the most significant route of workplace exposure. Chemicals can volatilize from the process or evaporate from work surfaces where they are deposited. Exposure to a high vapor pressure solvent can be evaluated solely from the rate of vaporization and the effectiveness of ventilation controls unless there is also significant skin contact with liquid or vapor. With lower vapor pressure chemicals, longer term volatilization of spills from work surfaces may be important. Dusty environments created by the generation of fine particles, like carbon dust, or in the cleaning of a manufacturing line can also contribute to inhalation exposure.

Figure 6.2-2 presents the framework for calculating exposure to chemical substances by the inhalation route. Exposure with units of mass is the product of the severity (mass/time) of exposure and the duration (time) of exposure. Severity is, in turn, the product of the environmental concentration (mass/volume) and the

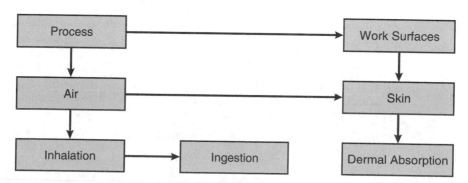

Figure 6.2-1   Exposure pathway model.

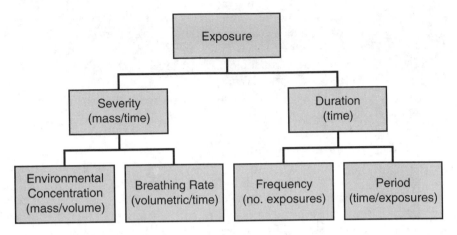

Figure 6.2-2   Inhalation exposure framework.

breathing rate (volume/time). Similarly, duration is the product of the frequency (number of exposures) and period (time/exposure) of exposure. A separate estimate of the rate of absorption of inhaled materials is necessary to calculate the intake of a chemical into the body.

Dermal contact can also represent an important route of exposure for some chemicals, particularly those that readily absorb through the skin in immediately toxic amounts and those that pass through the skin and accumulate in the body. This exposure pathway usually results from direct contact of chemicals with the skin. Exposure may also result from contact with work surfaces that are contaminated.

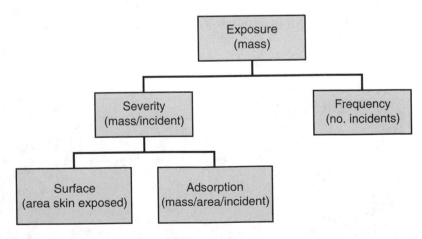

Figure 6.2-3   Dermal exposure framework.

Figure 6.2-3 presents the framework for calculating exposure to chemical substances by the dermal route. Exposure (mass) is the product of severity (mass
adsorbed per incident) and frequency (number of incidents). Severity of dermal exposure is, in turn, the product of the surface area of exposed skin and the mass adsorption per incident. Dermal intake into the body requires a separate estimate of
the rate of uptake of the chemical from the exposed skin surface.

Oral ingestion of chemicals is usually a relatively minor route of exposure in
the workplace, particularly when dining areas are separate from work areas and
employees practice a reasonable level of personal hygiene. However, this may be
an important route of exposure for chemicals that accumulate in the body over
long periods of exposure.

6.2.3   Monitoring Worker Exposure

Monitoring objectives can be grouped into three categories: baseline, diagnostic,
and compliance. Baseline monitoring is performed to evaluate the range of worker
exposures. The baseline data are used to determine the acceptability of exposures
to chemicals and the need for controls to reduce exposures. Diagnostic monitoring
is performed to identify principal sources and tasks contributing to exposure to
specific chemicals. The results of diagnostic monitoring are used to select appropriate control strategies for reducing exposure to known sources. Compliance monitoring is performed to demonstrate conformance with government regulations. The
sampling strategy for evaluating compliance is often to monitor the "most exposed" worker using a collection device attached to the worker near his breathing
zone.

Monitoring methods can be classified as either personal monitoring or area
monitoring. Personal monitoring is conducted to characterize the exposure of a
worker to the chemical substance of interest. The most common method of personal monitoring is a breathing zone measurement. A battery-powered pump is attached to the worker to draw air through a collection tube at a constant rate. The
inlet to the collection tube is connected to a flexible hose, which draws air from the
breathing zone of the worker. The sample is collected for a designated period and
the monitoring result is reported as a time-weighted average over the designated
period. Two common sample averaging times are 8 hours, a normal work shift, and
15 minutes, a common short-term exposure time limit. These durations correspond
to the averaging times of regulatory limits. Eight-hour sampling has the disadvantage that peak exposure information is usually lost. A mixture of full-shift and
short-term sampling is usually the best technique for evaluating worker exposures.

Personal monitoring of skin absorption is often difficult. Patch testing is conducted by affixing a patch of absorbent material to an exposed skin surface of a
worker for a known period of time. At the end of a specified time period, the patch
is removed and the chemical of concern is extracted from the patch and quantified.
Skin washes are used to remove the chemical of concern from the skin surface

using a suitable solvent. The quantity of chemical removed from a measured skin area is then quantified and reported as exposure per unit area.

Area monitoring of the ambient air is used to measure the background level of chemical contaminants when chronic conditions resulting from long-term exposure are of concern. Area monitoring is also used to warn of toxic concentrations of acutely hazardous substances. Monitoring the ambient atmosphere can also be used to demonstrate the effectiveness of ventilation controls by measuring the levels of chemical contaminants before and after the controls are installed.

Area monitoring also includes investigation of surface contamination by wipe test methods. Although not a direct method of exposure, wipe tests are useful for tracking levels of contamination, particularly for chemical substances readily absorbed by the dermal route. Wipe tests may be used to document trends in work practices and housekeeping procedures. They can also identify deficiencies in maintenance or operation of local exposure control systems, including safety hoods.

The number of samples collected in a monitoring program is determined by regulatory requirements or professional judgment. The cost and difficulty of sample collection and analysis will limit the number of samples collected. Conversely, a greater number of samples will decrease the likelihood of significant errors in the sample means and decrease the variance about the mean. The standard error of the mean decreases rapidly as the first few replicates are collected but the likely sample error decreases only slightly with each additional sample after 6 to 10 samples are collected. The variance about the sample mean is inversely proportional to the square root of the number of samples analyzed. Similarly, the variance about the sample mean decreases significantly over the initial samples and less so after 10 or more samples.

6.2.4 Modeling Inhalation Exposures

It is not always convenient or possible, in the case of a new or proposed process, to undertake a monitoring program to determine airborne concentrations of chemicals. In some instances, a more rapid estimate of potential worker exposures to chemical substances is needed. In this situation, the engineer may utilize models which simulate worker exposures.

6.2.4.1 The Mass Balance Model

A simple model often used to estimate the concentration of airborne contaminants in the workplace is the mass balance model also known as the box model. The work area is modeled as a box in which the contaminant is uniformly distributed. In this case, a mass balance can be written for the contaminant concentration within the work area.

$$V \frac{dC}{dt} = G - kQ(C - C_o) \qquad \text{(Eq. 6-1)}$$

where

C is the concentration of airborne contaminant in the work area (mass/length^3),
V is the volume of the work area (length^3),
t is the time during which the contaminant has been emitted,
G is the emission rate of the contaminant to the air (mass/time),
Q is the ventilation rate in the work area (length^3/time),
k is a mixing factor to account for incomplete mixing in the work area (unitless)
C_o is the concentration of the airborne contaminant entering the work area (mass/length^3).

If the emission rate and ventilation rate are constant, the concentration will reach a steady state and Equation 6-1 becomes:

$$C = C_o + \frac{G}{kQ}$$ (Eq. 6-2)

At times, emissions are episodic. Consider a work area that initially contains contaminant at concentration C_o. At some time, t=0, an emission source, releasing contaminant at rate G, is placed in the work area. In this case, the box model can be used to estimate the rise in concentration of the contaminant in the workplace. Again, assuming that the ventilation rate is constant, Equation 6-1 can be integrated to yield

$$C = C_o + \frac{G}{kQ}[1 - \exp(-kQt/V)]$$ (Eq. 6-3)

The mixing factor (k) typically ranges from 0.3 to 0.7 in small rooms without fans (Drivas 1972). Others have used mixing factors of 0.5 for work areas with average ventilation and 0.1 for poorly ventilated work areas (Fehrenbacher and Hummel, 1996).

The determination of G may be simple or complex, depending on the nature of the emission source. As an example, assume that the source is a pool of liquid that is evaporating at a constant rate. Estimating this emission rate requires input of the vapor pressure of the contaminant, the surface area of the evaporating liquid, and the relationship between the velocity of the air over the liquid surface and mass transfer from the liquid into the flowing air stream. The penetration model (Hummel et al., 1996) provides acceptable estimates of evaporation rates at low air speeds characteristic of indoor work areas:

$$G = 8.79 \times 10^{-5} \frac{(MW^{0.883})\,(VP)\,[(1/MW + 1/29)^{0.25}]\,(v^{0.25})}{(T^{0.05})\,(\Delta x^{0.5})\,(P^{0.5})}$$ (Eq. 6-4)

where

G is the evaporation rate (g/sec-cm^2)
MW is the molecular weight of the evaporating species (g/mole)
VP is the vapor pressure of the evaporating contaminant (atm)
v is the air velocity parallel to the surface of the evaporating liquid (cm/sec)

T is the surface temperature of the evaporating liquid ($^{\circ}K$)

Δx is the length of the evaporating pool in the direction of airflow (cm)

P is the ambient pressure (atm)

A survey and evaluation of other models used to estimate evaporation rates of volatile liquids are given by Lennert (1997).

Example 6.2-1

A cleaning bath for electronic parts emits 0.5 g/sec of CFC-12 into a small work room of dimensions 3 m $\times$ 3 m $\times$ 2.45 m high. Calculate the concentration in the room under average and poor ventilation conditions if the air velocity in the room is 0.3 m/s and compare the results to the OSHA PEL.

Solution: When the air speed is 0.3 m/s, the volume of air flowing through the room will be:

$$Q = 0.3 \text{ m/s} \times 3 \text{ m} \times 2.45 \text{ m} = 2.21 \text{ m}^3/\text{s},$$

and the concentration of CFC-12 in the air will be:

average ventilation $C = 0.5 \text{ g/sec}/(0.5 \times 2.21 \text{ m}^3/\text{s}) = 0.45 \text{ g/m}^3,$

poor ventilation $C = 0.5 \text{ g/sec}/(0.1 \times 2.21 \text{ m}^3/\text{s}) = 2.27 \text{ g/m}^3.$

A comparison with the permissible exposure limits given in Table 6.2-1 indicates that even under poor ventilation conditions, the OSHA PEL will not be exceeded and respiratory protection will not be needed to safe-guard the health of a person working in this room.

The simple mass-balance model does not account for all of the phenomena that influence the exposure to chemicals released into the workplace atmosphere. Exposure may be mitigated by adsorption of the chemical to walls and other surfaces in the work room. In this case, the mass balance on the airborne concentration is given by:

$$V \frac{dC}{dt} = G - kQ(C - C_o) - rC \qquad \text{(Eq. 6-5)}$$

where r is the nonventilatory removal coefficient of airborne contaminant (volume/time). If the ventilation and emission rates are constant, the box model predicts a steady state concentration of:

$$C = \frac{kQC_o + G}{(kQ + r)} \qquad \text{(Eq. 6-6)}$$

Solvents are often volatile and significant accumulations of their vapors may occur in the workplace air. In this instance, the concentration of solvent may exert a significant back pressure retarding the evaporation of additional solvent. Jayjock (1994) has published the solution of the mass balance equation when back pressure is significant.

The approach used in development of the mass-balance model, adjusting the ventilation rate to account for imperfect mixing and unventilated areas, has been criticized because the model is still used to describe an imperfectly mixed room. In addition, the mixing factor is an empirical adjustment that must be developed by experimental measurements. An alternative model divides the work area into two perfectly-mixed zones, one near the source of an airborne contaminant and the other removed from the source (Nicas, 1996). Mixing in the work area occurs as a result of ventilation between the two zones. The steady-state or upper bound on concentration in the zone of the work area nearest the source is given by:

$$C = \frac{G\,(B + Q)}{(BQ)}$$

(Eq. 6-7)

where

    C is the concentration of the contaminant in the work area near the source (mass/length^3)

    G is the rate of vaporization of the contaminant (mass/time)

    B is the rate of exchange of air between the zones located near and removed from the source (length^3/time)

    Q is the ventilation rate of the zone removed from the source (length^3/time)

Although this model does not require estimation of an empirical mixing factor, the air exchange rate, B, must be determined from the physical dimensions of the zones or other criteria.

Example 6.2-2

Calculate the concentration of freon in the cube, 1 m on a side, surrounding the top of the cleaning bath in Example 6.2-1 if the air exchange rate with the remainder of the room is 1 m^3/s. Repeat the calculation for an air exchange rate of 0.5 m^3/s.

Solution: From Example 6.2-1, the rate of release of freon from the cleaning unit is 0.5 g/s and the ventilation rate in the room is 2.21 m^3/s. Thus in the area closest to the cleaning bath, the concentration of freon can be calculated from Equation 6-7:

$$C = (0.5 \text{ g/s}) \times [1 \text{ m}^3/\text{s} + 2.21 \text{ m}^3/\text{s}] / [1 \text{ m}^3/\text{s} \times 2.21 \text{ m}^3/\text{s}] = 0.73 \text{ g/m}^3 = 730 \text{ mg/m}^3$$

or

$$C = (0.5 \text{ g/s}) \times [0.5 \text{ m}^3/\text{s} + 2.21 \text{ m}^3/\text{s}] / [0.5 \text{ m}^3/\text{s} \times 2.21 \text{ m}^3/\text{s}] = 1.23 \text{ g/m}^3 = 1,230 \text{ mg/m}^3$$

As would be expected, the local concentration of the chemical increases when the ventilation is less. Localized ventilation is an effective method of dispersing airborne chemicals and reducing exposures of workers to chemicals.

6.2.4.2   Dispersion Models

Diffusion of contaminants in workplace air results in the net movement of the contaminants from regions of higher concentration to regions of lower concentration. The spread of the contaminant is aided by the convective mass transfer driven by the ventilation system. The combination of these influences results in movement of contaminants away from their source into the surrounding room (Scheff et al., 1992).

The mass-balance model described above presumes a uniform concentration of the chemical contaminant in the work area. Dispersion models have a notable advantage; they describe the variation of contaminant concentration with distance from the source. The concentration gradient is described by the following equation when convection occurs in the x-direction only and dispersion occurs equally in all directions:

$$u \frac{dC}{dx} = \frac{D}{r^2} \frac{d}{dr} \left[\frac{r \, dc}{dr} \right] \qquad \text{(Eq. 6-8)}$$

where
   u is the wind velocity in the x direction (length/time)
   C is the concentration of airborne contaminant (mass/length^3)
   D is the diffusion coefficient (length^2/time)
   x is the distance downwind from the source (length)
   r is the distance from the source to the sampling point (length)

This equation has been solved for concentrations resulting from emissions into an infinite space:

$$C = \frac{G}{4 \pi Dr} \exp[(-u/2D)(r - x)] \qquad \text{(Eq. 6-9)}$$

where G is the contaminant emission rate from the source (mass/time).

The diffusion coefficient (D) can be derived from measurements at the sampling site or estimated from values available in the literature. Measurements of the diffusion coefficient in indoor industrial environments have ranged from 0.05 to 11.5 m^2/minute, with 0.2 m^2/min being a typical value (Jayjock, 1998).

Example 6.2-3

Freon is emitted from an open-top vapor degreaser at a rate of 0.74 g/min. Estimate the concentration in the air inhaled by a worker 3 m downwind from the degreaser if the air velocity is 0.79 m/min.

Solution: Since the worker is downwind of the degreaser, x = r in the diffusion-convection equation and

$$C = \frac{0.74 \text{ g/min}}{(4)(3.1416)(0.2 \text{ m}^2/\text{min})(3 \text{ m})} = 0.1 \text{ g/m}^3$$

Molecular diffusion theory strictly applies to vapors and gases; however, particulate matter with aerodynamic diameters less than 10 μm are distributed in workplace air in a similar manner. Using the dispersion models to describe the distribution of dusts and fumes is reasonable for small particles.

6.2.5   Assessing Dermal Exposures

Dermal hazards refer to chemicals that can cause dermatitis or otherwise damage the skin as well as to chemicals that can enter the body through the skin and cause toxic effects in other organs. Dermatitis refers to inflammation or damage to the skin which is localized and does not spread to other areas of the body. Acids, alkalis, and other irritating or corrosive chemicals damage skin which they contact. Repeated contact with epoxy resins may result in skin sensitization and dermatitis. The National Institute of Occupational Safety and Health has recognized allergic and irritant dermatitis as the second most common occupational disease (after hearing loss), accounting for 15 to 20 percent of all reported occupational diseases. Because of the often readily apparent reaction to chemicals causing dermatitis, exposures are usually quickly eliminated or protective clothing is used to preclude skin contact with toxic chemicals.

In contrast to contact dermatitis, toxic chemicals may be absorbed through the skin, mucous membranes, or eyes either by direct skin contact with the chemical or deposition of aerosols. This absorption can contribute to toxic effects on other organs. Some substances, such as amines and nitriles, pass through the skin so rapidly that the rate at which they enter the body is similar to rates of inhalation or ingestion. In 29 CFR 1910.1000, OSHA identifies nearly 100 chemicals which can enter the body through the skin and cause toxic effects elsewhere within the body. For these chemicals, the engineer should be alert to dermal exposures and should minimize contact of chemicals with the skin by process modifications or use of protective clothing.

The three mechanisms of dermal exposure are 1) direct contact between the worker's skin and a liquid or solid chemical as from splashing or immersion, 2) transfer of a chemical from a contaminated surface to the skin following direct contact, or 3) deposition or impaction on the skin as a vapor or aerosol. Aerosols are created when chemicals are applied by spraying or when fluids contact moving surfaces, e.g., metal working fluids interacting with machinery. Aerosols tend to settle rapidly, making an increased separation between the worker and operations using the chemical of concern a feasible means of controlling dermal exposures.

The amount of a chemical remaining on the skin depends on the processes of contamination, removal, and penetration through the skin. Possible removal processes are evaporation, incidental transfer to other surfaces, or intentional decontamination. Dermal exposure is often highly variable between workers over time and between different anatomical locations of the body. Since dermal penetration varies across the anatomical locations of the body, an overall average value for skin exposure is often insufficient.

Direct methods for measurement of skin exposure include collection of chemical contaminants on absorbent pads or clothing and wipe sampling of contaminated surfaces. The absorbent pad technique utilizes gauze pads, treated cloth, or alphacellulose pads which are attached to various sites on the worker's skin or outer clothing to capture chemicals that would have been deposited on the skin or clothing.

Table 6.2-2 Surface Area by Region of the Body for Adults in Square Centimeters.

Region of the Body	Men		Women	
	Median	5th to 95th Percentiles	Median	5th to 95th Percentiles
Head	1300	1190–1430	1110	1060–1170
Trunk	7390	5910–9350	5790	4900–7520
Arms	2910	2410–3540	2300	2100–2530
Hands	990	850–1170	817	730–966
Legs	6400	5390–7620	5460	4600–6830
Feet	1310	1140–1490	1140	1000–1340
Total	19,400	16,600–22,800	16,900	11,450–20,900

Source: U.S. Environmental Protection Agency, *Exposure Factors Handbook, Volume I* (EPA 600/P-95/002Fa), Washington, D.C. (1997).

Collection pads are exposed for a representative period of time to the work environment and are subsequently removed and analyzed for the chemical of concern. An estimate of the potential dermal exposure can be obtained by multiplying the amount of contaminant deposited on a unit area of the absorbent pad by the surface area of the body region that the pad is positioned to represent. Data on the surface area of the adult body is given in Table 6.2-2. This technique is generally used for sampling nonvolatile contaminants or compounds with low vapor pressure; charcoal-impregnated cloth has been used for sampling of volatile compounds.

Wipe samples are collected by washing the skin or clothing with water, surfactants, alcohol, acetone, or other solvents. Chemicals remaining on the skin or clothing are collected but those that have penetrated the skin are not collected by this technique. Wipe samples have been used to determine routes of dermal exposure to aromatic amines used as anti-oxidants, intermediates, and curatives in epoxy resins and urethanes. Wipe samples taken from the inside of protective gloves indicated that methylene dianiline used in aircraft composites had penetrated the gloves of workers engaged in the hand lay-up operations in aircraft and aerospace industries. The wipe samples revealed that chemical breakthrough of the protective clothing was the cause of elevated levels of the aromatic amine detected by biological monitoring (Groth, 1992). Wipe tests have also been used to identify significant exposure to toxic chemicals from handling contaminated tools (Klingner, 1992) and improper removal of contaminated clothing (Kusters, 1992).

Computerized image analysis techniques can be used together with fluorescent whitening agents to indirectly quantify exposure of the total body surface (Fenske 1997). Visual observation of fluorescent tracer deposition on skin has been used to characterize exposure in a variety of pesticide applications. The behavior of the fluorescent tracer in the application process must be similar to that of the chemical of concern. This technique provides a means of assessing exposures without use of toxic chemicals.

Methods to control dermal exposure to chemicals can take many forms. Substitution of a less toxic chemical is almost always a good option, unless the alternative chemical has a much higher vapor pressure and is likely to cause an inhalation hazard. Consideration should also be given to redesigning the work process to avoid splashes or immersion. Where that is not feasible, personal protection in the form of chemical protective gloves, an apron, or clothing may be selected. Performance characteristics of glove materials must be matched to the hazard to be avoided, i.e., cuts, abrasions, and dermal contact with toxic chemicals. Glove manufacturers can provide information on the ability of a variety of glove materials (natural rubber, polyvinyl chloride, neoprene, nitrile, butyl rubber, polyvinyl alcohol, viton, or norfoil) to preclude penetration of toxic chemicals.

Modeling Dermal Exposure

The quantity of a chemical contacting the skin during immersion, splashing, application to substrate, attachment of process lines, or weighing and transfer of chemicals can be estimated as the sum of the products of the exposed skin areas (in cm^2) and the amount of chemical contacting the exposed area of the skin (mg/cm^2/event). Usually, the amount of chemical transferred to the exposed area can only be measured after exposure has occurred. If the chemical of concern is absorbed rapidly, the amount of chemical contacting the skin will be greater than that estimated by direct measurement; indirect methods of measurement of exposure such as the fluorescent imaging described above may be required to obtain accurate estimates of dermal exposure.

During most dermal exposure events, exposure will be limited to a few areas of the body. For example, during sampling of a reactor, attachment of process lines, or manual weighing and dumping of powders, only the hands and perhaps the forearms would be exposed to the chemical of concern. Conversely, during the spray application of a paint, addition of a antimicrobial liquid to latex products, use of metal working fluids, or commercial pesticide applications, concern for aerosols, splashing of fluids, and general dispersion of the chemical will require that other areas of the body be protected from contact with process chemicals

The equation given below can be used to estimate the exposure to a chemical that is absorbed through the skin.

$$DA = (S)(Q)(N)(WF)(ABS)$$ (Eq. 6-10)

where

> DA is the dermal absorbed dose rate of the chemical (mass/time)
> S is the surface area of the skin contacted by the chemical ($length^2$)
> Q is the quantity deposited on the skin per event ($mass/length^2$/event)
> N is the number of exposure events per day (event/time)
> WF is the weight fraction of the chemical of concern in the mixture
> (dimensionless)
> ABS is the fraction of the applied dose absorbed during the event
> (dimensionless)

In the absence of monitoring data, the values given in Table 6.2-3 may be used to estimate dermal exposure to liquids during plant operations.

Table 6.2-3 Quantity of Chemical Deposited on the Skin per Exposure Event.

Activity	Quantity Transferred to the Skin per Event (mg/cm²)
Handling wet surfaces	6.0−10.3
Spray painting	6.0−10.3
Manual cleaning of equipment	0.7−2.1
Filling drums with liquid	0.5−1.8
Connecting transfer lines	0.7−2.1
Sampling	0.7−2.1
Ladling liquid/bench scale transfer	0.5−1.8

Source: US Environmental Protection Agency, *Occupational Dermal Exposure Assessment—A Review of Methodologies and Field Data,* Office of Pollution Prevention and Toxics, 1996.

Example 6.2-4

A worker is preparing an epoxy adhesive by adding the solvent (toluene) and the chemicals to produce the adhesive in a batch reactor. During the process, the reactor is sampled twice. At the end of the reaction, the worker fills drums with the epoxy adhesive and cleans the reactor. Estimate the dermal exposure of the worker to toluene. Assume the adhesive contains 20 percent toluene.

Solution: Equation 6-10 and the higher limits of dermal exposure given in Table 6.2-3 will be used to obtain a conservative estimate of dermal exposure. The skin surface areas for the hands are given in Table 6.2-2. Assume all of the toluene contacting the worker's hands is absorbed, that only one hand is exposed during sampling, and that both hands are exposed during other operations.

Connecting toluene inlet line: $(840 \text{ cm}^2) (2.1 \text{ mg/cm}^2/\text{event}) (1 \text{ event}) (1.0) (1.0) = 1{,}760 \text{ mg}$

Sampling reactor: $(420 \text{ cm}) (2.1 \text{ mg/cm/event}) (2 \text{ events}) (1.0) (0.2) = 350 \text{ mg}$

Filling drums with product: $(840 \text{ cm}) (1.8 \text{ mg/cm/event}) (1 \text{ event}) (1.0) (0.2) = 300 \text{ mg}$

Cleaning reactor: $(840 \text{ cm}) (2.1 \text{ mg/cm /event}) (1 \text{ event}) (1.0) (0.2) = \underline{350 \text{ mg}}$

Total potential exposure to toluene: 2,760 mg

Some chemicals will be absorbed through the skin during the exposure event, some will be absorbed after the exposure event, and some chemicals will be removed before absorption occurs. Fick's first law of diffusion has been used to characterize the rate of penetration of the skin. The skin is resistant to hydrophilic or water-soluble chemicals and the permeability constant is unlikely to exceed 0.001 cm/hr. Hydrophobic compounds are more readily absorbed and the penetration of organic solvents such as toluene and xylene may approach 1 cm/hr (US EPA, 1992). It has been recommended that the time during which absorption occurs be taken as four hours and that the fraction of the chemical remaining on the surface of the skin longer than four hours will be removed (Fehrenbacher, 1998).

Equation 6-11 can be used to estimate the uptake of a chemical that is absorbed through the skin when evaporation and organic solvent carrier effects are negligible.

$$DA = (S)(K_p)(ED)(WF)(\rho) \qquad \text{(Eq. 6-11)}$$

where

DA is the dermal absorbed dose of the chemical (mass)
S is the surface area of the skin contacted by the chemical (length^2)
K_p is the permeability coefficient for the chemical of concern (length/time)
ED is the exposure duration (time)
WF is the weight fraction of the chemical of concern in the mixture
    (dimensionless)
ρ is the density of the mixture (mass/length^3)

The following equation for the permeability coefficient was selected after independent statistical analysis of data for diffusion of organics in aqueous solution through the skin (US EPA 1992).

$$\log(K_p) = -2.72 + 0.71 \log(K_{ow}) \quad 0.0061\,(MW) \qquad \text{(Eq. 6-12)}$$

where

K_p is the permeability coefficient of the chemical of concern through the skin
    (cm/hr)
K_{ow} is the oil-water partition coefficient (dimensionless)
MW is molecular weight of the chemical of concern (mass/mole)

When the chemical of concern is dissolved in an organic solvent, the permeability of the skin to the organic solvent should be used to calculate the dermal absorption rate.

Example 6.2-5

A worker is dying cloth in a 15% by weight aqueous solution of the dye Red No. 19. The worker exposes his hands and forearms to the dye for 8 hours each work day. The density of the mixture is 1,030 kg/m^3. Physical properties of Red No. 19 include a K_{ow} of 1.0 and a molecular weight of 479 grams per gram-mole. Calculate the daily dermal uptake of Red No. 19 by the dye worker.

Solution: Equation 6-12 can be used to estimate the permeability coefficient for the dye Red No. 19; subsequently, Equation 6-11 can be used to calculate the absorbed dose. The median surface area of the hands and forearms of an adult male is 0.23 square meters.

$$\log(K_p) = -2.72 + 0.71 \log(1.0) - 0.0061\,(479) = -5.64; K_p = 2.28 \times 10^{-6}\,cm/hr$$

$$DA = (0.23\ m^2)\,(2.28 \times 10^{-8}\ m/hr)\,(8hr/workday)\,(0.15)\,(1,030\ kg/m^3)\,(10^6\ mg/kg)$$

$$= 6.5\ mg/workday$$

SECTION 6.2 QUESTIONS FOR DISCUSSION

1. For what types of chemicals would dermal exposures be more significant than inhalation exposures in the workplace?

2. For what types of processing operations would dermal exposures be more significant than inhalation in the workplace?

3. The simple exposure estimation procedures described in this section are useful primarily as screening tools. If these methods indicate potentially high exposures, more sophisticated models should be employed. Describe some of the chemical and physical processes important to inhalation and dermal exposure that more complex models should address.

6.3 EXPOSURE ASSESSMENT FOR CHEMICALS IN THE AMBIENT ENVIRONMENT

Exposure to chemicals in the ambient environment can occur through inhalation, ingestion, or dermal contact. Typically, exposure by ingestion is not as important as dermal or inhalation exposure. However, ingestion may be a significant route of exposure to chemical substances when animals used for food, such as fish or shellfish, accumulate and concentrate chemical contaminants. Ingestion may occur when particles are trapped and swallowed following respiration or when small children eat dust or soil. Ordinarily, exposure by inhalation or dermal absorption will accompany ingestion and result in more significant uptake of the chemicals of concern. In this section, only inhalation and dermal exposure will be considered.

Assessment begins with identification of all wastes and releases containing a chemical of concern and an estimate of the quantity of waste disposed from each source. Next, the concentration of the chemical of concern in the waste or release is measured or estimated and the characteristics of the waste matrix, such as whether it is a liquid, gas, or solid, are identified. The treatment and disposal practices associated with each waste are identified and the quantity of the chemical of concern released to the air, surface waters, groundwater, and land by the treatment and disposal practices are estimated. Finally, the transport and transformation of the chemical of concern through the air, surface waters, and ground water is modeled, along with the uptake through inhalation, ingestion, or dermal contact. In this section, exposure assessment is used to determine the amount of a chemical of concern potentially contacting a member of the general population.

6.3.1 Exposure to Toxic Air Pollutants

Exposure assessment for toxic air pollutants is a four-step process. The first step is to identify pollutants likely to be in the ambient air. Many chemicals found in factories, consumer goods, and waste treatment plants can be released to the air as toxic air pollutants. Some commonly released chemicals include perchloroethylene from dry cleaners, methylene chloride from industrial cleaning and consumer products such as paint strippers, and chromium from metal plating operations. The Toxic Release Inventory, available at the EPA website (www.epa.gov) and discussed elsewhere in this text, provides an extensive source of data on toxic chemical releases.

The second step in exposure assessment for toxic air pollutants is to estimate the quantities of pollutants released by point, area, and mobile sources. Point sources are sites with a specific, usually fixed, location. Point sources include chemical plants, steel mills, oil refineries, and hazardous waste incinerators. Pollutants can be released when equipment leaks, when chemicals are transferred from one area to another, or when pollutants are emitted from stacks. Area sources of toxic air pollutants are comprised of many small sources releasing pollutants to the outdoor air in a defined area. Examples include dry cleaners, small metal plating operations, and gas stations. Mobile sources include automobiles, trucks, buses, etc., which are important contributors of oxides of nitrogen and sulfur, hydrocarbons, carbon dioxide, and particulates in the air.

Routine releases, such as those from industry, cars, landfills, or incinerators, may follow regular patterns and occur continuously over time. Other releases may be routine but intermittent, such as when production is done in batches. Accidental releases can occur during an explosion, equipment failure, or a transportation accident; the timing and, often, the amount released during accidental releases are difficult to predict.

To estimate the amount of a routine or intermittent release, engineers will often sample the effluent from a facility as it is released. The sample is taken to a laboratory and analyzed to quantify the amount released during the collection period. The amount collected during the test is used to predict the amount released each operating day. For example, if 0.1 kilogram of sulfur dioxide is collected in one hour by a collector which samples 1 percent of the airflow from a stack, a plant operating 24 hours per day would be expected to emit 240 kg of sulfur dioxide per day.

Alternatively, engineers can use an emission factor to estimate the amount of toxic air pollutant released by a particular facility. Emission factors are averages of emission measurements from a few representative facilities that relate the quantity of a pollutant released to the level of production associated with the release of that pollutant. Emission factors are described in detail in later chapters. Example 6.3-1 illustrates the use of a particularly simple emission factor.

Example 6.3-1 Use of Emission Factors

An electrical power generating station with four electrical generating units burned 1,055,539 tons of coal, 22,122 thousand gallons of No. 6 fuel oil, and 606 million cubic feet of natural gas to generate electricity during 1998. Use the emission factors in the table below to estimate the releases of arsenic and mercury from the stacks at the power plant.

Pollutant	Coal Emission Factor (lb/ton)	Oil Emission Factor (lb/10³ gal)	Natural Gas Emission Factor (lb/10⁶ ft³)
Arsenic	4.1×10^{-4}	1.32×10^{-4}	2.30×10^{-4}
Cadmium	5.1×10^{-5}	3.98×10^{-4}	2.52×10^{-4}
Chromium	2.6×10^{-4}	8.45×10^{-4}	1.10×10^{-4}
Lead	4.2×10^{-4}	1.51×10^{-3}	2.71×10^{-4}
Mercury	8.3×10^{-5}	1.13×10^{-4}	7.15×10^{-4}
Nickel	2.8×10^{-4}	8.45×10^{-2}	3.61×10^{-3}

Source: "Study of Hazardous Air Pollutant Emissions from Electric Utility Steam Generating Units"—final Report to Congress, U.S. Environmental Protection Agency, Office of Air Quality Planning and Standards (453/R-98-004a), Washington, DC, February 1998.

Solution:

Arsenic: $(1,055,539 \text{ tons})(4.1 \times 10^{-4} \text{ lb/ton}) + (22,122 \times 10^3 \text{ gal})(1.32 \times 10^{-4} \text{ lb/10}^3 \text{ gal})$
$+ (606 \times 10^6 \text{ cf})(2.3 \times 10^{-4} \text{ lb/10}^6 \text{ cf}) = 435 \text{ lbs.}$

Mercury: $(1,055,539 \text{ tons})(8.3 \times 10^{-5} \text{ lb/ton}) + (22,122 \times 10^3 \text{ gal})(1.13 \times 10^{-4} \text{ lb/10}^3 \text{ gal})$
$+ (606 \times 10^6 \text{ cf})(7.15 \times 10^{-4} \text{ lb/10}^6 \text{ cf}) = 90.5 \text{ lbs}$

The third step is to estimate the concentration of the toxic pollutant at the location where exposure occurs. The concentration of a pollutant decreases as it disperses from the point of release. The decrease in concentration or dispersion of the toxic air pollutant is a function of the wind direction and speed and the terrain over which the air flows, whether flat or hilly, whether flowing over a mountain or through a valley. The location of the release, whether from a tall smokestack or a leak at ground level, will affect the distribution of the pollutant near the facility; a toxic air contaminant released from high stacks is dispersed and diluted while descending to ground level. Other factors that affect the concentration include the temperature and speed of the gas exiting the smoke stack and the location of the release within the facility.

The Gaussian dispersion model is most often used to characterize the dilution of toxic air pollutants with distance from the source. The model provides reasonable agreement with experimental data and is, in its simplest form, easy to perform calculations with. The mean concentration, C, resulting from emission at a continuous point source of strength Q at a height H above the totally reflecting earth along the plume centerline is given by

$$C = Q \left(\pi\, \sigma_x \sigma_y\, U \right)^{-1} \times \exp\left(-0.5\, H^2 \sigma_z^{-2} \right) \qquad \text{(Eq. 6-13)}$$

where
C is the concentration of toxic air pollutant ($\mu g/m^3$)
Q is the source release rate ($\mu g/s$)
U is the mean wind speed at the stack height (m/s)
H is the effective height of release above the earth (m)
y is the distance in a direction transverse to the wind (m)
z is the height at which the observation is made (m)

σ_y and σ_z are the standard deviations of the concentrations of plume transverse to the wind and perpendicular to the earth, respectively (m)

Published values of σ_y and σ_z are based on laboratory and field measurements of velocity fluctuations under a variety of atmospheric conditions. Atmospheric stability is used to represent the amount of mixing in the atmosphere and is generally classified as stable, neutral, or unstable. A stable atmosphere is characterized by temperatures that increase with distance from the surface of the earth and reduced vertical mixing; nighttime atmospheric conditions are generally represented as stable. More vigorous atmospheric mixing is expected as the sun warms the surface of the earth and the warmer, less dense air accumulates near the earth's surface; eventually, gravity will displace the warm air with cooler air from above. Daytime atmospheric conditions are typically represented as either neutral or unstable.

The product of the standard deviations has been represented by an equation of the form

$$\sigma_y\sigma_z = ax^b \qquad \text{(Eq. 6-14)}$$

where
   a and b are constants (nondimensional)
   x is the distance downwind from the source (length)

Kumar (1998, 1999) performed regression analysis to develop expressions for the constants, a and b, for urban and rural settings and neutral and stable atmospheric conditions. Urban settings are appropriate when there are many obstacles in the immediate area of the release; obstacles include buildings and trees. Rural settings are appropriate when there are no buildings in the immediate area of the release and the terrain is generally flat and unobstructed. Table 6.3-1 lists the results of the regression when the distance from the source is in meters.

Table 6.3-1   Regression Equations for Dispersion Coefficients.

a. Rural release, neutral atmosphere, x<500m:	$\sigma_y\sigma_z = 0.01082\ x^{1.78}$
b. Rural release, neutral atmosphere, x>500m:	$\sigma_y\sigma_z = 0.04487\ x^{1.56}$
c. Rural release, stable atmosphere, x<2000m:	$\sigma_y\sigma_z = 0.0049\ x^{1.66}$
d. Rural release, stable atmosphere, x>2000m:	$\sigma_y\sigma_z = 0.01901\ x^{1.46}$
e. Urban release, neutral atmosphere, x<500m:	$\sigma_y\sigma_z = 0.0224\ x^2$
f. Urban release, neutral atmosphere, x>500m:	$\sigma_y\sigma_z = 0.394\ x^{1.54}$
g. Urban release, stable atmosphere, x<500m:	$\sigma_y\sigma_z = 0.008\ x^2$
h. Urban release, stable atmosphere, x>500m:	$\sigma_y\sigma_z = 0.34\ x^{1.37}$

Example 6.3-2

Hydrogen sulfide is released from a low-level vent in a rural area at a rate of 0.025 kg/s. Calculate the concentration at the plant boundary located 300 m downwind from the vent during daytime conditions when the wind speed is 4 m/s and at night when the wind speed is 2.5 m/s. (a) If the concentration of concern for hydrogen sulfide is 42 mg/m^3, will this concentration be exceeded at the plant boundary? (b) Estimate the exposure to hydrogen sulfide of a person living 300 m downwind from the facility described during i) daytime and ii) nighttime if the individual at rest breathes 0.9 m^3/hr of the ambient air.

Solution:

(a) Equation 6-13 can be used to calculate the concentration at ground-level by setting the height above the earth equal to zero. The appropriate correlations for the dispersion coefficients are given by items a and c in Table 6.3-1 for daytime and nighttime conditions, respectively. For daytime conditions,

$$C = \frac{0.025 \text{ kg/s}}{(\pi)(0.01082)(300 \text{ m})^{1.78}(4 \text{ m/s})} = 7.17 \times 10^{-6} \text{ kg/m}^3 = 7.17 \text{ mg/m}^3$$

For nighttime conditions,

$$C = \frac{0.025 \text{ kg/s}}{(\pi)(0.0049)(300 \text{ m})^{1.66}(2.5 \text{ m/s})} = 5.02 \times 10^{-5} \text{ kg/m}^3 = 50.2 \text{ mg/m}^3$$

The concentration of hydrogen sulfide is below the concentration of concern under daytime conditions but exceeds the concentration of concern at night when atmospheric mixing is less.

(b) For daytime condition, $(7.17 \text{ mg/m}^3) (0.9 \text{ m}^3/\text{hr}) = 6.45 \text{ mg/hr}$
For nighttime conditions, $(50.2 \text{ mg/m}^3) (0.9 \text{ m}^3/\text{hr}) = 45.2 \text{ mg/hr}$

The last step in an exposure assessment is to estimate the number of persons exposed to a toxic air pollutant. Demographers can estimate the number of persons living in areas surrounding a source using census data. Combining the concentration estimates and the census data, engineers can estimate the numbers of people exposed to the pollutant at varied concentrations. To aid decision makers, these results can be compared to a selected benchmark such as an air quality standard or a level with a known health effect. Data on population densities in regions surrounding point sources are available at the Envirofacts section of the EPA website (http://www.epa.gov/enviro/index_java.html, see Appendix F).

6.3.2 Dermal Exposure to Chemicals in the Ambient Environment

Swimming in rivers, lakes, and streams is generally the only activity considered to cause significant dermal exposure. Although other activities—e.g., water skiing, fishing, standing in the rain—could lead to human dermal exposure, the frequency, duration, and the amount of skin surface available for exposure are small; therefore, for general and long-term assessments, these activities are considered negligible. Because swimming is an episodic activity, it is necessary to consider both frequency and duration of exposure. In addition, the surface area exposed is an important factor in dermal exposure calculation. These activity-related parameters, when coupled with data on the aquatic ambient concentration of a chemical toxin, yield an estimate of dermal exposure.

Frequency of swimming in natural surface water bodies can be defined from the number and duration of exposures occurring in a single year. A Department of Interior survey (USDOI 1973) found that 34% of the population swam in rivers, lakes, or oceans in the year surveyed. For these swimmers, the average frequency of swimming was seven days per year and the average duration was 2.6 hours.

Subsequent investigation of this survey found that the reported exposure time represented time on the shore as well as time in the water (EPA 1992). Furthermore, certain subpopulations, e.g., competitive swimmers, upwardly biased the average exposure frequency and time. Therefore, a reasonable average frequency for a recreational swimmer may be 5 days per year lasting 0.5 hour on each day when swimming occurs (EPA 1992).

An inherent assumption of many exposure scenarios is that clothing prevents dermal contact and subsequent absorption of contaminants. For swimming and bathing scenarios, past exposure assessments have assumed that 75% to 100% of the skin surface is exposed (Vandeven and Herrinton, 1989). Other studies have shown that dermal exposure may occur at sites covered by clothing (Maddy et al., 1983). Consequently, it is appropriate to assume that the entire body is exposed to the chemical of concern during swimming.

Data on the surface area of the body is given in the Exposure Factors Handbook (EPA 1997) and reproduced in Section 6.2. As shown in Table 6.2-1, total adult body surface area for males can vary from less than 1.7 square meters to over 2.3 m^2; for females, the range is from 1.45 m^2 to 2.1 m^2. For default purposes, the median skin surface areas, 1.94 m^2 for males and 1.69 m^2 for females, can be used.

Example 6.3-3

A man swims in a river downstream of a rubber processing plant that uses 1,1,1-trichloroethylene to clean molds used to shape the rubber parts. Wastewater discharged into the river results in contamination at a level of 3 µg per liter in the receiving stream. The man is of average stature and swims in the river about fifteen times each summer with each swim lasting one-half hour. Calculate the man's exposure to 1,1,1-trichloroethylene which results from swimming in the river.

Solution:

$$Exposure = (Mass/Event)*(15\ events/yr)$$

Mass/Event is obtained from Equations 6-11 and 6-12 where
$s = 1.94\ m^2$
$Kp = 10^{-1.8}\ cm/hr$
$ED = 0.5\ hr$
$WF = 3µg$ trichloroethylene per 1000g H_2O
$ρ = 10^6\ g/m^3$
$Exposure = 5 \times 10^{-7}\ (g\ TCE/event) \times (15\ events/yr) = 7.5 \times 10^{-6}\ (g\ TCE/yr)$

6.3.3   Effect of Chemical Releases to Surface Waters on Aquatic Biota

Wastewater generated in the manufacture, processing, or use of a chemical may contain a fraction of the chemical produced and the raw materials used in the manufacturing process. This loss may occur during reaction to produce the chemical, purification, blending, or cleaning of the reactors, piping, and equipment used to process the chemical substance. The wastewater must be either treated by facilities at the plant site or, more often, commingled with the wastes of others and treated at a publicly owned treatment works (POTW). Using physical-chemical property data and estimates of biodegradability, the effectiveness of the treatment can be estimated, so that the amount actually entering the receiving water body can be predicted. The receiving water body will dilute the discharge from the plant site or POTW so that the concentration in the receiving stream can be calculated if the flow in the stream is known. Stream in this context means the receiving body of water and, in this sense, can include creeks, rivers, lakes, bays, or estuaries.

Removal of chemicals during wastewater treatment is controlled by the physical and biological processes employed in the treatment works. The following processes are commonly used to remove chemicals during wastewater treatment:

1. Adsorption to suspended solids in the primary clarifier, aeration basin, and secondary clarifier;
2. Volatilization through surface vaporization in the primary and secondary clarifiers and through air-stripping in the aeration basin; and
3. Biodegradation by aerobic microorganisms, most commonly in an activated sludge aeration basin.

Under optimal conditions, a POTW will remove a large percentage, i.e., 70 to 99+%, of many organic pollutants from the wastewater, but treatment efficiency varies with the chemical and physical properties of the pollutant. The POTW is typically less efficient in removal of inorganic pollutants and many of these pass through the POTW unchanged.

Numerous models have been proposed to predict the fate of chemicals in a POTW consisting of a primary settling basin, an activated-sludge aeration basin, and a secondary clarifier. Clark, et al. (1995) proposed a simple fugacity analysis of the fate of organic chemicals in a POTW. The fugacity approach is predicated on equivalence of the chemical potential in phases in contact, in this case wastewater, solids suspended in the wastewater, and air in contact with the wastewater. The physical-chemical properties needed to model the fate of chemicals in a POTW by the fugacity analysis include water solubility, vapor pressure, octanol-water partition coefficient, and biodegradation half-life. These properties are discussed in Chapter 5 of this text. Sample removal efficiencies as calculated by Clark, et al. (1995) are shown in Table 6.3-2.

An important issue for surface water is the effect that a chemical may have on aquatic organisms including algae, freshwater crustaceans, and fish. A healthy stream with a wider variety of organisms will have a better ability to assimilate chemical releases than a stream whose quality is already compromised. If any link in the food chain in a stream is impacted, the effect can be deleterious to other organisms as well as the health of the stream. Organisms lower on the food chain, such as algae, have shorter lives; for these organisms short-term exposures to high concentrations of chemicals are critical. Consequently, the concentration of a chemical in the receiving body of water when the dilution is least is used to assess the impact of chemical releases on the aquatic biota. For this purpose, the historical stream flow representing the seven consecutive days of lowest flow over a ten-year

Table 6.3-2   Removal Efficiencies in a POTW Calculated by Clark, et al. (1995).

	Removal Efficiency	Volatilization	Biodegradation	Settled Solids	Effluent
1,1,1-trichloroethane	88%	73%	13%	1%	12%
1,1,2-trichloroethane	85%	69%	15%	1%	15%
toluene	87%	38%	48%	1%	13%
1,4-dichlorobenzene	72%	19%	46%	7%	28%
naphthalene	68%	7%	53%	7%	32%
anthracene	86%	<1%	47%	39%	14%
pyrene	87%	<1%	14%	73%	12%
dibutyl phthalate	81%	<1%	27%	54%	19%
2-ethyl hexyl phthalate	91%	<1%	27%	63%	9%
phenol	99%	<1%	99%	1%	1%
pentachlorophenol	87%	<1%	81%	6%	13%
2,4-D	83%	<1%	79%	4%	17%

period is often used to generate estimates of chronic concentrations of chemicals of concern for aquatic life. Data on historical stream flows is available from the U.S. Geological Survey or at the agency's Internet site at *http://www.usgs.gov/usa/nwis/sw*.

The following formula can be used to calculate surface water concentrations of the chemical of concern in free-flowing rivers and streams:

$$SWC = [\text{Release} \times (1 - WWT/100)]/\text{Stream flow}] \qquad \text{(Eq. 6-15)}$$

where

   SWC is the surface water concentration (mass/volume)
   Release is the quantity of chemical released in wastewater (mass/time)
   WWT is the percent removal in wastewater treatment (dimensionless)
   Stream flow is the measured or estimated flow of the receiving stream (volume/time)

Example 6.3-4

During periodic cleaning of reactors, a chemical plant using toluene as a process solvent discharges wastewater containing 32 kg/day of toluene to the Riverside POTW. The Riverside facility is an activated sludge plant with primary and secondary treatment for wastewater pollutants. The Riverside plant discharges its effluent into the Grande River which has a historical once-in-ten-years 7-day low-flow of 84 cfs. Assess the potential impact of the discharge on minnows in the river if the LC-50 for the minnows is 20 mg/l.

Solution: Using the estimated removal efficiency for toluene from Table 6.3-2 of 87%, the estimated concentration of toluene in the Grande River at low-flow conditions can be calculated using Equation 6-15.

$$SWC = [(32 \text{ kg/da})(1 - 87/100)(10^6 \text{ mg/kg})]/[(84 \text{ cfs})(86{,}400 \text{ s/da})(28.32 \text{ l/ft}^3)]$$

$$= 0.020 \text{ mg/l}$$

The estimated concentration of toluene is three orders of magnitude (1000 times) less than that which would be lethal to one-half of the minnows. This is a reasonable margin of safety for aquatic biota.

6.3.4 Ground Water Contamination

Industrial solid wastes are often sent to land disposal in municipal, industrial, or hazardous waste landfills. Although less common, surface impoundments and land treatment may be used to contain and treat industrial wastes. Chemicals may leach from the wastes, either in free liquid contained in the waste or in rainwater percolating through the waste, and be carried into the underlying soils. Chemicals entering the soil solubilize in water contained in the pore space between the soil particles. This interstitial water, called groundwater, may subsequently percolate downward into the water table, carrying the chemical contaminants with it.

Groundwater contamination is most common beneath urban areas, agricultural areas, and industrial complexes. Frequently, groundwater contamination is not discovered until long after the actions leading to the contamination have occurred. One reason for this is the slow movement of groundwater through soils and underlying rock strata; in fine-grained soils and low permeability rock strata, groundwater movement is often less than one foot per day. Contaminants in groundwater do not mix or spread quickly, but remain concentrated in slow-moving, localized plumes that may persist for many years. This often results in a delay in the detection of groundwater contamination. In some cases, groundwater contamination discovered today is the result of agricultural, industrial, and municipal practices several decades ago. This also means that the land disposal practices of today may have effects on groundwater quality many years from now.

Groundwater is a vital natural resource. It is used for public and domestic water supply, for irrigation of crops, and for industrial, commercial, mining, and thermoelectric power production purposes. In 1990, the United States Geological Survey reported that groundwater supplied 51% of the nation's total population with drinking water. Unfortunately, groundwater is vulnerable to contamination and, once contaminated, is difficult to remediate. Table 6.3-3 lists National Primary Drinking Water Standards prescribed by the US Environmental Protection Agency which must be met by all drinking water supplies after treatment, if any.

The transport of a chemical in the subsurface depends on physical-chemical properties of the chemical and the characteristics of the subsurface environment. Some of the more important properties influencing the spread of chemical contaminants in the subsurface include water solubility, soil organic carbon partition coefficient, and vapor pressure. A chemical that is readily soluble in water will be carried deeper into the subsurface by rainwater and once it reaches the water table it will mix intimately with the groundwater. A chemical that has an affinity for organic solvents is likely to be adsorbed onto soil organic matter which constitutes a range of less than 1% to 20% of topsoils with the concentration generally decreasing with increasing depth; adsorption from the groundwater retards the movement of dissolved chemicals. In addition, cationic species can be expected to attach to

Table 6.3-3   National Primary Drinking Water Standards for Maximum Contaminant Limits (MCL), US EPA 1994.

Contaminant	MCL (mg/L)	Potential Health Effects from Ingestion of Water	Sources of Contamination in Drinking Water
Benzene	0.005	Cancer	Gasoline, paint, plastics industry
Carbon Tetrachloride	0.005	Cancer	Solvents and degradation products
Chlorobenzene	0.1	Nervous system, liver	Metal degreasing processes
o-Dichlorobenzene	0.6	Liver, kidney damage	Paints, dyes, chemical wastes
p-Dichlorobenzene	0.075	Cancer	Room and water deodorants
1,2-Dichloroethane	0.005	Cancer	Leaded gas, fumigants, paints
1,1-Dichloroethylene	0.007	Cancer, liver, kidney damage	Plastics, dyes perfumes, paints
1,2-Dichloroethylene	0.07	Liver, kidney, nervous system	Waste industrial extraction solvents
Diethylhexyl phthalate	0.006	Cancer	Polyvinyl chloride, other plastics
Ethylbenzene	0.7	Liver, kidney, nervous system	Gasoline, chemical manufacturing
Pentachlorophenol	0.001	Cancer, liver, kidney damage	Wood preservative
PCBs	0.0005	Cancer	Transformer oils, plasticizers
Styrene	0.1	Liver, nervous system damage	Plastics, rubber, landfill leachate
Tetrachloroethylene	0.005	Cancer	Dry cleaning solvent, other solvents
Toluene	1.0	Liver, kidney, nervous system	Gasoline, chemical solvent
1,1,1-Trichloroethane	0.2	Liver, nervous system damage	Adhesives, paints, metal degreasing
1,1,2-Trichloroethane	0.005	Kidney, liver, nervous system	Rubber processing, chemical mfg.
Vinyl chloride	0.002	Cancer	PVC pipe, solvent degradation
Xylenes	10	Liver, kidney, nervous system	Gasoline refining, paints, inks

soil particles that are negatively charged. Chemicals with significant vapor pressure may vaporize to the atmosphere from shallow pore water before precipitation carries the chemical downward to the saturated zone.

Even the simplest descriptions of contaminant migration in groundwater often rely on numerical solutions of the equations governing flow, physical equilibrium, and chemical reaction. Analytical solutions are available for a variety of conditions when only a single spatial dimension is considered (van Genuchten and Alves, 1982). For example, the one-dimensional form of the analytical equation for convection and dispersion for dissolved, nonreactive constituents in a homogeneous sediment is

$$D \frac{\partial^2 C}{\partial x^2} - u \frac{\partial C}{\partial x} = \frac{\partial C}{\partial t} \qquad \text{(Eq. 6-16)}$$

where

   C is the concentration of dissolved solute in the groundwater (mass/volume)
   D is the hydrodynamic dispersivity in the direction of flow (length^2/time)
   u is the average interstitial groundwater velocity (length/time)
   x is the distance along the flow path (length)
   t is the temporal variable (time)

Hydrodynamic dispersion is due to mixing of the groundwater and molecular diffusion of the dissolved species. These components are combined to yield

$$D = \alpha u + D^* \qquad \text{(Eq. 6-17)}$$

where

   α is the dynamic dispersivity of the porous media (length, typical value for α
      is 0.1)
   D^* is the coefficient of molecular diffusion of the solute (length^2/time)

The boundary conditions for a step function input are described mathematically as

$$C(x,0) = 0 \qquad x \geq 0$$
$$C(0,t) = C_0 \qquad t \geq 0$$
$$C(\infty,t) = 0 \qquad t \geq 0$$

For these boundary conditions, the solution to Equation 6-16 for a saturated, homogeneous porous media is given by

$$C(x,t) = C_0 \{ \text{erfc}[(x - ut)/(4Dt)^{1/2}] + \exp(u/D)\, \text{erfc}[(x + ut)/(4Dt)^{1/2}]\} \qquad \text{(Eq. 6-18)}$$

where erfc is the complementary error function, which is tabulated in mathematical handbooks.

   One-dimensional expressions for the transport of dissolved constituents, such as Equation 6-18, are of limited utility in field problems because dispersion occurs in directions transverse to flow as well as in the direction of flow. Baetsle (1969) has described the concentration distribution in a plume of contamination originating as an instantaneous slug at the point $x = 0$, $y = 0$, $z = 0$. As the contamination is carried away from the source in the x-direction, the concentration distribution resulting from instantaneous release of a mass M is given by

$$C(x,y,z,t) = \frac{M}{8\,(\pi\, t)^{3/2}\,[D_x D_y D_z]^{1/2}} \exp\{-[(x - ut)^2/4D_x t] - [y^2/4D_y t] - [z^2/4D_z t]\}$$
$$\text{(Eq. 6-19)}$$

   The maximum concentration in the plume occurs at the center of mass of the contaminant cloud where $x = ut$, $y = 0$, $z = 0$, at which the exponential term is equal to unity. The zone in which 99.7% of the contaminant mass occurs is described by the ellipsoid with dimensions, measured from the center of mass, of $d_i = (2D_i t)^{1/2}$ where $i = x$, y, or z.

Example 6.3-5

A rupture of a storage tank containing liquid waste released 100 kg of dissolved arsenic into a shallow saturated groundwater zone in which the flow is horizontal. The average groundwater velocity is 0.5 m/day, the dynamic dispersivity is 0.1 m, and the coefficient of molecular diffusion of arsenic in water is 2×10^{-8} m^2/s. Arsenic is not removed from the groundwater by adsorption or chemical precipitation. Estimate the location and size of the waste plume 90 days after the rupture of the tank.

Solution: After 90 days, the center of gravity of the contaminant plume has moved (0.5 m/day)(90 days) = 45 meters from the site of the rupture in the direction of groundwater flow. The dispersivities in the coordinate directions are, in the absence of better data, estimated to be:

$$D_x = (0.1 \text{ m})(0.5 \text{ m/day}) + (2 \times 10^{-8} \text{ m}^2/\text{s})(86{,}400 \text{ s/day}) = 0.0517 \text{ m}^2/\text{day}$$

$$D_y = D_z = (2 \times 10^{-8} \text{ m}^2/\text{s})(86{,}400 \text{ s/day}) = 1.73 \times 10^{-3} \text{ m}^2/\text{day}$$

The extension of the plume from the center of gravity in the three dimensions are

$$d_x = [(2)(0.0517 \text{ m}^2/\text{day})(90 \text{ days})]^{1/2} = 3.05 \text{ m ahead and behind the center of the plume}$$

$$d_y = d_z = [(2)(1.73 \times 10^{-3} \text{ m}^2/\text{day})(90 \text{ days})]^{1/2} = 0.56 \text{ m above, below, and to either side}$$

The concentration at the center of the plume is

$$C = \frac{100 \text{ kg}}{(8)[(3.14159)(90 \text{ days})]^{3/2} (0.0517 \text{ m/day})^{1/2} (1.73 \times 10^{-3} \text{ m}^2/\text{day})} = 6.68 \text{ kg/m}^3$$

SECTION 6.3   QUESTIONS FOR DISCUSSION

1. What classes of chemicals will be highly mobile in groundwater and what classes of chemicals would be relatively immobile?
2. What classes of chemicals will be likely to be present at high concentrations in surface waters, even after treatment in POTWs and other wastewater treatment units?
3. The simple exposure estimation procedures described in this section are useful primarily as screening tools. If these methods indicate potentially high exposures, more sophisticated models should be employed. Describe some of the chemical and physical processes important to determining concentrations of contaminants in surface and groundwaters.

6.4   DESIGNING SAFER CHEMICALS

A challenge for chemical engineers is to use the general principles outlined in this chapter and in Chapter 5 in designing chemicals that will reduce toxicity. The remainder of this chapter presents semi-quantitative principles and guidelines that can be used in designing safer chemicals and is adapted from material presented by DeVito (1996).

In designing safer chemicals, it is useful to think about modifying properties so that

- persistence and dispersion in the environment are minimized, reducing exposures,
- uptake by the body is minimized, reducing dose, and
- toxicity is minimized.

This section will consider property modifications that can lead to reduced exposure, dose, and toxicity. Consider first the issue of dose.

6.4.1 Reducing Dose

Converting an exposure (e.g., inhaling a chemical) into a dose (e.g., absorption by the blood through the lung membrane) generally involves the transport of a chemical across a membrane. The three primary membranes of interest are the lung, which controls uptake of chemicals that are inhaled; the skin, which controls the uptake of compounds from dermal exposures; and the gastrointestinal tract, which controls the uptake of chemicals that are ingested. Some of the characteristics of these membranes are listed in Table 6.4-1.

From Table 6.4-1, it is apparent that the gastrointestinal tract has one of the greatest surface areas available for uptake of chemicals by the body. The uptake of chemicals across this membrane is controlled by lipid solubility, water solubility, dissociation constant, and molecular size.

High water solubility enhances uptake through the gastrointestinal tract because water soluble materials are more easily mobilized in the large and small intestine and the materials therefore experience less mass transfer resistance in migrating to the intestine wall. In contrast, high lipid solubility enhances uptake and transport across the membrane. Thus, the compounds that are likely to be transported from the gastrointestinal tract into the blood streams are compounds with moderate water solubility and moderate lipid solubility. Highly water soluble (lipid insoluble) and highly lipid soluble (log $K_{ow} > 5$, water insoluble) compounds are less likely to be taken up through the gastrointestinal tract.

Molecular weight also plays a role in determining uptake through the gastrointestinal tract. A general guideline is that molecules with molecular weights less than 300 that are both lipid and water soluble are well absorbed, and those with molecular weights in excess of 1000 are only sparingly absorbed.

The lung also provides a relatively large surface area for uptake of chemicals. The lung is a relatively thin membrane and because the membrane is so thin, lipid solubility plays less of a role in chemical uptake than for the gastrointestinal tract. High water solubility will promote uptake through the lung, as will the delivery of the compound on fine particles (less than 1 micron in diameter). Small particles can be inhaled deeply and will deposit deep in the lung, allowing the chemicals adsorbed on or dissolved in the particles to reside in the lung for very long periods.

The skin presents a formidable barrier to chemicals transport. For a chemical to be taken up through the skin, it must pass through multiple layers. As with the

Table 6.4-1 Characteristics of Membranes That Control Chemical Uptake by the Body (DeVito, 1996).

Membrane	Surface area (m^2)	Thickness of absorption barrier (μm)	Blood flow (L/min)
Skin	1.8	100–1000	0.5
Gastrointestinal tract	200	8–12	1.4
lung	140	0.2–0.4	5.8

gastrointestinal tract, moderate lipophilicity (log K_{ow} < 5) promotes absorption through the skin because transport must occur through both largely lipid and largely aqueous layers.

Finally, note that once a compound is absorbed into the blood stream, it must still reach a target organ. Many organs have their own barriers to uptake that may influence dose (e.g., the blood-brain barrier is more easily crossed by lipophilic materials). In addition, chemicals may be removed by the body through urine and feces before the target organ is reached (water solubility enhances elimination via this mechanism).

6.4.2   Reducing Toxicity

Designing safer chemicals by reducing toxicity requires a knowledge of the mechanisms by which compounds exert a toxic effect. While these mechanisms are not known in many cases, there are a few general mechanisms for toxicity that can be examined, leading to safer chemical designs.

One group of mechanisms associated with toxic effects are the reactions of electrophilic species with nucleophilic substituents of cellular macromolecules such as DNA, RNA, enzymes and proteins. Table 6.4-2 presents the possible effects of a number of common electrophiles.

Table 6.4-2   Examples of Electrophilic Substituents and the Reactions They Undergo with Biological Nucleophiles, and the Resulting Toxicity* (DeVito, 1996).

Electrophile	General Structure	Nucleophilic Reaction	Toxic effect
Alkyl halides	R−X where X= Cl,Br,I,F	substitution	Various; e.g., cancer
α,βunsaturated carbonyl and related groups	C−C−C−O C≡C−C=O C=C−C≡N	Michael addition	Various; e.g., cancer, mutations, hepatoxicity, nephrotoxicity, neurotoxicity, hematoxicity
γ diketones	R_1−C(=O)−CH_2−CH_2− C(=O)−R_2	Schiff base formation	Neurotoxicity
Terminal epoxides	−CH−CH_2 (O) −O−CH_2−CH−CH_2 (O)	addition	Mutagenicity, testicular lesions
Isocyanates	−N=C=O −N=C=S	addition	Cancer, mutagenicity, immunotoxicity

*The presence of these substituents in a substance does not automatically mean that the substance is or will be toxic. Other factors, such as bioavailability, and the presence of other substituents that may reduce the reactivity of these electrophiles can influence toxicity as well.

Ideally, the use of these groups would be avoided, however, in many cases the electrophilic groups are necessary to produce a desired property. For example, for the case of the unsaturated carbonyls, the Michael addition reaction that causes the toxic effect may be the desired commercial property. Nevertheless, the toxic effects can sometimes be reduced by introducing selected substituents. For example, the addition of a methyl substituent to ethyl acrylate reduces potential health effects:

$$CH_2=CH-C=O \atop | \atop O-CH_2-CH_3$$

ethyl acrylate (carcinogenic)

$$CH_3 \atop | \atop CH_2=C-C=O \atop | \atop O-CH_3$$

(methyl methacrylate, noncarcinogenic)

Isocyanates present another example. In this case, the electrophilic nature of the isocyanate can be masked in some applications by converting the material to a ketoxime derivative.

$$R-NH-C=O \atop | \atop O-N=C-CH_2-CH_2-CH_2-CH_3 \atop | \atop CH_3$$

The ketoxime derivative is then removed, *in situ,* during the use of the compound. This reduces potential exposures and the resultant toxicity.

Clearly, the identification of such structural modifications requires a detailed knowledge of the mechanism of the potential toxicity and the structural sensitivity of that mechanism. Case studies of structural modifications leading to reduced toxicities are available in the US EPA's Green Chemistry Expert System, which is available at *http://www.epa.gov/greenchemistry/gces.htm.* Such detailed knowledge is not available for all materials, but the examples cited above demonstrate that there is potential for designing materials with reduced toxicities.

REFERENCES

Baetsle, L.H., "Migration of radionuclides in porous media," *Progress in Nuclear Energy, Series XII, Health Physics,* A.M.F. Duhamel, ed., Pergamon Press, Elmsford, N.Y. (1969).

Clark, B., Henry, J.G., and Mackay, D., "Fugacity analysis and model of organic chemical fate in a sewage treatment plant," *Environ. Sci. Technol.,* 29, 1488–1494 (1995).

DeVito, S., "General Principles for the Design of Safer Chemicals: Toxicological Considerations for Chemists," *Designing Safer Chemicals,* American Chemical Society, Symposium Series 640, Washington, D.C., 1996.

Drivas, P.J., Simmonds, P.G., and Shair, F.H., "Experimental characterization of ventilation systems in buildings," *Environ. Sci. & Tech.,* 6, 609–614 (1972).

Fenske, R.A., and Birnbaum, S.G., "Second generation video imaging technique for assessing dermal exposure (VITAE system)," *Am. Ind. Hyg. Assoc. J.*, 58, 636–645 (1997).

Fehrenbacher, M.C. and Hummel, A.A., "Evaluation of the mass balance model used by the Environmental Protection Agency for estimating inhalation exposure to new chemical substances" *Am. Ind. Hyg. Assoc. J.*, 57, 526–536 (1996).

Fehrenbacher, M.C., "Dermal Exposure Assessments," *A Strategy for Assessing and Managing Occupational Exposures,* J.R. Mulhausen and J. Damiano, eds., AIHA Press, Fairfax, VA (1998).

Groth, K., "Assessment of dermal exposures to 4,4'-Methylene dianaline in aircraft maintenance operations involving advanced technology materials," Proceedings of the Conference on Advanced Composites, 113–118, ACGIH, Cincinnati, OH (1992).

Hummel, A.A., Braun, K.O., and Fehrenbacher, M.C., "Evaporation of a liquid in a flowing air stream," *Am. Ind. Hyg. Assoc. J.*, 57, 519–526 (1996).

Jayjock, M.A., "Back pressure modeling of indoor air concentrations from volatilizing sources," *Am. Ind. Hyg. Assoc. J.*, 55, 230–235 (1994).

Jayjock, M.A., "Estimating airborne exposure with physical-chemical models," *A Strategy for Assessing and Managing Occupational Exposures,* J.R. Mulhausen and J. Damiano, eds., AIHA Press, Fairfax, VA (1998).

Klingner, T., "New developments in surface contamination monitoring for aromatic amines," *Proceedings of the Conference on Advanced Composites,* ACGIH, 43–46 (1992).

Kumar, A., "Estimating hazard distances from accidental releases," *Chemical Engineering,* 121–128 (August 1998).

Kumar, A., "Estimate dispersion for accidental releases in rural areas," *Chemical Engineering,* 91–94 (July 1999).

Kusters, E., "Biological monitoring of MDA," *Brit. J. Ind. Med.,* 49, 72–79 (1992).

Lennart, A. Nielsen, F., and Breum, N.O., "Evaluation of evaporation and concentration distribution models-a test chamber study," *Ann. Occup. Hyg.*, 41, 625–641 (1997).

Nicas, M., "Estimating exposure intensity in an imperfectly mixed room," *Am. Ind. Hyg. Assoc. J.*, 57, 542–560 (1996).

OSHA, "Precautions and the order of testing before entering confined and enclosed spaces and other dangerous atmospheres," 29 CFR 1915.12, US Occupational Safety and Health Administration, Washington, D.C.

OSHA, "Occupational safety and health standards, air contaminants," 29 CFR 1910.1000, U.S. Occupational Safety and Health Administration, Washington, D.C.

Scheff, P.A., Friedman, R.L., Franke, J.E., Conroy, L.M., and Wadden, R.A., "Source activity modeling of freon emissions from open-top vapor degreasers," *Appl. Occup. Environ. Hyg.*, 7, 127–134 (1992).

US EPA 1991, *Chemical Engineering Branch Manual for the Preparation of Engineering Assessments,* US Environmental Protection Agency, Washington, D.C. (1991).

US EPA 1992, *Dermal Exposure Assessment: Principles and Applications* (EPA/600/8-91/011B), US Environmental Protection Agency, Office of Research and Development, Washington, D.C. (1992).

US EPA 1994, *National Primary Drinking Water Standards,* US Environmental Protection Agency, Office of Water, Washington, D.C. (1994).

US EPA 1996, *Occupational Dermal Exposure Assessment—A Review of Methodologies and Field Data,* US Environmental Protection Agency, Office of Pollution Prevention and Toxics (1996).

US EPA 1997, *Exposure Factors Handbook* (EPA/600/P-95/002), US Environmental Protection Agency, Office of Research and Development, Washington, D.C. (1997).

van Genuchten, M.Th. and Alves, W.J., "Analytical Solutions of the One-Dimensional Convective-Dispersive Solute Transport Equation," Technical Bulletin Number 1661, US Department of Agriculture, Washington, D.C. (1982).

PROBLEMS

1. Using Equation 6-4, calculate the rate of evaporation of a circular pool (1 m^2 surface area) of ethylbenzene into a room where the temperature is 80°F and the ventilation produces an air velocity of 0.5 m/s. Use the methods described in Chapter 5 to estimate the vapor pressure of the liquid.

2. Calculate the concentration in the room described in Problem 1 under average and poor ventilation conditions, and compare the results to the OSHA PEL. Assume the room has dimensions $4 \text{ m} \times 4 \text{ m} \times 2.45 \text{ m}$.

3. Confirm that Equation 6-3 is the solution to Equation 6-1 for a transient emission source.

4. Derive Equation 6-7, which describes ventilation of a two-compartment room. Describe how you would extend the analysis to a three-compartment room.

5. Assume that the room described in problems 1 and 2 is characterized as a cube, 1.5 meter on each side, centered over the pool of ethylbenzene, that exchanges air at a rate of 1 m^3/s with the rest of the room. Develop expressions for the transient and steady-state concentrations of ethylbenzene in the air immediately above the pool, and in the rest of the room.

6. Estimate the dermal and inhalation exposures that might be associated with collecting a sample of ethylbenzene from the pool described in problems 1, 2, and 5. Make reasonable assumptions about the time required to collect the sample and inhalation rates. Is dermal or inhalation the dominant exposure route?

7. A liquid transfer pump is leaking at the seals, releasing 1 milliliter per minute (about two ounces per hour) of an aqueous solution containing 4 percent acrolein (2-butenal). Nearby, a process tank has a leaking seam from which is weeping an aqueous solution containing 5 percent methyl ethyl ketone (2-butanone) at a rate of 30 milliliters per minute (about 2 quarts per hour). The ventilation rate in the process building where the leaking pump and weeping tank are located is 200 cubic meters per hour. Is either of these releases a potential health hazard to workers in the process building?

8. Use the Green Chemistry Expert System (*http://www.epa.gov/greenchemistry/gces.htm*) to identify structural modifications that can be made to nitriles to reduce their toxicity.

9. In designing chemicals that will minimize human uptake, you may wish to consider properties such as volatility, octanol-water partition coefficient, and water solubility.

For high, medium, and low values of each of these parameters, characterize whether exposure due to inhalation, ingestion, and dermal contact are likely to be important. For each of the properties, complete a table like the one shown below.

Exposure route	High water solubility	Moderate water solubility	Low water solubility
Inhalation			
Ingestion	Potentially high uptake	Potentially high uptake	Low uptake due to poor mass transfer within g.i. tract
Dermal contact			

10. The process line shown in Figure Problem 10 delineates the steps in the formulation and packaging of a primer coating for metal parts.
    (a) Identify all possible sources of occupational exposure to hazardous chemicals and indicate if exposure is due to inhalation, dermal contact, or other route.
    (b) Identify all possible sources of releases to the environment and indicate if the release is to the air, water, or land.

11. The process line shown in Figure Problem 11 delineates the steps in the coloring of leather with dyes.
    (a) Identify all possible sources of occupational exposure to hazardous chemicals and indicate if exposure is due to inhalation, dermal contact, or other route.
    (b) Identify all sources of releases of hazardous chemicals to the environment and indicate if the release is to the air, water, or land.

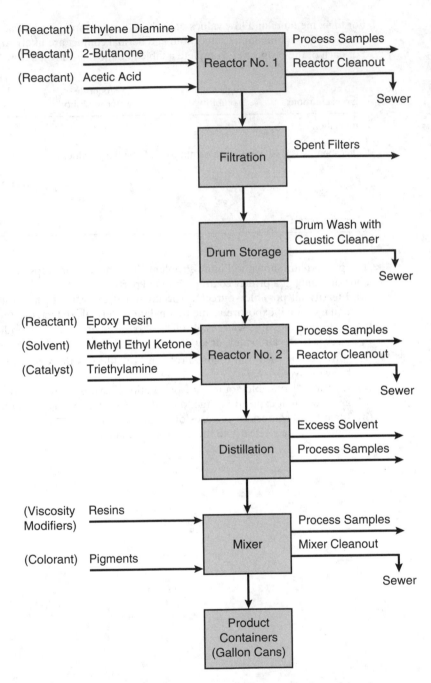

Figure Problem 10   Formulation and packaging of a primer coating for metal parts.

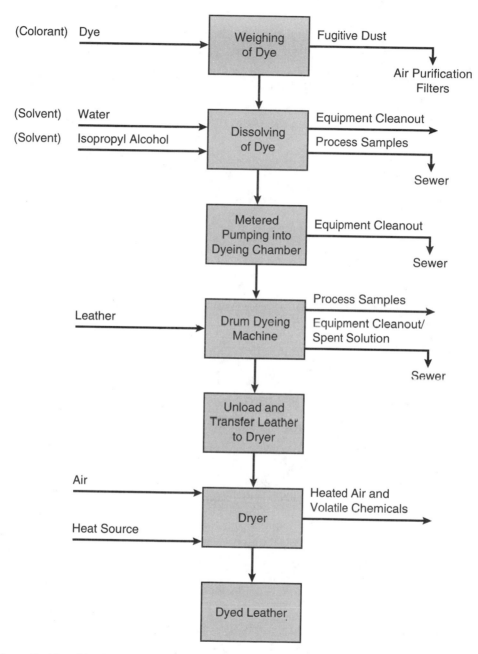

Figure Problem 11   Coloring of leathers and dyes.

CHAPTER 7

Green Chemistry

by
Paul T. Anastas and David Allen

7.1 GREEN CHEMISTRY

Chemical products can be manufactured using a wide variety of synthesis routes. The designer of a chemical process must choose from alternative raw materials, solvents, reaction pathways, and reaction conditions, and these design choices can have a significant impact on the overall environmental performance of a chemical process. Ideal chemical reactions would have attributes such as

- simplicity
- safety
- high yield and selectivity
- energy efficiency
- use of renewable and recyclable reagents and raw materials

In general, chemical reactions cannot achieve all of these goals simultaneously and it is the task of chemists and chemical engineers to identify pathways that optimize the balance of desirable attributes.

Identification of environmentally preferable pathways requires creative advances in chemistry as well as process design. Because the number of choices in selecting reaction pathways is so large and the implications of those choices are so complex, systematic, quantitative design tools for identifying green chemistries are not available. Nevertheless, an extensive body of knowledge concerning green chemistry exists and some qualitative and quantitative design tools are emerging.

Green chemistry, defined as the design of chemical products and processes that reduce or eliminate the use and generation of hazardous substances, is presented in two basic parts—one qualitative and the other quantitative. The first part describes qualitative principles to be used in developing alternatives—alternative solvents, alternative reactants, alternative chemistries—that may lead to environmental improvements. Section 7.2 provides this overview of qualitative principles that can be used to identify green chemistry alternatives. Section 7.3 describes quantitative, optimization-based approaches that have been used to identify environmentally preferable reaction pathways. Finally, Section 7.4 briefly describes the US Environmental Protection Agency's Green Chemistry Expert System, which provides case studies of many of the principles described in this chapter.

7.2 GREEN CHEMISTRY METHODOLOGIES

The design of a chemical manufacturing process involves feedstock selection, selection of solvents, catalysts and other materials, and selection of reaction pathways. This section describes some of the alternatives that are available in making these design decisions and suggests a set of principles that process designers can use to identify alternatives. Specifically, the following issues are addressed:

- alternative feedstocks
- green solvents
- synthesis pathways
- inherently safer chemistry

7.2.1   Feedstocks

The synthesis and manufacture of any chemical substance begins with the selection of a starting material from which the final product will be synthesized. In many cases, the selection of a starting material can be the most significant factor in determining the impact of a chemical manufacturing process on the environment. There are a number of criteria that can be used in evaluating the potential environmental impacts of materials and these will be discussed at length in Chapter 8. For now, note that criteria that may be important in evaluating the environmental performance of a material include its persistence in the environment, its bioaccumulation, potential, its ecotoxicity and human toxicity. The scarcity of the material and whether it is a renewable or non-renewable resource may also be considered. In addition to considering the persistence, bioaccumulation, and toxicity of materials, the designer should be cognizant of the environmental impacts associated with creating the feedstock material. If a material does not pose any hazard to human health or the environment, for example, but the retrieval or isolation of the substance causes significant risk, then this should be taken into account in the selection. These issues will be addressed more completely in Chapters 13 and 14.

Setting aside, for the moment, the issue of what criteria will be used to evaluate environmental performance of materials, the question remains—How do we identify alternative raw materials that might improve environmental performance? Case studies can illustrate a variety of approaches to answering this question.

Consider the case of adipic acid manufacturing shown in Figure 7.2-1. The traditional method for adipic acid manufacture uses benzene, a fossil-fuel based, carcinogenic feedstock. The designer may wish to consider feedstocks which are renewable or less toxic. One potentially environmentally preferable alternative uses glucose, a renewable feedstock which is innocuous. Thus, the adipic acid pathway using glucose (Draths and Frost, 1998), shown in Figure 7.2-1, has some advantages. A complete evaluation, however, would need to consider the environmental issues associated with glucose and benzene production and purification (for a more detailed discussion of these types of issues, see Chapters 13 and 14).

A second example of the use of less hazardous materials is provided by the synthesis of disodium iminodiacetate as shown in Figure 7.2-2. The traditional synthesis uses hydrogen cynanide, while an alternative route using diethanolamine avoids these substances.

These few examples can be expanded into a set of more general principles and guidelines, which are described below.

Innocuous

The selection of starting materials should start with an evaluation of the materials themselves, using the methods described in Chapters 5 and 6, to determine if they possess any hazardous properties. Inherent to this analysis is determining whether the process or reaction step requiring the hazardous material is necessary, or whether the final target compound could be obtained from an alternative pathway that uses a less hazardous material.

Generates Less Waste

An important consideration associated with the use of a particular raw material is whether it is responsible for the generation of more or less waste than other materials. The amount of waste either generated or eliminated, however, cannot be the only consideration. The type of waste generated must also be assessed. Just as all chemical products are not equal in terms of their hazard, neither are chemical waste streams. Waste streams therefore must also be assessed for any hazardous properties that they possess.

Selective

Utilizing a raw material or reaction pathway that is more selective means that more of the starting material will be converted into the desired product. High product selectivity does not always translate into high product yield (and less waste generated), however. Both high selectivity and high conversion must be achieved for a

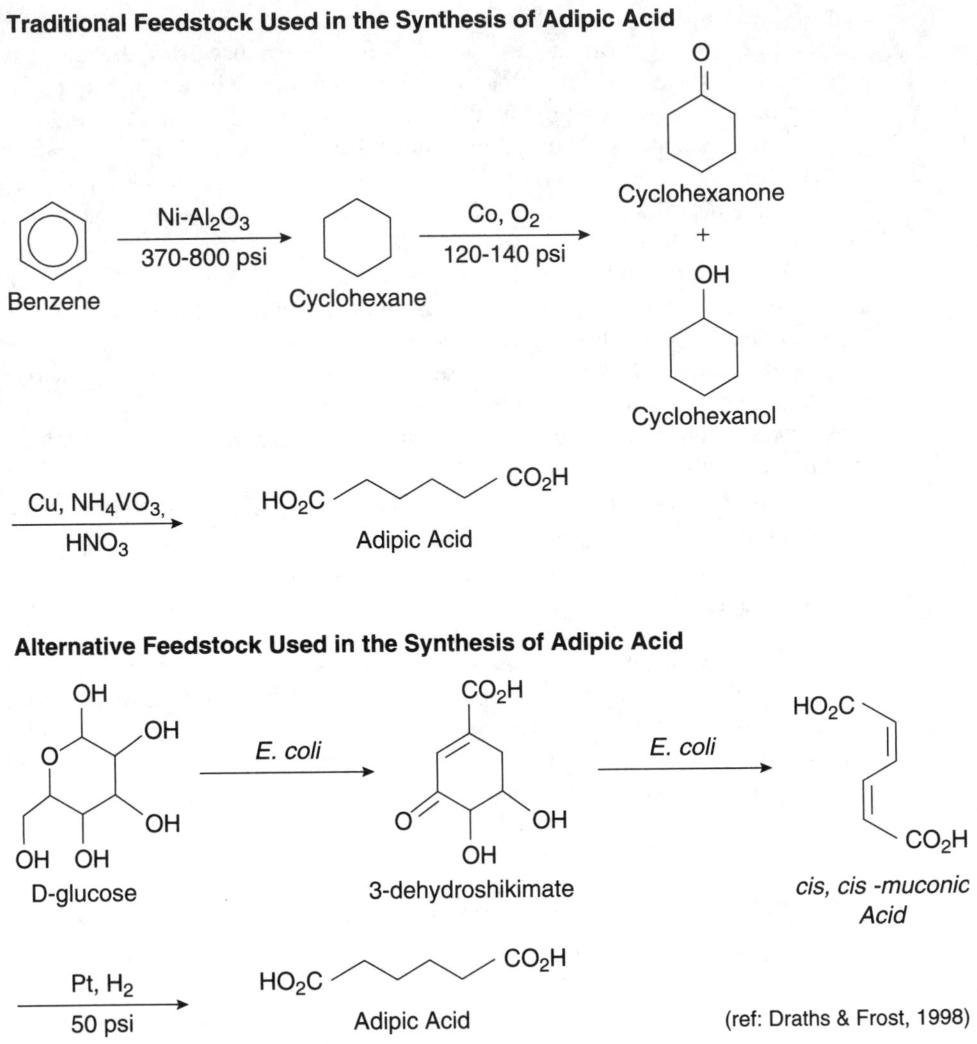

Traditional Feedstock Used in the Synthesis of Adipic Acid

Alternative Feedstock Used in the Synthesis of Adipic Acid

(ref: Draths & Frost, 1998)

Figure 7.2-1   Traditional and alternative synthesis pathways for adipic acid; the traditional pathway uses benzene, a fossil-fuel-based, carcinogenic feedstock. The alternative uses glucose, a renewable feedstock which is innocuous. (Draths and Frost, 1998.)

synthetic transformation to generate little or no waste. Using highly selective reagents can mean, however, that separation, isolation, and purification of the product will be significantly less difficult. Since a substantial portion of the burden to the environment that chemical manufacturing processes incur result from separation and purification processes, highly selective materials and reaction pathways are very desirable.

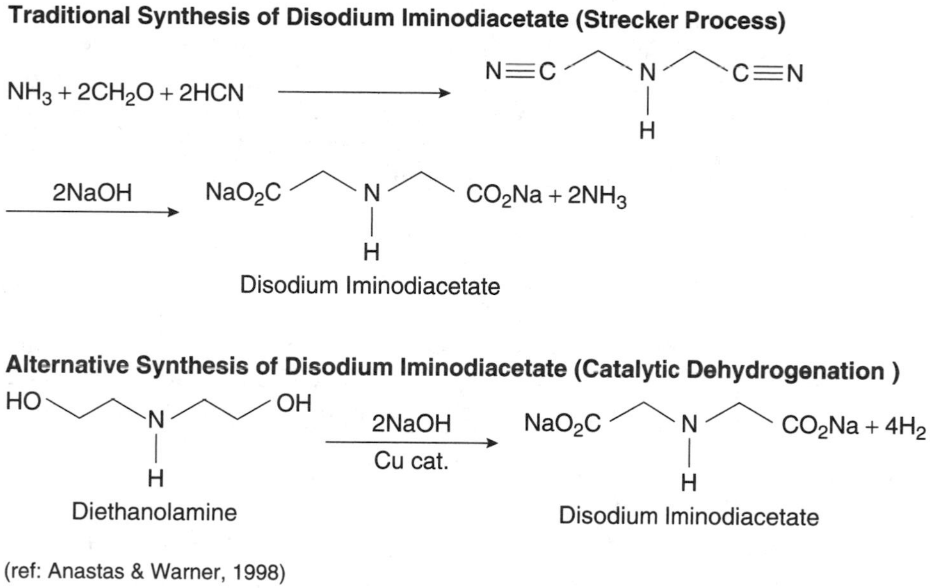

Traditional Synthesis of Disodium Iminodiacetate (Strecker Process)

$NH_3 + 2CH_2O + 2HCN$ $\longrightarrow$

Disodium Iminodiacetate

Alternative Synthesis of Disodium Iminodiacetate (Catalytic Dehydrogenation)

Diethanolamine   Disodium Iminodiacetate

(ref: Anastas & Warner, 1998)

Figure 7.2-2 Traditional and alternative synthesis pathways for disodium iminodiacetate. The traditional synthesis uses hydrogen cyanide, while an alternative route using diethanolamine avoids these substances. (Anastas and Warner, 1998.)

Efficient

Reaction efficiency, much like product selectivity, has long been a goal of synthetic design, and even prior to the advent of green chemistry principles, has offered benefits. When the overall yield of a process is increased by 10 or 20 percent, less material ends up in waste streams and more is converted into product. However, yield and selectivity are not entirely adequate as a measure of reaction efficiency. As Trost (1991) has outlined, a synthetic transformation can achieve 100 percent selectivity to product and still generate a substantial amount of waste if the transformation is not "atom economical." Atom economy, a ratio of the molecular weight of the starting materials and reagents and the molecular weight of the target molecule, provides a measure of the intrinsic efficiency of the transformation; its use is described in an example later in this chapter.

7.2.2   Solvents

The use of solvents in the chemical industry and the chemical-related industries is ubiquitous. In 1991, the production of the 25 most commonly used solvents was more than 26 million tons per year. According to Toxic Release Inventory (TRI) data, of the chemicals and chemical categories tracked by the program in 1994, 5 of the top 10 chemicals released or disposed were solvents and include methanol, toluene, xylene, methyl ethyl ketone, and dichloromethane. The total quantity of these chemicals released or disposed was over 687 million pounds, which accounts for 27 percent of the total quantity of TRI chemicals released and disposed in that year. With increasing regulatory pressure focusing on solvents, there is significant attention being paid to the use of alternatives to traditional solvents. General guiding principles in the selection of solvents are given below.

Less Hazardous

Solvents have been developed with an eye toward safety since they are used in such large volumes. The earliest and most obvious hazards that were addressed in the design of solvent molecules were their ability to explode or ignite. With the greater understanding of the health and environmental effects that could be caused by a large number of solvents, new solvents are being scrutinized for other hazards as well.

Human Health

Solvents are of particular concern because the likelihood of significant levels of exposure is high. Many solvents, by their nature, have high vapor pressure and, in combination with the volumes that are often used, can result in significant exposures. Halogenated solvents such as carbon tetrachloride, perchloroethylene, and chloroform have been implicated as potential and/or suspect carcinogens while other classes of solvents have demonstrated neurotoxicological effects. However, the direct toxicity to humans is only one aspect of the total hazards that solvents possess. There are a number of environmental implications of the use of large volumes of solvents.

Environment (Local and Global)

The use of solvents has caused both global and local environmental concerns. At the global level, the role of chlorofluorocarbons (CFCs) in stratospheric ozone depletion has led to a global phase-out of the substances from virtually all uses. Other solvents have been found to possess significant global warming potential and are thought to contribute to the overall greenhouse gas loading in the environment. At a more local level, the use of certain volatile organic compounds (VOCs) as solvents and in other applications has generated concern about their ability to elevate air pollution levels.

Case Studies

Some of the main alternatives to traditional solvents include supercritical fluids, aqueous applications, polymerized/immobilized solvents, ionic liquids, solvent-less systems and reduced hazard organic solvents. Some examples are provided below.

Consider first the use of supercritical carbon dioxide as a reaction medium. Supercritical CO_2 is non-toxic, non-flammable, renewable, and inexpensive. Further, because solubility of most solutes in supercritical fluids changes dramatically around the critical point, it is often possible to recover materials from supercritical CO_2 merely by reducing pressure below the critical point. Thus, a range of applications of supercritical CO_2, such as decaffeinating coffee (where supercritical CO_2 replaced methylene chloride), rely strictly on the physical, solvating properties of this solvent. Not all materials are soluble in supercritical CO_2, however. For materials such as high molecular weight hydrocarbons, which are not highly soluble in supercritical CO_2, the advantages of supercritical fluids can be obtained by adding a surfactant to the supercritical CO_2. Adding the right surfactant creates a micelle phase in which materials not normally soluble in supercritical CO_2 can be suspended. Figure 7.2-3 shows an example of this concept. In this case, a polymerization reaction, which produces high molecular weight materials, is conducted in a surfactant-supercritical CO_2 system, replacing the use of conventional solvents.

While in many cases solvents are used strictly for their physical, solvating properties, in some situations, such as chemical reactions occurring in solvents, the solvents play a role in the chemical synthesis. In cases where reactions occur in a solvent, the use of a supercritical fluid may enhance or inhibit the desired reaction. The effect of supercritical fluids on reaction chemistry is an active area of research, and Figure 7.2-3 shows examples of recent progress. As shown in the figure, a class of reactions, referred to as asymmetric catalytic reductions, have been conducted in supercritical CO_2. For this class of reactions, selectivity in supercritical CO_2 are comparable or superior to those achieved in conventional solvents (Burk, 1991; Burk, et al., 1993, 1995; U.S. EPA, 1996).

Consider next the use of water as an alternative solvent. As in the case of supercritical CO_2, water is non-toxic, non-flammable, renewable, and inexpensive. The limited solubility of many hydrocarbon reactants in water has often limited its utility, however. A number of case studies of innovative use of water as a reaction medium have been reported by Anastas and Williamson (1998) and by Li and Chan (1997). One example is shown in Figure 7.2-4 (Breslow and Zhu, 1995; Breslow, et al., 1996), where water with an alcohol co-solvent is used in Diels Alder reactions. Some Diels Alder reactions, such as the dimerization of 1,3 cyclopentadiene, are accelerated in water. This is due to favorable packing of hydrophobic surfaces in the reaction's transition state (Breslow, et al., 1995), a generally unanticipated result.

Many other organic reactions that have traditionally been carried out in organic solvents have now been carried out in aqueous media. Examples include the Barbier-Grignard reaction, pericyclic reactions, and transition metal catalyzed reactions (for a thorough discussion, see Li, 1998).

As a final case study, consider the use of derivatized, immobilized solvent materials. The concept behind this solvent replacement is to reduce the emissions and promote the recovery of hazardous solvents by attaching the solvent to a hydrocarbon backbone. The use of this concept in the replacement of tetrahydrofuran (THF) is shown in Figure 7.2-5 (Hurter and Hatton, 1992; U.S. EPA, 1996). In this case, the hazardous substance, THF, is attached to a polymeric backbone using a chlorinated styrene derivative. The THF remains relatively mobile, but because it

Catalytic Copolymerization of CO_2 with Epoxides

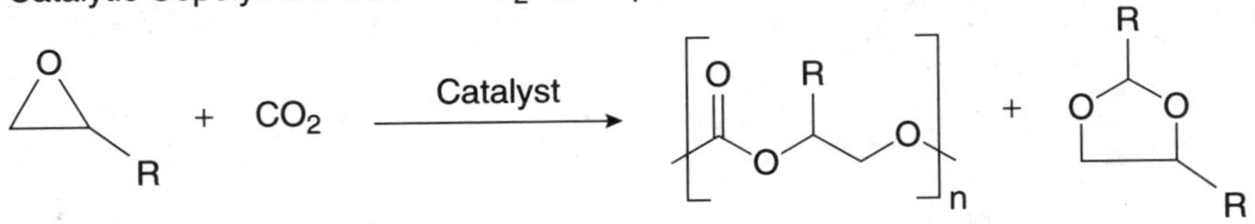

Assymetric Catalytic Hydrogenation of Enamides

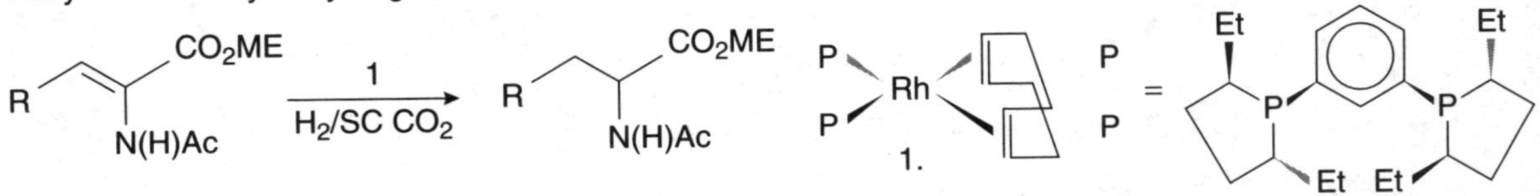

(ref: Buelow, et al., 1998)

Figure 7.2-3 Examples of the use of supercritical CO_2 to replace conventional solvents; in the first example, a co-solvent is used along with the supercritical CO_2 to allow a polymerization reaction to take place; in the second example, the use of supercritical CO_2 enhances the selectivity of a catalytic reaction (Buelow, et al., 1998).

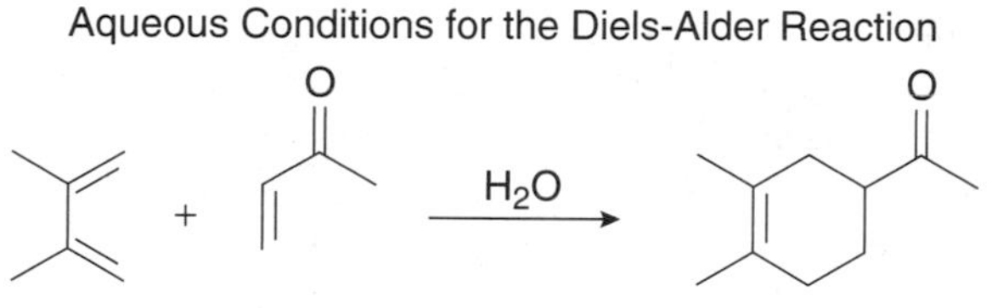

Aqueous Conditions for the Diels-Alder Reaction

Figure 7.2-4 When water is used as a solvent for certain Diels Alder reactions, reaction rates can be accelerated.

is attached to a polymeric backbone, it is less likely to volatilize and is easily recoverable using ultrafiltration or other methods.

These limited examples provide some indication of the variety of alternatives available for conventional solvents. More examples are described in the volume edited by Anastas and Williamson (1998) and in the Green Chemistry Expert System available from the US EPA.

7.2.3   Synthesis Pathways

Identifying chemical synthesis pathways that may lead to superior environmental performance is complex and relies on extensive knowledge of synthetic organic chemistry. It is beyond the scope of this chapter to provide a detailed review of this rapidly advancing field, but it is important for a process engineer to be able to identify classes of chemical reactions that have the potential for improvement. Addition reactions (A + B → AB), substitution reactions (AB + C → AC + B), and

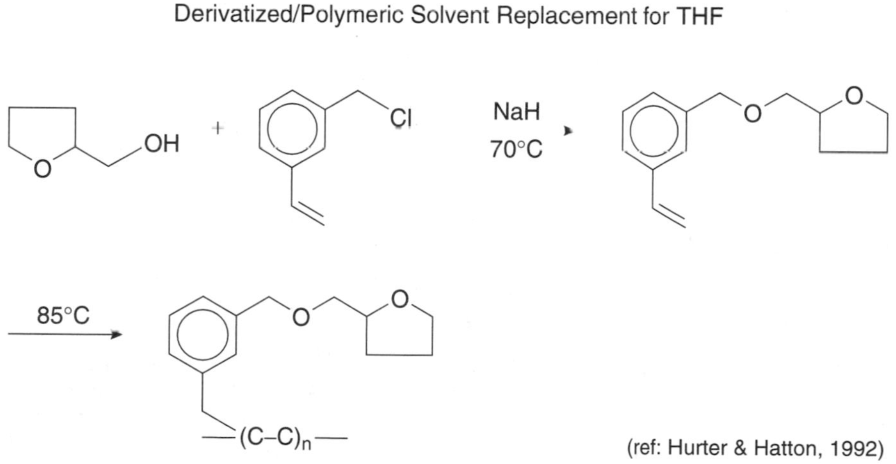

Derivatized/Polymeric Solvent Replacement for THF

(ref: Hurter & Hatton, 1992)

Figure 7.2-5   Use of a solvent functionality (tetrahydrofuran), attached to a large polymer backbone, to replace a volatile solvent (tetrahydrofuran) (Hurter and Hatton, 1992).

elimination reactions (AB → A + B), for example, can have different degrees of impact on human health and the environment. Addition reactions incorporate the starting materials into the final product and, therefore, do not produce waste that needs to be treated, disposed of, or otherwise dealt with. Substitution reactions, on the other hand, necessarily generate stoichiometric quantities of substances as byproducts and waste. Elimination reactions do not require input of materials during the course of the reaction other than the initial input of a starting material, but they do generate stoichiometric quantities of substances that are not part of the final target molecule. This guidance is qualitative. A semi-quantitative tool that a process engineer or chemist can use in evaluating synthetic pathways is the concept of atom efficiency. The atom efficiency characterizes the fraction of starting materials that are incorporated into desired products and is best illustrated through an example.

Friedel Crafts acylations have atom efficiencies that are relatively low (Clark, 1999). The process typically involves the substitution reaction of an acid chloride with an aromatic substrate. The reaction is frequently accomplished using an aluminum chloride catalyst. The product forms a complex with the catalyst, requiring a water wash, resulting in the formation of hydrochloric acid and salt wastes. The overall reaction is shown in Figure 7.2-6.

A simplistic overall atom and mass balance, outlined in Example 7.2-1 (Clark, 1999), suggests that only about 30% of the starting material ends up in the product. Thus, a simple calculation of atom efficiency identifies Friedel Crafts reactions as a potential target for environmental improvements. One type of improvement would be to retain the chemical pathway, but to regenerate and reuse the aluminum chloride catalyst. In the synthesis of ethylbenzene via the alkylation of benzene with ethylene, for example, approximately 1 ton of $AlCl_3$ waste is generated per

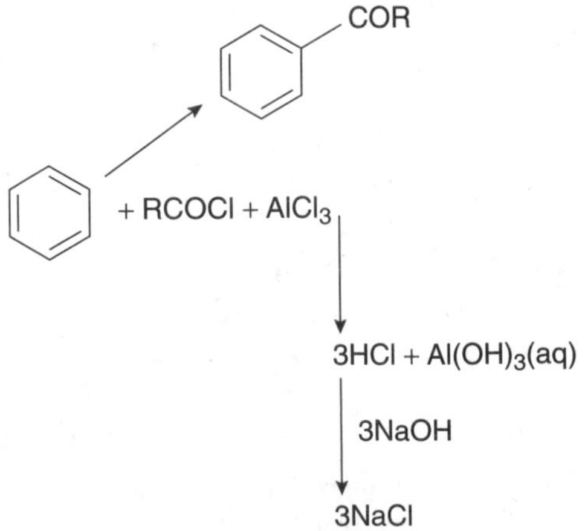

Figure 7.2-6 Friedel Crafts acylation generates a relatively large amount of waste material, even if the reaction is carried out at 100% yield and 100% selectivity, because of the dissipative use of the aluminum chloride catalyst.

100 tons of product (Davis, 1994), a significant reduction in waste generation relative to the case of no catalyst recovery. Another alternative for improving environmental performance would be to identify an alternative catalyst, such as highly
acidic zeolites (Davis, 1994).

Example 7.2-1

Calculate atom and mass efficiencies for the Friedel Crafts reaction shown in Figure
7.2-5. Assume that the substituent R on the organic chloride is a methyl group.

Solution: To calculate the atom efficiencies, determine the fraction of the carbon,
hydrogen, aluminum, chlorine, sodium, and oxygen atoms that emerge as product and
the fraction that emerges from the reaction as waste.

- Virtually all of carbon (100%) becomes product.
- Most of hydrogen (excluding water in the water wash) becomes product; however,
  if hydrogen used in the water wash is included, virtually all of the hydrogen be-
  comes waste.
- All of the aluminum becomes waste (0% efficiency).
- All of the chlorine becomes waste (0% efficiency).
- All of the sodium becomes waste (0% efficiency).
- One mole of oxygen in the organic chloride is incorporated into the product.
  Three moles of oxygen in the sodium hydroxide becomes waste. Therefore, ex-
  cluding oxygen in water, the atom efficiency is 25%.

The mass efficiency can be calculated using atomic weights (ignoring water use).

Mass in product = 8 moles carbon * 12 + 10 moles hydrogen * 1 + 1 mole oxygen * 16 = 122

Mass input = 8 moles carbon * 12 + 16 moles H * 1 + 4 moles oxygen * 16 + 1 mole aluminum
* 27 + 3 moles chlorine * 35.5 + 3 moles sodium * 23 = 378

Approximate mass efficiency = 122/378 = 0.32

Partial oxidations provide additional examples of industrially important reactions where atom utilization can be low. Figure 7.2-7 shows a number of commercially important partial oxidation reactions.

In the case of partial oxidations the poor atom utilization can be due to the
oxidizing agent. If molecular oxygen is used in the partial oxidation, then atom utilization may be high if selectivity is high. If, however, oxidizing agents such as
dichromate or permanganate are used, atom efficiencies can be low, as shown in
Table 7.2-1. Environmental performance might be improved in these cases by carefully selecting the oxidizing agents.

Wastes can sometimes be reduced by simplifying synthesis pathways. Consider the synthesis of Ibuprofen, an over-the-counter pain reliever. The traditional
synthesis method (note that it involves Friedel-Crafts chemistry in the first step) is
shown in Figure 7.2-8. An alternative synthesis, replacing the AlCl₃ acid catalyst

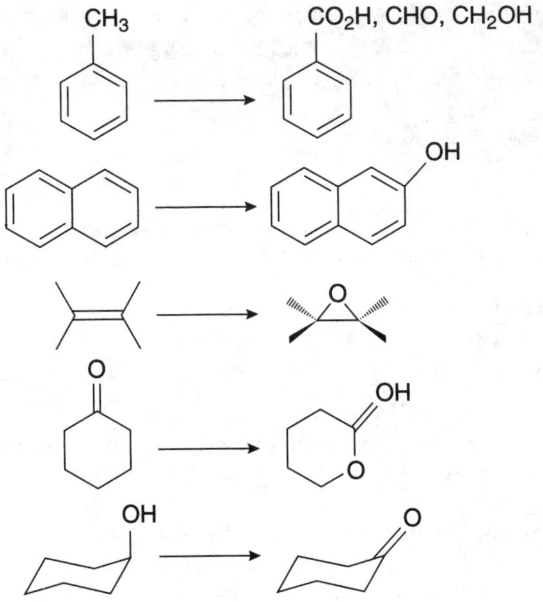

Figure 7.2-7 Commercially important partial oxidation reactions (Clark, 1999).

with HF and reducing the number of subsequent transformations and solvent usage, is also shown in the figure. Atom utilization increased from less than 40% in the traditional synthesis to approximately 80% using the new pathway (US EPA, 1998).

These few brief examples—Friedel Crafts reactions, partial oxidations, and reducing the number of steps in synthesis pathways—only scratch the surface of the rich variety of work in synthetic organic chemistry that can improve the

Table 7.2-1 Atom Utilization in Oxidizing Reagents (from Davis, 1994, adapted from Sheldon, 1993).

Molecule used as oxidizing agent in partial oxidation reaction	wt % active oxygen	Byproduct
MnO_2	18.4	MnO
PhIO	7.3	PhI
H_2O_2	47.0	H_2O
t-BuOOH	17.8	BuOH
NaOCl	21.6	NaCl
$K_2Cr_2O_7$	21.8	Cr_2O_3
$KMnO_4$	20.2	MnO_2

Traditional Synthesis of Ibuprofen

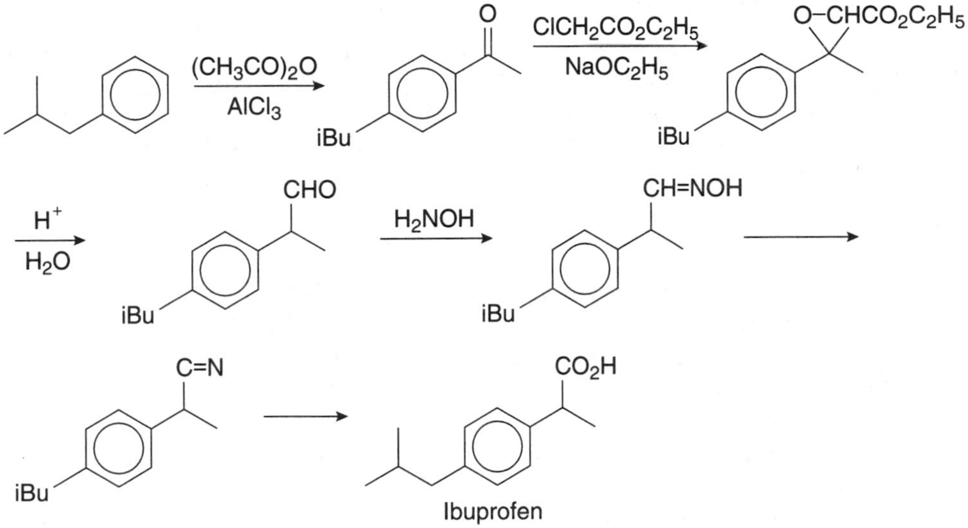

Alternative Synthesis of Ibuprofen

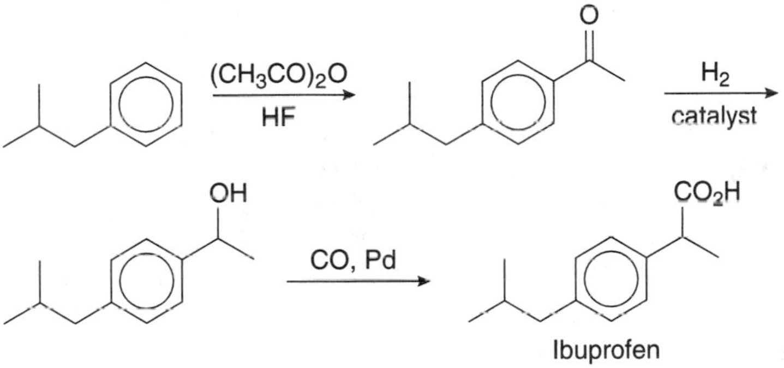

(ref: U.S. EPA, 1998)

Figure 7.2-8 The traditional synthesis of Ibuprofen involved a large number of steps, including a Friedel Crafts reaction that generated byproducts of the type shown in Figure 7.2-8. The new route is simpler and employs a recoverable strong acid as the catalyst. (U.S. EPA, 1998.)

performance of chemical manufacturing. In examining reaction pathways, the process engineer should employ tools, such as calculating atom and mass efficiencies, that help determine the magnitude of environmental improvements that may be possible. Additional methods for assessing reaction pathways that may lead to critical environmental improvements are described in Chapter 8. Later sections in this chapter describe some tools and analysis methods that can help to identify alternative chemical reagents and pathways.

7.2.4 Functional Group Approaches to Green Chemistry

A number of tools can be used in the design of more environmentally benign chemistries, including structure-activity relationships, identification and avoidance of toxic functional groups, reducing bioavailability, and designing chemicals for innocuous fate. These concepts rely on the same functional group principles introduced in Chapter 5, and are described qualitatively below.

Structure-Activity Relationship

Many times the mechanism of action may not be known, but structure-activity relationships can be used to identify structural modifications that may improve a chemical's safety. As an example, if the methyl-substituted analog of a substance has very high toxicity, and the toxicity decreases as the substitution moves from ethyl to propyl, it might be reasonable to increase the alkyl chain length to design a safer chemical. Even in cases where the reason for the effect the alkyl chain length has on decreasing toxicity is not known, if the results can ultimately be borne out empirically, then the structure-activity relationship is certainly a powerful design tool. Chapters 5 and 6 describe in detail the use of structure-activity relationships in evaluating a chemical's environmental fate, bioaccumulation, and toxicity.

Elimination of Toxic Functional Group

A class of chemicals is often defined by certain structural features, such as aldehyde, ketone, nitrile, or isocyante functional groups. If information is not available about the specific chemical's toxicity or the mechanism by which it produces that toxicity, the assumption that certain reactive functional groups will react similarly within the body or in the environment is often a good one. The assumption is especially good if there are data on other compounds in the chemical class that demonstrate a common toxic effect.

In cases such as this, the design of a safer chemical could proceed by removing the toxic functionality, which defines the class. In some cases this is not possible because the functionality is what gives the molecule the properties that are required for the chemical to perform in the desired way. In these cases, there are still options such as masking the functional group to a non-toxic derivative form and only releasing the parent functionality when necessary.

The masking of vinyl sulfones provides an interesting example of this technique. The vinyl sulfone functionality is highly electrophilic and reacts with cellulosic fibers, making it an effective component of dyes. There are, however, a variety of toxic effects associated with this functionality (DeVito, 1996). The sulfones can be made safer by masking the functional group. Rather than manufacturing, storing, and transporting the relatively hazardous sulfone, the sulfone can be generated when and as needed by converting a hydroxyethylsulfone into a vinyl sulfone, using the chemistry shown in Figure 7.2-9 (DeVito, 1996).

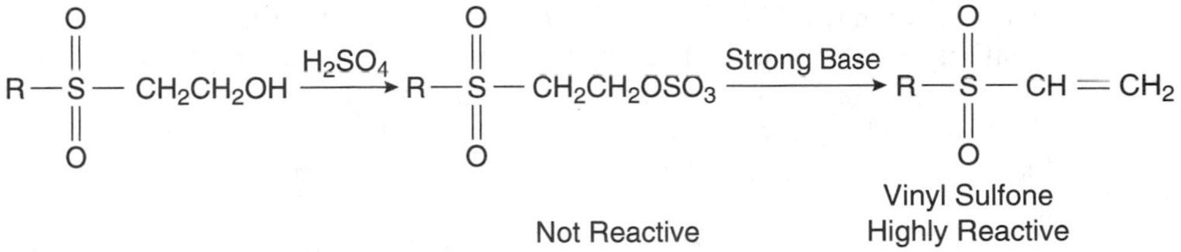

Figure 7.2-9 Masking of vinyl sulfones as hydroxyethylsulfones allows these relatively hazardous materials to be used more safely.

Reduce Bioavailability

If it is not known what structural features of the molecule need to be modified in order to make it less hazardous, then there is still the option of making the substance less bioavailable. If the substance is unable, due to structural design, to reach the target of toxicity, then it is in effect, innocuous. This can be done through a manipulation of the water-solubility/lipophilicity relationships that often control the ability of a substance to pass through biological membranes such as skin, lungs, or the gastrointestinal tract (see Chapter 6). The same principle applies to designing safer chemicals for the environment such as ozone depleting substances. For a substance to have a significant ozone depleting potential, it must be able to both reach the altitudes and have a sufficient lifetime in those altitudes in order to cause damage. Many substances are now being designed which have the same properties as substances which are known ozone depleters but without the ability to be available to the target of the hazard, in this case the stratospheric ozone layer.

Design for Innocuous Fate

It was often the goal of the chemist to design substances which were robust and could last as long as possible. This philosophy has resulted in persistent, and at times bioaccumulative and toxic substances. It is now known that it is more desirable to not have substances persist in the environment or a landfill forever, but that they should be designed to degrade after their useful life is over. Therefore, the design of safer chemicals cannot be limited to only hazards associated with the manufacture and use of the chemical but also that of its disposal and ultimate end of life cycle.

7.3   QUANTITATIVE/OPTIMIZATION-BASED FRAMEWORKS FOR THE DESIGN OF GREEN CHEMICAL SYNTHESIS PATHWAYS

One of the challenges associated with the design of green chemical synthesis pathways is identifying alternatives. Section 7.2 provided general guidelines and suggestions for improving the environmental performance of raw materials, reagents, and synthesis pathways. While useful, these guidelines still rely on the knowledge and creativity of chemists and chemical engineers in identifying specific alternatives. Identifying all possible alternatives is beyond the knowledge or experience of any single individual, so an increasingly popular method for rapidly searching for possible alternative materials or chemical synthesis routes relies on combinatorial approaches.

In a combinatorial approach to identifying green chemistry alternatives, the first step is to select a set of molecular or functional group building blocks from which a target molecule can be constructed. Next, a series of stoichiometric, thermodynamic, economic, and other constraints can be identified. These constraints serve to reduce the number of possibilities that might be considered. Finally, a set of criteria can be used to identify reaction pathways that deserve further examination.

The first step in the systematic construction of alternative chemical pathways is to select a set of functional group building blocks. Because the number and variety of pathways that are generated is a strong function of the starting materials that are used, this is a critical step. To keep the alternatives as varied as possible, it is desirable to include as many functional group building blocks as possible, yet to keep the search focused and tractable, the number of groups should be limited. Buxton, et al. (1997) have reported a number of rules and guidelines that can be effective in selecting a group of starting materials. They are:

- Include the groups present in the product.
- Include groups present in any existing industrial raw materials, coproducts or byproducts.
- Include groups which provide the basic building blocks for the functionalities of the product or of similar functionalities.
- Select sets of groups associated with the general chemical pathway employed (cyclic, acyclic, or aromatic).
- Reject groups that violate property restrictions.

These rules can be clarified through an example. Consider the synthesis of 1-naphthyl-methylcarbamate (Crabtree and El-Halwagi, 1995), manufactured by Union Carbide and known as carbaryl. In 1984, a catastrophic release of methylisocyanate, a reactant used in the synthesis of carbaryl, occurred at a carbaryl manufacturing facility in Bhopal, India, killing thousands. This incident, and other less catastrophic events, demonstrate the importance of identifying reaction pathways that minimize the use of hazardous materials. For the carbaryl synthesis used in 1984 in Bhopal, α-naphthol and methylisocyanate were used as reactants, as shown in Figure 7.3-1. An alternative chemistry for carbaryl is also shown.

Are other chemistries for the synthesis of carbaryl possible or desirable? The first step in identifying alternative pathways is to select a set of functional group building blocks that will be included in the analysis. Since the product molecule contains aromatic groups, it will be necessary to include a range of aromatic

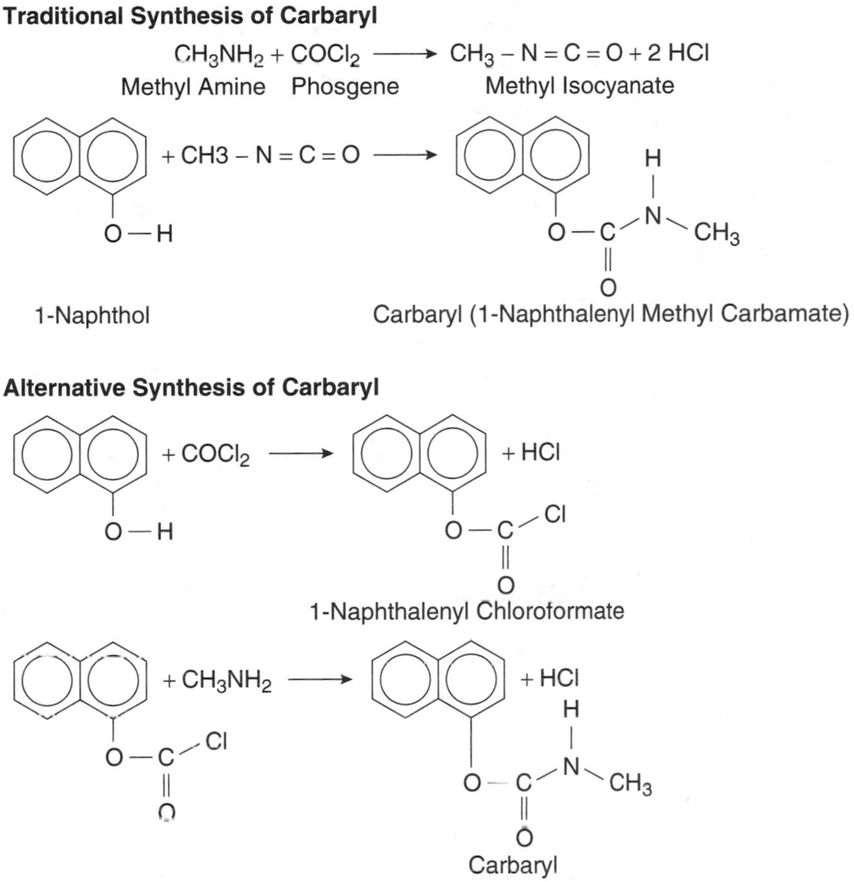

Traditional Synthesis of Carbaryl

$$CH_3NH_2 + COCl_2 \longrightarrow CH_3 - N = C = O + 2\ HCl$$

Methyl Amine   Phosgene   Methyl Isocyanate

1-Naphthol   Carbaryl (1-Naphthalenyl Methyl Carbamate)

Alternative Synthesis of Carbaryl

1-Naphthalenyl Chloroformate

Carbaryl

Figure 7.3-1   The synthesis of carbaryl can be accomplished with a methylisocyanate intermediate. An alternative route, not involving methylisocyanate, is also shown.

functionalities such as aromatic carbon bound to hydrogen (ACH in the notation of Chapter 5), aromatic carbon bound to other aromatic carbon (AC−), aromatic carbon bound to chlorine, and aromatic carbon bound to a hydroxl group (ACOH). More aromatic functionalities could be chosen, if desired. Other groups appearing in the product molecule, or related to the groups appearing in the product molecule, are (using the notation of Chapter 5) −CH₃, CH₃NH<, CH₃NH₂−, −COO−, −CHO, −CO₂H, −OH, −Cl.

These functional group building blocks can be used to identify a set of potential molecular reactants. Going from a set of functional group building blocks to potential molecular starting materials can generate very large numbers of potential reactants, so constraints, based on chemical intuition, are generally imposed. For example, in identifying alternatives to the carbaryl synthesis, Buxton, et al. (1997) assumed that

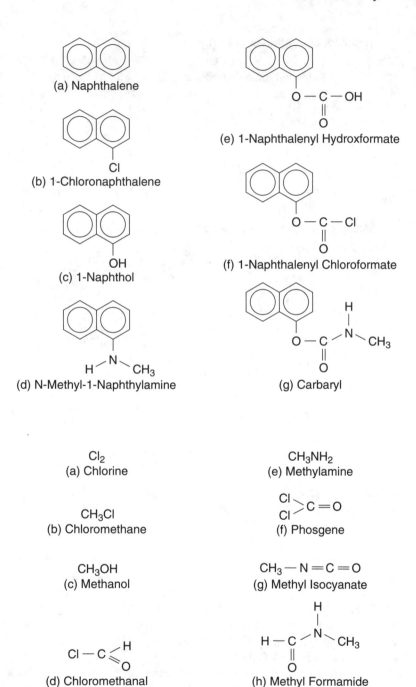

Figure 7.3-2 Potential reactants in carbaryl synthesis identified by Buxton, et al. (1997).

only monosubstituted aromatic molecules would be used, since the product is mono-substituted; they also assumed that reactants for which the carbon skeleton would need to be altered would not be used (for example, benzene would not be used as a reactant since forming the product would require a ring condensation reaction). Using these and other assumptions, a limited set of reactants can be identified. Potential reactants identified by Buxton, et al. (1997) are shown in Figure 7.3-2.

Once a set of potential reactants has been selected, a set of rules and constraints must be applied to describe how the reactants can interact to form molecules. The most obvious of these constraints are stoichiometric. For example, the product molecule contains 7 aromatic carbons bound to hydrogen and two aromatic carbons bound to other aromatic carbon. Thus, the reactants must provide sufficient aromatic carbons, of various types, to generate the product molecule. Similar stoichiometric constraints could be written for the other types of groups in the molecule. Some reaction pathway analysis methods assume that reactions, appropriately balanced for stoichiometry, can proceed with 100% selectivity and yield. Other methods include thermodynamic constraints on selectivity (see, for example, Crabtree and El-Halwagi, 1995).

Once constraints are established, pathways can be identified and ranked. Ranking schemes might include cost and environmental performance metrics. Buxton, et al. (1997) identified and ranked 13 different reaction pathways for the synthesis of carbaryl. The results are shown in Table 7.3-1. The economic ranking is based on the price differential between product and reactants. The environmental ranking is based on the assumption that a fixed percentage of the materials used is released to the environment.

While the results of Table 7.3-1 are intriguing, it would be inappropriate to suggest that this type of analysis will yield the optimal reaction pathway. Rather, the point of these analysis methods is to inject systematic decision rules into the search for alternative pathways. Sets of starting materials are identified based on stoichiometry and chemical intuition. Then, pathways can be identified and potential upper bounds for selectivity can be estimated using thermodynamics. Finally, alternatives can be quickly ranked using economic and environmental criteria. These systematic procedures may lead to a desirable alternative pathway, or they may merely lead to a clear definition of the constraints that should be considered in evaluating alternative pathways.

7.4 GREEN CHEMISTRY EXPERT SYSTEM CASE STUDIES

The concepts and examples described in this chapter provide only an introduction to the concepts of green chemistry. A mechanism for further exploring this area is the Green Chemistry Expert System (GCES), which is downloadable from the US EPA website (http://www.epa.gov/greenchemistry/tools.htm). This software provides more depth on many of the concepts and tools presented in this chapter; it also provides a searchable literature database on green chemistry.

Table 7.3-1 Alternative Pathways for the Synthesis of Carbaryl (Buxton, et al., 1997). The species numbers are listed below the Table and refer to the compounds shown in Figure 7.3-2; the profit is the difference in value between reactants and products and the environmental ranking is determined by assuming that a fixed fraction of the reactants and product are released to the environment.

1	2	3	4	5	6	7	8	9	10	11	12	13	14	15	16	17	18	19	Profit	Env. Rank
-1								-1	1	1	-2								1.45	9
	1	1		-1	-1					1		-1							1.03	7
		2		-1	-1					1		-1							1.00	2
		1		-1						1		-1				1	-1		1.00	12
-1				-1	1					1		-1							1.00	1
		1		-1				-1		1		-1							0.976	13
				-1						1		-1				1		-1	0.967	4
	1				-1			-1		1		-1							0.952	8
					-1					1		-1				1		-1	0.952	11
		2		-1	-1			-1		1									0.604	5
	1	1			-1			-1		1								-1	0.543	6
1				-1				-1		1									0.503	3
								-1	-1	1									0.451	10

1=Oxygen; 2=Hydrogen; 3=Hydrogen chloride; 4=Naphthol chloroformate; 5=Methyl formamide; 6=Water; 7=Methylamine; 8=Phosgene; 9=Methyl isocyanate; 10=Naphthol; 11=Carbaryl; 12=Naphthalene; 13=Chloronaphthalene; 14=methyl-naphthylamine; 15=Naphthenyl hydroxyformate; 16=Chlorine; 17=Chloromethane; 18=Methanol; 19=Chloromethanal

If, for example, the concept of supercritical solvents were to be explored in more detail, a search of the GCES online database would provide literature citations for a number of case studies. More detail on alternative pathways for partial oxidation reactions or Friedel Crafts reactions could be found. A database of solvents could be searched, or the design of inherently safer chemicals could be explored. Some of the problems at the end of this chapter involve using the GCES to explore green chemistry alternatives.

QUESTIONS FOR DISCUSSION

1. Suggest quantitative metrics that could be used to rank the impact that a synthesis method would have on the environment. How do your suggestions compare to those in Chapter 8?

2. What are the strengths and weaknesses of combinatorical approaches to identifying reaction pathways?

REFERENCES

Anastas, P.T. and Williamson, T.C., eds., *Green Chemistry: Frontiers in Benign Chemical Syntheses and Processes,* Oxford University Press, New York, 1998.

Anastas, P.T. and Warner, J.C., *Green Chemistry: Theory and Practice*, Oxford University Press, New York, 1998.

Breslow, R., Connors, R. and Zhu, Z., "Mechanistic studies using antihydrophobic enantio-selective hydrogenation reactions," *Pure Appl. Chem.,* 68, 1527–33.

Breslow, R. and Zhu, Z., "Quantitative antihydrophobic effects as probes for transition state structures. 2. Diels Alder reactions," *Journal of American Chemical Society,* 117, 9923–4 (1995).

Burk, M.J., "C₂-symmetric bis(phospholanes) and their use in highly enantioselective hydrogenation reactions," *Journal of American Chemical Society,* 113, 8518–8519 (1991).

Burk, M.J., Feaster, J.E., Nugent, W.A. and Harlow, R.L., "Preparation and use of C₂-symmetric bis(phospholanes): Production of α-amino acid derivatives," *Journal of American Chemical Society,* 115, 10125–10138 (1993).

Burk, M.J., Gregory, T., Harper, P., and Kalberg, C.S., "Highly enantioselective hydrogenation of β-keto esters under mild conditions," *Journal of American Chemical Society,* 117, 4423–4424 (1995).

Buelow, S., Dell-Orco, P., Morita, D., Pesiri, D., Birnbaum, E., Borkowsky, S., Brown, G., Feng, S., Luan, L., Morganstern, D., and Tumas, W., "Recent advances in chemistry and chemical processing in dense phase carbon dioxide at Los Alamos," in *Green Chemistry: Frontiers in Benign Chemical Syntheses and Processes,* P.T. Anastas and T.C Williamson, eds., Oxford University Press, New York, 1998 (pp. 265–285).

Buxton, A., Livingston, A.G., and Pistikopoulos, E.N. "Reaction Path Synthesis for Environmental Impact Minimization," *Computers in Chemical Engineering,* 21, S959–964 (1997).

Clark, J.H., "Green Chemistry: challenges and opportunities," *Green Chemistry,* 1 (1), 1–11, 1999.

Crabtree, E.W. and El-Halwagi, M.M., "Synthesis of Environmentally Acceptable Reactions," in *Pollution Prevention via Process and Product Modifications,* M.M. El-Halwagi and D.P. Petrides, eds., *AIChE Symposium Series,* 90(303), 117–127, 1995.

Dartt, C.B. and Davis, M.E., "Catalysis for Environmentally Benign Processing," *Industrial and Engineering Chemistry Research,* 33, 2887–2899 (1994).

DeVito, S.C., "General Principles for the Design of Safer Chemicals: Toxicological Considerations for Chemists," in *Designing Safer Chemicals,* S.C. DeVito and R.L. Garrett, eds., ACS Symposium Series 640, American Chemical Society, Washington, D.C. (1996).

Draths, K.M. and Frost, J.W., "Improving the environment through process changes and product substitutions," *Green Chemistry: Frontiers in Benign Chemical Syntheses and Processes,* P.T. Anastas and T.C Williamson, eds., Oxford University Press, New York, 1998 (pp. 150–165).

Hurter, P.N. and Hatton, T.A., *Langmuir,* 8, 1291–9 (1992).

Li, C.-J., "Water as a benign solvent for chemical synthesis," *Green Chemistry: Frontiers in Benign Chemical Syntheses and Processes,* P.T. Anastas and T.C Williamson, eds., Oxford University Press, New York, 1998 (pp. 234–249).

Li, C.-J. and Chan, T.H., *Organic Reactions in Aqueous Media,* Wiley, New York, 1997.

US Environmental Protection Agency, The Presidential Green Chemistry Challenge Awards, Summary of 1996 Award Entries and Recipients. EPA 744-K-96-001 (July, 1996).

US Environmental Protection Agency, The Presidential Green Chemistry Challenge Awards, Summary of 1997 Award Entries and Recipients. EPA 744-S-97-001 (April, 1998).

Trost, B.M., "The Atom Economy—A Search for Synthetic Efficiency," *Science,* 254, 1471–7 (1991).

PROBLEMS

1. The text noted that the atom and mass efficiencies for addition reactions are generally higher than for substitution or elimination reactions. To illustrate this concept, calculate the mass and atom efficiencies for the following reactions:

   (a) Addition reaction

   Isobutylene + methanol → methyl,tert-butyl ether
   $$C_4H_8 + CH_3OH \rightarrow (C_4H_9)-O-CH_3$$

   Calculate mass, carbon, hydrogen and oxygen efficiencies.

   (b) Substitution reaction

   Phenol + ammonia → aniline + water
   $$C_6H_5-OH + NH_3 \rightarrow C_6H_5-NH_2 + H_2O$$

   Calculate mass, carbon, hydrogen, nitrogen, and oxygen efficiencies.

   (c) Elimination reaction

   Ethylbenzene → styrene + hydrogen
   $$C_6H_5-C_2H_5 \rightarrow C_6H_5-C_2H_3 + H_2$$

   Calculate mass, carbon, and hydrogen efficiencies.

   (d) Identify additional industrially significant examples of addition, substitution, and elimination reactions; calculate atom and mass efficiencies for these reactions.

2. In Table 7.2-1, atom economies are presented for a variety of oxidation agents. Confirm these calculations.

3. Use the US EPA's Green Chemistry Expert System to identify a new solvent replacement technology. Review the original scientific literature on the technique and write a one-page summary of the new technology and the solvent it replaces.

4. Use the US EPA's Green Chemistry Expert System to identify a new chemical synthesis method. Review the original scientific literature on the chemistry and write a one-page summary of the new reaction pathway and the pathway it replaces.

CHAPTER 8

Evaluating Environmental Performance During Process Synthesis

by
David T. Allen, David R. Shonnard, and Scott Prothero

The design of chemical processes proceeds through a series of steps, beginning with the specification of the input-output structure of the process and concluding with a fully specified flowsheet. Traditionally, environmental performance has only been evaluated at the final design stages, when the process is fully specified. This chapter presents methodologies that can be employed at a variety of stages in the design process, allowing the process engineer more flexibility in choosing design options that improve environmental performance.

8.1 INTRODUCTION

The search for "greener chemistry," described in the previous chapter, can lead to many exciting developments. New, simpler synthesis pathways could be discovered for complex chemical products, resulting in a process that generates less toxic byproducts and lowers the overall risk associated with the process. Toxic intermediates used in the synthesis of commodity chemicals might be eliminated. Benign solvents might replace more environmentally hazardous materials. However, these developments will involve new chemical processes as well as Green Chemistry.

The art and craft of creating chemical processes is the topic of a number of excellent textbooks (see, for example, Douglas, 1988). A fundamental theme that arises in each of these texts is that the design process proceeds through a series of steps, each involving an evaluation of the process performance. At the earliest stages of a design, only the most basic features of a process are proposed. These include the raw materials and chemical pathway to be used, as well as the overall material balances for the major products, byproducts, and raw materials. Large numbers of design alternatives are screened at this early design stage, and the screening tools used to evaluate the alternatives must be able to handle efficiently large numbers of alternative design concepts. As design concepts are screened, a select few might merit further study. Preliminary designs for the major pieces of equipment to be used in the process need to be specified for the design options that merit further study. Material flows for both major and minor byproducts are estimated. Rough emission estimates, based on analogous processes, might be considered. At this development stage, where fewer design alternatives are considered, more effort can be expended in evaluating each design alternative, and more information is available to perform the evaluation. If a design alternative appears attractive at this stage, a small-scale pilot plant of the process might be constructed and a detailed process flow sheet for a full-scale process might be constructed. Very few new design ideas reach this stage, and the investments made in evaluating design alternatives at this level are substantial. Therefore, process evaluation and screening tools can be quite sophisticated.

Traditionally, evaluations of environmental performance have been restricted to the last stages of this engineering design process, when most of the critical design decisions have already been made. A better approach would be to evaluate environmental performance at each step in the design process. This would require, however, a hierarchy of tools for evaluating environmental performance. Tools that can be efficiently applied to large numbers of alternatives, using limited information, are necessary for evaluating environmental performance at the earliest design stages. More detailed tools could be employed at the development stages, where potential emissions and wastes have been identified. Finally, detailed environmental impact assessments would be performed as a process nears implementation.

This chapter and Chapter 11 present a hierarchy of tools for evaluating the environmental performance of chemical processes. Three tiers of environmental performance tools will be presented. The first tier of tools, presented in Section 8.2, is appropriate for situations where only chemical structures and the input-output structure of a process is known. Section 8.3 describes a second tier of tools which is appropriate for evaluating the environmental performance of preliminary process designs. This tier includes tools for estimating wastes and emissions. Finally, Section 8.4 introduces methods for the detailed evaluation of flowsheet alternatives, which will be discussed in Chapter 11.

8.2   TIER 1 ENVIRONMENTAL PERFORMANCE TOOLS

At the earliest stages of a process design, only the most elementary data on raw materials, products, and byproducts of a chemical process may be available and large numbers of design alternatives may need to be considered. Evaluation methods, including environmental performance evaluations, must be rapid, relatively simple, and must rely on the simplest of process material flows. This section describes methods for performing environmental evaluations at this level.

8.2.1   Economic Criteria

As a simple example, consider two alternative processes for the manufacture of methyl methacrylate. Billions of pounds of methyl methacrylate are manufactured annually. Methyl methacrylate can be manufactured through an acetone-cyanohydrin pathway:

$$(CH_3)_2\,C{=}O + HCN \rightarrow HO{-}C(CH_3)_2{-}CN$$

(Acetone + hydrogen cyanide → acetone cyanohydrin)

$$HO{-}C(CH_3)_2{-}CN + H_2SO_4 \rightarrow CH_3{-}(C{=}CH_2){-}(C{=}O){-}NH_2(H_2SO_4)$$

(acetone cyanohydrin → methacrylamide sulfate)

The methacrylamide sulfate is then cracked, forming methacrylic acid and methyl-methacrylate:

$$CH_3{-}(C{=}CH_2){-}(C{=}O){-}NH_2(H_2SO_4) + CH_3OH \rightarrow CH_3{-}(C{=}CH_2){-}(C{=}O){-}OH$$
$$\rightarrow CH_3{-}(C{=}CH_2){-}(C{=}O){-}O{-}CH_3$$

Alternatively, methyl methacrylate can be manufactured with isobutylene and oxygen as raw materials.

$$CH_3{-}(C{=}CH_2){-}CH_3 + O_2 \rightarrow CH_3{-}(C{=}CH_2){-}(C{=}O)H + H_2O$$

isobutylene + oxygen → methacrolein

$$CH_3{-}(C{=}CH_2){-}(C{=}O)H + 0.5\,O_2 \rightarrow CH_3{-}(C{=}CH_2){-}(C{=}O){-}OH$$

methacrolein → methacrylic acid

$$CH_3{-}(C{=}CH_2){-}(C{=}O){-}OH + CH_3OH \rightarrow CH_3{-}(C{=}CH_2){-}(C{=}O){-}O{-}CH_3 + H_2O$$

methacrylic acid + methanol (in sulfuric acid) → methylmethacrylate

What would be an appropriate method for evaluating these alternatives for synthesizing methyl methacrylate? The first step in answering this question is to select a set of criteria to be used in the evaluation. In traditional methods of process synthesis, cost is the most common screening criterion. To evaluate alternative processes, such as the two processes used in the synthesis of methyl methacrylate, the value of the product could be compared to the cost of the raw materials. Such an evaluation would require data on the raw material input requirements, product and byproduct output, and market values of all of the materials. Approximate stoichiometric and cost data for the methyl methacrylate processes (Chang, 1996; Rudd, et al., 1981) are provided in Table 8.2-1.

Table 8.2-1   Stoichiometric and Cost Data for Two Methyl Methacrylate Synthesis Routes.

Compound	Pounds produced or pounds of raw material required per pound of methyl methacrylate*	Cost per pound[1]
Acetone-cyanohydrin route		
Acetone	− .68	$0.43
Hydrogen cyanide	− .32	$0.67
Methanol	− .37	$0.064
Sulfuric acid	−1.63	$0.04
Methyl methacrylate	1.00	$0.78
Isobutylene route		
Isobutylene	−1.12	$0.31
Methanol	−0.38	$0.064
Pentane	−0.03	$0.112
Sulfuric acid	−0.01	$0.04
Methyl methacrylate	1.00	$0.78

*A negative stoichiometric index indicates that a material is consumed; a positive index indicates that it is produced in the reaction.
[1]Data from Chang (1996)

The raw material costs per pound of methyl methacrylate are simply the stoichiometric coefficients, multiplied by the cost per pound. For the first pathway, the raw material costs per pound of methyl methacrylate are:

$$0.68 \times \$0.43 + 0.32 \times \$0.67 + 0.37 \times \$0.064 + 1.63 \times 0.04 = \$0.60 \text{ pound of methyl methacrylate}$$

For the isobutylene route, a similar calculation leads to a cost of $0.37 per pound of methyl methacrylate. From this simple evaluation, it is clear that the isobutylene route has lower raw material costs than the acetone-cyanohydrin route, and is probably economically preferable. It is important to note, however, that raw material costs are not the only cost factor. Different reaction pathways may lead to very different processing costs. A reaction run at high temperature or pressure may require more energy or expensive capital equipment than an alternative pathway with more expensive raw materials. Or, raw materials may be available as byproducts from other processes at a lower cost than market rates. So, simple evaluations of raw material costs should only be used in a qualitative fashion. Nevertheless, they provide a simple screening method for chemical pathways and may lead to rapid elimination of alternatives where the raw material inputs are more valuable than the products.

8.2.2 Environmental Criteria

In addition to a simple economic criterion, simple environmental criteria should be available for screening designs, based on input-output data. Selecting a single criterion or a few simple criteria that will characterize a design's potential environmental impacts is not a simple matter. As noted elsewhere in this text, a variety of impact categories could be considered, ranging from global warming to human health concerns. Not all of these potential impacts can be estimated effectively. Further, if only input-output data are available, there may not be sufficient information to estimate some environmental impacts. For example, estimates of global warming impacts of a design would require data on energy demands, which are often not available at this design stage.

One set of environmental criteria that can be rapidly estimated, even at the input-output level of design, are the persistence, bioaccumulation, and toxicities of the input and output materials. Chapter 5 described, in some detail, how these parameters can be estimated based on chemical structure. Consider how this might be applied to the problem of evaluating the methyl methacrylate reaction pathways. Persistence and bioaccumulation for each of the compounds listed in Table 8.2-1 are listed in Table 8.2-2.

The values for persistence and bioaccumulation reported in Table 8.2-2 were calculated using the EPISUITE software package (see Appendix F), which is based on the methods described in Chapter 5. In Chapter 5, classification schemes, based on the values of persistence and bioaccumulation factors, are presented. These classifications are partially reproduced in Table 8.2-3.

Comparing these classifications to the values presented in Table 8.2-2 leads to the conclusion that none of the reactants or products in either scheme bioaccumulate or are persistent in the environment. This is a qualitative assessment. Later in this section, quantitative evaluations are discussed, and for the purposes of those quantitative

Table 8.2-2 Bioaccumulation and Persistence Data for Two Synthesis Routes.

Compound	Persistence (atmospheric half life[1])	Aquatic half-life (biodegradation index)	Bioaccumulation (bioconcentration factor)
Acetone-cyanohydrin route			
Acetone	52 days	weeks	3.2
Hydrogen cyanide	1 year	weeks	3.2
Methanol	17 days	days-weeks	3.2
Sulfuric acid[2]			
Methyl methacrylate	7 hours	weeks	2.3
Isobutylene route			
Isobutylene	2.5 hours	weeks	12.6
Methanol	17 days	days-weeks	3.2
Pentane	2.6 days	days-weeks	81
Sulfuric acid[2]			
Methyl methacrylate	7 hours	weeks	2.3

[1]The atmospheric half life is based on the reaction with the hydroxyl radical and assumes an ambient hydroxyl radical concentration of $1.5*10^6$ molecules per cubic centimeter and 12 hours of sunlight per day.
[2]The group contribution method does not estimate an atmospheric reaction rate for sulfuric acid; however, its lifetime in the atmosphere is short due to reactions with ammonia.

Table 8.2-3   Classification Schemes for Persistence and Bioaccumulation.

Persistence		
Rapid	>60% degradation over 1 week	Rating index = 0
Moderate	>30% degradation over 28 days	Rating index = 1
Slow	<30% degradation over 28 days	Rating index = 2
Very Slow	<30% degradation over more than 28 days	Rating index = 3
Bioaccumulation		
High Potential	$8.0 > \text{Log } K_{ow} > 4.3$ or BCF>1000	Rating index = 3
Moderate Potential	$4.3 > \text{Log } K_{ow} > 3.5$ or 1000>BCF>250	Rating index = 2
Low Potential	$3.5 > \text{Log } K_{ow}$ or 250>BCF	Rating index = 1

assessments, the numerical ratings given in Table 8.2-3 are useful. In this case, all of the compounds would have persistence ratings of 1 and bioaccumulation ratings of 1.

While persistence and bioaccumulation can generally be evaluated using the structure-activity methods described in Chapter 5, toxicity is more problematic. Some structure-activity relationships exist for relating chemical structures to specific human health or ecosystem health endpoints, but often the correlations are limited to specific classes of compounds. The ideal toxicity parameter would recognize a variety of potential human and ecosystem health endpoints and would be readily accessible. No such parameter exists. A variety of simple toxicity surrogates have been employed, however, including Threshold Limit Values, Permissible Exposure Limits, Recommended Exposure Limits, inhalation reference concentrations, and oral response factors. Each of these are described below.

8.2.3   Threshold Limit Values (TLVs), Permissible Exposure Limits (PELs), and Recommended Exposure Limits (RELs)

These parameters were developed to address the problem of establishing workplace limits for concentrations of chemicals. TLVs, PELs, and RELs are the estimated concentrations of chemicals that workers can be safely exposed to in occupational settings. Threshold Limit Values (TLVs), Permissible Exposure Limits (PELs), and Recommended Exposure Limits (RELs) reflect the different health impacts of chemicals and variations in exposure pathways. They are defined as follows:

Threshold Limit Value (TLV). The TLV is one type of airborne concentration limit for individual exposures in the workplace environment. The concentration is set at a level for which no adverse effects would be expected over a worker's lifetime. A number of TLVs can be cited for a chemical, depending on the length of the exposure. In this chapter, the TLVs will be time-weighted averages for an 8-hour workday and a 40-hour workweek. The concentration, again, is the level to which nearly all workers can be exposed without adverse effects. TLVs are established by a nongovernmental organization, the American Conference of Governmental Industrial Hygienists (ACGIH) (http://www.acgih.org).

Permissible Exposure Limits (PELs). The United States Occupational Safety and Health Administration (OSHA) has the legal authority to place limits on exposures to chemicals in the workplace. The workplace limits set by OSHA are referred to as PELs, and are set by OSHA in a manner similar to the setting of TLVs by ACGIH.

Recommended Exposure Limits (RELs). The National Institute for Occupational Safety and Health (NIOSH), under the Centers for Disease Control and Prevention (CDC), publishes RELs based on toxicity research. As the research complement to OSHA, NIOSH sets RELs that are intended to assist OSHA in the setting and revising of the legally binding PELs. Because no rule-making process is required for NIOSH to set RELs, these values are frequently more current than the OSHA PELs.

The TLV, PEL and REL values in Table 8.2-4 are generally quite similar, but some of the differences are worthy of comment. TLV values represent a scientific and professional assessment of hazards, while PEL values have legal implications in defining workplace conditions. Because of these legal implications, PELs are directly influenced by political, economic and feasibility issues. NIOSH, as the research complement to OSHA, is not affected by these external issues and can set their limits in a purely research environment. Because RELs do not face the same practicality issues as the PELs, NIOSH has chosen not to set safe levels of exposure for potential carcinogens, but instead recommends minimizing exposures to these substances. It is not unusual for a TLV or REL value to be established before a PEL value. Because of the greater number of chemicals for which there are reported values, there is a tendency to use TLV or REL data in screening methodologies rather than PEL values.

One method of using TLV and PEL values to define a toxicity index is to use the inverse of the TLV (see, for example, Horvath, et al., 1995).

$$\text{Environmental Index} = 1/(\text{TLV}) \qquad \text{(Eq. 8-1)}$$

The concept is simple. Higher TLVs imply that higher exposures can be tolerated with no observable health effect, implying a lower health impact. A simple way to express this relationship mathematically is with an inverse relationship, as shown in Equation 8-1.

Using the TLV (or PEL, REL) as a surrogate for all toxicity impacts is a gross simplification. The TLV only accounts for direct human health effects via inhalation, and even for this purpose, it is dangerous to use the TLV as a measure of relative health impact. Figure 8.2-1 illustrates one of the pitfalls of using TLV as an indicator of relative human health impact.

Figure 8.2-1 shows the toxic response of two chemicals, A and B, as a function of dose. Chemical A has a higher threshold concentration, at which no toxic effects are observed, than chemical B. Once the threshold dose is exceeded, however, chemical A has a greater response to increasing dose than chemical B. If the TLV were based on the dose at which 10% of the population experienced health effects, then chemical B would have a lower TLV than chemical A. In contrast, if the TLV were based on the dose at which 50% of the population experienced a health

Table 8.2-4 Threshold Limit Values, Permissible Exposure Limits and Recommended Exposure Limits for Selected Compounds (adapted from Crowl and Louvar, 1990. Updated with 2001 data. Note that these values continue to be periodically updated. Readers interested in current values of these parameters should consult the appropriate reference. See Appendix F.)

Compound	TLV (ppm)	PEL (ppm)	REL (ppm)
Acetaldehyde	25	200	Potential carcinogen—minimize exposure
Acetic acid	10	10	10
Acetone	500	1000	250
Acrolein	0.1	0.1	0.1
Ammonia	25	50	25
Arsine	0.05	0.05	0.0002
Benzene	0.5	1	0.1
Biphenyl	0.2	0.2	0.2
Bromine	0.1	0.1	0.1
Butane	800	None est.	800
Carbon monoxide	25	50	35
Chlorine	0.5	1	0.5
Chloroform	10	50	2
Cyclohexane	300	300	300
Cyclohexene	300	300	300
Cyclopentane	600	None est.	600
1,1 Dichloroethane	100	100	100
1,2 Dichloroethylene	200	200	200
Diethyl ketone	200	None est.	200
Dimethylamine	5	10	10
Ethylbenzene	100	100	100
Ethyl chloride	500	1000	Potential carcinogen—minimize exposure
Ethylene dichloride	10	50	1
Ethylene oxide	1	1	0.1
Formaldehyde	0.3	0.75	0.016
Gasoline	300	None est.	Potential carcinogen—minimize exposure
Heptane	400	500	85
Hexachloroethane	1	1	1
Isobutyl alcohol	50	100	50
Isopropyl alcohol	400	400	400
Maleic anhydride	0.1	0.25	0.25
Methyl ethyl ketone	200	200	200
Naphthalene	10	10	10
Nitric acid	2	2	2
Nitric oxide	25	25	25
Nitrogen dioxide	3	5	1
Phosgene	0.1	0.1	0.1
Sulfur dioxide	2	5	2
Trichloroethylene	50	100	Potential carcinogen—minimize exposure
Vinyl chloride	1	1	Potential carcinogen—minimize exposure

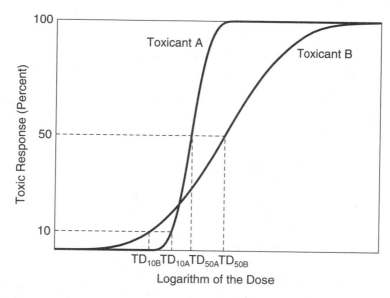

Figure 8.2-1   Dose response curves for two compounds that have different relative threshold limit values (TLVs), depending on how the effect level is defined (Crowl and Louvar, 1990).

impact, chemical A would have the lower TLV. So, which chemical is more toxic? The answer depends on the precise definition of toxicity and the specifics of the dose-response relationship.

   This conceptual example is designed to illustrate the dangers of using simple indices as precise, quantitative indicators of environmental impacts. There is value, however, in using these simple indicators in rough, qualitative evaluations of potential environmental impacts.

8.2.4   Toxicity Weighting

An additional limitation of TLV values is that they do not consider ingestion pathways. An alternative measure of potential toxicities might incorporate both inhalation and ingestion exposure pathways. Such a system has been developed by the US EPA using data available from the EPA's IRIS (Integrated Risk Information System) database. IRIS compiles a wide range of available data on individual compounds (http://www.epa.gov/ngispgm3/iris/subst/index.html). Three data elements that are of use in assessing potential toxicities are the inhalation reference concentration, the oral ingestion slope factor, and the unit risk. As defined in the IRIS documentation, a reference concentration is "an estimate (with uncertainty spanning perhaps an order of magnitude) of a continuous inhalation exposure to the human population (including sensitive subgroups) that is likely to be without an appreciable risk of deleterious noncancer effects during a lifetime." The inhalation reference concentration is in some ways related to the TLV, and ratios of the TLVs of different compounds would be expected to be similar to the ratios of the inhalation reference concentrations.

An oral slope factor characterizes response to ingestion of a compound and is defined as "the slope of a dose response curve in the low dose region. When low dose linearity cannot be assumed, the slope factor is the slope of the straight line from 0 dose (and 0 excess risk) to the dose at 1% excess risk. An upper bound on this slope is usually used instead of the slope itself. The units for the slope factor are usually expressed as $(mg/kg\text{-}day)^{-1}$" (US EPA, IRIS, 1999).

The unit risk is "the upper bound excess lifetime cancer risk estimated to result from continuous exposure to an agent at a concentration of 1 microgram/L in water and 1 microgram/cubic meter in air."

A simple example may clarify the meaning of these indicators of toxicity. Consider the data available on IRIS (August, 1999) for acrylonitrile. IRIS lists acrylonitrile as a probable human carcinogen. Non-carcinogenic effects include inflammation of nasal tissues. The reference concentration for inhalation is given as 0.002 mg/m^3. Lifetime exposure to this concentration is likely to be without an appreciable risk of nasal tissue inflammation and degeneration. The oral slope factor for carcinogenic risk is given as 0.54 $(mg/kg\text{-}day)^{-1}$. A 100 kg person exposed to 100 mg per day would have a 0.54% excess risk. The potential individual excess lifetime cancer risk (i.e., unit risk) is 6.8×10^{-5} per microgram/m^3. For a region with a population of 100,000, this corresponds to approximately 6.8 potential excess cancer cases based on a lifetime exposure of 1 microgram/m^3 of acrylonitrile (i.e., an upper bound of the lifetime risk is 6.8 in 100,000). Note that 6.8 represents an upper bound and the actual risk may be much less.

The US EPA has used data such as reference concentrations, oral slope factors, and unit risk factors to determine toxicity weighting for approximately 600 compounds reported through the Toxic Release Inventory. A complete description of the methodology and the toxicity weights are available at http://www.epa.gov/opptintr/env_ind/index.html. To briefly summarize, the EPA assembled up to four preliminary human health toxicity weights for each compound: cancer-oral, cancer-inhalation, non-cancer-oral, and non-cancer-inhalation. For each exposure pathway (oral and inhalation) the greater of the cancer and non-cancer toxicity weights was chosen. If data on only one exposure pathway were available, then the toxicity weight for that pathway was assigned to both pathways; however, if there is evidence that no exposure occurs through one of the pathways, then the toxicity weight for that pathway was assigned a value of 0.

The toxicity weights were based on the values for unit risks and slope factors. A sample of the scheme used to assign toxicity weights is given in Table 8.2-5.

For the acrylonitrile, a probable carcinogen with an oral slope factor of 0.54, the oral toxicity weight would be 10,000. The toxicity weight for inhalation, based on a unit risk of 6.8×10^{-5} per (microgram/m^3) or .068 per (milligram/m^3), would be 1000. The overall toxicity weight would be based on the larger of the two values. Table 8.2-6 provides a sampling of toxicity weights. The compounds listed are the same compounds for which TLV data were listed in Table 8.2-3. The data are somewhat more sparse than the TLV data.

Table 8.2-5   Assignment of Toxicity Weights for Chemicals with Cancer Health Effects.

Range of oral slope factor (SF) (risk per mg/kg-day)	Range of inhalation unit risk factor (UR) (risk per mg/m³)	Known or probable carcinogen	Possible carcinogen
SF<0.005	UR<0.0014	10	1
0.005<SF<0.05	0.0014<UR<0.014	100	10
0.05<SF<0.5	0.014<UR<0.14	1000	100
0.5<SF<5	0.14<UR<1.4	10,000	1000
5<SF<50	1.4<UR<14	100,000	10,000
50>SF	UR>14	1,000,000	100,000

8.2.5   Evaluating Alternative Synthetic Pathways

As a case study of the use of TLVs and toxicity weights in evaluating toxicity, consider once again the two routes for producing methyl methacrylate. Stoichiometric, TLV, and toxicity weight data for the two pathways are shown in Table 8.2-7.

Both the TLVs and toxicity weights in Table 8.2-7 indicate that the major health concerns associated with the two reaction pathways are due to sulfuric acid, and to a lesser extent, hydrogen cyanide.

Once these data, together with data on persistence and bioaccumulation, are known for the reactants and products, some composite index for the overall input-output structure could be established. Ideally, the index would be based on the emission rates, weighted by measures of persistence, bioaccumulation, and toxicity. In preliminary screenings, however, it is highly unlikely that detailed information will be available on emission rates. Therefore, approximations for emission rates are required. One possible approach is to use flow rate, based on stoichiometry, as a surrogate for emissions. This surrogate for emissions can then be weighted by an appropriate index.

In choosing weighting factors and an overall index for assessing environmental performance at this early stage of a design, it is important to recognize that there is no single correct choice. Many different indices have been employed. This chapter will illustrate two types of approaches that have appeared frequently in the literature. One approach is to use toxicity as a weighting factor. In this approach, the overall environmental index for a reaction is typically calculated as:

$$\text{Environmental index} = \Sigma \, |v_i| \times (\text{TLV}_i)^{-1} \qquad \text{(Eq. 8-2)}$$

Where $|v_i|$ is the absolute value of the stoichiometric coefficient of reactant or product i, TLV_i is the threshold limit value of reactant or product i, and the summation is taken over all reactants and products. For the acetone-cyanohydrin route:

$$\text{Index} = 0.68 \times (1/750) + 0.32 \times (1/10) + 0.37 \times (1/200)$$
$$+ \, 1.63 \times (1/2) + 1 \times (1/100) = 0.86$$

For the acetone-cyanohydrin process, the index calculated using Equation 8-2 is 0.86, and for the isobutylene process, the index is 0.01, indicating a preference for

Table 8.2-6   Selected Toxicity Weights Drawn from the U.S. EPA's Environmental Indicators Project.

Compound	Overall inhalation toxicity factor	Overall oral toxicity factor
Acetaldehyde	1000	1000
Acetic acid		
Acetone		
Acrolein	100000	100000
Ammonia	100	100
Arsine		
Benzene	100	100
Biphenyl	100	100
Bromine		
Butane		
Carbon Monoxide		
Chlorine	10	10
Chloroform	1000	100
Cyclohexane		
Cyclohexene		
Cyclopentane		
1,1 Dichloroethane	1000	1000
1,2 Dichloroethylene	100	100
Diethyl ketone		
Dimethylamine		
Ethylbenzene		
Ethyl chloride		
Ethylene dichloride		
Ethylene oxide	10000	10000
Formaldehyde	100	10
Gasoline		
Heptane		
Hexachloroethane	10	1000
Isobutyl alcohol		
Isopropyl alcohol		
Maleic anhydride	10	10
Methyl ethyl ketone	10	1
Naphthalene		
Nitric acid		
Nitric oxide		
Nitrogen dioxide		
Phosgene		
Sulfur dioxide		
Trichloroethylene		
Vinyl chloride	10000	10000

the isobutylene process. This is because the indices are dominated by the contribution of sulfuric acid, which is used at a lower rate in the isobutylene process.

Alternatively, the toxicity factors developed by the US EPA could be used, rather than the TLVs. In this case:

$$\text{Environmental index} = \Sigma \, |v_i| \times (\text{maximum of oral and inhalation weighting factor})$$

$$(\text{Eq. 8-3})$$

Table 8.2-7   Stoichiometric, TLV, and Toxicity Weight Data for Two Methyl Methacrylate Synthesis Routes.

Compound	Pounds produced or pounds of raw material required per pound of methyl methacrylate*	1/TLV (ppm)	Overall inhalation toxicity factor	Overall oral toxicity factor
Acetone-cyanohydrin route				
Acetone	−.68	1/750	NA	NA
Hydrogen cyanide	−.32	1/10	1000	100
Methanol	−.37	1/200	10	10
Sulfuric acid	−1.63	1/2(est.)	10,000	1
Methyl methacrylate	1.00	1/100 (PEL)	10	10
Isobutylene route				
Isobutylene	−1.12	1/200 (est)	NA	NA
Methanol	−0.38	1/200	10	10
Pentane	−0.03	1/600	NA	NA
Sulfuric acid	−0.01	1/2 (est)	10,000	1

*A negative stoichiometric index indicates that a material is consumed; a positive index indicates that it is produced in the reaction.

Using this approach, the index for the acetone-cyanohydrin process would be:

$$Index = 0.68 \times (0) + 0.32 \times (1000) + 0.37 \times (10)$$
$$+ 1.63 \times (10,000) + 1 \times (10) = 16,600$$

For the isobutylene process, the index is 100, again indicating a preference for the isobutylene process.

Another approach that appears in preliminary environmental assessments employs persistence, bioaccumulation, and toxicity factors. Combining these factors into a composite environmental index requires that the factors be placed in a common unit system. This is generally done by assigning ratings to the persistence, bioaccumulation, and toxicity parameters. Table 8.2-2 gave rating factors for persistence and bioaccumulation for the two methyl methacrylate pathways. Ratings for human toxicity are more difficult to assign. In the evaluation of chemicals under the Toxic Substances Control Act, the US EPA employs three levels of concern for human toxicity (Wagner, et al., 1995):

- High concern
    Evidence of adverse effects in human populations
    Conclusive evidence of severe effects in animal studies
- Moderate concern
    Suggestive animal studies
    Data from close chemical analogue
    Compound class known to produce toxicity
- Low concern
    Chemicals that do not meet the criteria for moderate or high concern

Based on these criteria, the human toxicity concerns of the two methyl methacrylate pathways would be dominated by the concerns associated with sulfuric acid. Thus, the two pathways would have very similar levels of toxicity concern unless the relative amounts of sulfuric acid used were incorporated into the evaluation. As noted earlier, the bioaccumulation and persistence of the compounds associated with the two pathways were also identical; therefore, the overall environmental performance of the two pathways could be viewed as virtually identical.

Table 8.2-8 provides a set of three ratings for each pathway. These three ratings could be combined into a single index, or they could be retained in the matrix format shown in the table.

To summarize, the environmental performance of the two pathways for manufacturing methyl methacrylate was evaluated based on economics, toxicity, and a combined assessment of persistence, bioaccumulation, and toxicity. All of the approaches indicate a preference for the isobutylene pathway. A similar case study with a different result is given in Example 8.2-1.

Example 8.2-1

Acrylonitrile can be produced via the ammoxidation of propylene or via the cyanation of ethylene oxide. Stoichiometric, TLV, persistence, bioaccumulation, toxicity, and cost data for the two reactions are given below.

(a) Estimate the persistence and bioaccumulation potential of the two pathways
(b) Evaluate the toxicity potential of the two pathways
(c) Suggest which pathway is preferable based on environmental and economic criteria

ammoxidation of propylene:

$$C_3H_6 + NH_3 + 1.5\ O_2 \rightarrow C_3H_3N + 3\ H_2O$$

cyanation of ethylene oxide

$$C_2H_4 + 0.5\ O_2 \rightarrow C_2H_4O$$

$$C_2H_4O + HCN \rightarrow HOC_2H_4CN \rightarrow C_3H_3N + H_2O$$

Table 8.2-8   Evaluation of Methyl Methacrylate Pathways Based on Persistence, Bioaccumulation, and Toxicity.

Pathway	Persistence of raw materials and products	Bioaccumulation potential of raw materials and products	Toxicity of raw materials and products
Acetone-cyanohydrin route	All raw materials and products on a time scale of weeks; rating index =1	Bioaccumulation potential of all raw materials and products is low; rating index = 1	Toxicity is dominated by sulfuric acid, which is a respiratory toxicant and a suspected carcinogen; rating index = 2
Isobutylene route	All raw materials and products on a time scale of weeks; rating index =1	Bioaccumulation potential of all raw materials and products is low; rating index = 1	Toxicity is dominated by sulfuric acid, which is a respiratory toxicant and a suspected carcinogen; rating index = 2

Table 8.2-9   Bioaccumulation and Persistence Data for Two Acrylonitrile Synthesis Routes.

Compound	Persistence (atmospheric half life[1])	Aquatic half-life (Biodegradation index)	Bioaccumulation
Ammoxidation of propylene			
Proplyene	4.9 hours	weeks	4.6
Ammonia	NA[2]	weeks	3.2
Acrylonitrile	30.5 hours	weeks	3.2
Hydrogen cyanide	1 year	weeks	3.2
Acetonitrile	1 year	weeks	3.2
Cyanation of ethylene oxide			
Ethylene	15 hours	weeks	1.1
Hydrogen cyanide	1 year	weeks	3.2
Acrylonitrile	30.5 hours	weeks	3.2
Carbon dioxide	—	—	—

[1]The atmospheric half-life is based on the reaction with the hydroxyl radical and assumes an ambient hydroxyl radical concentration of $1.5*10^6$ molecules per cubic centimeter and 12 hours of sunlight per day.
[2]The group contribution method does not estimate an atmospheric reaction rate for ammonia; however, its lifetime in the atmosphere is short due to reactions with acid gases.

Solution:

(a) Estimate the persistence and bioaccumulation potential of the two pathways

Based on the data in the table below, the materials used in the two pathways have comparable, relatively low persistence and bioaccumulation potentials.

The values for persistence and bioaccumulation were calculated using the EPISUITE™ software package, which is based on the methods described in Chapter 5.

Table 8.2-10   Stoichiometric, TLV, and Toxicity Weight Data for Two Acrylonitrile Synthesis Routes.

Compound	Pounds produced or pounds of raw material required per pound of acrylonitrile*	TLV (ppm)	Overall inhalation toxicity factor	Overall oral toxicity factor
Ammoxidation of propylene				
Proplyene	−1.1	>10,000	1	1
Ammonia	−0.4	25	100	100
Acrylonitrile	1	2	1000	10,000
Hydrogen cyanide	0.1	10	1000	100
Acetonitrile	0.03	40	100	100
Cyanation of ethylene oxide				
Ethylene	−0.84	>10,000	1	1
Hydrogen cyanide	−0.6	10	1000	100
Acrylonitrile	1	2	1000	10,000
Carbon dioxide	0.3	5000		

*A negative index indicates that a material is consumed; a positive index indicates that it is produced

Table 8.2-11   Stoichiometric, TLV and Cost Data for Two Acrylonitrile Synthesis Routes.

Compound	Stoichiometry*	1/TLV (ppm)$^{-1}$	Cost per pound[1]
Ammoxidation of propylene			
Proplyene	−1.1	1/10,000	$0.13
Ammonia	−0.4	1/25	$0.07
Acrylonitrile	1	1/2	$0.53
Hydrogen cyanide	0.1	1/10	$0.68
Acetonitrile	0.03	1/40	$0.65
Cyanation of ethylene oxide			
Ethylene	−0.84	1/10,000	$0.23
Hydrogen cyanide	−0.6	1/10	$0.68
Acrylonitrile	1	1/2	$0.53
Carbon dioxide	0.3	1/5,000	

*A negative stoichiometric index indicates that a material is consumed; a positive index indicates that the material is produced in the reaction
[1]Data from Chang (1996)

(b) Evaluate the toxicity potential of the two pathways

As shown in the table and calculations below, the toxicity is dominated by the product, acrylonitrile, so the two pathways have very similar environmental performance indices.

For the acetone-cyanohydrin process, the environmental index based on the TLV and the index based on EPA's toxicity weights are given by:

$$\text{TLV Index} = 1.1/10,000 + 0.4/25 + 1/2 + 0.1/10 + 0.03/40 = 0.53$$

$$\text{EPA Index} = 1.1 \times 1.0 + 0.4 \times 100 + 1.0 \times 10,000 + 0.1 \times 1,000 + 0.03 \times 100 = 10,144$$

For the cyanation of ethylene oxide the indices are :

$$\text{TLV Index} = 0.84/10,000 + 0.6/10 + 1/2 + 0.3/5000 = 0.56$$

$$\text{EPA Index} = 0.84 \times 1.0 + 0.6 \times 1000 + 1.0 \times 10,000 = 10,600$$

Based on these criteria, the human toxicity concerns of the two acrylonitrile pathways would be dominated by the concerns associated with acrylonitrile. Thus, the two pathways would have very similar levels of toxicity concern. As noted earlier, the bioaccumulation and persistence of the compounds associated with the two pathways were also identical; therefore, the overall environmental performance of the two pathways could be viewed as virtually identical.

(c) Suggest which pathway is preferable based on environmental and economic criteria

A simple economic evaluation considers the raw material costs. For the ammoxidation of propylene, the economic index is given by:

$$\text{Index} = 1.1 \times (\$0.13) + 0.4 \times (\$0.07) = \$0.17$$

Table 8.2-12   Evaluation of Acrylonitrile Pathways Based on Persistence, Bioaccumulation and Toxicity.

Pathway	Persistence of raw materials and products	Bioaccumulation potential of raw materials and products	Toxicity of raw materials and products
Ammoxidation of propylene	All raw materials and products on a time scale of weeks; rating index =1	Bioaccumulation potential of all raw materials and products is low; rating index = 1	Toxicity is dominated by the product, acrylonitrile, which is a probable carcinogen; high concern rating
Cyanation of ethylene oxide	All raw materials and products on a time scale of weeks; rating index =1	Bioaccumulation potential of all raw materials and products is low; rating index=1	Toxicity is dominated by the product, acrylonitrile, which is a probable carcinogen; high concern rating

Alternatively, an index could include raw material costs minus the value of salable byproducts:

$$\text{Index} = 1.1 \times (\$0.13) + 0.4 \times (\$0.07) - 0.1 \times (\$0.68) - 0.03 \times (\$0.65) = \$0.14$$

For the cyanation of ethylene oxide, the economic index is:

$$\text{Index} = \$0.84 \times \$0.23 + 0.6 \times \$0.68 = \$0.60$$

Thus, the ammoxidation of propylene is preferable to the cyanation of ethylene oxide on a cost basis; the pathways have comparable environmental characteristics.

SECTION 8.2   QUESTIONS FOR DISCUSSION

1. What criteria would you suggest for evaluating the environmental performance of reaction pathways?
2. Can you suggest alternatives to stoichiometric coefficients for weighting environmental indices in evaluating reaction pathways?
3. What are the strengths and limitations of the environmental performance criteria described in this section?

8.3   TIER 2 ENVIRONMENTAL PERFORMANCE TOOLS

Once the basic input-output structure of a flow sheet is determined, a preliminary process flowsheet is developed. Typically, storage devices, reactors, and separation devices might be identified, and some information would be available about equipment sizes or process stream flow rates. This level of process specification is an appropriate time to re-examine environmental performance. At this stage of analysis, it still may be necessary to screen large numbers of design alternatives, but more information about the process is available and should be incorporated into the environmental performance evaluation. This section describes methods for performing environmental evaluations at this intermediate level. A first step in this analysis is to use the information available on the process units to estimate the magnitude and composition of emissions and wastes. Some of these emission estimation tools are described in Section 8.3.1. Once the emissions, wastes, and other process flows are characterized, any of a number of environmental performance evaluation methods can be employed. Environmental performance evaluation tools, suitable for this level of analysis, are described in Section 8.3.2.

8.3.1 Environmental Release Assessment

8.3.1.1 Basics of Releases

Releases include any spilling, leaking, pumping, pouring, emitting, emptying, discharging, injecting, escaping, leaching, dumping, or disposing into the environment (including the abandonment or discarding of containers and other closed receptacles) of any chemical or chemical mixture. The term "environment" includes water, air, and land, the three media to which release may occur. Related to releases are transfers of chemical wastes off-site for purposes other than making a salable product. Such purposes could include treatment or disposal.

8.3.1.2 Release Assessment Components

Release assessments are documents that contain information on release rates, frequencies, media of releases, and other information needed to characterize to the fullest possible extent the issues related to the releases. The audience for the document determines the amount of information about the methods used to estimate releases that should be presented. The steps required in making release assessments are:

1. Identify purpose and need for release assessment.
2. Obtain or diagram a process flowsheet.
3. Identify and list waste and emissions streams.
4. Examine the flowsheet for additional waste and emission streams.
5. For each release point identified in steps 3 and 4, determine the best available method for quantifying the release rate.
6. Determine data or information needed to use the quantification methods determined in step 5.
7. Collect data and information to fill gaps.
8. Quantify the chemical's release rates and frequencies and the media to which releases occur.
9. Document the release assessment, including a characterization of the uncertainties in the estimates.

The purpose of the release assessment determines the information and data needed to complete the assessment. If a release assessment is to be used as part of a screening-level risk assessment, less detail and accuracy will be required than if the release assessment were being used as part of a detailed risk assessment. Knowing the purpose of the assessment can also refine the scope of the analysis. For example, if only aquatic impacts of release are of concern, then only releases to water may need to be addressed.

8.3.1.3   Process Analysis

A release assessment begins after one or more processes have been selected for analysis. At this point, the basic features (e.g., mass balances, unit operations, operating conditions) of the design are available. A flow diagram showing process streams is often a key tool in beginning the analysis.

From the flow diagram, process output streams that are not usable or salable products can be identified. List these output streams as potential releases. Often, the flowsheet does not identify other potential releases for various reasons: some are not directly attributable to process equipment, some result from process inefficiencies that are not normally considered significant, some may be infrequent, some may be difficult to quantify, and some may be overlooked. If the flowsheet development was not rigorous, many opportunities to identify potential releases, not included on the flowsheet, may exist.

Common sources of releases that are often missing in a flowsheet are:

- Fugitive emissions (including leaks, which are defined later in this section)
- Venting of equipment (e.g., breathing and displacement losses, etc.)
- Periodic equipment cleaning (may be frequent or infrequent)
- Transport container residuals (e.g., from drums, totes, tank trucks, rail cars, barges, etc.)
- Incomplete separations (e.g., distillation, gravity phase separation, filtration, etc.)

The manner in which a chemical is released is a crucial factor in assessing environmental impact. In characterizing the manner in which a chemical is released, it is convenient to first determine whether the release is expected to occur on-site or from some extension of the site to an off-site location, such as a pipe extending into a water body. On-site releases to the environment include emissions to the air, discharges to surface waters, and releases to land and underground injection wells. Both routine releases, such as fugitive air emissions, and accidental or non-routine releases, such as chemical spills, are part of the on-site releases. On-site releases do not include transfers or shipments of chemicals from the facility for sale or distribution in commerce, or of wastes to other facilities for disposal, treatment, energy recovery, or recycling. Chemical wastes that are transferred or shipped to an off-site location, such as a publicly owned treatment works (POTW), where the waste may be fully or partially released, are called "off-site transfers."

Once emissions and wastes have been characterized as on-site or off-site, the on-site releases are classified by the medium or media to which the chemical is released. Releases to common classes of media are described below.

Air Releases. Releases to air (often called emissions) can be categorized into primary or secondary emissions. Primary emissions occur as a direct consequence of the production or the use within an industrial process of the compound under consideration. These primary emissions may come from either point (often called stack) or non-point (often called fugitive) sources. Stack releases of chemicals to the air occur through stacks, vents, ducts, pipes, or other confined gas streams. Stack releases include storage tank and unit operation vent emissions and, generally, air releases from air pollution control equipment. Unit operations of importance as emission sources include those which are relatively few in number within a process and are easily identifiable, such as pressure relief vents on reactors, and vents on distillation column condensers, absorption and stripping columns vents, and feed or product storage tank vents.

Fugitive air emissions are not releases through stacks, vents, ducts, pipes, or any other confined gas streams. These releases include fugitive equipment leaks from valves, pump seals, flanges, compressors, sampling connections, open-ended lines, etc.; releases from building ventilation systems; and any other fugitive or non-point air emissions. Fugitive emissions occur from process sources that are not easily identifiable and are of relatively large number within the process. Within a typical industrial process there may be tens to hundreds of thousands of fugitive sources (Berglund, et al., 1989).

Secondary emissions occur indirectly as a result of the production or use of a specific compound. These emission sources include utilities consumption, evaporative losses from surface impoundments and spills, and industrial wastewater collection systems.

Because emissions to air can be difficult to measure and emission sources can be difficult to locate, some resources for preparing plant-wide emissions inventories have been developed. Several references for these resources are shown in Table 8.3-1.

Water Releases. Releases of chemicals from discharge points in a process can be to a receiving stream or water body. These releases include process outfalls such as pipes and open trenches, releases from on-site wastewater treatment systems, and, sometimes, the contribution from storm-water runoff. Water releases do not include discharges to a POTW or other off-site wastewater treatment facilities. These are off-site transfers.

Underground Injection Releases. Some chemicals may be injected into wells at a facility. US EPA regulations apply to underground wells, which are classified by the type of material injected into the well. The Underground Injection Control Program of the Federal Safe Drinking Water Act is found in 40 CFR Parts 144-147.

Releases to Land. Some chemicals may be released to land within the boundaries of a facility. Some facilities may have on-site landfills for chemical disposal.

Table 8.3-1   Resources for Preparing Plant-Wide Emission Inventories (Allen and Rosselot, *Pollution Prevention for Chemical Processes* © 1997, This material is used by permission of John Wiley & Sons, Inc.)

Reference	Information Content	Location of Reference
"Compilation of Air Toxics Emissions Inventories," Lahre, T.F., US EPA Publication number EPA/450/4-86-010, July 1986	Preparing Inventories	NTIS as PB86238086
"How to Develop Your Toxic Emissions Inventory: Approaches, Problems and Solutions," Walther, E.G. et al. in Proceedings of the National Research and Development Conference on the Control of Hazardous Materials, Anaheim, CA Feb. 1991	Preparing Inventories	Through the Hazardous Materials Control Research Institute, Greenbelt, MD, (301)-982-9500
"Prepare Now for the Operating Permit Program," Van Wormer, M.B. and Iwamchuck, R.M.	Compendium of guides for estimation estimation and measurement techniques	*Chemical Engineering Progress*, April 1992.

Land treatment/application farming is a disposal method in which a waste containing chemicals is applied onto or incorporated into soil. While this disposal method is considered a release to land, any volatilization of chemicals into the air occurring during the disposal operation is a fugitive air release.

Chemicals may also be disposed to a surface impoundment. A surface impoundment is a natural topographic depression, man-made excavation, or diked area formed primarily of earthen materials (although some may be lined with man-made materials) that is designed to hold an accumulation of liquid wastes or wastes containing free liquids. Examples of surface impoundments are holding, settling, storage, and elevation pits; ponds; and lagoons. If the pit, pond, or lagoon is intended for storage or holding without discharge, it would be considered to be a surface impoundment used as a final disposal method.

If a volatile chemical (e.g., benzene) that is in waste sent to a surface impoundment partially evaporates, that part of the release is a fugitive air emission. Chemicals released to surface impoundments that are used merely as part of a wastewater treatment process generally are not releases to land. However, if the impoundment accumulates sludges containing chemicals, this accumulation is a land release unless the sludges are removed and otherwise disposed (storage tanks are not considered to be a type of disposal).

Other land disposal includes chemicals released to land that does not fit the categories of landfills, land treatment, or surface impoundment. This other disposal would include any spills or leaks of chemicals to land.

8.3.2 Release Quantification Methods

Once the process has been analyzed to determine the points and sources of releases, the amounts of release can be estimated. These methods can be characterized and a general hierarchy of preferences can be used to determine how releases may best be quantified. The following list shows the potential methods for quantifying releases, in order, from the most preferred to the least preferred.

a. Measured release data for the chemical or indirectly measured release data using mass balance or stoichiometric ratios.

b. Release data for a surrogate chemical with similar release-affecting properties and used in the same (or very similar) process. Surrogate data may be measured, indirectly measured, modeled, or some combination of these. Some emission factors would be considered to be surrogate data.

c. Modeled release estimates:
   1. Mathematically modeled (e.g., process design software, mass transfer models, etc.) release estimates for the chemical or by analogy to a surrogate chemical.
   2. Rule-of-thumb release estimates, or those developed using engineering judgment.

This order of preference is expected to apply generally to most cases of release assessment. However, judgment may dictate that, in some cases, the order within the hierarchy should be changed. Examples of such a change of hierarchy order may include when data are judged to be unreliable or unrepresentative. Also, some estimates may be based on a combination of two or more of these methods. In other cases, the method used to generate some estimates may not be known due to lack of documentation (e.g., industry survey results). Judgment may also include factors such as how rigorous the modeling is in various modeling methods.

This textbook is intended to apply primarily to situations in which chemical manufacturing processes are being designed. For process design, measured release data are generally not available, requiring the use of emission factors and modeled estimates. However, for the sake of completeness, the material below will examine all methods for quantifying releases.

8.3.2.1 Measured Release Data for the Chemical

Measured release data for a chemical of interest are generally not applicable to processes in design but rather to existing processes. The following examples illustrate how data may be used to generate estimates of releases. For a continuous process, a release can be estimated by calculating the product of three measures: a chemical's average concentration, the average volumetric flow rate of the release stream containing that chemical, and the density of the release stream. Similarly, a release could be estimated using the chemical's weight fraction and the mass flow rate of the release stream.

Example 8.3-1

A wastewater pretreatment plant runs every day and averages 1.5 million gallons per day. The following chromium concentrations were measured. How much chromium does the POTW receive annually?

Sample Number	Cr(III) in ppm (mg/kg)
1	2.7
2	0.9
3	4.1
4	3.4
5	5.1
6	2.3
7	3.8

Solution: The average Cr(III) concentration of the seven samples is 3.2 mg/kg. We can assume that the effluent has a density very close to that of water, 3.78 kg/gal. The average mass effluent flow is 1,500,000 gal/day effluent × 3.78 kg/gal = 5,670,000 kg/day. Multiplying the average total mass daily flow by the average concentration and operating days per year yields the annual estimate: 5,670,000 kg/day effluent × 3.2 kg Cr(III) / 1,000,000 kg effluent × 365 days/yr = 6,600 kg/yr Cr(III).

8.3.2.2   Release Data for a Surrogate Chemical

Release data for analogous or surrogate chemicals from existing processes can sometimes be used to estimate releases of chemicals of interest either in processes in design or in existing processes. To use surrogate chemical data, similarities must exist in some physical/chemical properties of the chemicals, unit processes and their operating conditions, and quantities of chemical throughput. For instance, in Example 8.3-1, if an estimate of the release rate for Cr(VI), which is a different oxidation state of chromium than Cr(III), was desired, then Cr(III) might be used as a surrogate. If data were available indicating that a typical ratio of Cr(III) to Cr(VI) were 1000:1, then the release rate of Cr(VI) might be estimated as 1/1000 of the release rate of Cr(III).

8.3.2.3   Emission Factors

Emission factors are commonly used to estimate releases to air. A number of unit operation-specific emission factor databases have been compiled for the US EPA. For air pollutants regulated through the National Ambient Air Quality Standards (NAAQS) (SO_2, NO_2, CO, O_3, hydrocarbons, particulates), the AP-42 compilation with supplements provides emission factors for several industrial sectors, including stationary and mobile combustion units, refuse incineration, storage tank emissions, and units in the chemical, metallurgical, minerals, food and agriculture, and wood products industries (US EPA, 1998).

The emission factors in the AP-42 database are not, in general, compound-specific. Several databases are available from the EPA for estimating emissions of specific organic and inorganic compounds from a variety of processes in the chemical manufacturing industry. The earliest of these documents (*Organic Chemical Manufacturing*) is a ten-volume report which summarizes emission factors of 39 compounds for several production and use processes (US EPA, 1980). These emission factors are tabulated on a process-unit-by-process-unit basis and include reactors, separation columns, storage tanks, fugitive emission, transfer and handling operations, and wastewater treatment units. A more recent compilation of emission factors for hazardous compounds is provided in documents titled *Locating and Estimating Air Emissions from Sources* (L&E) (US EPA, 1998). The L&E database contains much of the data in the previously mentioned volumes of *Organic Chemical Manufacturing* (OCM).

The most recent and comprehensive emission factor document from the US EPA is titled the *Factor Information Retrieval* (FIRE) System. It contains EPA's recommended criteria and hazardous air pollutant (HAP) emission estimation factors. The above-mentioned emission factor sources are included in the Air CHIEF CD-ROM available from the US EPA (US EPA, 1998). In addition to emission factors, the Air CHIEF software contains source classification codes (SCC) for over 9000 chemicals; a program to calculate controlled emissions for particulate matter (PM-2.5 and PM-10); a biogenic emission inventory system; a spreadsheet for estimating volatile organic compound (VOC) emissions from treatment, storage and disposal facility processes; and a menu-driven computer program for estimating emissions from wastewater treatment systems. (See Appendix F for more information on these resources.)

8.3.2.4   Emissions from Process Units and Fugitive Sources.

The rate of emission (E, mass/time) from process units and operations such as reactors, distillation columns, storage tanks, transportation and handling operations, and fugitive sources can be calculated using the formula

$$E = m_{VOC}\, EF_{av}\, M \qquad\qquad \text{(Eq. 8-4)}$$

where

m_{VOC} is the mass fraction of the volatile organic compound in the stream or process unit

EF_{av} (kg emitted/10^3 kg throughput) is the average emission factor ascribed to that stream or process unit

M is the mass flow rate through the unit (mass/time)

Table 8.3-2 shows average emission factors obtained from the US Environmental Protection Agency L&E database for several process units found in chemical plants.

Table 8.3-3 lists the average emission factors for fugitive sources found in refineries, gas plants and synthetic organic chemical manufacturing operations.

Table 8.3-2 Average Emission Factors for Chemical Process Units Calculated from the US EPA L&E Database (Shonnard, et al. 1995).

Process unit	EF_{av}; (kg emitted/10^3 kg throughput)
Reactor vents	1.50
Distillation columns vents	0.70
Absorber units	2.20
Strippers	0.20
Sumps/decanters	0.02
Dryers	0.70
Cooling towers	0.10

Note that the form of the emission factors given in Table 8.3-3 is slightly different from the form shown in Equation 8-4. This is not unusual and consequently, emission factor equations should be used with care and attention to the details of units. Also note that in Table 8.3-3, liquid streams are classified into light and heavy service. A light liquid is defined as a stream in which the most volatile component (present > 20% by weight) has a vapor pressure at the stream temperature of > 0.04 lb/in^2.

Table 8.3-3 Average Emission Factors for Estimating Fugitive Emissions

Source	Service	Emission Factor (kg/hour/source)		
		$SOCMI$[a]	$Refinery$[b]	$Gas\ Plant$[a]
Valves	Hydrocarbon gas	0.00597	0.027	
	Light liquid	0.00403	0.011	
	Heavy liquid	0.00023	0.0002	
	Hydrogen gas		0.0083	
	All			0.02
Pump Seals	Light liquid	0.0199	0.11	
	Heavy liquid	0.00862	0.021	
	Liquid			0.063
Compressor Seals	Hydrocarbon gas	0.228	0.63	
	Hydrogen gas		0.05	
	All			0.204
Pressure-relief Valves	Hydrocarbon gas	0.104	0.16	
	Liquid	0.007[c]	0.007[c]	
	All			0.188
Flanges and other connections	All	0.00183	0.00025	0.0011
Open-ended lines	All	0.0017	0.002	0.022
Oil/water separators (uncovered)	All		14,600[d]	
Sampling connections	All	0.015		

[a]Synthetic Organic Chemical Manufacturing Industries, US EPA (1993) except as noted.
[b]US EPA(1998) except as noted.
[c]US EPA(1985b).
[d]Based on limited data (330,000 bbl/day capacity) (US EPA, 1998).

Example 8.3-2

A company wants to sell a newly-developed substitute for dimethylsulfoxide (DMSO) in pharmaceutical manufacturing. 300,000 lb/yr of this new drop-in substitute for DMSO could be sold to two customer pharmaceutical plants that will each purchase the same volume of the substitute. US EPA requires submission of a Premanufacture Notice (PMN) for this new chemical. The PMN form requests estimated releases to all media from downstream customers (the pharmaceutical plants) in units of kg/day. Generate these estimates.

Solution: Assume that the physical/chemical properties of the new chemical match the properties of DMSO very closely. The customer plants are reluctant to divulge proprietary process information, but one detail they give is that the facilities operate five days each week for 50 weeks of the year. Based on a search of several release estimation resources, you find that the US EPA document entitled "Compilation of Air Pollutant Emission Factors" (often referred to as AP-42) contains emission factors for DMSO for the pharmaceutical industry. The reference table notes that these data are based on an industry survey. Because both the release-affecting properties and the industrial use of the new chemical and DMSO are so similar, you have concluded that these DMSO emission factors are suitable surrogates for estimating releases of your company's new chemical. The AP-42 DMSO emission factors for the pharmaceutical industry are 1% to air emissions, 28% to sewer, and 71% to incineration (AP-42, Section 6.13). You contacted the potential customers, and they acknowledged that these AP-42 emission factors reasonably represent their facilities.

Calculation: First calculate the amount used per day at each site, then apply the emission factors to calculate the release to the media.

1. Daily average amount (mass) used at each site =

   300,000 lb/yr/2.20 kg/lb/2 sites/(5 days/week $\times$ 50weeks/yr) = 273 kg/day

2. Partition the daily use amount to the media based on the emission factors:

   release to air = 1% of 273 kg/day = 2.7 kg/day

   release to water = 28% of 273 kg/day = 76 kg/day

   release to incineration = 71% of 273 kg/day = 194 kg/day

Example 8.3-3

Estimate the fugitive emissions from a chemical manufacturing facility that contains 1400 valves (168 in gas service), 3048 flanges and other connectors, 27 pumps (all in liquid service), 20 pressure relief valves, and 20 sampling connections (Berglund, et al., 1989) Determine the total fugitive emissions in pounds per year using the Synthetic Organic Chemical Manufacturing Industry (SOCMI) emission factors given in Table 8.3-3. Assume that the process fluids are composed almost entirely of volatile compounds. (Allen and Rosselot, Pollution Prevention for Chemical Processes, © 1997, This material is used by permission of John Wiley & Sons, Inc.)

Solution: For the valves in gas service, the emissions are:

(168 valves) $\times$ (0.00597 kg VOC/(hr − valve))(2.2 lb/kg)(24 hr/day)(365 day/year)
           = 19,300 lb/yr

For the valves in light liquid service, the emissions are:

$$((1400 - 168)\text{valves}) \times (0.00403 \text{ kg VOC}/(\text{hr} - \text{valve})(2.2 \text{ lb/kg})$$
$$(24 \text{ hr/day})(365 \text{ day/year})) = 95{,}700 \text{ lb/yr}$$

Summing these values gives the total estimated emissions from valves, which is 115,000 lb/year. Results for the other component types are given below.

Equipment type	Emissions, lb/yr	% of emissions, by equipment type
Valves	115,000	41
Flanges	108,000	38
Pumps	10,300	3.7
Pressure relief valves	40,200	14
Open-ended lines	700	0.25
Sampling connections	5,700	2.1
Total	276,000	100

8.3.2.5   Losses of Residuals from Cleaning of Drums and Tanks

Some limited data are available to help estimate the amounts of material that may be lost from cleaning drums, tanks, and similar vessels for containing chemicals. To estimate the amount of material that may be lost from cleaning of drums and tanks, the nature of the cleaning process should be considered; the capacities, shapes, and materials of construction of the vessels to be cleaned; the cleaning schedule; the residual quantity of the chemical in the vessels; the type and amount of solvent used (aqueous or organic); the solubility/miscibility of the chemical in the solvent; and, if applicable, any treatment of wastewater containing the chemical. Vessels may be either rinsed or flushed with aqueous or organic solvent, which may depend in part upon the solubility of the chemical in various solvent options. If an aqueous solvent is used, the waste will typically be sent to wastewater treatment or reworked into a new batch. If an organic solvent is used, it will typically either be recycled or incinerated. Some of these issues are discussed in further detail below.

One approach to quantifying the loss is to assume a certain residual fraction of the chemical based on approximate vessel capacity. The first step in this approach is to determine the type and volumetric capacity of the vessels to be cleaned (or volume of chemical per batch processed through those vessels). Once all vessels to be cleaned are identified, the amount of chemical in those vessels during operation can be determined. This will be the batch volume. Adjustments should be made for the concentration of the chemical in the mass contained in the equipment. If the size of the vessels is not known, values can be assumed based on information from literature or best estimates. Factors that could significantly impact residual volumes should also be considered. Such differences could be packing in a reaction vessel that would have a significant liquid hold up.

The likely frequency of cleaning of the vessels must be determined. Clean-out after every batch may occur if quality of product demands it. Other reasons for frequent clean-out are changes in the type of batch being run (e.g., color change in paint mixing), possible solidification of product within a reactor, or proper operation of mechanical equipment (e.g., a plate and frame filter may not close if not cleaned after each use). Clean-out after every batch should only be assumed if a specific reason for such cleaning can be identified. Otherwise, cleaning only after one week's run or at the completion of a campaign may be assumed.

Another factor to be considered is the possible recycle of cleaning effluent back to the process. Although such flushing may occur after every batch, it may not result in a release (i.e., the residue may be added to the product stream or used in the next batch). This can occur when mixing vessels are rinsed with water that is subsequently reworked into batches of similar product, or when product is to be subsequently isolated from the cleaning solvent by distillation. Cleaning that results in a release may be very infrequent in these cases.

The amount of the chemical that remains in equipment prior to cleaning is the amount available for loss. Many parameters affect this amount, including the design configuration of the equipment, the method of removing or unloading the chemical from the equipment, the viscosity of the chemical, and the material of construction or lining of the equipment. Sometimes the amount of chemical available for loss is calculated as a fraction or percent of the total amount of the material in the equipment during normal operation of a batch.

Table 8.3-4 presents factors for estimating percent chemical remaining in drums and tanks after unloading. These factors were derived from a pilot scale research project investigating the effect of the design configuration of the equipment, the method of removing or unloading the chemical from the equipment, the viscosity of the chemical, and the material of construction or lining of the equipment on residue quantities (PEI, 1986). It was concluded that the amount of residue is generally influenced most by the method of unloading. The viscosity of the chemical and the design configuration of the equipment appeared to affect residue quantities to a lesser degree. Material of construction or lining of the equipment appeared to have little effect on residue quantities. The values listed in the table represent residue quantities as a weight percent of vessel capacity (pounds chemical residue per 100 pounds of chemical). The values presented in Table 8.3-4 may be used to represent typical residues, which should only be applied to similar vessel types, unloading methods, and bulk fluid materials. The research was performed with materials with viscosities below 100 cp. For materials with significantly higher viscosities (>200 cp), estimates of percent residue were made based on engineering judgment.

Example 8.3-4

A facility purchased 42,500 pounds of hydrazine this year. To determine the value of the hydrazine lost as residual, the company requests us to provide estimated releases from cleaning the emptied 55-gallon drums.

Solution: You know that the hydrazine is received in steel drums and is pumped into a process vessel. A chemical handbook shows that the viscosities of hydrazine are nearly the same as water at ambient temperatures. Using Table 8.3-4, the loss fraction

Table 8.3.4 Summary of Residue Quantities from Pilot-Scale Experimental Study, Weight Percent

Unloading method	Vessel type	Material				
		Surfactant solution[a]	Water[b]	Kerosene[c]	Motor oil[d]	Material[e] with viscosity >200 cp
Pumping	Steel drum	3.06	2.29	2.48	2.06	3
Pumping	Plastic drum	Not available	3.28	2.61	2.30	4
Pouring	Bung–top steel drum	0.485	0.403	0.404	0.737	1
Pouring	Open–top steel drum	0.089	0.034	0.054	0.350	0.5
Gravity drain	Slope–bottom steel tank	0.048	0.019	0.033	0.111	0.1
Gravity drain	Dish–bottom steel tank	0.058	0.034	0.038	0.161	0.2
Gravity drain	Dish–bottom glass-lined tank	0.040	0.033	0.040	0.127	0.2

[a]Surfactant solution viscosity = 3 centipoise, surface tension = 31.4 dynes/cm^2.

[b]For water, viscosity = 1 centipoise, surface tension = 77.3 dynes/cm^2.

[c]For kerosene, viscosity = 5 cent poise, surface tension = 29.3 dynes/cm^2.

[d]For motor oil, viscosity = 97 centipoise, surface tension = 34.5 dynes/cm^2.

[e]Residue quantities for high viscosity material were not defined by the study; thus, the quantities presented are estimates of a reasonable worst case scenario based on engineering judgment.

Source: PEI 1986.

for water pumped from steel drums is 2.29%. We can use data for water residual as surrogate data for estimating hydrazine residual. Calculation: 42,500 lb/yr × 2.29 lb loss/100 lb delivered = 973 lb hydrazine estimated to be lost as drum residue.

8.3.2.6   Secondary Emissions from Utility Sources

Utility consumption in chemical processes is a large generator of environmental impact. Table 8.3-5 lists emission factors for uncontrolled releases for residual and distillate oil combustion. Table 8.3-6 shows similar emission factors for the combustion of natural gas. Each emission factor listed in Tables 8.3-5 and 8.3-6 is based on the volume of fuel burned. In order to relate the emissions of the pollutants to the energy demand in the process, we must first know the fuel value (energy/volume of fuel burned). The efficiency of the boiler supplying the energy transfer agent (steam) needs to be incorporated. The emissions for fuel and natural gas combustion are given by the following general equation:

$$E \, (\text{kg/unit/yr}) = (ED)(EF \, (FV)^{-1}(BE)^{-1} \qquad \text{(Eq. 8-5)}$$

where
   ED is the energy demand of a process unit (energy demand/unit/yr)
   EF is the emission factor for the fuel type (kg/volume of fuel combusted)
   FV is the fuel value (energy/volume fuel combusted)
   BE is the boiler efficiency (unitless; 0.75 to 0.90 is a typical range of values)

A typical value for the fuel value of natural gas is shown in the footnotes of Table 8.3-6. Typical heating values for solid, liquid, and gaseous fuels are provided in Table 8.3-7. If the boiler efficiency has already been accounted for in the process simulation or analysis, the factor BE is not necessary in the equation above.

A reasonable approximation for criteria pollutant emission factors (short tons emitted/kW hr) for electricity use in a chemical process can be derived from the data shown in Table 8.3-8 by dividing the emissions by the power generated (bottom row). An average factor (derived from the right column) will be used if the fuel utilized to generate the electricity is not known. The electricity generated by coal, petroleum, and gas fired units does not add to the total electricity generated reported in the table because this total also includes other sources like nuclear power and renewable energy. If not specified in the mass and energy balances for the process, an efficiency for the electric motor or other device must be included. Efficiencies can range from 0.75 to 0.95, depending upon the size and type of the motor. Estimating emissions from electricity consumption in processes is given by;

$$E \, (\text{kg/unit/yr}) = (ED)(EF) \, (ME)^{-1} \qquad \text{(Eq. 8-6)}$$

where ED is the electricity demand of the unit per year and ME is the efficiency of the device.

Table 8.3-5   Criteria Pollutant Emission Factors (*EF*) for Uncontrolled Releases from Residual and Distillate Oil Combustion (EPA, 1998).

Firing Configuration (SCC)[a]	SO_2[b] kg/ 10^3 L	SO_3 kg/ 10^3 L	NO_x[c] kg/ 10^3 L	CO[d,e] kg/ 10^3 L	Filterable PM kg/ 10^3 L	TOC[f] kg/ 10^3 L	CO_2 kg/ 10^3 L
Utility boilers							
No. 6 oil-fired, normal firing	19S	0.69S	8	0.6	g	0.125	3,025
No. 6 oil-fired, tangential firing	19S	0.69S	5	0.6	g	0.125	3,025
No. 5 oil-fired, normal firing	19S	0.69S	8	0.6	g	0.125	
No. 5 oil-fired, tangential firing	19S	0.69S	5	0.6	g	0.125	
No. 4 oil-fired, normal firing	18S	0.69S	8	0.6	g	0.125	
No. 4 oil-fired, tangential firing	18S	0.69S	5	0.6	g	0.125	
Industrial boilers							
No. 6 oil-fired (1-02-004-01/02/03)	19S	0.24S	6.6	0.6	g	0.154	3,025
No. 5 oil-fired (1-02-004-04)	19S	0.24S	6.6	0.6	g	0.154	3,025
Distillate oil-fired (1-02-005-01/02/03)	17S	0.24S	2.4	0.6	g	0.03	
No. 4 oil-fired (1-02-005-04)	18S	0.24S	2.4	0.6	g	0.03	
Commercial/institutional/residential combustors							
No. 6 oil-fired	19S	0.24S	6.6	0.6	g	0.193	3,025
No. 5 oil-fired	19S	0.24S	6.6	0.6	g	0.193	
Distillate oil-fired	17S	0.24S	2.4	0.6	g	0.067	
No. 4 oil-fired	18S	0.24S	2.4	0.6	g	0.067	
Residential furnace (No SCC)	17S	0.24S	2.2	0.6	0.3	0.299	

[a]SCC = Source Classification Code.

[b]S indicates that the weight % of sulfur in the oil should be multiplied by the value given.

[c]Expressed as NO_2. Test results indicate that at least 95% by weight of NOx is NO for all boiler types except residential furnaces, where about 75% is NO. For utility vertical fired boilers use 12.6 kg/10^3 L at full load and normal (>15%) excess air. Nitrogen oxides emissions from residual oil combustion in industrial and commercial boilers are related to fuel nitrogen content, estimated by the following empirical relationship: kg NO_2/10^3 L = 2.465 + 12.526(N), where N is the weight percent of nitrogen in the oil.

[d]CO emissions may increase by factors of 10 to 100 if the unit is improperly operated or not well maintained.

[e]Emission factors for CO_2 from oil combustion should be calculated using kg CO_2/10^3 L oil = 31.0 C (distillate) or 34.6 C (residual), or use data in far right column.

[f]Filterable PM is that particulate collected on or prior to the filter of an EPA Method 5 (or equivalent) sampling train. PM-10 values include the sum of that particulate collected on the PM-10 filter of an EPA Method 201 or 201A sampling train and condensable emissions as measured by EPA Method 202.

[g]Particulate emission factors for residual oil combustion are, on average, a function of fuel oil grade and sulfur content:

No. 6 oil: 1.12(S) + 0.37 kg/10^3 L, where S is the weight % of sulfur in oil.

No. 5 oil: 1.2 kg/10^3 L

No. 4 oil: 0.84 kg/10^3 L

No. 2 oil: 0.24 kg/10^3 L

Table 8.3-6 Emission Factors for Sulfur Dioxide (SO_2), Nitrogen Oxides (NOx), and Carbon Monoxide (CO) from Natural Gas Combustion[a] (US EPA, 1998).

Combustor Type	SO_2[b] kg/ $10^6 m^3$	lb/ $10^6 ft^3$	NOx[c] kg/ $10^6 m^3$	lb/ $10^6 ft^3$	CO kg/ $10^6 m^3$	lb/ $10^6 ft^3$	CO_2 kg/ $10^6 m^3$	lb/ $10^6 ft^3$
Utility/Large Industrial Boilers								
Uncontrolled	9.6	0.6	3040	190[d]	1344	84	1.9×10^6	1.2×10^5
Controlled—Low NOx burners	9.6	0.6	2240	140[d]	1344	84	1.9×10^6	1.2×10^5
Controlled—Flue gas recirculation	9.6	0.6	1600	100	1344	84	1.9×10^6	1.2×10^5
Small Industrial Boilers								
Uncontrolled	9.6	0.6	1600	100	1344	84	1.9×10^6	1.2×10^5
Controlled—Low NOx burners	9.6	0.6	800	50[d]	1344	84	1.9×10^6	1.2×10^5
Controlled—Flue gas recirculation	9.6	0.6	512	32	1344	84	1.9×10^6	1.2×10^5
Commercial Boilers								
Uncontrolled	9.6	0.6	1600	100	330	21	1.9×10^6	1.2×10^5
Controlled—Low NOx burners	9.6	0.6	270	17	425	27	1.9×10^6	1.2×10^5
Controlled—Flue gas	9.6	0.6	580	36	ND	ND	1.9×10^6	1.2×10^5
Residential Furnaces								
Uncontrolled	9.6	0.6	1500	94	640	40	1.9×10^6	1.2×10^5

[a]Units are kg of pollutant/10^6 cubic meters natural-gas-fired and lb. of pollutant/10^6 cubic feet natural-gas-fired. Based on an average natural-gas-fired higher heating value of 8270 kcal/m^3 (1000 Btu/scf). The emission factors in this table can be converted to other natural gas heating values by multiplying the given emission factor by the ratio of the specified heating value to this average heating value. ND = no data.
[b]Based on average sulfur content of natural gas, 4600 g/10^6 Nm^3 (2000 gr/10^6 scf).
[c]Expressed as NO_2. For tangentially fired units, use 4400 kg/10^6 m^3 (275 lb/10^6 ft^3). Note that NOx emissions from controlled boilers will be reduced at low load conditions.
[d]Emission factors apply to packaged boilers only.

Table 8.3-7 Typical Heating Values for Solid, Liquid, and Gaseous Fuels (Perry and Green, 1997).

Fuel Oil, Btu/US gal	
No. 1	137,000
No. 2	139,600
No. 4	145,100
No. 5	148,800
No. 6	152,400
Propane, Btu/US gal	91,500
Natural gas, Btu/Standard ft.^{3}	1,035
Coal, Btu/lb	
Bituminous	11,500–14,000
Subbituminous	8,300–11,500
Lignite	6,300–8,300

Table 8.3-8   Emissions from Fossil-Fueled Steam-Electric Generating Units (EFacts, 1992).

Emission (thousands of short tons[a])	Coal Fired	Petroleum Fired	Gas Fired	Total[b]
Carbon dioxide	1,499,131	87,698	156,748	1,747,418
Sulfur dioxide	14,126	637	1	14,766
Nitrogen oxides	6,879	208	599	7,690
Power generated (billion kW hr)	1,551	111	264	2,796

[a]1 short ton equal to 2,000 pounds or 0.8929 metric tons.
[b]Also include light oil, methane, coal/oil mixture, propane gas, blast furnace gas, wood, and refuse.

8.3.3   Modeled Release Estimates

Many process design software programs contain rigorous methods for calculating contents of output streams from process equipment, and some of these output streams are points of release of the process. However, process design software often does not account for all of the releases from a process or is not the most efficient method for generating some release estimates. The material below provides guidance and methods for calculating some of the release points not normally included in process design software and conventional methods.

8.3.3.1   Loading Transport Containers

AP-42 (US EPA, 1985), a US EPA document on emissions issues and estimation methods, discusses the estimation of losses due to vapors generated from bulk loading of petroleum products. Additional information on AP-42 can be found in Appendix F. This model and related information should apply to other non-petroleum organic chemicals that can generate vapors. The following discussion of the model and issues contains many excerpts from AP-42.

Loading losses are a primary source of evaporative emissions from rail tank car, tank truck, and similar bulk loading operations. Loading losses occur as vapors in "empty" cargo tanks are displaced to the atmosphere by the liquid being loaded into the tanks. These vapors are a composite of (1) vapors formed in the empty tank by evaporation of residual product from previous loads (if applicable), (2) vapors transferred to the tank in vapor balance systems (discussed below) as product is being unloaded (if applicable), and (3) vapors generated in the tank as the new product is being loaded. The quantity of evaporative losses from loading operations is, therefore, a function of the following parameters:

- Physical and chemical characteristics of the previous cargo.
- Method of unloading the previous cargo.
- Operations to transport the empty carrier to a loading terminal.
- Method of loading the new cargo.
- Physical and chemical characteristics of the new cargo.

The principal methods of cargo carrier loading are illustrated in Figures 8.3-1, 8.3-2, and 8.3-3. In the splash loading method, the fill pipe dispensing the cargo is lowered only part way into the cargo tank. Significant turbulence and vapor/liquid contact occur during the splash loading operation, resulting in high levels of vapor generation and loss. If the turbulence is great enough, liquid droplets will be entrained in the vented vapors.

A second method of loading is submerged loading. Two types are the submerged fill pipe method and the bottom loading method. In the submerged fill pipe method, the fill pipe extends almost to the bottom of the cargo tank. In the bottom loading method, a permanent fill pipe is attached to the cargo tank bottom. During most of submerged loading by both methods, the fill pipe opening is below the liquid surface level. Liquid turbulence is controlled significantly during submerged loading, resulting in much lower vapor generation than encountered during splash loading. The recent loading history of a cargo carrier is just as important a factor in loading losses as the method of loading. If the carrier has carried a nonvolatile liquid such as fuel oil, or has just been cleaned, it will contain vapor-free air. If it has just carried gasoline and has not been vented, the air in the carrier tank will contain volatile organic vapors, which will be expelled during the loading operation along with newly generated vapors.

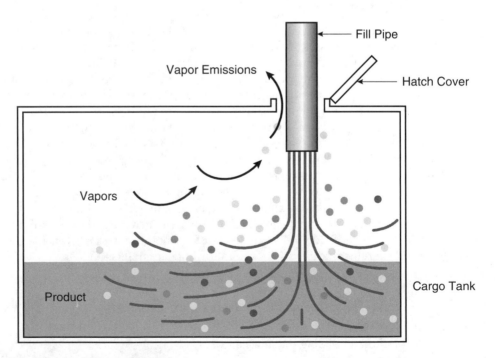

Figure 8.3-1   Splash loading method.

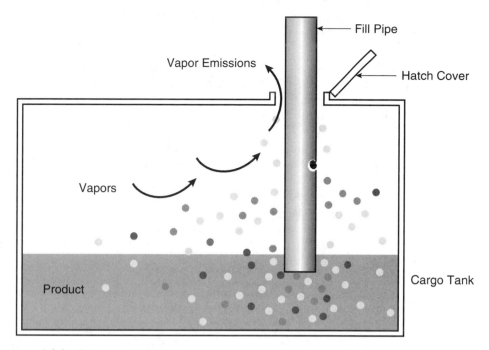

Figure 8.3-2 Submerged fill pipe.

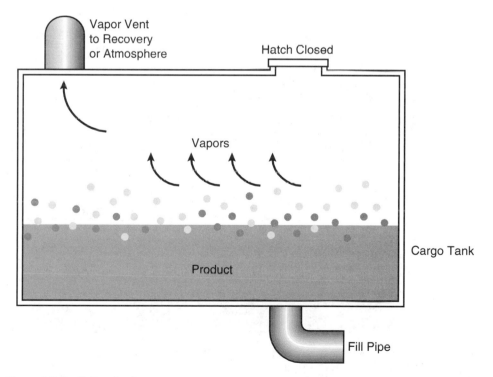

Figure 8.3-3 Bottom loading.

233

Cargo carriers are sometimes designated to transport only one product, and in such cases are practicing "dedicated service." Dedicated gasoline cargo tanks return to a loading terminal containing air fully or partially saturated with vapor from the previous load. Cargo tanks may also be "switch loaded" with various products, so that a nonvolatile product being loaded may expel the vapors remaining from a previous load of a volatile product such as gasoline. These circumstances vary with the type of cargo tank and with the ownership of the carrier, the liquids being transported, geographic location, and season of the year.

One control measure for vapors displaced during liquid unloading at bulk plants or service stations is called "vapor balance service". The cargo tank on the truck retrieves the vapors displaced, then the truck transports the vapors back to the loading terminal. Figure 8.3-4 shows a tank truck in "vapor balance service" filling an underground tank and taking on displaced gasoline vapors for return to the terminal. A cargo tank returning to a bulk terminal in "vapor balance service" normally is saturated with organic vapors, and the presence of these vapors at the start of submerged loading of the tanker truck results in greater loading losses than encountered during non-vapor balance, or "normal", service. Vapor balance service is usually not practiced with marine vessels, although some vessels practice emission control by means of vapor transfer within their own cargo tanks during

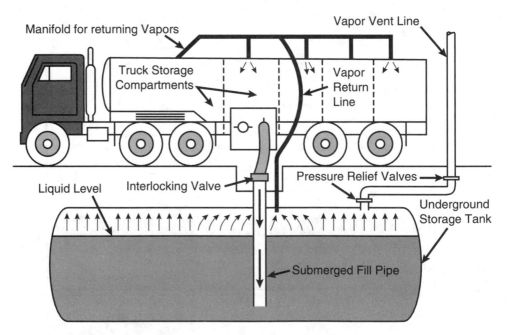

Figure 8.3-4   Tank truck unloading into a service station underground storage tank, practicing "vapor balancing."

ballasting operations. More information about ballasting losses may be found in AP-42 section 5.2.2.1.2.

If the evaporation rate is negligible in comparison to the displacement rate, emission losses from loading liquid, given in units of pounds per 1000 gallons of liquid loaded, can be estimated (with a probable error of 30 percent) using Equation 8-7 from AP-42 (US EPA 1985):

$$L_L = 12.46 \times SPM/T \qquad \text{(Eq. 8-7)}$$

where

L_L = loading loss [(lbs/10^3 gal) of liquid loaded]
S = saturation factor [dimensionless] (see Table 8.3-9)
P = true vapor pressure of liquid loaded [psia]
M = molecular weight of vapors [lb/lb-mole]
T = temperature of bulk liquid loaded [°R (°F + 460)]

This equation can be used for liquids with vapor pressures below 0.68 psia (35 torr). Equation 8-7 can be used to estimate the loading loss of a single chemical in a liquid mixture of two or more chemicals. In such a case, the true vapor pressure of the single chemical may overestimate the loss. If the mole fraction of the chemical in the mixture is expected to be less than unity, we may wish to account for this decrease in the calculated loss. For mixtures, the vapor pressure of a chemical component (P_a in atmospheres) can be calculated using Raoult's Law as shown in Equation 8-8:

$$P_a = P \times X_a \qquad \text{(Eq. 8-8)}$$

where

P = Vapor pressure of pure substance, atm
X_a = Mole fraction of component of component a

Raoult's Law may be too simplistic in certain circumstances. For chemicals that are solid at ambient temperature and volatilize by sublimation, simple techniques for estimating physical properties are not adequate for predicting vapor pressure.

Saturation factors for tank truck loading of petroleum liquids are expected to range from 0.5 to 1.45 (US EPA, 1985). If complete saturation of the vapor space within a vessel is assumed, the saturation factor is equal to 1. Table 8.3-9 lists typical saturation factors by mode of loading for tank trucks, rail cars, drums, and small containers.

In some cases, emission rates, as well as emissions per filling event, must be calculated. The loading loss calculated in Equation 8-7 for vapor being displaced from a container may be converted to a generation rate, which is shown in Equation 8-9. The differences between Equations 8-7 and 8-9 are simply converting units to metric units, separating the universal gas constant R from the equation coeffi-

Table 8.3-9 Saturation Factors for Loading Operations

Mode of operation	Saturation factor (S), dimensionless
Submerged loading:	
Clean cargo vessel	0.50
Normal dedicated service	0.60
Dedicated vapor balance service	1.00
Drums and small containers	0.50
Splash loading	
Clean cargo vessel	1.45
Normal dedicated service	1.45
Dedicated vapor balance service	1.00
Drums and small containers	1.00

Sources: US EPA 1985 and US EPA 1991.

cient, and multiplying by the amount of volume displaced over time (represented by container volume (V) and fill rate (r)).

$$G = SMVrP/(3600RT_L) \qquad \text{(Eq. 8-9)}$$

where

> G = vapor generation of component [g/sec]
> S = saturation factor [dimensionless]
> M = molecular weight of vapors [g/g-mole]
> V = volume of container [cm^3]
> r = fill rate [containers/hr]
> P = vapor pressure of component [atm, at T_L]
> R = universal gas constant [82.05 atm*cm^3/g-mol*K]
> T_L = liquid temperature [K]

EPA has chosen some default parameters for use when information is not available to determine the actual or expected values of these parameters for a given situation. These default values are shown in Table 8.3-10. Some typical and conservative values for fill rates associated with various transfer operations, with accompanying values for other parameters, are given in Table 8.3-10.

Example 8.3-5

ABC Chemical Company plans to produce and sell 50,000 pounds of n-butyl lactate (NBL) this year. All of this product will be shipped in 55-gallon drums. ABC will produce 5,000 lb/day of NBL for 10 days, and each day's production is drummed in 30 minutes. How much of the NBL product will be emitted daily as fugitive vapors from ABC's drumming operation?

Solution: The fugitive emission may be calculated using Equation 8-9, and the parameters needed for the calculation follow (information/data source for each parameter is shown in parentheses).

Table 8.3-10   Transfer Operation Default Parameters for Equation 8-9.

Vessel / Parameters	container fill rate, [hr^{-1}]	fill rate, [gal/min]	volume, V [cm^3]	saturation factor, S [dimensionless]
Drumming (55 gal.)				
Conservative Case	30	27.5	2.1×10^5	1.0
Typical Case	20	18.3	2.1×10^5	0.5
Cans/Bottles (5 gal.)				
Conservative Case	30	2.5	1.9×10^4	1.0
Typical Case	20	1.7	1.9×10^4	0.5
Tank Truck (5000 gal.)				
Conservative Case	2	167	1.9×10^7	1.0
Typical Case	2	167	1.9×10^7	1.0
Tank Car (20,000 gal.)				
Conservative Case	1	333	7.6×10^7	1.0
Typical Case	1	333	7.6×10^7	1.0

Source: US EPA, 1991, Table 4-11.

S (saturation factor) = 0.5 (Table 8.3-9, typical value chosen)
M (molecular weight of vapor in g/g-mole) = 146.2
V (volume of container in cm^3) = 2.1×10^5 (Table 8.3-10 for 55-gal drum)
r (fill rate as containers/hr) = 5,000 lb/day / [0.98 (specific gravity, NIOSH, 1997) × 8.33 lb/gal] / 55 gal/drum = 11 drums/day / 0.5 hr/day = 22 drums/hr
P (vapor pressure of NBL in atm at T$_L$) = 0.4 mm (NIOSH, 1997) / 760 mm/atm = 0.0005 atm
R (universal gas constant in atm × cm^3/g-mol × K) = 82.05
T$_L$ (liquid temperature in K) – 68EF – 20EC – 293 K

These parameters may be placed in the equation:

$$G = 0.5 \times 146.2 \times 210,000 \times 22 \times 0.0005 / (3.600 \times 82.05 \times 293) = 2.05 \times 10^{-3}\,\text{g/sec}$$
$$= 2.05 \times 10^{-3}\,\text{g/sec} \times 3,600\,\text{sec/hr} \times 0.5\,\text{hr/day} = 3.7\,\text{g/day} = 3.7 \times 10^{-3}\,\text{kg/day}$$

8.3.3.2   Evaporative Losses from Static Liquid Pools

Vapors may be generated from evaporation from pools of liquid that are open to the air. Routine emissions may occur from open surface operations, which would include work related to open vats or tanks, solvent dip tanks, open roller coating, and cleaning or maintenance activities. More sporadic emissions may occur from liquid pools caused by events such as unintentional spills. A number of models are available to estimate air emissions from open liquid pools, and these models can be used to make very conservative estimates for open tanks used in routine operations. Only one of these models, which is a correlation developed by US EPA, will be examined to demonstrate a typical approach.

To develop this correlation for EPA, the evaporation rates of 16 different pure compounds in a test chamber were measured to determine an empirical model to describe the relationship between evaporation rate and physical chemical properties. The compounds were studied at different air velocities and temperatures, and the data were curve-fitted. Based on mass balance of a differential element above a liquid pool, the evaporation rate was derived:

$$G = 13.32 \, M \, P \, A \, T^{-1} \, (D_{ab} \, v_z \, \Delta z^{-1})^{0.5} \qquad \text{(Eq. 8-10)}$$

where
      G = Generation rate, lb/hr
      M = Molecular weight, lb/lb mole
      P = Vapor pressure, in. Hg
      A = Area, ft^2
      D_{ab} = Diffusion coefficient, ft^2/sec of a through b (in this case b is air)
      v_z = Air velocity, ft/min
      T = Temperature, °K
      Δz = Pool length along flow direction, ft

Gas diffusivities of volatile compounds in air are available for some chemicals. More often, however, the diffusion coefficient will not be known. An equation to estimate diffusion coefficients has been developed. The expression for the diffusion coefficient is:

$$D = 4.09 \times 10^{-5} \, T^{1.9} \, (29^{-1} + M^{-1})^{0.5} \, M^{-0.33} \, P_t^{-1} \qquad \text{(Eq. 8-11)}$$

where
      D_{ab} = Diffusion coefficient, cm^2/sec
      T = Temperature, °K
      M = Molecular weight, g/g-mole
      P_t = Pressure, atm

8.3.3.3 Storage Tank Working and Breathing Losses

Storage tanks are units common to almost every chemical process. They provide a buffer for raw materials availability in continuous processes and allow for storage of finished product before delivery is taken. Tanks have the potential to be major contributors to airborne emissions of volatile organic compounds from chemical facilities because of the dynamic operation of these units. There are two major losses mechanisms from tanks; *working losses* and *standing losses*. Working losses originate from the raising and lowering of the liquid level in the tank as a result of raw material utilization and production of product. The gas space above the liquid must expand and contract in response to these level changes. During tank emptying, air from the outside or an inert gas will enter the tank. Volatile organic vapors from the liquid will evaporate into the gas in an attempt to achieve an equilibrium condition between the concentrations of each component in the liquid and gas phases. When the tank is filled again, these vapors in the gas will exit the unit via the vent to be dispersed into the atmosphere unless pollution control devices are installed. Even if the tank level is static, standing losses from the tank will occur as a result of daily temperature and ambient pressure fluctuations which cause a pressure difference between the gas inside the tank and the outside air.

There are four major types of storage tanks; fixed-roof, floating-roof, variable-vapor-space, and pressurized tanks. Equations for estimating emissions from fixed-roof and floating-roof storage tanks are provided in Appendix C. Example problems are also included. Software that performs these calculations (the TANKS program) is available from the EPA CHIEF website (http://www.epa.gov/ttn/chief/).

8.3.4   Release Characterization and Documentation

Estimating releases often requires judgment, and the reliability of emission estimates based on judgment is often difficult to assess. The uncertainty depends on how well we know the process, how well we understand the estimation method and its data and parameters, and how well the method and parameters seem to match up with those expected for the actual process. Uncertainties are inherent in making estimations, and the issues of uncertainty are complex. Some uncertainties can be quantified using established methods, but, more frequently, uncertainties cannot be quantified. Often, the factor quality rating system of the US EPA is used in assessing the accuracy and representativeness of emission data. This rating system assigns a quality index of A through E and a U for unratable. Detailed explanations for the ratings of A–D can be found elsewhere (EPA, 1998). A factor rating of A is excellent and is assigned for factors developed from A-rated source tests taken from many randomly chosen facilities in industry. The source category is specific enough to minimize variability within the source population (i.e., one type of reactor or separation device). A factor rating of B is above average and is taken from A-rated source tests from a reasonably large number of facilities, but does not necessarily represent a random sampling of industry. A factor rating of C is average and is the same as a B rating except that the source tests include B-rated source test results. A rating of D is below average and is similar to C except that only a small number of facilities were sampled, there is reason to believe that the factor is not representative of industry, and there appears to be evidence of variability within the source category population. A rating of E is poor because the factor is developed from C- and D-rated test data from a small number of facilities. There is reason to believe that the facilities tested do not represent a random sampling of industry and there is variability within source category population. A rating of U may apply to gross mass balance estimation, deficiencies found with C- and D-rated test data, and use of engineering judgment.

An initial screening of the FIRE database showed that out of the approximately 650 emission factors for process vents and other units, only 10 were rated as A (Shonnard, et al., 1995). These factors were all for secondary emission of trichloroethylene from wastewater treatment plants. There were a total of 22 B-rated emission factors, and except for 3, all were from secondary emissions from wastewater treatment plants. A much broader range of processes and process units was related to the 17 C-rated factors found in FIRE. There were only 43 D-rated factors, and these also covered a range of processes and units. The vast majority of emission factors were rated U, unratable. Clearly, many of the emission estimation procedures developed in this chapter provide only order of magnitude estimates of actual or expected process emissions.

After making release estimates, it can be valuable to look at the estimates and ask whether the estimates seem realistic relative to the process flow streams and relative to one another. The assumptions used to make estimates can be evaluated and the sensitivity of the estimates to selected variables can be assessed.

Case Study

The emission estimation procedures described in this section represent only a small fraction of the emission estimation procedures relevant to chemical processes. Nevertheless, they are representative of the types of emission estimation tools used in chemical process designs. To conclude this section, the collection of emission estimation tools will be applied to the chemical process flowsheet shown in Figure 8.3-5.

This is a process in which cyclohexane is oxidized, producing cyclohexanone and cyclohexanol (a ketone/alcohol mixture). This mixture is used in the manufacture of adipic acid, which in turn is used in the production of nylon.

The first step in estimating the emissions for this flowsheet, as outlined in this section, is to identify major emission sources. Among the major sources of emissions from this process are:

- venting from the feed and product storage tanks
- off-gases from the scrubbers
- liquid wastes from the scrubbers
- emissions from the decanting and purification columns
- emissions from the boilers
- fugitive emissions
- feed and product loading and off-loading emissions

Each of these emissions can be calculated, at varying levels of detail, using the methods described in this chapter.

Consider first the emissions from reactors and the emissions from the stripper and the decanting and purification columns. Since no direct process data have been provided for these units, the emissions should be estimated from the general emission factors listed in Table 8.3-2. The relevant factors are 1.5 kg/10^3 kg throughput for reactor vents, 0.2 kg/10^3 kg throughput for the stripper, 0.02 kg/10^3 kg throughput for decanters, and 0.7 kg/10^3 kg throughput for distillation column vents. For the reactor, we might assume that half of the emissions are reactants (cyclohexane), and half are products (ketone and aldehyde). For the stripper, decanter, and distillation columns, it can be assumed that all of the emissions are product.

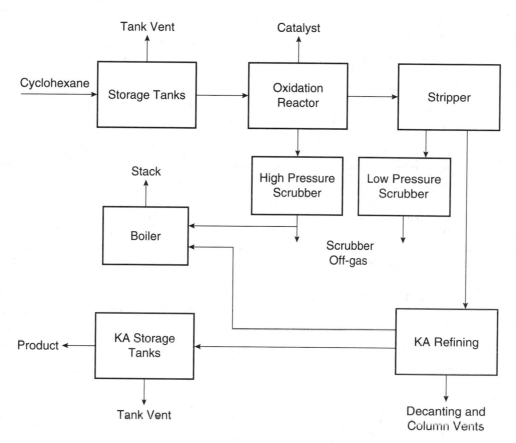

Figure 8.3-5   Chemical process flowsheet for case study.

This leads to total emissions for this section of 0.8 kg/10^3 kg throughput for cyclo-hexane and 1.6 kg/10^3 kg throughput for the ketone and alcohol.

A next step in the emission estimation process would be to consider emissions from boilers. To estimate these emissions, an estimate of energy consumption per kg product is required. Rudd, et al. (1981) provide estimates of energy consumption for a number of processes and suggest a value of 0.5 metric tons of fuel oil equivalent per metric ton (10^3 kg) of product. Assuming that #6 fuel oil with 1% sulfur is used and that no emission controls are in place leads to estimates (based on Table 8.3-5) of:

SO_2 emissions:

   19 kg/10^3 L fuel oil × 0.8 kg/L × 500 kg fuel oil/10^3 kg product
   = 7.6 kg SO_2/10^3 kg product

SO_3 emissions:

   0.69 kg/10^3 L fuel oil × 0.8 kg/L × 500 kg fuel oil/10^3 kg product
   = 0.3 SO_3/10^3 kg product

NO_x emissions:

$$8 \text{ kg}/10^3 \text{ L fuel oil} \times 0.8 \text{ kg/L} \times 500 \text{ kg fuel oil}/10^3 \text{ kg product}$$
$$= 3.2 \text{ kg } NO_x/10^3 \text{ kg product}$$

Particulate Matter emissions:

$$1.5 \text{ kg}/10^3 \text{ L fuel oil} \times 0.8 \text{ kg/L} \times 500 \text{ kg fuel oil}/10^3 \text{ kg product}$$
$$= 0.6 \text{ kg PM}/10^3 \text{ kg product}$$

Accurately estimating fugitive emissions requires a count of valves, flanges, fittings, pumps, and other devices that are used in the process. Such counts are not generally available for preliminary process designs; however, rough estimates can be made based on experience. Typically, fugitive emissions for chemical processes total 0.5–1.5 kg per 10^3 kg product (Berglund and Hansen, 1990). In this case, we have probably already accounted for some of the fugitive emissions through the emission factors for the reactors and distillation column; therefore, an estimate of 0.5 kg per 10^3 kg product is appropriate, with the emissions evenly split between products and reactants.

The remaining emissions include emissions from tanks, emissions from loading and off-loading emissions and the off-gases and liquid wastes from the scrubbers. The loading and off-loading emissions could be estimated using Equation 8-7. Assuming a saturation factor of 0.6 (Table 8.3-9), a vapor pressure of 4.1 mm Hg for the ketone (estimated using the methods described in Chapter 5), a molecular weight of 98, and a temperature of 530 R gives a loading loss of :

$$L_L = 12.46\,(0.6 \times 4.1 \times (14.7/760) \times 98/530) = 0.1 \text{ lb}/10^3 \text{ gal} = 0.15 \text{ kg}/10^3 \text{ kg product}$$

Losses from tanks can be estimated using the methods described in Appendix C. Without a detailed flowsheet, exact specifications for the tank are not available. A rough estimate of tank dimensions, however, can be derived from annual production rates. If we assume a production rate of 100 million pounds per year, and that a typical tank should hold 2–3 days of production capacity, an approximate tank volume can be calculated. For this production rate, a tank 35 ft in diameter and 20 feet high with a fixed roof, is reasonable. We will also assume that, to minimize emissions, the tank is painted white, the paint is in good condition and the tank is generally kept 80% full. If the facility is located in Houston, Texas, the data and procedures described in the appendix lead to an estimate of order 0.5 kg emitted /10^3 kg product for standing and working losses. We will assume that these are emissions of the feed material, cyclohexane.

Finally, emission rates should be estimated for the off-gases and liquid wastes from the scrubbers. These emissions depend strongly on the assumed efficiency of the scrubbers. If data are not available for the process of interest, it is generally a sound practice to obtain data from similar processes. For this part of the adipic acid process, AP-42 reports the gas emissions from the scrubbers shown in Table 8.3-11.

Hedley, et al., report rates of liquid waste generation from the combined scrubbers. They suggest approximately 200 kg of organic sodium salts are generated in the scrubbers per 10^3 kg product.

Table 8.3-11   Gas Phase Emissions from the Scrubbers in Cyclohexane Partial Oxidation (AP-42).

Emission type	Emissions from high pressure scrubber (kg emitted /10^3 kg product)	Emissions from low pressure scrubber (kg emitted /10^3 kg product)
Total non-methane hydrocarbons	7.0	1.4
CO	25	9.0
CO_2	14	3.7
CH_4	0.08	0.05

The data summarized in Table 8.3-12 provide a reasonable starting point for estimating the environmental impacts of this chemical process. The next section describes how these emission and other data can be converted to a set of environmental performance metrics.

Table 8.3-12   Preliminary Emission Estimates for the Cyclohexane Partial Oxidation Process (kg emitted /10^3 kg product).

Source	Cyclohexane air emissions	Ketone and aldehydye air emissions	Criteria pollutant emissions	Organic liquid wastes
Venting from the feed storage tanks	0.5			
Off-gases from the scrubbers	8.4		34	
Liquid wastes from the scrubbers				200
Emissions from reactor, and the decanting and purification columns	0.8	1.6		
Emissions from the boilers			11.7	
Fugitive emissions	0.25	0.25		
Feed and product loading and off-loading emissions		0.15		
Total	**10**	**2**	**46**	**200**

8.3.5   Assessing Environmental Performance

Once preliminary estimates of material flows, energy requirements, wastes and emissions have been made for a flowsheet, the overall environmental performance of the flowsheet can be evaluated. Two types of assessments have been commonly employed. One type of assessment evaluates the treatability or costs of treatment of the waste streams. Douglas and co-workers (Schultz, 1998) have suggested approximate costs for the treatment of waste streams. A second method for assessing environmental performance of flowsheets is to evaluate a set of relatively simple environmental performance indicators. The performance indicators that will be used in this chapter were developed by the Canadian National Roundtable on the Environment and the Economy (NRTEE, 1999). These indicators have been tested and evaluated by a number of organizations throughout North America (including the American Institute of Chemical Engineer's Center for Waste Reduction Technologies; see CWRT, 1998, 1999). The indicators are:

- Energy consumed from all sources within the manufacturing or delivery process per unit of manufactured output (with electricity consumption converted to equivalent fuel use, based on an average efficiency of converting energy to electricity in power plants).
- Total mass of materials used directly in the product, minus the mass of the product, per unit of manufactured output.
- Water consumption (including water present in waste streams, contact cooling water, water vented to the atmosphere, and the fraction of non-contact cooling water lost to evaporation) per unit of manufactured output.
- Emissions of targeted pollutants (those listed in the Toxic Release Inventory) per unit of manufactured output.
- Total pollutants (including acidifying emissions, eutrophying emissions, salinity, and ozone depleting substances) per unit of manufactured output.

Taken together, these cost and environmental performance metrics provide additional guidance on the performance of flowsheets. Consider how the these performance evaluations would be applied to the cyclohexane oxidation process described in the previous section.

Material use for this process can be determined based on the data of Rudd, et al. (1981) and Hedley, et al. (1975). Rudd reports that the manufacture of 1 ton of cyclohexanol requires 1.64 tons of cyclohexane and 0.13 ton of sodium hydroxide. Cyclohexanone is produced at a rate of 0.38 tons per ton of cyclohexanol. If we assume that both cyclohexanone and cyclohexanol are desirable products, then the material intensity (excluding water) is:

$$\text{Material intensity}/[(1.64 + 0.13) \text{ tons raw materials} - 1.38 \text{ pounds product}]/$$
$$(1 + 0.38) \text{ tons product} = 0.28$$

Water use has been estimated by Hedley, et al. as 5,000 gpm of cooling water and 10 gpm of process water (used in the scrubber) for a 85,000,000 pound per year facility. This leads to an estimate of approximately 30 gallons of water per pound of product, with most of this water use dedicated to cooling.

Energy intensity is approximately 0.4 L fuel oil per kg product; assuming 150,000 BTU per gallon for fuel oil, this leads to an energy intensity of 7 kBTU per pound of product.

Pollutant generation, which is dominated by organic liquid wastes, is estimated to be approximately 0.3 lb/lb product (Table 8.3-12).

What do these indices mean? Are the values high? Are they low? The best way to evaluate the indices is to consider how other chemical manufacturing processes fare. Table 8.3-13 is a listing of environmental performance indices for a number of chemical manufacturing processes. These were derived by E. Beaver and colleagues in an analysis of chemical manufacturing processes conducted for the Department of Energy (Bridges to Sustainability, 2000).

The sample results shown in Table 8.3-13 reveal that different processes can have dramatically different characteristics. For example, water use, material use, and energy use per pound of product are significantly different for acetic acid, acrylonitrile, and maelic anhydride. This implies that the metrics are sensitive to process designs and chemistry. Data for two different routes for producing sulfuric acid, also shown in Table 8.3-13, reveal that the metrics are sensitive to process design.

With this as background, the metrics obtained for the cyclohexane oxidation can be put in context. Recall that the preliminary estimates for the cyclohexane oxidation flowsheet were:

Material use:	0.28 lb/lb prod.
Energy use:	7 kBTU/lb prod.

Table 8.3-13 Representative Environmental Performance Metrics for Chemical Manufacturing Processes (Bridges to Sustainability, 2000).

Compound	Process	Material Intensity/lb prod. (lb/lb)	Energy/lb prod. (10^3 BTU/lb)	Water/lb prod. (gal./lb)	Toxics/lb prod. (lb/lb)	Pollutants/lb prod. (lb/lb)	Pollutants + CO_2/lb prod. (lb/lb)
Acetic acid	from methanol by low pressure carbonylation	0.062	1.82	1.24	0.00011	0.0000	0.133
Acrylonitrile	by ammoxidation of propylene	0.493	5.21	3.37	0.01514	0.00781	0.966
Maelic anhydride	from n-butane by partial oxidation	0.565	0.77	1.66	0.000	0.000	2.77
Sulfuric acid	from pyrometallurgical sulfur dioxide	0.002	0.073	0.57	−0.65	−0.63	−0.04
Sulfuric acid	from sulfur	0.001	−0.87	0.70	0.00195	0.00195	0.002

Note: Negative values for material use indicate that waste materials from other processes are used as raw materials; air and water used as raw materials are not included in the material use; negative values for energy use indicate that the process is a net energy generator.

Water use: 30 gal/lb prod.

Pollutants: 0.3 lb/lb prod.

These values are at the high end of the range reported in Table 8.3-13, suggesting that a number of process improvements may be possible. Chapter 9 will describe approaches to identifying process improvements.

SECTION 8.3   QUESTIONS FOR DISCUSSION

1. What are the major sources of emissions from chemical manufacturing processes?
2. Describe the sources of uncertainty in estimating emissions from process flowsheets.
3. Describe how you might use the benchmark data on environmental metrics for chemical processes. Specifically, consider what types of processes can be compared. Does it make sense to compare the data from a partial oxidation reaction to the data from a cracking reaction?

8.4   TIER 3 ENVIRONMENTAL PERFORMANCE TOOLS

Once the basic structure of the process flowsheet is determined, detailed specifications of reactor and separator sizes, stream compositions, energy loads, and other process variables can be established. This level of process specification is, once again, an appropriate time to examine environmental performance. At this stage of analysis, a relatively limited number of design alternatives will be screened, but more information about the process is available and should be incorporated into each environmental performance evaluation. Chapter 11 examines methods for evaluating environmental performance at this final level. Before those methods are presented, Chapters 9 and 10 examine methods for improving (as opposed to assessing) environmental performance at the level of a conceptual process flowsheet.

REFERENCES

Allen, D.T. and Rosselot, K.S., *Pollution Prevention for Chemical Processes,* John Wiley & Sons, New York, NY, 434 pages, 1997.

American Institute of Chemical Engineers, Center for Waste Reduction Technologies (CWRT), "Sustainability Metrics Interim Report #1," AIChE CWRT, New York, 1998.

American Institute of Chemical Engineers, Center for Waste Reduction Technologies (CWRT), "Sustainability Metrics Interim Report #2," AIChE CWRT, New York, 1999.

Berglund, R. L., Wood, D. A. and Covin, T. J. "Fugitive emissions from the acrolein production industry," in the Air and Waste Management Association International Specialty Conference: SARA Title III, Section 313, Industry Experience in Estimating Chemical Releases, April 6, 1989.

Bridges to Sustainability, "Sustainability Metrics: Making Decisions for Major Chemical Products and Facilities," Houston, Texas (2000).

Chang, D., M.S. Thesis, University of California, Los Angeles, 1996.

Crowl, D. A., and Louvar, J.F., *Chemical Process Safety: Fundamentals with Applications,* Prentice Hall, Englewood Cliffs, New Jersey, 1990.

Douglas, J.M., *Conceptual Design of Chemical Processes,* McGraw Hill, New York, 1988.

Energy Facts 1992, Energy Information Administration, U.S. Department of Energy, DOE/EIA-0469 (1992).

Hedley, W.H., Mehta, S.M., Moscowitz, C.M., Reznik, R.B., Richardson, G.A., and Zanders, D.L., *Potential Pollutants from Petrochemical Processes,* Technomic, Westport, Conn., 1975.

National Roundtable on the Environment and the Economy (NRTEE), *Measuring Eco-efficiency in Business: Feasibility of a Core Set of Indicators,* Renouf Publishing, Ottowa Canada, 1999 (ISBN 1-895643-98-8).

PEI, 1986, "Releases during Cleaning of Equipment," Report prepared under Contract 68-02-4248 to US EPA, Office of Toxic Substances.

Perry, R. H. and Green, D. W., *Perry's Chemical Engineers' Handbook,* 7th Edition, McGraw-Hill Book Company, New York, NY, 1997.

Rudd, D.F., Fathi-Afshar, S., Trevino, A., and Stahtherr, M.A., *Petrochemical Technology Assessment,* Wiley-Interscience, New York, 1981.

Schultz, M. A., "A Hierarchical Design Procedure for the Conceptual Design of Pollution Prevention Alternatives for Chemical Processes," Ph.D. Thesis, University of Massachusetts, 1998.

Shonnard, D. R., Markey, J. M., and Deshpande, P. A., "Evaluation of unit operation-specific emission databases," the *Proceedings of the 1995 American Institute of Chemical Engineers Summer National Meeting,* Boston, Mass., August, 1995.

US EPA, "Organic Chemical Manufacturing: Selected Processes," US Environmental Protection Agency, Research Triangle Park, NC, 1980, EPA-450/3-80-028c.

US EPA, "Compilation of Air Pollutant Emission Factors, Volume I: Stationary Point and Area Sources," 4th edition, with Supplements A-D, US Environmental Protection Agency, Research Triangle Park, NC, Publication AP-42, 1985.

US EPA, "Estimating Releases and Waste Treatment Efficiencies for the Toxic Chemical Release Inventory Form" (EPA/560/4-88-002), US Environmental Protection Agency, Washington, D.C. (1987).

US EPA, "Toxic Air Pollutant Emission Factors - A Compilation of Selected Air Toxic Compounds and Sources" (EPA/450/2-88-006a), US Environmental Protection Agency, Washington, D.C. (1988).

US EPA, "Locating and Estimating Air Emissions," US Environmental Protection Agency, Research Triangle Park, NC, Report Number EPA-450/4-, 1979-1991.

US EPA, "Protocol for Equipment Leak Emission Estimates," US Environmental Protection Agency, Research Triangle Park, NC, available through NTIS (PB93-229219, June, 1993.

US EPA, Integrated Risk Information System (IRIS), http://www.epa.gov/ngispgm3/iris/Substance_List.html, US Environmental Protection Agency, 1997.

US EPA, "Compilation Of Air Pollutant Emission Factors, Volume I, Fifth Edition," AP-42, Air CHIEF CD-ROM, ClearingHouse For Inventories And Emission Factors, Air CHIEF CD-ROM, (EFIG/EMAD/OAQPS/EPA), Version 6.0, U.S. Environmental Protection Agency, Research Triangle Park, NC, EPA-454/F-98-007, 1998 (see also http://www.epa.gov/ttn/chief/).

Wagner, P.M., Nabholz, J.V., and Kent, R.J., "The new chemicals process at the Environmental Protection Agency (EPA): structure activity relationships for hazard identification and risk assessment," *Toxicology Letters,* 79, 67-73 (1995).

PROBLEMS

1. Compare the carbonylation of dinitrotoluene and the amine-phosgene routes for the production of toluene diisocyanate (TDI) using a Tier 1 economic and environmental performance evaluation. The amine-phosgene route involves the reaction of phosgene with toluenediamine in a chlorobenzene solvent. The carbonylation route has been demonstrated in laboratories, but is not presently a commercial technology. Data from the patent literature (see the Green Chemistry Expert System, Appendix F) indicate that the reaction of 2,4 dinitrotoluene with carbon monoxide occurs over a mixed oxide catalyst. Conversion approaches 100% with selectivity to the desired product ranging from 70–99%. Laboratory data indicate that the reaction can be performed in a chlorobenzene and pyridine solvent. Approximate stoichiometric data, based on the patent data, are given in the table below.

Amine-phosgene route:

$$C_6H_3(CH_3)(NH_2)_2 + 2\ COCl_2 \rightarrow C_6H_3(CH_3)(-N=C=O)_2 + 4\ HCl$$

Carbonylation of dinitrotoluene:

$$C_6H_3(CH_3)(NO_2)_2 + 6\ CO \rightarrow C_6H_3(CH_3)(-N=C=O)_2 + 4\ CO_2$$

Compound	Pounds produced or pounds of raw material required per pound of TDI*	Cost ($/lb)**	PEL (mg/m^3)	Overall inhalation toxicity factor	Overall oral toxicity factor
Amine-phosgene route					
toluene diamine	−0.76	0.576	0.1 (est.)	NA	NA
chlorobenzene	−0.01	0.550	350	100	100
hydrochloric acid	0.4 (est.)	0.027	7	100	100
phosgene	−1.26	0.610	0.4	NA	NA
TDI	1.00	1.340	0.14	100,000	100
Carbonylation route					
dinitrotoluene	1.04 (est.)	0.365	1.5	1,000	1,000
carbon monoxide	1.0 (est.)	0.040	55	NA	NA
TDI	1.00	1.340	0.14	100,000	100
carbon dioxide	1.0 (est.)		9000	NA	NA

*A negative stoichiometric index indicates that a material is consumed; a positive index indicates that it is produced in the reaction.
**Chang, 1996

2. Perform a Tier 2 environmental assessment for the production of maleic anhydride from n-butane. A conceptual process flowsheet for the process is given below.
   (a) Identify major sources of emissions.
   (b) Using the methods described in this chapter and data available in AP-42 documents available on-line (http://www.epa.gov/ttn/chief/), estimate wastes and emissions for this process.

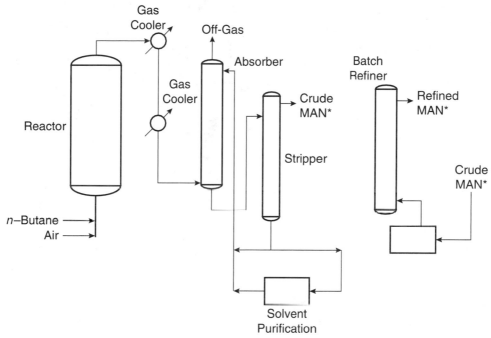

Schematic Flow Diagram of the Huntsman Fixed-bed Maleic Anhydride Process
*MAN = Maleic Anhydride

CHAPTER 9

Unit Operations
and Pollution Prevention

by
David R. Shonnard

9.1 INTRODUCTION

In developing a flowsheet for the production of a chemical, it is desirable to consider the environmental ramifications of each unit operation in the process rather than postponing this consideration until the flowsheet is finished. This "front end" environmental assessment is more likely to result in a chemical process that has less potential to cause environmental harm. In many instances, this environmentally benign design will also be more profitable, because the improved design will require lower waste treatment and environmental compliance costs and will convert a higher percentage of raw materials into salable product.

In considering pollution prevention for unit operations in the design of chemical processes, the following considerations are important.

1. Material Selection: Many of the environmental concerns can be addressed by reviewing material properties and making the correct choice of unit operation and operating conditions. The materials used in each unit operation should be carefully considered so as to minimize the human health impact and environmental damage of any releases that might occur.
2. Waste Generation Mechanisms: Often, a careful evaluation of the mechanisms of in-process waste generation can direct the process designer toward environmentally sound material choices and other pollution prevention options.
3. Operating Conditions: The operating conditions of each unit should be optimized in order to achieve maximum reactor conversion and separation efficiencies.
4. Material Storage and Transfer: The best material storage and transfer technologies should be considered in order to minimize releases of materials to the environment.

5. Energy Consumption: Energy consumption in each unit should be carefully reviewed so as to reasonably minimize its use and the associated release of utility-related emissions.

6. Process Safety: The safety ramifications of pollution prevention measures need to be reviewed in order to maintain safe working conditions.

In the following sections, we apply this framework for preventing pollution in unit operations by considering choices in materials, technology selection, energy consumption, and safety ramifications. In Section 9.2, material choices that are generic to most chemical processes, like process water and fuel type, are analyzed with respect to in-process waste generation and emission release. Other process materials that are more specific to various unit operations are discussed in subsequent sections of this chapter. Chemical reactors are the topic in Section 9.3. The environmental issues related to the use of reactants, diluents, solvents, and catalysts are discussed first. Then the effects of reaction type and order on product yield and selectivity are covered. The effects of reaction conditions (temperature and mixing intensity) on selectivity and yield are illustrated. Finally, the benefits of additional reactor modifications for pollution prevention are tabulated. In Section 9.4, the most important topics include the choice of material (mass separating agent) to be used in separations, design heuristics, and examples of the use of separation technologies for recovery of valuable components from waste streams, leading eventually to their reuse in the process. Separative reactors are the topic in Section 9.5. These hybrid unit operations have special characteristics to help achieve higher conversions and yields in chemical reactors compared to conventional reactor configurations. Section 9.6 discusses methods for reducing emissions from storage tanks and fugitive sources. The safety aspects of pollution prevention and unit operations are the topic of Section 9.7. It is shown that many pollution prevention efforts tend to make chemical processes more complex, necessitating a higher level of safety awareness.

In making pollution prevention decisions that include choices of materials, unit operations technologies, operating conditions, and energy consumption, it is very important to consider health and environmental risk factors. It is also important to incorporate cost factors and to be aware of safety ramifications. In Section 9.8, we review a method for minimizing the potential environmental impact of unit operations by considering the optimum reactor operating conditions as an example application. Although no generally accepted method exists for these assessments, the method outlined in Chapter 8 and applied in Section 9.8 is useful for incorporating multiple risk factors into decisions regarding unit operations.

Finally, it is also important to introduce the concept of "risk shifting." Pollution prevention decisions that are targeted to reduce one kind of risk may increase the level of risk in other areas. For example, a common method of conserving water resources at chemical manufacturing facilities is to employ cooling towers. Process water used for cooling purposes can be recycled and reused many times. However, there is an increased risk for workers who may be exposed to the biocides used to control microbial growth in the cooling water circuit. Also, in some cooling water processes, hazardous waste is created by the accumulation of solids— for example, from the use of hexavalent chromium (a cancer causing agent) as a corrosion inhibitor.

Another example of shifting risk from the environment and the general population to workers involves fugitive sources (valves, pumps, pipe connectors, etc.). One strategy for decreasing fugitive emissions is to reduce the number of these units by eliminating backup units and redundancy. This strategy will decrease routine air releases but will increase the probability of a catastrophic release or other safety incidents. Simply put, the objective of pollution prevention is to reduce the overall level of risk in all areas and not to shift risk from one type to another.

9.2　POLLUTION PREVENTION IN MATERIAL SELECTION FOR UNIT OPERATIONS

One very important element of designing and modifying process units for pollution prevention is the choice of materials that are used in chemical processes. These materials are used as feedstocks, solvents, reactants, mass separating agents, diluents, and fuels. In considering their suitability as process components, it is not sufficient to consider only material properties that are directly related to processing; it is increasingly important to consider the environmental and safety properties as well. Use of materials that are known to be persistent, bioaccumulative, or toxic should be avoided as they are under increasing regulatory scrutiny and many manufacturers are moving away from their use. Questions regarding material selection include:

a) What are the environmental, toxicological, and safety properties of the material?

b) How do these properties compare to alternative choices?

c) To what extent does the material contribute to waste generation or emission release in the process?

d) Are there alternative choices that generate less waste or emit less while maintaining or enhancing the overall yield of the desired product?

If processing materials can be found which generate less waste and if the hazardous characteristics of those wastes are less problematic, then significant progress may be made in preventing pollution from the chemical process.

Materials are involved in a wide range of processing functions in chemical manufacturing, and depending on the specific application, their environmental impacts vary greatly. For example, reactants for producing a particular chemical can vary significantly with respect to toxicity and inherent environmental impact potential (global warming, ozone depletion, etc.), and they can exhibit various degrees of selectivity and yield toward the desired product. In addition, the properties of reaction byproducts can vary widely, similarly to reactants. Some catalysts are composed of hazardous materials or they may react to form hazardous substances. For example, catalysts used for hydrogenation of carbon monoxide can form volatile metal carbonyl compounds, such as nickel carbonyl, that are highly toxic (Gates, 1993). Many catalysts contain heavy metals, and environmentally safe disposal has become an increasing concern and expense. After the deposition of inhibitory substances, the regeneration of certain heterogeneous catalysts releases significant amounts of SOx, NOx, and particulate matter. For example, the regenerator for a fluid catalytic cracking unit (FCCU) is a major source of air pollutants at refineries (Upson and Lomas, 1993). Note that the removal of the FCCU would result in very low yields and consequently unacceptable waste generation at facility level.

The choice of a mass separating agent for solids leaching or liquid extraction applications can affect the environmental impacts of those unit operations. Agents that are matched well to the desired separation will consume less energy and release less energy-related pollutants than those that are not well suited for the application. Typically, agents that have lower toxicity will require less stringent clean up levels for any waste streams that are generated in the process.

The choice of fuel for combustion in industrial boilers will determine the degree of air pollution abatement needed to meet environmental regulations for those waste streams. As an illustration, using fuel types having lower sulfur, nitrogen, and trace metals levels will yield a flue gas with lower concentrations of acid rain precursors (SOx, and NOx) and particulate matter, as shown by the following example.

Example 9.2-1

Compare the emission of sulfur dioxide (SO_2) resulting from the combustion of three fuel types that will satisfy an energy demand of 10^6 BTU. The fuel types are No. 6 fuel oil (F.O.), No. 2 fuel oil, and natural gas. The elemental compositions by weight for each fuel are listed below along with the density and lower heating value.

	No. 6 F.O.	**No. 2 F.O.**	**Natural Gas**
Density (lb/ft^3)	61.23	53.66	.0485
Lower Heating Value (BTU/gal)	148,000	130,000	1060 BTU/ft^3
Carbon (wt%)	87.27	87.30	74.8
Hydrogen (wt%)	10.49	12.60	25.23
Sulfur (wt%)	0.84	0.22	
Oxygen (wt%)	0.64	0.04	0.0073
Nitrogen (wt%)	0.28	0.006	
Ash (wt%)	0.04	<0.01	

Solution: **No. 6 F.O.**

The volume of No. 6 F.O. needed is (10^6 BTU/148,000 BTU/gal) = 6.76 gal

The mass of No. 6 F.O. needed is $(6.76 \text{ gal})(1 \text{ ft}^3/7.48 \text{ gal})(61.23 \text{ lb/ft}^3) = 55.18 \text{ lb}$

Therefore, the mass of SO_2 generated is $(55.31 \text{ lb})(.0084 \text{ lb S/lb})(64.06 \text{ lb } SO_2/32.06 \text{ lb S}) = .928 \text{ lb } SO_2$

No. 2 F.O.; similarly as for No. 6 F.O.
The mass of SO_2 is .243 lb SO_2

Natural Gas
The mass of SO_2 is 0.0 lb SO_2

The percent reduction in SO_2 generated for No. 2 F.O. compared to No. 6 F.O. is $(0.928 \text{ lb } SO_2 - 0.243 \text{ lb } SO_2)/0.928 \text{ lb } SO_2 \times 100 = 73.81\%$.

Discussion: Focussing on the sulfur content of each fuel is a useful approach for reducing acid rain. However, there are other risk factors that need to be considered. Other considerations could include a) toxicological properties of each fuel, b) difference in emission rates to air from storage and transfer operations, c) smog formation potential of fuel components, and d) the cost of the fuel.

(Adapted from a problem by John Walkinshaw in *Motivating Pollution Prevention Concepts: Homework Problems for Engineering Curricula,* editors M. Becker, I. Farag, and N. Hayden, 1996.)

Less toxic materials, such as water and air, can still have important environmental implications due to the waste streams that are generated in their use. Air is often used in chemical reactions either as a diluent or as a source of oxygen. For certain high temperature reactions, the nitrogen and oxygen molecules in the air react, forming oxides of nitrogen. Upon release, NOx will participate in photochemical smog reactions in the lower atmosphere. Therefore, it is important to consider alternative sources of oxidants, such as enriched air or pure oxygen, and diluents, such as carbon dioxide or other inert reaction byproducts. Water is used for many purposes in chemical processes; as boiler feed, a cooling medium, reactant, or a mass separating agent. The following example illustrates that the quality of the feed water can have a profound influence on the generation of hazardous waste in a refinery

Example 9.2-2

Figure 9.2-1 shows the many uses of process water in a refinery. Water is brought into contact with crude oil in order to remove salts and other solid contaminants that could disrupt the operation of downstream equipment. The spent water from this operation is sent to the wastewater treatment facility to recover residual oil and to remove toxic constituents. Water that is used as feed to the boilers is softened in an ion exchanger. Steam generated in the boilers is used for process heating and a fraction is returned to the boiler as condensate.

Problem
Solids accumulate in the boiler and excessive levels of suspended solids lead to fouling of heat transfer surfaces in the process, a decrease in heat transfer efficiency, and requires periodic shut-down and cleaning of these surfaces to restore normal operation.

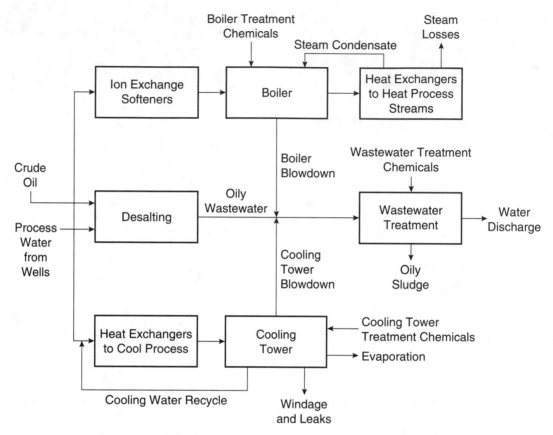

Figure 9.2-1 Conceptual diagram of process water at a refinery. *Source:* Allen and Rosselot, *Pollution Prevention for Chemical Processes* © 1997. This material is used by permission of John Wiley & Sons, Inc.

To control solids accumulation, the contents of the boiler are sent to wastewater treatment when the dissolved solids content is above a cut-off level, in a step termed "blowdown." Similarly, in the cooling system of the refinery, dissolved solids accumulate because the mechanism of cooling in the cooling tower is evaporation, which effectively retains solids in the process. When the high calcium solids from the cooling tower blowdown meet the alkaline boiler blowdown, precipitation occurs. This precipitate can clog wastewater treatment equipment and can form oily sludges upon being blended with the wastewater from the desalter unit. It has been shown that every pound of solids precipitate in oily wastewater creates about 10 pounds of oily sludge. The oily sludge is a RCRA hazardous waste and is costly because of expensive disposal fees and the oil lost from the process that could have been made into products.

Pollution Prevention Solution: At a southwest United States petroleum refinery, the solution to this waste disposal problem involved the pretreatment of all process water using reverse osmosis to separate dissolved solids from the feed water, thereby eliminating the source of solids for the oily sludge (Rosselot and Allen, 1996). This solution proved to be cost effective because the savings in disposal costs alone was enough to pay for the pretreatment equipment and its operation. Additional savings were realized because fewer boiler and cooling tower treatment chemicals were needed (90% reduction in use). Also, maintenance costs were lower since scale build-up on the cooling waste heat exchanger surfaces was reduced.

Additional examples of material use and pollution prevention in unit operations will be provided in the sections of this chapter on chemical reactors and separation equipment. A summary table relating unit operations, materials selection, and risk factors is presented in Table 9.2-1.

9.3   POLLUTION PREVENTION FOR CHEMICAL REACTORS

From an environmental perspective, reactors are the most important unit operation in a chemical process. The degree of conversion of feed to desired products influences all subsequent separation processes, recycle structure for reactors, waste treatment options, energy consumption, and ultimately pollutant releases to the environment. Once a chemical reaction pathway has been chosen, the inherent product and byproduct (waste) distributions for the process are to a large extent established. However, the synthesis must be carried out on an industrial scale in a particular reactor configuration and under specified conditions of temperature, pressure, reaction media (or solvent), mixing, and other aspects of the reactor operation.

In designing chemical reactors for pollution prevention, there are many important considerations. The raw materials, products, and byproducts should have a relatively low environmental and health impact potential. This means that the environmental and toxicological properties of the chemicals involved should indicate that they are relatively benign. In addition, the conversion of reactants to desired products should be high and their conversion toward byproducts should be low. In other words, the reaction *yield* and *selectivity* for the desired product should be as high as possible. Finally, energy consumption for the reaction should be low. Another consideration that is beyond the scope of this chapter is that the life cycle impacts of reactants, products, and byproducts should be relatively low (see Chapters 13 and 14). For example, cumulative emissions and impacts of raw materials should be relatively low, environmental impacts during subsequent use by consumers should be small, and if possible the reaction products should be recyclable. Engineers must balance all of these considerations. For the discussion in this chapter, these reactor considerations will be classified as

1) material use and selection,
2) reaction type and reactor choice, and
3) reactor operation.

Table 9.2-1 Summary of Material Selection Issues for Unit Operations in Chemical Processes.

Unit Operation	Materials	Risk and Environmental Impact Issues	Chapter Sections
Boilers	Fuel type.	• Emission of criteria pollutants. • High efficiency and low emissions boilers.	9.2
Reactors	Feedstocks, reactants, products, byproducts, diluents, oxidants, solvents, catalysts.	• Environmental and toxicological properties. • Reaction yield, conversion, and selectivity. • Waste generation and release mechanisms. • Catalyst reuse or disposal.	9.3
Separators	Mass separating agents, extraction solvents, solid adsorbents.	• Environmental and toxicological properties. • Process properties (relative volatility, etc.). • Energy consumption. • Regeneration of solid adsorbents.	9.4
Storage Tanks	Feedstocks, products, solvents.	• Environmental and toxicological properties. • Air emissions. • Vapor pressure of liquids.	9.6
Fugitive Sources	Feedstocks, products, solvents.	• Same as storage tanks.	9.6
Cooling Towers	Water, biocides.	• Environmental/toxicological properties of biocides. • Waste generation by dissolved solids.	
Heat exchangers	Heat transfer fluids.	• Environmental and toxicological properties.	

The following discussion proceeds from the most general to the more specific topics on preventing pollution for chemical reactors.

9.3.1 Material Use and Selection for Reactors

Issues involving the use of materials in a chemical reactor include the choice of feed entering the reactor, the catalyst if one is needed, and solvents or diluents. Many of these material choices were already made in previous design steps in the generation of a flowsheet, particularly using the assessments methods described in Chapters 5–8. However, it is important to mention material selection here in light of their influence on the environmental impacts of reactors in chemical processes.

Raw Materials and Feedstocks

Raw materials used in chemical reactions can be highly toxic or can cause undesirable byproducts to form. Although some of these raw materials may be converted to relatively benign chemicals through chemical reactions in the process, their presence may be a concern because of the potential for uncontrolled release and exposure to humans in the workplace and also in the environment. An important strategy for environmental risk reduction for chemical processes is to eliminate as many of these toxic raw materials, intermediates, and products as possible.

The elimination of a raw material or the use of a more benign substitution may necessitate the adoption of new process chemistry. For example, phosgene is used in large volumes all over the world in the manufacture of polycarbonates and urethanes. Phosgene – $COCl_2$ is highly toxic and may pose risks for workers at manufacturing facilities and to the surrounding population if large releases occur. In the phosgene process for producing polycarbonates, polycarbonate is produced from bisphenol-A monomer and phosgene in the presence of two solvents, methylene chloride and water. A new process for polycarbonate synthesis has been demonstrated using solid-state polymerization in the absence of both phosgene and methylene chloride (also toxic) (Komiya et al., 1996), by including diphenyl carbonate (DPC) and phenol instead. Similarly, alternative phosgene-free routes to urethanes have recently been developed (see citations in the Green Chemistry Expert System (GCES) available at http://www.epa.gov/greenchemistry/). The "Tier 1" environmental performance tools outlined in Chapter 8 (Section 8.2) are very useful for evaluating these and other alternative reactions chemistries.

In the production of fuels for transportation, petroleum refineries are required to remove sulfur from their products. If not removed, SO_2 is released to the atmosphere upon combustion of the fuels in automobiles, trucks, or stationary combustion sources. Exposure to SO_2-contaminated air causes lung irritation and other more serious health effects, and SO_2 emissions contribute to acidification of surface water and ecosystem damage. Choosing a crude oil raw material with lower sulfur content (sweet crude) reduces the amount of sulfur that needs to be removed and reduces operating costs, but is *considerably* more expensive to purchase. Another option to consider would be to use, and therefore incur the associated costs with, a hydrodesulfurization or a hydrotreating unit to remove the sulfur. The sulfur can then become a salable product.

In partial oxidation reactions of hydrocarbons to form alcohols or other oxygenated organics, air has traditionally been the source of oxygen in the reaction, and the nitrogen in the air has acted as a heat sink agent (diluent) to help control temperature rise for the exothermic reaction. Some CO_2 and H_2O are produced, and due to the presence of N_2, some NOx is formed. NOx is a precursor in the formation of photochemical smog in urban atmospheres and its emission from industrial facilities is regulated under the Clean Air Act. One method to reduce or eliminate the formation of NOx in partial oxidation reactions is to use pure oxygen or enriched air as the oxidizing agent, thus preventing NOx formation. Carbon dioxide that is recovered and recycled from the reactor effluent or water vapor could be used as the heat sink instead of nitrogen. Another method is to install NOx control equipment on the original process. An important issue in this case is whether the costs associated with purchasing and operating the CO_2 recovery equipment are lower than operating NOx control equipment. Another consideration is whether the additional pollutant releases associated with NOx prevention equipment are lower than the releases in the original process.

Solvents

Another important class of raw materials used in chemical reactors is solvents. This is especially true in "solution" and "emulsion" polymerization reactions in which the reaction of monomers to create high molecular weight polymers occurs either within the solvent phase or within dispersed droplets of monomer in the solvent phase. In some polymerizations, addition of solvents can enhance precipitation of polymer in solid form, co-solubilize monomer and initiator, and act as a diluting medium to modulate the rate of reaction and rate of heat removal (Elias, 1984). The highest production polymers in the United States are low-density polyethylene (LDPE), high-density polyethylene (HDPE), polyvinyl chloride (PVC), polypropylene, and polystyrene, with approximately 20%, 15%, 15%, 13% and 8% of annual mass production, respectively (Aggarwal and Caneba, 1993). Solvents used in the production of some of these high volume polymers include xylene, methanol, lubricating oil, hexane, heptane, and water. Solvents are a concern due to their high volatility and potential to cause low-level ozone during smog formation reactions in the atmosphere. They may also be a health concern for workers and the general population in the vicinity of the facility. Candidate substitute solvents having similar solubility parameters can be found in standard references and handbooks (Hansen, 2000; Sullivan, 1996; Barton, 1983; Flick, 1985). In addition, there are several on-line resources for evaluating substitute solvents (see Appendix F). The solubility, toxicological, cost, and environmental properties of the candidate solvents can be compared with each other and with the original solvent, using the methods in Chapter 8.

Supercritical carbon dioxide is being studied as a substitute solvent in many reaction systems (Morgenstern et al., 1996). Applications include both homogeneous and dispersed phase polymerization reactions (DeSimone et al., 1992, 1994) in which the supercritical CO_2 replaces volatile organic compounds and chlorofluorocarbons as traditional solvents in the reaction mixture.

Catalysts

A catalyst is a substance that is added to a chemical reaction mixture in order to accelerate the rate of reaction. Catalysts are either homogeneous, being dissolved in the reaction mixture, or heterogeneous, typically existing as a solid within a reacting fluid mixture. The choice of catalyst has a large impact on the efficiency of the chemical reactor and ultimately upon the environmental impacts of the entire chemical process. Advances in catalysts can improve the environmental performance of a chemical reactor in several ways. Catalysts can allow the use of more environmentally benign chemicals as raw materials, can increase selectivity toward the desired product and away from unwanted byproducts (wastes), can convert waste chemicals to raw materials (Allen, 1992), and can create more environmentally acceptable products directly from the reactions (Absi-Halabi, et al., 1997).

The production of reformulated gasoline (RFG) and diesel fuels from crude oil is a clear example of how improved catalysts can create chemicals that are better for the environment. Because of recent trends in the petroleum refining industry, improved catalysts are being used in several reaction processes within modern refineries. These trends include

   a) increased processing of crude oils with lower quality (higher percentages of sulfur, nitrogen, metals, and carbon residues),
   b) more demand for lighter fuels and less for heavy oils, and
   c) environmental regulations that limit the percentages of sulfur, heavy metals, aromatics, and volatile organic compounds in transportation fuels.

In particular, the inclusion of RFG in the Clean Air Act (CAA) has prompted many changes in catalyst formulation and reactor configurations. Table 9.3-1 is a summary of conventional and improved catalytic reaction processes for RFG and diesel production. As seen in the table, the major emphasis is on catalyst improvements for sulfur and nitrogen removal from heavier crude fractions, reduced aromatics content, and increased production of branched C5-C7 alkanes for octane enhancement.

In this section, we presented examples of how material selection in chemical reactors can impact the environment and reviewed cases where waste generation or toxicity were reduced. In the next part of Section 9.3, we investigate the effects of reaction type and reactor choice on waste generation in chemical reactors.

9.3.2   Reaction Type and Reactor Choice

The details of any chemical reaction mechanism, including the reaction order, whether it has series or parallel reaction pathways, and whether the reaction is reversible or irreversible, influences pollution prevention opportunities and strategies for chemical reactors. These details will determine the optimum reactor temperature, residence time, and mixing. In addition, reactor operation influences the degree of reactant conversion, selectivity, and yield for the desired product, byproduct formation, and waste generation. As a general rule, in designing chemical reactors for pollution prevention, one would like a reaction with a very high conversion of the reactants, high selectivity toward the desired product, and low selectivity toward any byproducts. A typical reactor efficiency measure pertaining to

Table 9.3-1 Summary of Conventional and Improved Catalysts for Reformulated Gasoline and Diesel Production (Adapted from Absi-Halabi, et al., 1997).

Process	Objective	Conventional Catalyst	Improved Catalyst	Benefits of Improved Catalyst
Reformulated Gasoline				
FCC	Conversion of heavy oils to gasoline	• Zeolites • ReY zeolites	• USY zeolites • USY + ZSM-5 • USY/matrix GSR	• Increased gasoline yield • Reduced coking • Increased light olefins/selectivity • Gasoline sulfur reduction
Reforming	Gasoline octane enhancement	• Pt/Al_2O_3	• $Pt-Ir/Al_2O_3$ • $Pt-Re/Al_2O_3$ • $Pt-Re/Al_2O_3$ + zeolite • $Pt-Sn/Al_2O_3$	• Low-pressure operation • Reduced coking • Increased octane • Improved catalyst stability
Alkylation	Production of branched alkanes for gasoline octane enhancement	• H_2SO_4 • HF	• Supported BF_3 • Modified SbF_3 • Supported liquid-acid catalysts	• Less corrosive • Safe handling • Fewer environmental problems
Isomerization	Conversion of C5/C6 alkanes into high-octane branched isomers	• Pt/Al_2O_3 • $Pt/SiO2-Al_2O_3$ • Modernite (zeolite)	• Solid super acid catalysts (e.g., sulfonated zirconia)	• Low temperatures • Increased conversion • Less cracking
Diesel Production				
Middle distillate hydrotreating	Diesel desulfurization	• $Co-Mo/Al_2O_3$	• High metal Co-Mo/Al_2O_3 with modified support pore structure	• Sulfur removal to < 500 ppm.
Middle distillate aromatics hydrdogenation	Production of low aromatics diesel	• $Ni-Mo/Al_2O_3$	• $Ni-Mo/Al_2O_3$ • Noble metal-zeolite combination in two-stage process	• Aromatics hydrogenation to acceptable low aromatics levels in diesel
VGO hydrotreating	FCC feed pretreatment to reduce sulfur and nitrogen levels	• $Co-Mo/Al_2O_3$	• $Co-Mo/Al_2O_3$ with improved formulation & pore structure • $Ni-Mo/Zeolite$ + amorphous SiO_2-Al_2O_3	• Increased activity for N & S removal • Improved cycle length • Increased throughput • Mild hydrocracking • Increased middle distillate selectivity
Gas oil hydrocracking	Conversion of heavy gas oils to lighter products (gasoline & diesel)	• $Ni-Mo/Al_2O_3$ • $Ni-W/Al_2O_3$	• Ni-W/modified Al_2O_3 • $Ni-W/SiO_2-Al_2O_3$ • Ni-W/Zeolite + amorphous SiO_2-Al_2O_3	• Increased middle distillate selectivity • superior quality middle distillates • Increased catalyst life

FCC: Fluidized Catalytic Cracker
GSR: Gasoline Sulfur reduction
RE: Rare Earth
US: Ultra Stable
VGO: Vacuum gas oil
Y & ZSM-5: Crystalline forms y zeolite catalyst

reactant conversion is the reaction yield, defined as the ratio of the exiting concentration of product to inlet reactant ($[P]/[R]_o$). Reaction selectivity is defined as the ratio of exiting product concentration to the undesired byproduct concentration (Fogler, 1992). We define a *modified* selectivity as the ratio of exiting product concentration to the sum of product and byproduct (waste) concentrations ($[P]/([P]+[W])$ = $[P]/[\text{Reactant consumed}]$). This allows us to display both yield and selectivity on the same scale, from 0 to 1. Yields and selectivity values that are very close to unity indicate an efficient reaction, with little waste generation or reactant to separate in downstream unit operations.

Parallel reaction pathways are very common in the chemical industry. An example of an industrial parallel reaction is the partial oxidation of ethylene to ethylene oxide, whereas the parallel reaction converts ethylene to byproducts, carbon dioxide, and water.

We will begin our discussion of reaction types and their implications for pollution generation with the simple irreversible first-order parallel reaction mechanism shown below.

$$R \xrightarrow{k_p} P$$

$$R \xrightarrow{k_w} W \qquad \qquad \text{(Eq. 9-1)}$$

R is the reactant, P the product, W a waste byproduct, and k_p and k_w are the first-order reaction rate constants for product formation and waste generation (time^{-1}), respectively. The relative concentrations of products and waste components are significantly affected by the ratio of the first order reaction rate constants, k_p/k_w. Figure 9.3-1 illustrates the dependence of the reactor effluent concentrations of products and waste as a function of reactor residence time for several values of these rate constant ratios. In order to achieve maximum reactor yields, the residence time must be about 5 times the reaction time constant $(k_p+k_w)^{-1}$. The reaction selectivity is constant and independent of reactor residence time for first-order irreversible, isothermal parallel reactions. As shown next, in a series reaction, selectivity is affected by reactor residence time, and therefore this parameter must also be considered for pollution prevention in chemical reactors.

In a series reaction, the rate of byproduct (waste) generation depends on the rate of product formation, as shown by the first-order irreversible series reaction below.

$$R \xrightarrow{k_p} P \xrightarrow{k_w} W \qquad \qquad \text{(Eq. 9-2)}$$

Longer reactor residence times lead to not only more product formation but also more byproduct generation. The amount of waste generation for a series reaction depends on the ratio of product formation rate constant (k_p) to the byproduct generation rate constant (k_w) and also on the residence time in the reactor. Figure 9.3-2 illustrates the effect of reactor residence time on reactant, product, and byproduct concentrations for several reaction rate constant ratios (k_p/k_w). For

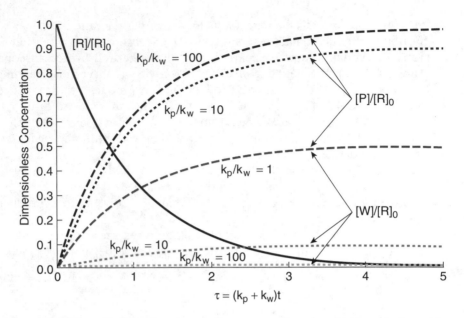

Figure 9.3-1 Effect of k_p/k_w on the reactor outlet concentrations of products, reactants, and byproducts (waste) for a simple irreversible first-order parallel reaction mechanism.

each ratio, there is an optimum reactor residence time that maximizes the product concentration. Figure 9.3-3 shows the product yield ($[P]/[R]_o$) and modified selectivity ($[P]/([P]+[W])$) over a range of reactor residence time for several reaction rate ratios. For irreversible series reactions, the modified selectivity continues to decrease with time. At longer residence times, the rate of waste generation is greater than the rate of product formation. To minimize waste generation in series reactions, it is important to operate the reactor so that the ratio k_p/k_w is as large as possible and to control the reaction residence time. Also, if there is a way to remove the reaction product as it is being formed and before its concentration builds up in the reactor, then byproduct generation can be minimized. We discuss this point more when the topic of separative reactors is covered later in this chapter.

Reversible reactions are another important category of chemical reactions. Figure 9.3-4 shows the reactant, product, and byproduct concentrations profiles in parallel and series reversible reactions for a wide range of reaction rate constants. It is evident that reversible reactions inhibit full conversion of reactants to products. Also, the reactor residence time is a key operating parameter for reversible reactions. Selectivity improvements for reversible reactions, operated at equilibrium, can be achieved by utilizing the concept of *recycle to extinction*. As an example of this concept, consider the steam reforming of methane to form synthesis gas ($CO + H_2$) for methanol production.

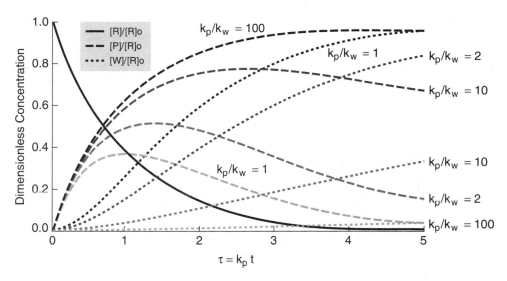

Figure 9.3-2 Effect of product and waste reaction rate constants on product and waste concentrations in a first-order irreversible series reaction. The reactor residence time has been made dimensionless using the product reaction rate constant.

$$CH_4 + H_2O \leftrightarrow CO + 3H_2$$

$$CO + H_2O \leftrightarrow CO_2 + H_2$$

Both reactions are reversible and at equilibrium. When CO_2 is recovered and recycled back to the reactor, it decomposes in the reactor as fast as it forms, and no net conversion of methane to CO_2 occurs. This requires additional operating costs, but there is no selectivity loss of reactant, the process is cleaner, and it may be the lowest cost option overall (Mulholland and Dyer, 1999).

Figure 9.3-5 shows a process flow diagram for a reactor combined with a separator that recycles reactants and byproducts back to the reactor. This configuration can be operated such that all reactants fed to the reactor are converted to product with no net waste generation from the process. Selectivity improvements for reversible reactions can also be realized by employing separative reactors, as discussed later.

More complicated chemical reactions, compared to the few simplistic first-order reactions mentioned above, are common in the chemical industry, and their pollution-generating potential must be evaluated on a case-by-case basis. However, the general trends discussed are expected to hold for more complex reaction networks.

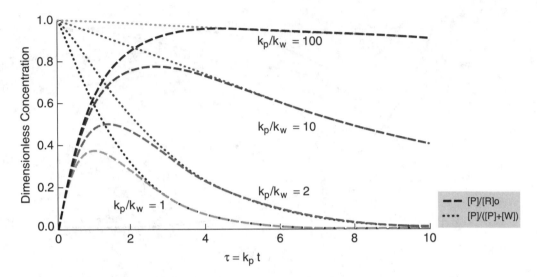

Figure 9.3-3 Effect of product and waste reaction rate constants on product yield ($[P]/[R]_o$) and product modified selectivity ($[P]/([P]+[W])$) for a first-order irreversible series reaction. The reactor residence time has been made dimensionless using the product reaction rate constant.

The choice of chemical reactor type within which the reaction is carried out is also an important issue for process design and pollution prevention. A continuous-flow stirred-tank reactor (CSTR) is not always the best choice. A plug flow reactor has several advantages in that it can be staged and each stage can be operated at different conditions to minimize waste formation (Nelson, 1992). In a novel application of a plug flow reactor, DuPont developed a catalytic route for the in-situ manufacture of methyl isocyanate (MIC) using a pipeline reactor, resulting in only a few pounds of MIC being inventoried in the process at any one time. This strategy minimizes the chance of a catastrophic release of MIC, such as happened at Bophal, India, in 1984 (Menzer, 1994; Mulholland, 2000).

When hot spots are a problem for highly exothermic reactions carried out in a fixed-bed catalytic reactor, a fluidized-bed catalytic reactor will likely avoid the unwanted temperature excursions. Good temperature control is critical for reducing byproduct formation reactions that are highly temperature-sensitive. An example where a fluidized-bed reactor succeeded in reducing waste formation is in the production of ethylene dichloride, an intermediate in the production of polyvinyl chloride (PVC) (Randall, 1994). The prior fixed-bed design operated with a temperature range of 230–300°C while the newer fluidized-bed design was able to run at between 220–235°C.

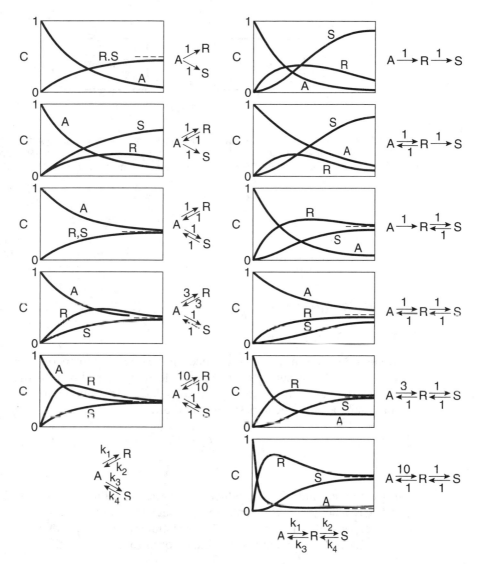

Figure 9.3-4 Product and byproduct profiles for reversible parallel and series reactions. *Source:* Levenspiel, *Chemical Reaction Engineering,* III © 1999. This material is used by permission of John Wiley & Sons, Inc.

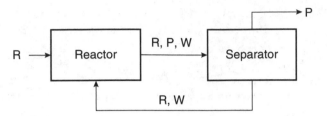

Figure 9.3-5 Process flow diagram illustrating the concept of recycle to extinction for reversible reactions (R = reactant, P = product, and W = waste). *Source:* Allen and Rosselot, *Pollution Prevention for Chemical Processes* © 1997. This material is used by permission of John Wiley & Sons, Inc.

9.3.3 Reactor Operation

Reaction Temperature

Reaction temperature can influence the degree of conversion of reactants to products, the product yield, and product selectivity. We illustrate the effects of temperature on reaction selectivity by considering the simple irreversible first-order parallel reaction mechanism shown below.

$$R \xrightarrow{k_p} P$$

$$R \xrightarrow{k_w} W \qquad \text{(Eq. 9-3)}$$

where R is the reactant, P the product, W a waste byproduct, and k_p and k_w are the first-order reaction rate constants for product formation and waste generation (time^{-1}), respectively. The ratio of the reaction rates for product formation to byproduct generation is an important indicator of reaction selectivity.

$$\frac{k_p[R]}{k_w[R]} = \frac{k_p}{k_w} = \frac{A_p e^{-(E_p/RT)}}{A_w e^{-(E_w/RT)}} \qquad \text{(Eq. 9-4)}$$

where A_p and A_w are the frequency factors (time^{-1}) and E_p and E_w are the activation energies (kcal/mole) for product and waste respectively, R is the gas constant (1.987x10^{-3} kcal/(mole•K)), and T is absolute temperature. Because the reaction rate constants, k_p and k_w, are functions of temperature, their ratio is also a function of temperature. For the purpose of illustration, we can calculate the change in this ratio ($\Delta(k_p/k_w)$) as the temperature is changed to a new value (T_1) above or below a given initial temperature (T_o).

$$\Delta \frac{k_p}{k_w} = \frac{e^{-(E_p/RT_1)} / e^{-(E_w/RT_1)}}{e^{-(E_p/RT_o)} / e^{-(E_w/RT_o)}} = \frac{e^{-(E_p-E_w)/RT_1}}{e^{-(E_p-E_w)/RT_o}} \qquad \text{(Eq. 9-5)}$$

Figure 9.3-6 shows the expected change in the ratio of product/byproduct rate constants when temperature is changed (ΔT) above and below T_o. When $E_p > E_w$, the ratio increases with increasing temperature and decreases with decreasing temperature. Therefore, pollution can be prevented in parallel (and also series) reactions by increasing reactor temperature when $E_p > E_w$. The opposite holds true for when $E_p < E_w$. Also, as the difference between E_p and E_w increases, temperature has a more pronounced influence on the change in the rate constant ratios.

Mixing

When a reactant in one inlet stream is added to another reactant that already exists in a well-stirred reactor, the course of complex multiple reactions can be affected by the intensity of mixing in the vessel. For irreversible reactions, the reaction yield and selectivity may be altered compared to the case where the reactants are mixed instantaneously to a molecular level. This may lead to a greater amount of waste byproduct generation. In addition, the rate of reaction can be reduced because of diffusional limitations between segregated elements of the reaction mixture. The complications that arise from imperfect mixing are particularly evident

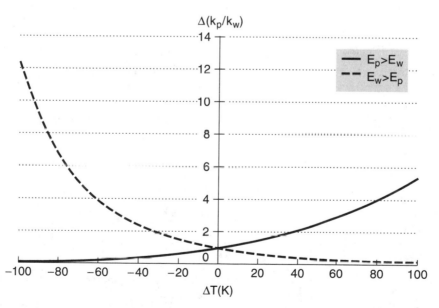

Figure 9.3-6   Effect of reaction temperature on the ratio of rate constants for a first-order parallel and irreversible reaction. For $E_p > E_w$, E_p was set to 20 kcal/mole and E_w to 10 kcal/mole. For $E_p < E_w$, E_p was set to 10 kcal/mole and E_w to 20 kcal/mole.

for rapidly reacting systems. In these situations, reactants are significantly converted to products and byproducts before mixing is complete.

To illustrate the effects of mixing, it is illustrative to examine the *competitive-consecutive* reaction carried out in a constant stirred tank reactor (CSTR), using the reaction mechanisms shown below (Paul and Treybal, 1971).

$$A + B \xrightarrow{k_1} R$$

$$R + B \xrightarrow{k_2} S \qquad \text{(Eq. 9-6)}$$

This reaction is also sometimes referred to as the *series-parallel* reaction. This reaction type is a good kinetic representation of the nitration and halogenation of hydrocarbons and saponification of polyesters, among its many industrially-relevant examples (Chella and Ottino, 1982). Reactant A is initially charged in the reactor and B is added as a solution through a feed pipe in a continuous manner until a stoichiometric amount of B is added. Species R is the desired product and S is a byproduct. If the reactions are first order, mixing will not affect selectivity. However, if the reactions are second order, the presence of local excess B concentrations can cause overreaction of R to S via the second reaction. This effect of mixing occurs for both homogeneous and heterogeneous reaction systems and for batch or semi-batch reactors (B added to an initial charge of A as described above).

A detailed experimental study of a homogeneous liquid phase second-order competitive-consecutive reaction was conducted to determine the effects of mixing on yield of reactants A and B to product R in a CSTR (Paul and Treybal, 1971). The reaction involved the iodization of L-tyrosine (A) in aqueous solution, as shown in Figure 9.3-7.

The authors investigated the effects of reaction temperature, initial concentration of reactant A (A_o), rate of addition of B, agitation rate of the vessel impeller, and presence or absence of baffling within the reactor. A correlating equation for all of these parameters was found between the ratio of measured reactor yield to expected yield (Y/Y_{exp}) versus the dimensionless quantity $(k_1 B_o \tau)(A_o/B_o)$, where

k_1 = product reaction rate constant (liters/(gmole $\cdot$ sec))

k_2 = byproduct reaction rate constant (liters/gmole $\cdot$ sec))

A_o = initial concentration of species A in the feed (gmole/liter)

B_o = initial concentration of species B in the feed (gmole/liter)

τ = microtime scale for mixing of eddies of pure B with bulk liquid (sec)

Y = measured yield = R/A_o

$(k_1 B_o \tau)$ = extent of conversion of A and B under conditions of partial segregation

Y_{exp} = "expected yield" (perfect mixing) = $\dfrac{R}{A_o} = \dfrac{1}{(k_2/k_1 - 1)}\left[\dfrac{A}{A_o} - \left(\dfrac{A}{A_o}\right)^{k_2/k_1}\right]$

A/A_o = fraction of reactant A remaining at the end of the reaction.

$$\text{(Eq. 9-7)}$$

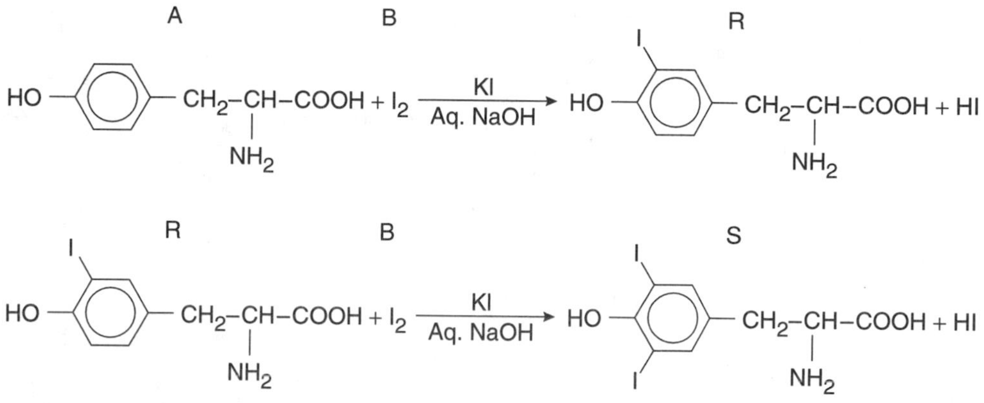

Figure 9.3-7   Iodization of L-tyrosine in aqueous solution.

To give an idea of the range of observed yields in the experiments, values of Y/Y_{exp} were measured from 0.66 to 0.98, depending upon mixing intensity and other parameters. A correlation fitting the data is presented in Figure 9.3-8. It was found that when the quantity $(k_1 \, B_o \, \tau)(A_o/B_o)$ was less than or equal to 10^{-5}, $Y \approx Y_{exp}$. This criterion allows us to "set" the mixing intensity for any second-order competitive-consecutive reaction.

$$10^{-5} = (k_1 \, B_o \, \tau)(A_o \, / \, B_o) = (k_1 \, \tau \, A_o) \qquad \text{(Eq. 9-8)}$$

Rearranging for τ from the above equation, and incorporating the Kolomogoroff universal equilibrium theory for turbulent motion (Kolmogoroff, 1941), we get

$$\tau = \frac{10^{-5}}{A_o k_1} = \frac{0.882 v^{3/4} \, L_f^{3/4}}{(u')^{7/4}} \qquad \text{(Eq. 9-9)}$$

where
L_f = a characteristic length scale of the vessel (ft)
u' = fluctuating turbulent velocity (ft/sec)
v = kinematic viscosity (ft^2/sec)

We can rearrange the equation above for u' and incorporate a correlation for turbulent fluctuation velocity in an agitated CSTR (Cutter, 1966) for feed entering at the impeller ($u' = 0.45 \, \pi \, D \, N$). We arrive at

$$u' = \left[\frac{0.882 \, v^{3/4} \, L_f^{3/4}}{\left(\dfrac{10^{-5}}{A_o \, k_1} \right)} \right]^{4/7} = 0.45 \, \pi \, D \, N \qquad \text{(Eq. 9-10)}$$

Thus, with this equation, we can establish the necessary impeller agitation speed (N, revolutions per second) to ensure that mixing will not adversely affect the yield, given k_1, v, L_f, and D, the impeller diameter (ft).

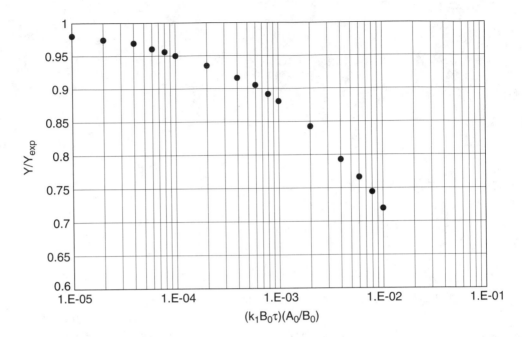

Figure 9.3-8 Correlation of reaction yield efficiency with the mixing parameters of Paul and Treybal (1971) for an irreversible consecutive-competitive second-order reaction.

Example 9.3-1   CSTR Mixer Design to Maximize Yield

A second-order competitive-consecutive reaction is being carried out in an industrial reactor. Use the correlation derived above based on the work of Paul and Treybal to determine the required impeller rotation rate (N) to eliminate mixing effects and achieve the expected yield.

Data: $k_1 = 35$ (liter/(gmole•sec))
$L_f = 1.5$ ft
$\nu =$ kinematic viscosity of mixture $= 1.08$ cs $= 1.16 \times 10^{-5}$ ft^2/sec
$A_o =$ initial concentration of A in the CSTR $= .1$ (gmole/liter)
$D =$ impeller diameter $= 0.5$ ft

Solution: The equation above can be re-arranged to solve for *N*.

$$N = \frac{1}{0.45\,\pi\,D}\left[\frac{0.882\,\nu^{3/4}\,L_f^{3/4}}{\left(\dfrac{10^{-5}}{A_o\,k_1}\right)}\right]^{4/7} = \frac{1}{0.45\,\pi(0.5)}\left[\frac{0.882(1.16\times 10^{-5})^{3/4}(1.5)^{3/4}}{\left(\dfrac{10^{-5}}{(.1)(35)}\right)}\right]^{4/7}$$

$$N = 17.69 \text{ rps} = 1{,}062 \text{ rpm}$$

Example 9.3-2   Estimate the magnitude of the mixing effect on reaction yield.

A second-order competitive-consecutive reaction is being carried out in a CSTR. The initial concentration of reactant A in the vessel is 0.2 gmole/liter and the feed containing reactant B is introduced into the reactor at the impeller. The volume of the vessel is 100 liters, the impeller diameter is 0.5 ft, k_1 is 35 liter / (gmole•sec), impeller speed is 200 rpm. Additional data are shown below. Estimate the reaction yield as a fraction of the expected yield.

Additional Data:

$L_f = 1.5$ ft

v = kinematic viscosity of mixture = 1.08 cs = 1.16×10^{-5} ft^2/sec

Solution: The x-axis in Figure 9-3.8 requires that τ be calculated, and τ requires u'.

$$u' = 0.45 \, \pi \, D \, N = 0.45 \, \pi (0.5)(200 / 60) = 2.36 \text{ ft/sec}$$

$$\tau = 0.882 \frac{v^{3/4} L_f^{3/4}}{u'^{7/4}} = 0.882 \frac{(1.16 \times 10^{-5})^{3/4}(1.5)^{3/4}}{(2.36)^{7/4}} = 5.29 \times 10^{-5} \text{ sec}$$

$$(k_1 B_o \tau)(A_o / B_o) = (k_1 \tau A_o) = (35)(2.32 \times 10^{-5}(0.2) = 3.70 \times 10^{-4}$$

From Figure 9-3.8, the estimated value of Y/Y_{exp} is approximately 0.92. Thus, the mixing in this reactor is almost sufficient to achieve the expected yield. Byproduct generation is not affected to a large degree by mixing in this CSTR, but could be improved slightly by operating the mixer at higher speeds.

Effect of Reactant Concentration

The selectivity of series and parallel chemical reactions can be sensitive to the initial concentration, since the rates of product formation and byproduct generation are dependent on concentration. For a parallel irreversible reaction, the rates of product formation and waste generation can be expressed as

$$\text{Rate of Product Formation} = k_p [R]^{n_p} \qquad \text{(Eq. 9-11)}$$

and

$$\text{Rate of Waste Generation} = k_w [R]^{n_w} \qquad \text{(Eq. 9-12)}$$

where [R] is the concentration of reactant and n_p and n_w are the orders of the reaction. The ratio of these rates is an indicator of the reaction selectivity toward product formation.

$$\frac{k_p [R]^{n_p}}{k_w [R]^{n_w}} = \frac{k_p}{k_w} [R]^{(n_p - n_w)} \qquad \text{(Eq. 9-13)}$$

If $n_p > n_w$, then increasing the concentration of reactant will increase the reaction selectivity toward the product and away from the waste byproduct. Conversely, if $n_p < n_w$, then increasing reactant concentration will decrease selectivity toward the desired product.

Summary of Other Methods

There are numerous operational modifications for improving the environmental and economic performance of reactors. A summary of many important modifications is shown in Table 9.3-2, along with a short description of the nature of the problem addressed, the modification, and the observed benefit. Additional case studies involving improvement in reactor operation can be found elsewhere (US EPA, 1993a).

9.4 POLLUTION PREVENTION FOR SEPARATION DEVICES

Separation technologies are some of the most common and most important unit operations found in chemical processes. Because feedstocks are often complex mixtures and chemical reactions are not 100% efficient, there is always a need to separate chemical components from one another prior to subsequent processing steps. Separation unit operations generate waste because the separation steps themselves are not 100% efficient, and require additional energy input or waste treatment to deal with off-spec products. In this section, we discuss the use of separation devices with respect to pollution prevention in chemical processes. First, the importance of the choice of material (mass separating agent) to be used in the separation step is presented. Next, design heuristics regarding the use of separation technologies in chemical processes are covered. Finally, we present examples of the use of separation technologies for recovery of valuable components from waste streams, leading eventually to their reuse in the process.

9.4.1 Choice of Mass Separating Agent

The correct choice of mass separating agent to employ in a separation technology is an important issue for pollution prevention. A poor choice may result in exposure to toxic substances for not only facility workers but also consumers who use the end product. This is especially important in food products, where exposure to residual agents is by direct ingestion into the body with the food. For example, decaffeinated coffee beans and instant coffee used to be extracted with a chlorinated solvent. While the solvent was extremely effective in extracting caffeine from the bean, residuals in the final product posed an unacceptable health risk to consumers. Caffeine is now extracted using supercritical carbon dioxide (among other benign agents), whose residuals pose no health risk. Edible oils are extracted from plant material using volatile solvents. The oil is recovered from the solvent using distillation while the solvent is recycled back to the process. Residuals can be present in the final product and therefore, it is important to use the lowest toxicity mass separating agent in these applications. In addition to these toxicological issues, a poor choice of mass separating agent may lead to excessive energy

Table 9.3-2   Additional Reactor Operation Modifications Leading to Pollution Prevention (Nelson, 1992; Mulholland and Dyer, 1999).

Improve Reactant Addition

Problem: Non-optimal reactant addition can lead to segregation and excessive byproduct formation.
Solution: Premix liquid reactants and solid catalysts before their introduction into a reactor using static in-line mixers.
Benefits: This will result in more efficient mixing of reactants and reduced waste generation by side-reactions for 2nd order or higher competitive-consecutive reactions.
Solution: Improve dip tube and sparger designs for tank reactors. Do not add low-density material above the liquid surface of a batch reactor. Control residence time of gases added to liquid reaction mixtures.
Benefits: Improved bottom-nozzle dip-tube design and improved gas residence time; control strategy reduced hazardous waste generation by 88% and saved $200,000 per year.

Catalysts

Problem: Homogeneous catalysts can lead to heavy metal contamination of water and solid-waste streams.
Solution: Consider using a heterogeneous catalyst where the metals are immobilized on a solid support.
Problem: Old catalyst designs emphasized conversion of reactants over selectivity to the desired product
Solution: Consider a new catalyst technology that features higher selectivity, and better physical characteristics (size, shape, porosity, etc.).
Benefits: Lower downstream separation and waste treatment costs for byproducts—for example, a new catalyst for making phosgene ($COCl_2$) minimized for formation of carbon tetrachloride and methyl chloride, saving $1 million and eliminating an end-of-pipe treatment device.

Distribute Flows in Fixed-Bed Reactors

Problem: Reactants entering a fixed-bed reactor are poorly distributed. The flow preferentially travels down the center of the reactor. The residence time of the fluid in the center is too short and at the reactor walls is too long. Yield and selectivity suffer.
Solution: Install a flow distributor at the reactor entrance to ensure uniform flow across the reactor cross-section.

Control Reactor Heating/Cooling

Problem: Conventional heat exchange design is not optimum for controlling reactor temperature.
Solution: For highly exothermic reactions, use cocurrent flow of cooling fluid on the external surface of tubular reactors at the inlet where reaction rates and heat generation rates are highest. Use countercurrent flow of cooling fluid near reactor exit where reaction rates and heat generation rates are smallest.
Problem: Diluents added to gas phase reactions, often nitrogen or air, help to dissipate heats of reaction but can result in the generation of wastes, such as oxides of nitrogen in partial oxidation reactions.
Solution: Use a non-reactive substitute diluent, such as carbon dioxide in partial oxidation reactions or even water vapor. Carbon dioxide will need to be efficiently separated from product streams, cooled, and recycled back to the reactor. If water vapor is used, it can be condensed but might result in a wastewater stream for certain reactions.

Additional Reactor Operation Issues

- Improve measurement and control of reactor parameters to achieve optimum state.
- Provide a separate reactor for recycle streams.
- Routinely calibrate instrumentation.
- Consider using a continuous rather than a batch reactor to avoid cleaning wastes.

consumption and the associated health impacts of the emitted criteria pollutants (CO, CO_2, NOx, SOx, particulate matter).

Choice of a mass separating agent in an adsorption application can be illustrated using a simple example. Adsorption is a technology whereby a chemical dissolved in a liquid or a gas phase will preferentially become immobilized on the surface of a solid matrix (adsorbent) packed within a column. Separation and recovery of toxic metal ions from aqueous streams is one very important application of adsorption. Granular activated carbon (GAC) is a very common type of adsorbent, but for the recovery of metals, it has been found that typical strong cation exchange resins have approximately a 20-fold higher capacity to adsorb Cu^{2+} (Mulholland and Dyer, 1999) than GAC. The metal must be recovered from the regenerated adsorbent using a strong acid. In this case, the use of GAC would require more energy consumption and would generate more acid waste than the cation exchange resin.

9.4.2 Process Design and Operation Heuristics for Separation Technologies

A typical chemical process might be depicted as shown in Figure 9.4-1, where a reactor converts feed materials into products and byproducts that must be separated from each other by the additional input of energy. There are waste streams leaving the process and entering the air, the water, and the soil compartments of the environment. While it may be difficult or impossible to eliminate all waste streams, there is every reason to believe that wastes can be minimized by the judicious choice of mass separating agent, by the correct choice and sequencing of separation technologies, and the careful control of system parameters during operation.

The first step in minimizing wastes generated from separation units in chemical processes is to choose the correct technology for the separation task. Making the correct choice, based on the physical and chemical properties of the molecules to be separated, will lead to processes that generate less waste and use less energy per unit of product. The separation task may pertain to a process stream or a waste stream. Table 9.4-1 shows unit operation choices and property differences between the components to be separated.

After selecting the best separation technology, it is worth considering several pollution prevention heuristics to guide the design of the flowsheet and operation of the units. Table 9.4-2 shows several design and operation heuristics for separation processes. Streams of similar enough composition can be combined in order to reduce the number of unit operations and their associated capital costs and emission sources. Corrosive materials should never be added unless necessary and if added or generated in the process, should be separated immediately. Their removal can minimize investment and the generation of trace metals. Unstable materials should be removed, preferably at low temperatures, to reduce the formation of undesirable waste products, such as tars. Removing the highest volume components first in a process will minimize downstream equipment investment, associated energy-related costs, and the addition of materials required for processing that could become another source of waste. If component properties are very close (difficult separation) or product purity requirements from the separation are extremely high, reducing the number of components involved will make the separation easier. Therefore, these cases should be left until the end in a separation sequence. Whereas raw materials and products add value to the design, mass separating agents only increase investment, operating cost, and waste loads. Therefore, mass separating agents should not be added to the process unless necessary. If a mass separating agent needs to be added, it should be removed (preferable recycled) in the next step of the process using an energy separating agent technology. The process should avoid separation technologies that operate far from ambient temperature and pressure. If departures from ambient are required, it is more economical to operate above rather than below ambient.

Distillation accounts for over 90% of the separation applications in chemical processing in the United States (Humphrey, 1995). Because of its prominence, we will present a number of pollution prevention techniques that are specific to distillation. Distillation columns contribute to process waste in four major ways;

a) by allowing impurities to remain in a product,
b) by forming waste within the column itself,
c) by inadequate condensing of overhead product (Nelson, 1992), and
d) by excessive energy use.

Product impurities above allowable levels must eventually be removed, leading to additional waste streams and energy consumption. Waste is formed within distillation columns in the reboiler where excessive temperatures and unstable materials combine to form high molecular weight tars or polymers on the heat transfer

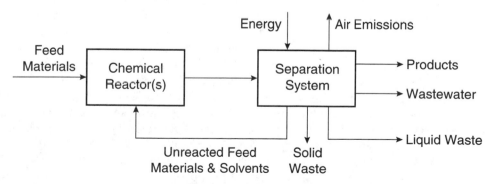

Figure 9.4-1   Typical chemical process (Adapted from Mulholland and Dyer, 1999).

Table 9.4-1 Unit Operations for Separations and Property Differences (Null, 1987).

Unit Operation	Property Difference
Adsorption	Surface sorption
Chromatography	Depends on stationary phase
Crystallization	Melting point or solubility
Dialysis	Diffusivity
Distillation	Vapor pressure
Electrodialysis	Electric charge and ionic mobility
Electrophoresis	Electric charge and ionic mobility
Gel Filtration	Molecular size and shape
Ion Exchange	Chemical reaction equilibrium
Liquid-liquid extraction	Distribution between immiscible liquid phases
Liquid membranes	Diffusivity and reaction equilibrium
Membrane gas separation	Diffusivity and solubility
Reverse osmosis	Molecular size
Micro- and ultrafiltration	Molecular size

surfaces. The condenser vent must be open to the external environment to relieve non-condensable gases that build up in the column. If the condenser duty is insufficient for the internal vapor load in the column, excess vapor will exit the vent as a waste stream. Energy use leads to the direct release of criteria pollutants (CO, NOx, SOx, particulate matter, volatile organic compounds) and global warming gases (primarily CO_2).

The most common way to increase product purity in distillation is to increase the reflux ratio. However, this increases the pressure drop across the column, raises the reboiler temperature, and increases the reboiler duty. But for stable materials, this may be the easiest way to decrease waste generation due to inadequate product purity. If a column is operating close to flooding (increasing reflux ratio is not an option), then adding a section to the column leads to higher-purity products. Replacing existing column internals (trays or packing) with high-efficiency packing results in greater separation for an existing column, and results in both lower pressure drop and reboiler temperature. Changing the feed location to the optimum may increase product purity without changing any other system parameters. In one documented case, relocating the feed to the optimum position reduced the loss of product to waste from 30 to 1 lb/hr, increased column capacity by 20%, and decreased the refrigeration cooling load by 10% (Mulholland and Dyer, 1999). The net benefit of this single step was greater than $9,000,000 per year. Additional ways to increase column separation efficiency are to insulate the column and reduce heat losses; improve feed, reflux, and liquid distribution; and preheat the column feed, employing cross-exchange with other process streams. Finally, if the overheads product contains a light impurity, it may be possible to withdraw the product from a side stream near the top of the column. A bleed stream from the overhead condenser can be recycled back to the process to rid the column of the light component.

Table 9.4-2   Separation Heuristics to Prevent Pollution (Adapted from Mulholland and Dyer, 1999).

1. Combine similar streams to minimize the number of separation units.
2. Remove corrosive and unstable materials early.
3. Separate highest-volume components first.
4. Do the most difficult separations last.
5. Do high-purity recovery fraction separations last.
6. Use a sequence resulting in the smallest number of products.
7. Avoid adding new components to the separation sequence.
8. If a mass separating agent is used, recover it in the next step.
9. Do not use a second mass recovery agent to recover the first.
10. Avoid extreme operating conditions.

Example 9.4-1   Energy Savings in Ethanol-Water Distillation: Side Stream Case

When a product from a distillation is needed with a composition between the distillate (x_D) and bottoms (x_B) products, a side stream with this composition collected from the column will always save energy compared to combining the top and bottom product streams. Consider the distillation column with a side stream of composition x_S and flow rate of S (moles/hr), as shown in Figure 9.4-2. Using a McCabe-Thiele analysis, demonstrate energy savings can be achieved using a side stream. Mole fractions are ethanol.

Solution: It can be easily shown, using graphical methods presented in any standard textbook (Wankat, 1988), that the required separation for this column, feed conditions, and separation requirements can be achieved using a column with 12 equilibrium stages operating with a reflux ratio of L/D = 2.5. The side stream is taken from the 5th stage from the top of the column. Similarly, for a 12-stage column without a side stream, the required separation for this feed can be accomplished using a reflux ratio of only 2.0. Nonetheless, the side stream has clear energy savings as shown in the table below.

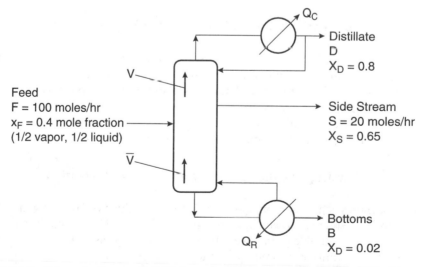

Figure 9.4-2   Schematic diagram of distillation column with side stream.

Column Design	L/D	D	V	Q_R (cal/hr)
Side Stream	2.5	32.56	63.96	63.96×10^4
No Side Stream	2.0	48.72	96.16	96.16×10^4

Using a side stream design, the energy savings are $(96.16\text{-}63.96)/96.16 \times 100 =$ 33.5%. As shown in Chapter 11, the energy savings will translate to reduced impacts for global warming and acid rain/deposition.

Note: $D = F\dfrac{x_F - x_B}{x_D - x_B}; Q_R = \overline{V}\lambda_W; \overline{V} = D\left(\dfrac{L}{D} + 1\right) - F(1 - q)$

In this example, the feed quality (q) is taken to be 1/2 (1/2 vapor and 1/2 liquid) and λ_W is the latent heat of vaporization of water (in the reboiler; 10^4 cal/mole).

There are several ways to decrease the generation of tars in the reboiler of the column. One way is to reduce the column pressure, resulting in lower reboiler temperatures. Caution must be taken, as this affects the condenser temperature and efficiency. The above step should be coupled with steps to reduce the steam temperature. Reboiler temperature can be reduced by desuperheating the steam, by using a lower pressure steam, by installing a thermocompressor, or by using an intermediate heat transfer fluid. The existing reboiler may be retrofitted with high flux tubes allowing for the use of lower pressure steam.

If the overheads condenser is undersized relative to the vapor loading in the column, it can be re-tubed or replaced with a larger capacity unit. This step will reduce the likelihood that fluctuations of column operation will result in hot vapor being expelled from the column vent. An additional way to reduce distillation column emissions and waste generation is by improving the process control technology. This step will assure that product purity specifications will be met and reduce the possibility that off-spec product will be created.

9.4.3 Pollution Prevention Examples for Separations

Pollution can be prevented using separation processes by selective recovery and reuse of valuable components from waste streams. Often, a good knowledge of the capabilities of separation technologies combined with markets for components recovered from waste streams can result in processes that are not only profitable but that also prevent pollution from entering the environment. Table 9.4-3 summarizes many successful applications of separation technologies for pollution prevention.

9.4.4 Separators with Reactors for Pollution Prevention

Separators can be combined with reactors to reduce byproduct generation from reactors and increase reactant conversion to products. These combinations of separators and reactors can involve either distinct units, as illustrated in Section 9.3, or integrated units as outlined in Section 9.5.

Table 9.4-3 Pollution Prevention Examples for Separation Processes

Separation Technology	Stream Type	Description	References
Distillation	Liquid	Solvent recovery from wastewater. A wastewater stream from a solution polymerization process contains organic solvents that are regulated under RCRA. The wastewater stream was previously incinerated. A re-evaluation found that distillation followed by extraction could be used to recover more than 10 million lb/yr of solvent and reduce incineration loads by 4 million lb/yr, and had a payback period of only 2 years.	Mulholland and Dyer, 1999
Distillation	Liquid	Ink and solvent recycle. Waste ink from newspaper printing contains organic solvent (20%), water (15%) and ink (65%). A flash distillation is used to separate the high-boiling ink from the solvent/ water solution and binary distillation is used to separate the solvent from the water. Solvent and ink are reused in the process.	Palepu et al., 1995
Distillation	Liquid	Batch distillation of used antifreeze. Pure ethylene glycol is recovered and blended with water plus other additives to make new antifreeze.	Palepu et al., 1995
Distillation	Liquid	Acid recovery from spent acid streams. In the electroplating industry, spent acids from etching tanks, cleaning tanks, and pickling tanks can be processed using distillation to recover pure acids (HCl, HNO_3, etc.).	Jones, 1990
Distillation	Liquid	Solvent recovery and reuse in automobile paint operations. A closed-loop solvent utilization system has been established for cleaning out paint lines between color changes. The collected paint/ solvent mixture is transported to a central reprocessing facility, the solids are separated, and pure solvent is recovered by distillation. The solvent is reused in automobile painting.	Gage Products Inc.
Extraction	Liquid	Extraction of a batch process residue. A batch process has difficulties using distillation, resulting in about 1/3 of the production run being incinerated. A low-boiling-point material recovered from the batch residue was found to be an effective extraction solvent to recover more product from the residue.	Mulholland and Dyer, 1999
Extraction	Liquid and Sludges	Hydrocarbon recovery from refinery wastewater and sludge. Triethylamine is used as a solvent to recover hydrocarbons from refinery wastewater and sludges. The hydrocarbons are recycled back to the process.	Tucker and Carson, 1985

(continued)

Table 9.4-3 Pollution Prevention Examples for Separation Processes (*continued*)

Separation Technology	Stream Type	Description	References
Reverse Osmosis	Liquid	Closed-loop rinsewater for process electroplating. Reverse osmosis is able to return pure water and a concentrated metals-containing stream to the plating bath. There are over 200 documented industrial applications.	Werschulz, 1985
Reverse Osmosis	Liquid	Recovery of homogeneous metal catalysts. $300,000 per year was saved by using reverse osmosis rather than chemical precipitation agents.	Radecki et al., 1999
Ultrafiltration	Liquid	Polymer recovery from wastewater. Cleaning of polymerization reactors generates a stream from which polymers such as latex can be recovered. Also, polyvinyl alcohol can be recovered in the manufacture of synthetic yarn.	Bansal, 1976
Adsorption	Gas	Natural gas dehydration. Molecular sieve adsorbents are being used to dehydrate natural gas, thereby eliminating the use of a solvent (triethyleneglycol).	Mulholland and Dyer, 1999
Adsorption	Liquid	Replacement of azeotropic distillation. Azeotropic solvents, such as benzene and cyclohexane, can be eliminated by contacting azeotropes (ethanol/water or isopropanol/water) with molecular sieve adsorbents.	Radecki et al., 1999
Membrane	Gas	Recovery and recycle of high-value volatile organic compounds. Examples include recovery of olefin monomer from polyolefin processes, gasoline vapor recovery from storage facilities, vinyl chloride recovery from PVC reactor vents, and chlorofluorocarbons (CFCs) recovery from process vents and transfer operations. Emerging applications included also.	Radecki et al., 1999
Membrane	Liquid	Recovery of organic compounds from wastewater streams. Pervaporation is a membrane process used to recover organics from low flow (10–100 gal/min) and moderate concentration (0.02 to 5% by wt.) wastewater.	Radecki et al., 1999
Membranes (RO, NF, UF, MF, ED)*	Liquid	Metal ion recovery from aqueous waste streams.	Radecki et al., 1999

*RO–reverse osmosis, NF – nanofiltration, UF – ultrafiltration, MF – microfiltration, ED – electrodialysis.

9.5 POLLUTION PREVENTION APPLICATIONS FOR SEPARATIVE REACTORS

An exciting new reactor type that has a very high potential for reducing waste generation is the *separative reactor*. These hybrid systems combine chemical reaction and product separation in a single process unit. When chemical reaction and separation occur in concert, the requirements for downstream processing units are reduced, leading to lower capital costs. The key feature allowing for the prevention of waste generation and maximizing product yield is the ability to control the addition of reactant and the removal of product more precisely than in traditional designs. Unfavorable chemical equilibrium can be shifted to maximize reactant conversion and product yield. Unwanted byproduct generation can be minimized in series reactions by the removal of the desired product within the reaction zone and before significant secondary reactions can occur. Separation units that have been integrated with reaction include distillation, membrane separation, and adsorption. A recent review has been written on emerging uses of separative reactors for pollution prevention employing membranes and solid adsorbents (Radecki et al., 1999).

A good demonstration of reaction coupled with adsorption is oxidative coupling of methane (OCM). Methane reacts with oxygen in the presence of metal oxide catalysts at a temperature of about 1000K to form ethane and ethylene.

$$2CH_4 + 1/2O_2 \rightarrow C_2H_6 + H_2O$$

$$2CH_4 + O_2 \rightarrow C_3H_4 + 2H_2O$$

A parallel path in which methane is completely oxidized is shown below.

$$CH_4 + 2O_2 \rightarrow CO_2 + 2H_2O$$

There is also a concern that the ethylene product can be oxidized to carbon dioxide, thereby reducing product yield and increasing waste generation. Successful application of OCM in industry would allow the use of methane, a high-production chemical that is difficult to transport, as a feedstock for ethylene, an important intermediate in polymer production. The difficulty with traditional OCM in a fixed-bed or fluidized-bed reactor is that the feed ratio of CH_4/O_2 must be kept around 50 or more to limit complete oxidation reactions from occurring. This results in relatively high selectivity (80–90%) but limits the yield of C_2s to less than 20%. A separative reactor composed of a series of reactor/adsorber sections substantially improved the yield of C_2 (50–65%) (Tonkovich, et al., 1993; Tonkovich and Carr, 1994). Each section contained a fixed-bed catalytic reactor operating at high temperature (1000K), followed immediately by a cooler adsorption bed. Figure 9.5-1 shows the arrangement of a four-section simulated countercurrent moving-bed chromatographic reactor (SCMCR) in which the smaller columns are the catalytic fixed-bed reactors and the larger columns are for adsorption. In section 1, the carrier gas (N_2) sweeps unreacted adsorbed CH_4 into the next section (feed section 2).

A small make-up feed stream comprised of CH_4/O_2 in stoichiometric amounts (to make up for consumption in reaction) is combined with a carrier gas stream from section 1, which then enters the reactor in the feed section. Reaction products (C_2s) and unreacted CH_4 are adsorbed in the large column of section 2, with the C_2 products being retained in the upper portion and the more mobile CH_4 in the bottom. Section 3 is isolated from flow, yet contains unreacted CH_4 in the adsorption column. Section 4 is the product removal section in which the C_2 products are swept off the adsorption bed and into a side stream roughly midway in the column. After maintaining the SCMCR in this configuration for a prescribed time interval, the flow configurations are advanced one section to the left so that section 1 is the product removal section, section 2 is the carrier section, section 3 is the feed section, and section 4 is the isolated section.

Another application of the coupled catalytic reactor with adsorbent beds is the partial oxidation of methane to methanol. Methanol is in demand as a fuel oxygenate, is a feedstock for other oxygenates in reformulated gasoline, and is being investigated as an alternative fuel for gasoline and diesel engines. Reformulated gasoline reduces emissions of CO, NOx, volatile organic compounds, and benzene from automobiles, as required by the Clean Air Act of 1990. The current process for methanol production is a costly two-step process, consisting of steam reforming of methane to produce CO and H_2, followed by methanol formation by passing CO and H_2 over a metal oxide catalyst. The overall reaction of CO and H_2 to form methanol is endothermic by 125 kJ/mole, requiring significant energy input. In contrast, the partial oxidation reaction of methane has the overall reaction

$$CH_4 + 1/2O_2 \rightarrow CH_3OH$$

and is exothermic by 126 kJ/mole. However, in order to minimize over-oxidation of methanol to CO_2 in a series reaction, the feed CH_4/O_2 ratio must be kept high, leading to disappointingly low per-pass methanol yields of less than 10%. In recent

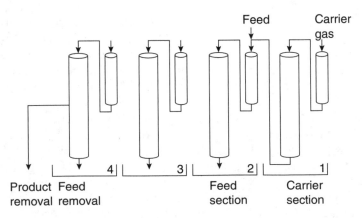

Product Feed   Feed   Carrier
removal removal   section   section

Figure 9.5-1 Schematic of a four-section SCMCR for the oxidative coupling of methane. Adapted from Tonkovich et al., 1993.

experiments, methanol yield increased from 3–4% in a tubular reactor to 17% when the reactor was interfaced with adsorber beds and operated in the SCMCR mode (Bjorklund and Carr, 1996). This demonstrated that the SCMCR mode of reaction is useful for increasing performance of low conversion per pass reactions.

Reaction coupled with membrane separation is another often-studied configuration for increasing the efficiency of chemical reactions. Much like adsorption-based separative reactors, the equivalent membrane-based unit can be used to selectively remove either products or byproducts from the reaction zone, thereby overcoming low conversions in equilibrium-limited reactions and reduce waste generation in series reactions. However, membrane-based separative reactors can also be used to selectively permeate reactant into the reaction zone in order to control excessive byproduct formation (e.g., permeation of O_2 in partial oxidation or oxidative coupling reactions). Both of these modes of operation are shown in Figure 9.5-2. Membrane materials can be organic, porous inorganic, or nonporous (dense) inorganic, and either can be constructed of inert (non-reactive) material or can contain catalysts in various configurations.

Applicable reaction types that can be improved by membrane separative reactors include

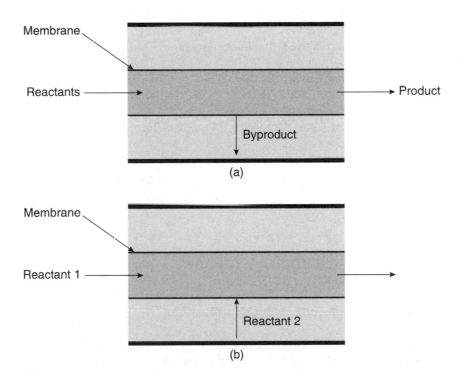

Figure 9.5-2   Main modes of operation for membranes Separative reactors: (a) Selective removal of products / byproducts and (b) Selective permeation of reactants.

a) thermodynamically-limited reactions (e.g., $C_6H_{12} \leftrightarrow C_6H_6 + 3H_2$),

b) parallel reactions in which product formation has a lower reaction order than byproduct generation,

c) series reactions such as selective dehydrogenations and partial oxidations, and

d) series-parallel reactions.

Applications of membrane separative reactors for partial oxidation have shown encouraging results. The test reaction of ethane oxidative dehydrogenation to ethylene showed that per pass yields increased from 12 to 52% (Tonkovich et al., 1996). Positive results were also demonstrated for membrane separative reactors for dehydrogenation of ethylbenzene to produce styrene, where conversions as high as 70% were observed, approximately 15% higher than conventional methods, and an increase from 2–5% in styrene selectivity (Radecki et al., 1999).

The potential for membrane separative reactors or other means to improve the environmental and economic performance of some of the top 50 commodity chemicals in the U.S. chemical industry formed through partial oxidation or by dehydrogenation reactions was recently studied (Tonkovich, 1994; Tonkovich and Gerber, 1995). The maximum energy saving was estimated to be 0.25 quadrillion BTU if every commodity chemical formed through either selective oxidation or dehydrogenation achieved maximum efficiency. Maximum annual savings in feedstocks were estimated to be $1.4 billion.

Additional challenges remain before commercial application of membrane separative reactors can be realized. These include

a) economical manufacture of thin, defect-free selective membrane layers over large surface areas,

b) leak-free reaction systems with high temperature seals,

c) elimination or reduction of sweep gases which dilute product streams, and

d) enhanced membrane and catalyst performance, including resistance to fouling and deactivation.

9.6 POLLUTION PREVENTION IN STORAGE TANKS AND FUGITIVE SOURCES

9.6.1 Storage Tank Pollution Prevention

Storage tanks are very common unit operations in several industrial sectors, including petroleum production and refining, petrochemical and chemical manufacturing, storage and transportation, and other industries that either use or produce organic liquid chemicals. Tanks are used for many purposes, including storage of fuels and for feedstock or final product buffer capacity. The main environmental impact of storage tanks is the continual occurrence of air emissions of volatile organic compounds from roof vents and the periodic removal of oily sludges from tank bottoms.

Tank bottoms are solids or sludges composed of rusts, soil particles, heavy feedstock constituents, and other dense materials that are likely to settle out of the liquid being stored. There are various methods of dealing with these materials once they are present. They may be periodically removed and either treated via land application or disposed of as hazardous waste. As long as the bottoms components are compatible with downstream processes, they may be prevented from settling to the tank bottom by the action of mixers that keep the solid particles suspended in the liquid (API, 1991a). Another method is to use emulsifying agents that keep water and solids in solution and out of the tank bottoms. A concern with the use of this method is the potential to generate oily waste downstream in the refinery processes from the presence of the emusifiers (API, 1991b).

Air emissions of volatile organic compounds from storage tanks are a major source of airborne pollution from petroleum and chemical processing facilities. These emissions stem from the normal operation of these units in response to the changes in liquid level within the tank and the action of ambient changes in temperature and pressure. These loss mechanisms are termed *working losses* and *standing losses*, respectively. The emissions are dependent upon the vapor pressure of the stored liquids, tank characteristics such as tank type, paint color and condition, and also the geographic location of the tank. There are six major types of storage tanks. A listing of these tank types, short descriptions, a summary of emission mechanisms, and pollution reduction measures are listed in Table 9.6-1.

The following example will illustrate the emission reduction that is possible when substituting a floating-roof tank for a fixed-roof tank in a process design.

Example 9.6-1   Storage tank emissions

A toluene product stream (516,600 gal/yr) exits from a VOC recovery process for a gaseous waste stream at a facility in the vicinity of Detroit, MI. Using the TANKS (see Chapter 8) software (US EPA TTN, 1999), calculate and compare the uncontrolled annual emissions for a new tank design having the following dimensions and conditions:

Fixed-Roof Tank: Height = 20 ft, Diameter = 12 ft, Working Volume = 15,228.53 gallons, Maximum Liquid Level = 18 ft, Average Liquid Level = 10 ft, no heating, domed roof of height 2 ft and diameter of 12 ft, vacuum setting of −.03 psig and pressure setting of .03 psig.

Internal-Floating Roof Tank: Height = 20 ft, Diameter = 12 ft, Working Volume = 15,228.53 gallons, self-supporting roof, internal shell condition of light rust, floating roof type is pontoon, primary seal is a mechanical shoe, secondary shoe is shoe-mounted, deck type is welded, deck fitting category is typical.

Domed External Floating-Roof Tank: same as internal floating-roof tank.

Table 9.6-1   Storage Tank Types and Pollution Reduction Strategies.

Storage Tank Type	Description	Loss Mechanisms	Pollution Reduction
Fixed Roof	Cylindrical shell with permanent roof (flat, cone, or dome), freely vented or with pressure/vacuum vent.	Working losses – VOCs in headspace above liquid are expelled when tank is filled. Standing losses – headspace gas expands/contracts by ambient T and P.	Pressure / vacuum vents reduce standing losses, heating the tanks reduces standing losses, pollution control equipment on vent (adsorption, absorption, cooling) reduce emissions 90–98%. Vapor balancing approach.
External Floating Roof	Cylindrical shell without a fixed roof, a deck floats on the liquid surface and rises and falls with liquid level, deck has flexible seals on shell inner wall to scrape liquid off shell wall.	Working losses – evaporation from wetted shell wall or columns as liquid is withdrawn. Standing losses – small annular space between deck system and shell wall is source of these losses.	Little reduction can be accomplished to control or prevent the wind–driven emissions from the shell wall. Emissions actually greater than fixed-roof tanks.
Internal Floating Roof	Same as external floating roof with a permanent fixed roof above. Roof is either column or self-supported.	Same as external floating roof tank. Permanent roof blocks wind and reduces working losses.	60–99% emission reduction compared to a fixed-roof tank.
Domed External Floating Roof	Similar to an internal floating roof tank but has a self-supported domed roof.	Similar to self-supported permanent roof.	60–99% emission reduction compared to a fixed-roof tank.
Variable Vapor Space	Roof telescopes to receive expelled vapors. Diaphragm used to accept expelled vapors.	Working losses occur when liquid level is raised. Standing losses are eliminated.	No data available on emissions reduction.
Pressure Tanks	Low pressure (2–15 psig) and high pressure (> 15 psig)	No losses from high pressure tanks. Working losses from low pressure tanks during filling operations. No standing losses.	No data available on emissions reductions.

Vapor balancing involves routing the expelled vapors during tank filling to another tank that is supplying the liquid.

T are daily changes in ambient temperature.

P are changes in barometric pressure.

Reference: (US EPA, 1998).

Solution: The TANKS program allows the user to quickly calculate the annual emission rate for all three tank types. The results are

Fixed-Roof Tank: 337.6 lb/yr
Internal Floating-Roof Tank: 66.2 lb/yr.
Domed External Floating-Roof Tank: 42.8 lb/yr.

Discussion: The reduction in emissions for the floating-roof tanks compared to the vertical fixed roof tank are:

Internal Floating-Roof Tank: % reduction = (337.6-66.2)/337.6 x 100 = 80.4%
Domed External Floating-Roof Tank: % reduction = (337.6-42.8)/337.6 x 100 = 87.3%

The reductions are significant, and may help the facility achieve emission reduction targets established by local, state, and federal regulations. Floating-roof tanks are more expensive than fixed-roof tanks and this consideration would have to be incorporated into any design decision. Pollution control on fixed-roof tank vents can achieve even higher removal percentages (90–98%), but would require annual operating costs.

9.6.2   Reducing Emissions from Fugitive Sources

Fugitive emission sources in chemical processes include valves, pumps, piping connectors, pressure relief valves, sampling connections, compressor seals, and open-ended lines. There may be thousands of these components in a typical synthetic organic chemical manufacturing industry (SOCMI) facility and tens to hundreds of thousands in a large petroleum refinery. These emission sources are significant contributors to air pollution from SOCMI facilities, as estimates have shown that as much as one third of air emissions occur from fugitive sources (US EPA, 1986).

Within individual components, leaks are localized near seals, valve packings, and gaskets. Components in good working order rarely leak quantities of process fluids that are of significant concern. When leaks occur due to a seal, packing, or gasket failure, the exact timing, location, and rate of release is difficult to predict. These leaks are of two types—either low-level leaks that may persist for long periods of time until detected, or sudden episodic failures resulting in a large release. However, the leaks can be prevented or repaired, and leakless technologies are available for situations where even small rates of release cannot be tolerated.

In this section, we identify which of the components from fugitive sources listed above have the greatest potential for emission reductions as a result of pollution prevention efforts. Next, established methods for reducing or preventing fugitive emissions are presented. Finally, a study summarizing the emissions reductions that are possible in chemical manufacturing facilities is presented.

Fugitive Emission Profiles

The average rate of emission of volatile organic compounds (VOCs) from fugitive components of different types can vary significantly within a given facility. To demonstrate this, we estimate the emission rate from all fugitive sources within two processing units at a refinery, a cracking unit and a hydrogen plant. The average emission factors presented in Table 8.3-3 will be used along with a knowledge of the numbers of sources within a given component type and the mass fraction of VOC in the stream serviced by the component. The equation used to calculate the emission rate for each component is

$$E = m_{VOC} f_{av}$$

where E is the emission rate (kg/hr/source), m_{VOC} is the mass fraction of VOC in the stream, and f_{av} is the average emission factor. The numbers of fugitive sources and their contributions to the emissions from the two processing units at a refinery are shown in Table 9.6-2. Valves in all service are by far the largest source for emissions from these process units, comprising 55.3% and 63.4% of the total for the cracker unit and H$_2$ plant, respectively. These emissions are disproportionately large for the relative number of valves in the processes—22.5% and 22.8% for the cracker unit and H$_2$ plant, respectively. The component present in the largest number is connectors in all service, being 74.4% and 75.1% of the total for the cracker unit and H$_2$ plant, respectively. Relief valves appear to be significant emission sources, as do seals on pumps and compressors.

Table 9.6-2 Distribution of Fugitive Components and Emission Rates from a Cracking Unit and a Hydrogen Plant at a Refinery (Allen and Rosselot, 1997).

| Component | Service[a] | Cracker | | | | Hydrogen Plant[b] | | |
		Equipment Count	m_{VOC}	Emissions, kg/hr	(%)	Equipment Count	Emissions, kg/hr	(%)
Pump seals	LL	6	0.75	0.51	4.0	2	0.22	1.9
	HL	9	0.55	0.1	0.81	2	0.042	0.36
Compressor seals	HC gas	4	1.0	2.60	20	0	0	0
	H$_2$ gas	0		0	0	6	0.30	2.6
Valves	HC gas	200	1.0	5.3	42	70	1.9	16
	H$_2$ gas	0		0	0	80	0.66	5.7
	LL	196	0.75	1.6	13	427	4.7	41
	HL	294	0.55	0.037	0.29	427	0.85	0.73
Connectors	All	2277	0.75	0.42	3.3	3313	0.83	7.2
Relief valves	Gas	11	1.0	1.8	14	15	2.4	21
	Liquid	15	0.63	0.066	0.52	2	0.014	0.12
Open-ended lines	All	32	0.75	0.054	0.43	42	0.084	0.72
Sampling taps	All	17	0.75	0.19	1.5	24	0.36	3.1
Total	—	—	—	13	100	—	12	100

[a]HL: heavy liquid, LL: light liquid, HC: hydrocarbon
[b]$m_{VOC} = 1.0$ for all.

Methods to Reduce Fugitive Emissions

There are two methods for reducing or preventing emissions and leaks from fugitive sources. They are

1) leak detection and repair (LDAR) of leaking equipment, and
2) equipment modification or replacement with emissionless technologies.

Both methods can be effective in reducing low-level as well as large episodic leaks of process fluid.

In an LDAR program, equipment such as pumps and valves are monitored periodically using an organic vapor analyzer (OVA). The wand of the OVA is directed towards the suspected source of leakage on each piece of equipment, i.e. at a packing nut on a valve, at a shaft seal on a pump, or a gasket or weld on a flange or connector. Guidance documents are available from the US EPA on detailed procedures to monitor for leaks and estimate emissions for fugitive sources (US EPA 1993b). If the source registers an OVA reading over a threshold value (>10,000 ppm), the equipment is said to be leaking and repair is required. Progress towards achieving desired fugitive emission reduction targets can be measured by using the OVA screening values and US EPA emission correlations for fugitive sources, as shown in Table 9.6-3. The nature of the repairs depends upon the piece of equipment, but may involve something as simple as tightening a packing nut on a valve or it may require replacement of a seal on a pump or a gasket in a connector. Repairs may require shut-down of the process, and would be conducted during regularly-scheduled shut-down times in order to minimize the number of upsets in process operation and reduce repair-related emissions.

Industrial LDAR programs vary greatly in their frequency of monitoring and in their effectiveness. Constant monitoring of emissions using area monitors is possible when contaminants are detectable in very low concentrations. For cases where constant monitoring is either technically impossible or too expensive, periodic monitoring on a monthly, quarterly, or annual basis using an OVA is the preferred approach. Monitoring on a more frequent basis may be more costly, but has been shown to be more effective in reducing emissions. For example, monitoring and repairing valves in light liquid service at monthly intervals is a third more effective in reducing emissions compared to monitoring quarterly, and is three times as effective as monitoring every six months (US EPA 1982).

Equipment modification to reduce fugitive emissions might involve redesigning a process so that it has fewer pieces of equipment and connections, replacing leaking equipment with new conventional equipment, or the inclusion of new emissions-reducing technology, and sealant injection. In this discussion, we focus only on equipment replacement for the major fugitive emission sources, valves, connectors/flanges, compressors, and pumps. For a more complete treatment, the reader is referred to other textbooks on the subject (Allen and Rosselot 1997, Chapter 7). We discuss the types of equipment, where the leaks are likely to occur, and what equipment changes can be made to reduce or eliminate releases.

Table 9.6-3 Correlations for Estimating Fugitive Emissions and Their Default Values (US EPA, 1993b).

Leak Rate from Correlation, kg/hr/source[a]				Default Emissions,[b] kg/hr/source
Equipment	*Service*	*SOCMI*	*Refinery*	
Valves	Gas	$1.87 \times 10^{-6}\ C^{.873}$	$2.18 \times 10^{-7}\ C^{1.23}$	6.56×10^{-7}
	Light liquid	$6.41 \times 10^{-6}\ C^{.797}$	$1.44 \times 10^{-5}\ C^{.80}$	4.85×10^{-7}
Pump seals	Light liquid	$1.9 \times 10^{-5}\ C^{.824,c}$	$8.27 \times 10^{-5}\ C^{.83,d}$	$7.49 \times 10^{-6,c}$
	Heavy liquid		$8.79 \times 10^{-6}\ C^{1.04}$	
Compressor seals	Gas		$8.27 \times 10^{-5}\ C^{.83}$	
Pressure-relief valves	Gas		$8.27 \times 10^{-5}\ C^{.83}$	
Flanges / other connectors	All	$3.05 \times 10^{-6}\ C^{.885}$	$5.78 \times 10^{-6}\ C^{.88}$	6.12×10^{-7}

[a] C: screening value in ppm.
[b] These values are applicable to all source categories.
[c] This correlation/default-zero value can be applied to compressor seals, pressure-relief valves, agitator seals, and heavy liquid pumps.
[d] This correlation can be applied to agitator seals.

Connectors are the most numerous pieces of equipment and are used to connect pipe to other pipes or to equipment and vessels. Connectors on smaller piping (<2 inches in diameter) are either threaded pipes and couplers or nut-and-ferrule connectors. Flanges are connectors with a flexible seal junction that are used for pipes greater than 2 inches in diameter. Another connector type is welded pipe. Leaks may occur from connectors due to thermal deformation on correct assemblies or may result from cross-threading and incorrect assembly of nut-and-ferrule types. Monitoring of these pieces of equipment would occur near threaded junctions, gaskets, and welds.

Seals around moving parts are common locations of leaks for fugitive sources such as valves, pumps, and compressors. For valves, the moving part is the stem that connects the internal components of the valve with the outside. The packing is subject to degradation or the stem may have surface defects, both of which may promote leaks. Valve packing technologies that use rings to keep the packing from extruding and springs to maintain the packing under constant pressure and contact with the stem have been developed. These systems can reduce leak rates and reduce maintenance requirements for 10–50 times as long as conventional packing (Brestel et al., 1991). There are two main types of "leakless" valves that have no emissions through the stem. They are bellows valves, which are expensive and are mostly used in the nuclear power industry, and diaphragm valves, in which a physical barrier (diaphragm) exists between the process fluid and the valve stem.

Seals around pumps generally occur where a rotating shaft meets the stationary casing. Two main types of seals are used; packed seals and mechanical seals. Mechanical seals that are well-maintained are superior to packed seals, but because they are expensive and time-consuming to repair, a second packed or mechanical seal is commonly used. Sealless designs for pumps include canned motor pumps, where the bearings are in the process fluid, and diaphragm pumps, where a moving diaphragm pumps the fluid. Magnetic pumps are also available in which the impeller is driven by magnets. Mechanical seals for pumps have improved greatly over the last 10–20 years, making them a viable alternative for leakless pumps for many applications (Adams, 1991).

Compressors are similar to pumps in that they move fluid, but in this case the fluid is a gas. Both packed seals and mechanical seals are used, but packed seals are found only on reciprocating devices. Mechanical seals are not necessarily of the contact design used for pumps, but rather include carbon rings, labyrinth-type, and oil film seals.

The emissions control effectiveness of various emission reduction measures is shown in Table 9.6-4 as a percentage reduction. Leakless technologies are 100% effective in eliminating emissions for properly functioning equipment, but are expensive to purchase and maintain. For example, pumps with dual mechanical seals are estimated to have an amortized annual cost roughly 10 times quarterly or monthly LDAR. Compared to facilities that do not have emissions reduction programs, fugitive emissions from a moderately-sized petroleum refinery can be reduced by approximately 70% by using the most effective reduction techniques shown in Table 9.6-4 (Allen and Rosselot, 1997). Similarly, for SOCMI facilities, fugitive emission reductions are expected to be around 60–70%.

Table 9.6-4   Effectiveness of Various Fugitive Emission Reduction Techniques.

Equipment	Control Technique	Control Effectiveness (%)	
		SOCMI	*Petroleum Refinery*
Pumps, light liquid service	Dual mechanical seals	100	100
	Monthly leak detection and repair	60	80
	Quarterly leak detection and repair	30	70
Valves, gas/light liquid service	Monthly leak detection and repair	60	70
	Quarterly leak detection and repair	50	60
Pressure-relief devices	Tie to flare; rupture disk	100	100
	Monthly leak detection and repair	50	50
	Quarterly leak detection and repair	40	40
Open-ended lines	Caps, plugs, blinds	100	100
Compressors	Mechanical seals, vented to degassing reservoirs	100	100
Sampling connections	Closed purge sampling systems	100	100

Source: Dimmick and Hustvedt (1984).

9.7　POLLUTION PREVENTION ASSESSMENT INTEGRATED WITH HAZ-OP ANALYSIS

The hazard and operability study (HAZ-OP) is a formal procedure that can be applied to individual chemical process units to identify potential hazards (AIChE, 1985). Hazards are not only identified, but their possible causes are also investigated, the consequences of those hazards are defined, and actions to mitigate the hazard are summarized. It can be applied to a new process design before construction begins or can be applied to existing process units to improve safety performance. The study is typically conducted by a team of engineering and operations personnel who are familiar with the process. During the study, team members identify potential hazards through a structured examination of the design. One of the members of the team is assigned the task of recording the hazards and their suggested solutions. In this text, we demonstrate a limited use of HAZ-OP for the purpose of identifying hazards when process changes are made for pollution prevention. We do not identify actions to mitigate those hazards. Examples of complete HAZ-OP analyses can be found in standard textbooks on process safety (Crowl and Louvar, 1990).

The methodology for a HAZ-OP study is to apply a series of guide words to the process design intention. The design intention relates to what the certain steps or units in the process are intended to do. Examples of process intentions are a) cooling water flow through a reactor or distillation condenser, b) inerting system for a reactor, separator, or storage tank, or c) air supply for pneumatically-driven process control valves. The guide words associated with process intentions are shown in Table 9.7-1. For example, the use of the guideword "NO" in the process intention "cooling water flow through a reactor" would indicate a loss of cooling water. The consequences of this might be an overheated reactor, a run-away reaction, or, potentially, an explosion. If we apply each guideword to a process intention, we arrive at a set of possible consequences. The HAZ-OP approach is applied to each unit and to each pipeline into or out of each unit, and continues until every unit in the processes has been analyzed.

For large and complex processes, this approach can be very time-consuming and often tedious. Nonetheless, this procedure is finding increasing use in the chemical industry.

Table 9.7-1　HAZ-OP Procedure Guide Words (Adapted from Crowl and Louvar, 1990).

Guide Words	Meaning
NO or NOT	The complete negation of the intention
MORE	Quantitative increases
LESS	Quantitative decreases
AS WELL AS	A qualitative increase
PART OF	A qualitative decrease
REVERSE	The logical opposite of the intention
OTHER THAN	Complete substitution

Example 9.7-1   Safety Aspects of Storage Tank Pollution Prevention

Consider the storage tanks shown in Figure 9.7-1. A facility is considering replacing the existing external floating-roof tank with an internal floating-roof tank in order to reduce uncontrolled volatile organic compound emissions. Calculate the toluene emission reductions for the tank configurations outlined in Example 9.6-1 using the TANKS software. Conduct a limited HAZ-OP analysis for each tank configuration to evaluate the hazards for explosion of accumulated vapors within each tank. The inlet and outlet pumps are intended to fill and then empty the tanks to the desired levels. The inerting system on the internal floating-roof tank is intended to maintain an oxygen-free environment in the headspace above the floating roof and below the tank roof.

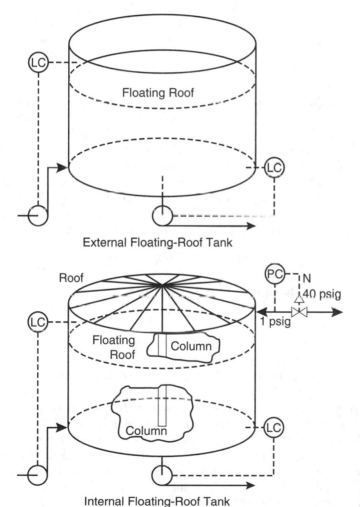

External Floating-Roof Tank

Internal Floating-Roof Tank

Figure 9.7-1   Schematic diagram of storage tank alternatives.

Solution:

Emissions Analysis:

The TANKS program allows the user to quickly calculate the annual emission rate for both tank types. The results are

External Floating-Roof Tank: 1,102.7 lb/yr.
Internal Floating-Roof Tank: 66.2 lb/yr.

The Internal floating-roof tank shows nearly a 20-fold reduction in toluene emission rate.

HAZ-OP Analysis (limited):

The results of the HAZ-OP analysis of both storage tank configurations is shown in Table 9.7-2. The hazards associated with the pumps and streams entering and exiting each tank are identical. The internal floating-roof storage tank has additional hazards associated with the interting gas because of the need to maintain a non-flammable mixture above the tank surface. The flash point for toluene is only 40°F, therefore, it is necessary to preclude air from the space above the floating roof. Vapor accumulation sufficient for ignition in air might be created by the action of the floating roof and the possibility of faulty roof seals.

The storage tank pollution prevention example above exhibited a higher level of safety hazard for the internal floating-roof tank compared to the external floating-roof tank. The additional complexity of the internal floating-roof tank introduced more possible modes of failure. Although the safety hazards associated with any process modification for pollution prevention would have to be evaluated on a case-by-case basis, many would result in a more complicated process. The rule of thumb in avoiding safety hazards in processes is to "keep it simple." The additional complexity of many pollution prevention applications makes safety assessment an important component of waste and risk reduction efforts in chemical process designs.

9.8 INTEGRATING RISK ASSESSMENT WITH PROCESS DESIGN— A CASE STUDY

Thus far in Chapter 9, we have incorporated environmental, health, and safety concerns into the design and operation of unit operations. We have used quantitative assessment measures only to a limited extent. It is very useful to provide a more quantitative risk assessment capability for the evaluation and optimization of unit operations. To this end, a screening-level risk assessment methodology is presented in this section using the Tier 1 assessment method presented in Chapter 8.

One important application could be screening of byproducts generated in a chemical reactor. Decisions regarding optimum reactor operation can then be made based on the risks posed by the individual byproducts generated rather than on just the mass rate of generation for each component. The case study deals with choosing residence time in a fluidized bed reactor for the production of acrylonitrile.

Table 9.7-2   Limited HAZ-OP Analysis of storage tank pollution prevention.

Guide Words	Deviation	EFRT	IFRT	Possible Cause	Consequences
NO	Inlet pump fails to stop or outlet pump fails to start	√	√	1. Level gauge failure 2. Pump failure	1. Toluene spills out top of tank 2. Soil and ground water contamination 3. Site personnel exposure
NO	Floating roof does not move	√	√	1. Seals binding to tank wall 2. Pontoon failure	1. Floating roof failure 2. Possible failure of level control system 3. Toluene spills out of tank top
NO	Inert N_2 stops		√	1. Pressure control failure 2. N_2 supply interrupt	1. Introduction of air into headspace of tank 2. Possible flammable mixture of toluene and air.
MORE	Inert N_2 fails to stop		√	1. Pressure control failure	1. Overpressure of storage tank and tank roof failure 2. Tank rupture and spill of toluene
LESS	Inert N_2 insufficient		√	1. Covered under "NO"	1. Covered under "NO"
AS WELL AS	Water in tank	√	√	1. Floating roof leak in EFRT 2. External roof leak in IFRT	1. Contamination of toluene product and generation of additional waste
PART OF	Inert N_2 insufficient		√	1. Covered under LESS	1. Covered under LESS
REVERSE	Pumps reverse	√	√	1. Impossible	1. Level control failure and spill of toluene
REVERSE	Inert system evacuated		√	1. Inert system mistaken for vacuum system	1. Collapse of tank with spillage of toluene
OTHER THAN	Another liquid than toluene	√	√	1. Mix up in supply to tanks	1. Contamination of toluene product 2. Reprocess tank contents 3. Waste tank contents
OTHER THAN	Another inerting gas		√	1. Another gas is used	1. If O_2 is used by mistake, a flammable mixture is created.

EFRT – External floating-roof tank
IFRT – Internal floating-roof tank

Case Study

Acrylonitrile Reactor (Hopper et al., 1992) Risk-Based Input-Output Analysis of a Reactor.

Acrylonitrile is produced in a fluidized-bed reactor containing a catalyst (Bi-Mo-O). The main reaction for acrylonitrile is ammonoxidation represented by

$$CH_2=CH-CH_3 + NH_3 + 3/2\,O_2 \to CH_2=CH-CN + 3H_2O$$

propylene   ammonia oxygen   acrylonitrile   water

In addition there are five other possible side reactions including

$$CH_2=CH-CH_3 + O_2 \to CH_2=CH-CHO + H_2O$$
acrolein

$$CH_2=CH-CH_3 + NH_3 + 9/4\,O_2 \to CH_3-CN + 1/2\,CO_2 + 1/2\,CO + 3H_2O$$
acetonitrile

$$CH_2=CH-CHO + NH_3 + 1/2\,O_2 \to CH_2=CH-CN + 2H_2O$$

$$CH_2=CH-CN + 2\,O_2 \to CO_2 + CO + \quad HCN \quad + 3H_2O$$
hydrogen cyanide

$$CH_3-CN + 3/2\,O_2 \to CO_2 + CO + HCN + H_2O$$

Hopper and coworkers (Hopper et al., 1992) constructed a set of reactor models for the above set of chemical reactions assuming first-order reaction kinetics with respect to the reactant, product, and byproduct species. The model also included mole balance and energy balance equations for the reactor. The model was used to predict the effects of reaction temperature, residence time, and reactor type—continuous flow stirred tank reactor (CSTR), plug flow reactor (PFR), and fluidized bed reactor (FBR)—on the generation of reaction byproducts in the acrylonitrile reaction. Here we illustrate the use of the FBR model predictions in determining the optimum residence time for minimum waste generation and acceptable economic performance. The evaluation is based on both mass generation as well as risk generation approaches.

The predicted concentrations of feed, product, and byproduct species from the reactor as a function of reactor residence time are shown in Figure 9.8-1. These results show that acrylonitrile concentration increases with residence time up to about 10 seconds. Thereafter, the increase in acrylonitrile concentration is slower and after 15 seconds, there is no further increase. Reactants (propylene and ammonia) continue to decline with increasing reactor residence time due to conversion of the reactant to product and byproduct species. Byproducts, hydrogen cyanide (HCN) and acetonitrile, exhibit complex profiles with respect to residence time. HCN is generated in significant amounts only above about 5 seconds residence time. HCN is the dominant reaction byproduct on a mass basis at higher residence times. Acetonitrile is generated in higher amounts than HCN at low residence times, but tends to remain at a constant concentration as residence time increases to 20 seconds. Based on these results, the authors (Hopper et al., 1992)

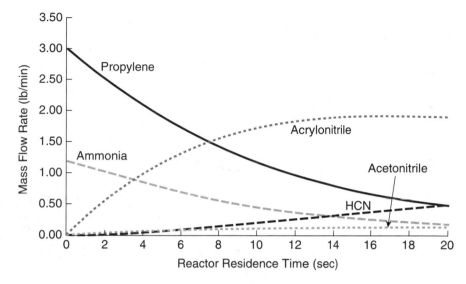

Figure 9.8-1   Effect of reactor residence time on the conversion of propylene and ammonia to product (acrylonitrile) and byproducts (hydrogen cyanide (HCN), and acetonitrile). The model is of a fluidized-bed reactor at 400 °C. Byproduct generation is shown on a mass basis.

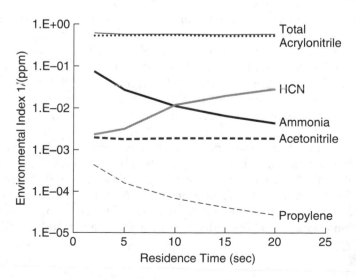

Figure 9.8-2 Effect of reactor residence time on the conversion of propylene to product (acrylonitrile) and byproducts (hydrogen cyanide (HCN), and acetonitrile). The model is of a fluidized-bed reactor at 400 °C. Byproduct generation is shown on a risk basis.

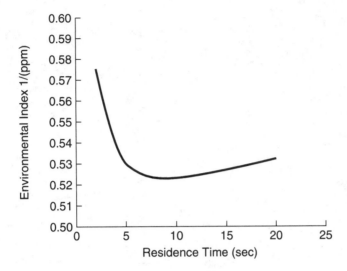

Figure 9.8-3 The total environmental index (Equation 8-2) as a function of reactor residence time for acrylonitrile production via the ammonoxidation pathway.

recommended operating the reactor at a temperature of 400–480 °C, with a reactor residence time of 2–10 seconds, and to use a fluidized-bed reactor.

A presentation of the same reactor results on a risk basis is shown in Figure 9.8-2. In generating this figure, stoichiometric coefficients at each residence time were calculated using the data shown in Figure 9.8-1 by taking the ratio of the mass flow rates of reactants and byproducts with respect to acrylonitrile. In addition, Equation 8-2 was employed using the TLV values shown in Table 8.2-7. The environmental index of the product acrylonitrile dominates the other environmental indexes shown in Figure 9.8-2, regardless of the reactor residence time. The total environmental index for this reactor is shown in greater detail in Figure 9.8-3. There is a minimum in the total index at a residence time of 10 seconds. The raw material cost per mass of acrylonitrile produced is shown in Figure 9.8-4 as a function of reactor residence time. It is apparent that the costs of raw materials is less than the cost of acrylonitrile when the residence time in the reactor is greater than 5 seconds.

This case study demonstrates that the "Tier 1" environmental assessment from Chapter 8 combined with a screening economic analysis provides valuable insights into the overall performance of the reactor design (residence time). There is little economic benefit in operating at residence times greater than 10 seconds, and the total environmental risk index is a minimum at 10 seconds. Therefore, a residence time of 10 seconds is a logical operating point for this reactor and reaction system.

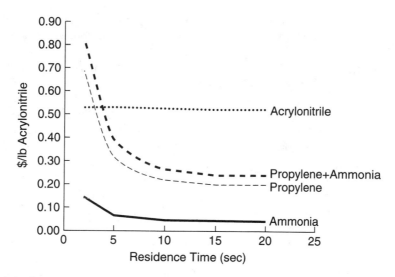

Figure 9.8-4 Raw materials costs per mass of acrylonitrile produced in ammonoxidation of propylene.

QUESTIONS FOR DISCUSSION

1. This chapter has considered pollution prevention alternatives for a variety of types of unit operations—reactors, separation devices, storage tanks and others. For a typical chemical process, which of these unit operations will be responsible for the majority of the emissions? Does your answer depend on the type of pollutant or the environmental medium to which the pollutant is released? (Refer to the emission inventories in Chapter 8.)

2. The pollution prevention alternatives identified in this chapter frequently result in reduced energy use and reduced material use. If the environmental improvements result in a design that uses less energy and less materials, why might a design engineer not choose to implement these options?

3. In this chapter, pollution prevention alternatives have been considered for individual unit operations. Is it possible that a pollution prevention strategy implemented for one unit operation may increase wastes and emissions in another unit operation? (Consider, as one example, the addition of emulsifiers in tanks to reduce solids formation. Can you think of other examples?)

REFERENCES

Absi-Halabi, W., Stanislaus, A., and Qabazard, H., "Trends in catalysis research to meet future refining needs," Hydrocarbon Processing, Feb. 1997, p. 45-55.

Adams, W.V., "Control fugitive emissions from mechanical seals," Chemical Engineering Progress, 87(8), 36-41, August, 1991.

Aggarwal, A. and Caneba, G.T., "Environmental Aspects of Polymers," report submitted to the Center for Clean Industrial and Treatment Technologies, Michigan Technological University, Houghton, MI, 1993.

AIChE, American Institute of Chemical Engineers, "Guidelines for Hazard Evaluation Procedures," New York, NY, 1985.

Allen, D.T., "The role of catalysis in industrial waste reduction," in *Industrial Environmental Chemistry,* ed. D.T. Sawyer and A.E. Martell, Plenum Press, New York, NY, p. 89-99, 1992.

Allen, D.T. and Rosselot, K.S., *Pollution Prevention for Chemical Processes,* John Wiley and Sons, New York, NY, 1997.

API, American Petroleum Institute, "Waste Minimization in the Petroleum Industry: A compendium of Practices," API Publication 849-00020, Washington, DC, 1991a.

API, American Petroleum Institute, "The Generation and Management of Wastes and Secondary Materials in the Petroleum Refining Industry: 1987-1988," API Publication 4530, Washington, DC, 1991b.

Bansal, I.K., "Concentration of oily and latex waste waters using ultrafiltration inorganic membranes," Industrial Water Engineering, October-November 1976.

Barton, A.F.M., *CRC Handbook of Solubility Parameters and Other Cohesion Parameters,* CRC Press Incorporated, Boca Raton, FL, 1983.

Becker, M, Farag, I.H., Hayden, N. (1996) (editors), *Motivating Pollution Prevention Concepts: Homework Problems for Engineering Curricula,* Northeast Waste Management Officials' Association (NEWMOA), 129 Portland Street, Suite 502, Boston, MA 02114, phone: 617-367-8558.

Bjorklund, M.C. and Carr, R.W., presented at the AIChE National Meeting, Chicago, IL, paper 195a, 1996.

Brestel, R. et al., "Minimize fugitive emissions with a new approach to valve packing," Chemical Engineering Progress, 87(8), 42-47, August 1991.

Chella, R. and Ottino, J.M., "Conversion and selectivity modifications due to mixing in unpremixed reactors," Chemical Engineering Science, Vol. 39, No. 3, pp. 551-567, 1982.

Crowl, D.A. and Louvar, J., *Chemical Process Safety: Fundamentals with Applications,* Prentice Hall, Englewood Cliffs, NJ, 1990.

Cutter, L., AIChE Journal, "Flow and turbulence in a stirred tank," Vol. 12, p. 35, 1966.

DeSimone, J.M, Guan, Z.G., and Elsbernde, C.S., "Synthesis of fluoropolymers in supercritical carbon dioxide," Science, Vol. 257, 14 Aug. 1992, p. 945 - 947.

DeSimone, J.M. Maury, E.E., Menceloglu, Y.Z., McClain, J.B., Romack, R.J., and Combes, J.R., "Dispersion polymerizations in supercritical carbon dioxide," Science, Vol. 265,15 July, 1994, p. 356–359.

Dimmick, W.F. and Hustveldt, K.C., "Equipment Leaks of VOC: Emissions and Their Control," paper No. 84-62.1 presented at the 77th Annual Meeting of the Air Pollution Control Association, San Francisco, June 24-29, 1984.

Elias, H-G., *Macromolecules, Volume 2: Synthesis, Materials, and Technology,* Plenum Press, New York, 1984.

Flick, E.W. ed, *Industrial Solvents Handbook,* 3rd ed., Noyes Data Corporation, Park Ridge, N.J., 1985.

Fogler, H.S., *Elements of Chemical Reaction Engineering,* 2nd Edition, Prentice Hall, Englewood Cliffs, NJ, 1992.

Gates, B., "Catalysis," in *Encyclopedia of Chemical Technology,* 4th edition, Volume 5, Editor Kroschwitz, J.I., John Wiley and Sons, New York, NY, p. 350, 1993.

Hansen, C.M., *Hansen Solubility Parameters: A Users Handbook*, CRC Press, Boca Raton, Florida, 2000.

Hopper, J.R., Yaws, C.L., Ho, T.C., Vichailak, M., and Muninnimit, A., "Waste minimization by process modification," in *Industrial Environmental Chemistry,* Sawyer, D.T., and Martell, A.E. ed., Plenum Press, New York, NY, p. 25–43, 1992.

Humphrey, J.L., "Separation processes: Playing a critical role," Chemical Engineering Progress, 91(10), 31-41, October 1995.

Jones, E.O., "Treating metal-bearing spent acids with a transportable test system," paper presented at the 11th AESF/EPA Conference on Environmental Control for the Metal Finishing Industry, Orlando, FL, 1990.

Kolmogoroff, A.N., Compt. Rend. Acad. Sci. U.R.S.S. (N.S.), vol. 30, p. 301, 1941, "The local structure of turbulence in incompressible" and "On degradation of isotropic turbulence in an incompressible viscous liquid," Vol. 31, p. 538-540 1941.

Komiya, K. et al., "New process for producing polycarbonate without phoshene and methylene chloride," in *Green Chemistry: Designing Chemistry for the Environment,* Anastas, P.T. and Williamson, T.C., Eds., ACS Symposium Series 626, American Chemical Society, Washington, DC, 1996, p 20-32.

Levenspiel, O., *Chemical Reaction Engineering*, 3rd Edition, John Wiley & Sons, New York, NY, 1999.

Manzer, L.E., "Chemistry and catalysis, keys to environmentally safer processes," in *Benign by design alternative: Synthetic Design for Pollution Prevention,* Anastas, P.T. and Farris, C.A., eds., ACS Symposium Series 577, American Chemical Society, Washington, DC, pp. 144-154, 1994.

Morgenstern, D.A., LeLacheur, R.M., Morita, D.K., Borkowsky, S.L., Feng, S., Brown, G.H., Luan, L., Gross, M.F., Burk, M.J., and Tumas, W., et al., "Supercritical carbon dioxide as a substitute solvent for chemical synthesis and catalysis," in *Green Chemistry: Designing Chemistry for the Environment,* Ed. Anastas, P.T. and Williamson, T.C., ACS Symposium Series 626, American Chemical Society, Washington, DC, 1996, p. 132-151.

Mulholland, K.L. and Dyer, J.A., *Pollution Prevention. Methodology, Technologies and Practices,* American Institute of Chemical Engineers, New York, 1999.

Mulholland, K.L., personal communication, 2000.

Nelson, K.E., "Practical techniques for reducing waste," in *Industrial Environmental Chemistry,* ed. D.T. Sawyer and A.E. Martell, Plenum Press, New York, NY, p. 3-17, 1992.

Null, H.R., "Selection of a separation process," in *Handbook of Separation Process Technology*, Rousseau, R.W., Ed., Wiley and Sons, New York, NY, p. 982-995, 1987.

Palepu, P.T., Chauhan, S.P., and Ananth, K.P., "Separation technologies," in *Industrial Pollution Prevention Handbook,* ed. Freeman, H.M., McGraw-Hill, New York, NY, p. 935, 1995.

Paul, E.L. and Treybal, R.E., "Mixing and product distribution for a liquid-phase, second-order, competitive-consecutive reaction," AIChE Journal, Vol. 17, No. 3, pp. 718-724, 1971.

Perry, R.H. and Green, D.W., Eds., *Perry's Chemical Engineers' Handbook,* 6th Ed., McGraw-Hill, New York, 1984.

Radecki, P.P., Crittenden, J.C., Shonnard, D.R., and Bulloch, J.L., Eds., *Emerging Separator and Separative Reaction Technologies for Process Waste Reduction,* Center for Waste Reduction Technologies, American Institute of Chemical Engineers, New York, NY, pg. 319, 1999.

Randall, P.M., "Pollution prevention strategies for the minimizing of industrial wastes in the VCM-PVC industry," Environmental Progress, Vol. 13(4), 269-277, 1994.

Rosselot, K.S. and Allen, D.T., "Upgrading process water to prevent pollution in petroleum refining," *Pollution Prevention Review,* vol. 6(1), 95–99, 1996.

Sullivan, D.A., Solvents - Industrial, in *Encyclopedia of Chemical Technology,* Vol. 22, Kroschwitz, J.I. Executive Editor, John Wiley and Sons, 529–571, 1996.

Tonkovich, A.L.Y., Carr, R.W., and Aris, R., "Enhanced C_2 yields from methane oxidative coupling by means of a separative chemical reactor," Science, 262, 221-223, 1993.

Tonkovich, A.L.Y. and Carr, R.W., "A simulated countercurrent moving bed chromatographic reactor for the oxidative coupling of methane: Experimental results," Chemical Engineering Science, 49, 4645-4656, 1994.

Tonkovich, A.L.Y., "Impact of Catalysis on the Production of the Top 50 U.S. Commodity Chemicals," PNL-9432, report prepared for the U.S. Department of Energy, EE-Office of Industrial Technology, 1994.

Tonkovich, A.L.Y. and Gerber, M.A., "The Top 50 Commodity Chemicals: Impact of Catalytic Process Limitations on Energy, Environment, and Economics," PNL-10684, report prepared for the U.S. Department of Energy, EE#-Office of Industrial Technology, 1995.

Tonkovich, A.L.Y., Zilka, J.L., Jimenez, D.M., Roberts, G.L., and Cox, J.L., "Experimental investigations of inorganic membrane reactors: a distributed feed approach for partial oxidation reactions," Chemical Engineering Science, 51(5), 789-806, 1996.

Tucker, S.P. and Carson, G.A., "Deactivation of hazardous chemical wastes," *Environmental Science and Technology,* 19(3), 215-220, 1985.

Upson, L.L. and Lomas, D.A., "Catalysts (Regeneration)," in *Encyclopedia of Chemical Technology,* 4th edition, Volume 5, Editor Kroschwitz, J.I., John Wiley and Sons, New York, NY, p. 431, 1993.

US EPA, United States Environmental Protection Agency, *Fugitive Emission Sources of Organic Compounds—Additional Information on Emissions, Emission Reductions, and Costs,* Publication EPA-450/3-82-010, Research Triangle Park, NC, 1982.

US EPA, United States Environmental Protection Agency, *Emission Factors for Equipment Leaks of VOC and HAP,* Publication EPA-450/3-86-002, Research Triangle Park, NC, January 1986.

US EPA, United States Environmental Protection Agency, *DuPont Chambers Works Waste Minimization Project,* EPA/600/R-93/203, 1993a.

US EPA, United States Environmental Protection Agency, "Protocols for Equipment Leak Emission Estimates," available through National Technical Information Service (NTIS) as PB93-229219, 1993b.

US EPA, United States Environmental Protection Agency, *Waste Minimization Prioritization Tool (WMPT),* EPA 530-R-97-019, 1997.

US EPA, *Compilation Of Air Pollutant Emission Factors,* Volume I, Fifth Edition, AP-42, Air CHIEF CD-ROM, ClearingHouse For Inventories And Emission Factors, Air

CHIEF CD-ROM, (EFIG/EMAD/OAQPS/EPA), Version 6.0, U.S. Environmental Protection Agency, Research Triangle Park, NC, EPA-454/F-98-007, 1998.

US EPA, Technology Transfer Network (TTN), United States Environmental Protection Agency, Office of Air Quality Planning and Standards, Technology Transfer Network website, TANKS storage tank emission estimation program, http://www.epa.gov/ttn/chief/tanks.html, June 1, 1999.

Wankat, P.C., *Equilibrium Staged Separations,* Prentice Hall, Englewood Cliffs, NJ, p. 707, 1988.

Werschulz, P., "New membrane technology in the metal finishing industry," in *Toxic and Hazardous Waste,* ed. Klugman, I.J., Technomic Publishing, Lancaster, PA, 1985.

PROBLEMS

1. **Solvent Choice for Caffeine Extraction from Coffee Beans**

   About 20% of the coffee consumed in the United States is decaffinated. There are many solvents and processes developed over the past century to accomplish this step. Critical issues in choosing a solvent are the caffeine/solvent affinity, the cost of the solvent, the ease of caffeine recovery from the solvent, safety aspects, and environmental impacts. The original process used a synthetic organic solvent to extract caffeine from un-roasted coffee beans. These solvents included trichloroethylene (C_2HCl_3) and methylene chloride (CH_2Cl_2). Today, caffeine is extracted using "natural" solvents including supercritical carbon dioxide, ethyl acetate (naturally found in coffee), oils extracted from roasted coffee, and water. Using Material Safety Data Sheets as a source of information, rank order these solvent candidates based only on their toxicological properties. Do not consider the "extracted oils" since the identity of these is not available. Use PEL and/or LD_{50} (rat) toxicological data.

   Information Source on Coffee Extraction: *Encyclopedia of Chemical Technology,* Volume 6, Coffee chapter, John Wiley and Sons, 1991 (4th edition) and 1978 (3rd edition).

2. **Optimum Plug Flow Reactor Design for a Series Reaction**

   A compound A reacts with S to form a desired product B. B also reacts with S to give a hazardous waste byproduct C that must be disposed of at great cost. The series reaction can be described as

$$A \xrightarrow{k_1} B \xrightarrow{k_2} C \qquad (2a)$$

   The reactions are irreversible and because they occur in large excess of S, are first order,

$$r_A = -k_1 C_A$$
$$r_B = k_1 C_A - k_2 C_B \qquad (2b)$$
$$r_C = k_2 C_B$$

   where $k_1 = 0.2$ min^{-1} and $k_2 = 0.1$ min^{-1}. A feed stream of volumetric flow rate $F = 100$ gal/min contains reactant A at a concentration of $C_{Ao} = 0.1$ lbmole/gal.

   (a) Determine the reactor volume in ft^3 that will maximize the yield of B and minimize the generation of C, and calculate the concentrations of A, B, and C in the reactor effluent stream. Assume that the density of the fluid is unaffected by the reaction.

(b) Describe the situation if S were not present in large excess. For example, would the potential be greater or less for creating the hazardous waste component, C?

(Adapted from a problem by Alfred Donatelli in *Motivating Pollution Prevention Concepts: Homework Problems for Engineering Curricula,* editors M. Becker, I. Farag, and N. Hayden, 1996).

3. **Energy Efficient Extraction Coupled with Distillation**

Background:

There are many examples of liquid extraction coupled with distillation in the chemical process industry. One important example is the extraction of aromatic components (benzene, toluene, xylenes) from paraffins in the refining of petroleum using a suitable solvent. A listing of potential solvents for this purpose is shown in Perry and Green (1984, pp. 15.9 to 15.13), along with values for the distribution ratio, $K_d = y/x$, where y and x are the mass fractions of solutes in the extract and raffinate phases, respectively. For this application, values for K_d tend to vary from 0.1 to 1.0. The choice of extraction solvent will affect the operation of the downstream distillation column in the following ways: a) by affecting the feed flow rate and solute composition that is fed to the distillation column and b) by affecting the relative volatility of the distillate compared to the bottoms product.

Problem:

In this problem, we wish to investigate the effects of extraction solvent choice on the consumption of energy in this process. Refer to the following diagram in solving this problem. We will assume that the major energy-consuming element is the reboiler of the distillation column.

(a) Determine the change in reboiler duty, $Q_R = \lambda \bar{V}$, for K_d values of 0.3, 0.5, and 0.7 in the extractor for a constant relative volatility in the distillation column of

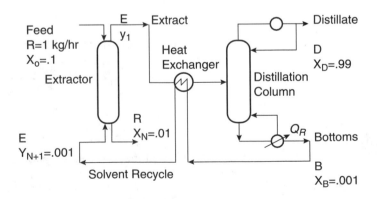

$\alpha_{P,S} = 3$. The subscripts P and S refer to product and solvent, respectively and where λ is the latent heat of vaporization of the liquid in the reboiler and $\bar{V}$ is the stripping section vapor flow rate.

(b) Determine the change in reboiler duty for $\alpha_{P,S} = 3, 5$, and 7 in the distillation column for a constant value of $K_d = 0.4$ in the extractor.

(c) Which parameter, K_d or $\alpha_{P,S}$, has the greatest influence on Q_R?

(d) Would higher or lower values of K_d or $\alpha_{P,S}$ result in a more energy-efficient process?

For this problem, make the following assumptions. The flow rates of raffinate and extract phases in the extractor are constant. The ratio of raffinate to extract flow rate must be 0.8 times the minimum ratio: $(R/E) = 0.8 \, (R/E)_{min}$. Assume that the feed to the distillation column is a saturated liquid. Use a reflux ratio (L/D) of 1.2 times the minimum $(L/D)_{min}$. The distillate mass fraction is $x_D = 0.99$ and the bottoms is $x_B = 0.001$. Use a feed flow rate to the extractor of 1 kg/hr as a basis for these calculations. Use a latent heat of vaporization for the liquid in the reboiler of $\lambda = 100$ kcal/kg.

4. **Effect of Storage Tank Type on Volatile Organic Compound (VOC) Emission Rates.**

A liquid benzene product stream, having an annual throughput of 516,600 gal/yr, is to be recovered from a gaseous waste stream at a facility in the vicinity of Detroit, MI. Using the TANKS software (Appendix F, US EPA TTN, 1999, http://www.epa.gov/ttn/chief),

(a) Calculate the uncontrolled annual emissions for a new tank design. Each tank is colored white and the paint condition is good.

(b) What is the percentage reduction in benzene emission rates for both of the floating-roof tanks compared to the vertical fixed-roof tank?

(c) Compare answers to Example 9.6-1.

Each of the tank types have the following dimensions and conditions.

Vertical Fixed-Roof Tank: Height = 20 ft, Diameter = 12 ft, Working Volume = 15,228.53 gallons, Maximum Liquid Level = 18 ft, Average Liquid Level = 10 ft, no heating, domed roof of height 2 ft and diameter of 12 ft, vacuum setting of $-.03$ psig and pressure setting of .03 psig.

Internal Floating-Roof Tank: Height = 20 ft, Working Volume = 15,228.53 gallons, self-supporting roof, internal shell condition of light rust, primary seal is a mechanical shoe, secondary shoe is shoe-mounted, deck type is welded, deck fitting category is typical.

Domed External Floating-Roof Tank: same as Internal Floating-Roof Tank.

5. **Net VOC Emissions Reduction from Painting an Existing Storage Tank.**

You wish to reduce the "standing" or "breathing" losses of toluene from an existing vertical fixed-roof storage tank that has a grey/medium color paint in poor condition. One pollution prevention strategy would be to repaint the existing tank a lighter color. In answering the following questions, use the TANKS software (Appendix F, US EPA TTN, 1999, http://www.epa.gov/ttn/chief).

The tank has the following dimensions and conditions.

Vertical Fixed-Roof Tank: Height = 20 ft, Diameter = 12 ft, Working Volume = 15,228.53 gallons, Maximum Liquid Level = 18 ft, Average Liquid Level = 10 ft, no heating, domed roof of height 2 ft and diameter of 12 ft, vacuum setting of $-.03$ psig and pressure setting of .03 psig.

(a) What is the annual toluene emission rate for the original tank having the poor condition paint and what is the percentage of the annual emission rate caused by standing losses?

(b) How much net emission reduction can be expected if the grey/medium color (poor condition) vertical fixed-roof tank (same as in Example 9.6-1) is painted white? Assume that the paint is an oil-based paint with 50% by volume toluene solvent, that all of the toluene in the paint is emitted after its application, and that one gallon covers 100 square feet of tank external surface. Assume that the dome roof is flat for this calculation.

CHAPTER 10

Flowsheet Analysis
for Pollution Prevention

by
Kirsten Sinclair Rosselot
David T. Allen

10.1   INTRODUCTION

The environmental performance of a process flowsheet depends on both the performance of the individual unit operations that make up the flowsheet and on the level to which the process streams have been networked and integrated. While Chapter 9 describes methods for improving the performance of individual unit operations, this chapter examines methods for assessing and improving the degree to which the unit operations are integrated. Specifically, Section 10.2 examines process energy integration and Section 10.3 examines process mass integration. The methods presented in these sections, and the case study presented in Section 10.4, demonstrate that improved process integration can lead to improvements in overall mass and energy efficiency.

Before examining process integration in detail, however, it is useful to review the methods that exist for sytematically assessing and improving the environmental performance of process designs. A number of such methods are available. Some are analogous to Hazard and Operability (HAZ-OP) Analyses (e.g., see Crowl and Louvar, 1990).

Section 9.7 briefly describes how a Haz-Op analysis is performed; to summarize, the potential hazard associated with each process stream is evaluated qualitatively (and sometimes quantitatively) by systematically considering possible deviations in the stream. Table 10.1-1 gives the guide words and examples of deviations used in HAZ-OP analysis. Each guide word is applied to each relevant stream characteristic, the possible causes of the deviation are listed, and the consequences of the deviation are determined. Finally, the action(s) required to prevent the occurrence of the deviation are determined.

Table 10.1   Guide Words and Deviations in HAZ-OP Analysis.

Guide Word	Example Deviations
NO or Not	No flow for an input stream.
MORE	Higher flow rate, higher temperature, higher pressure, higher concentrations.
LESS	Lower flow rate, lower temperature, lower pressure, lower concentrations.
AS WELL AS	Extra phase present, impurity present.
PART OF	Change in ratio of components, component missing.
MORE THAN	Extra phase present, impurity present.
REVERSE	Pressure change causes a vent to become an INLET.
OTHER THAN	Conditions that can occur during startup, shutdown, catalyst changes, maintenance.

For a single pipeline taking fluid from one storage tank to another, there may be several possible deviations, such as:

- no flow
- more flow
- more pressure
- more temperature
- less flow
- less temperature
- high concentration of a particular component
- presence of undesirable compounds

Note that each deviation may have more than one possible cause so that this set of deviations would be associated with dozens of possible causes. It would be difficult to consider all the deviations and their consequences without a structured system for analyzing the flowsheet. A similar analysis framework has been employed in a series of case studies to identify environmental improvements in process flowsheets (DuPont, 1993). In these case studies, a series of systematic questions are raised concerning each process stream or group of unit processes. Typical questions include:

- What changes in operating procedures might reduce wastes?
- Would changes in raw materials or process chemistry be effective?
- Would improvements in process control be effective?

Process alternatives, such as those defined in Chapter 9, can be identified, and in this way the environmental improvement opportunities for the entire flowsheet can be systematically examined. (See, for example, the cases from the DuPont report described by Allen and Rosselot, 1997.)

Other methods for systematically examining environmental improvement opportunities for flowsheets have been developed based on the hierarchical design methodologies developed by Douglas (1992). The hierarchical levels are shown in Table 10.2-1. Note that Level 1 in this table applies only to processes that are being designed, not to existing processes. The hierarchy is organized so that decisions that affect waste and emission generation at each level limit the decisions in the levels below it.

As an example of the use of hierarchical analysis procedures, consider a case study drawn from the AMOCO/US EPA Pollution Prevention Project at AMOCO's refinery in Yorktown, Virginia (Rossiter and Klee, 1995). In this example (adapted from Allen and Rosselot, Pollution Prevention for Chemical Processes, © 1997. This material is used by permission of John Wiley & Sons, Inc.), the flowsheet of a fluidized-bed catalytic cracking unit (FCCU) is evaluated for pollution prevention options. A flowsheet of the unit is shown in Figure 10.1-1.

Beginning with Level 2 of the hierarchy listed in Table 10.1-2 (input-output structure), the following pollution prevention strategies were generated:

1) Improve quality of the feed to eliminate or reduce the need for the vapor line washing system shown in the upper right-hand corner of Figure 10.1-1.

2) Reduce steam consumption in the reactor so that there is less condensate to remove from the distillation system.

3) Within the catalyst regeneration system, the loss of fines (upper left hand corner of Figure 10.1-1) is partly a function of the air input rate. A reduction in air flow (e.g., by using oxygen enrichment) is a possible means of reducing the discharge of fines.

Two ideas were generated during review of the recycle structure (level 3):

1) The reactor uses 26,000 lb/hr of steam. This is provided from the utility steam system. If this could be replaced with steam generated from process water, the liquid effluent from the unit would be reduced. Volatile hydrocarbons

Table 10.2-1 Levels for Hierarchical Analysis for Pollution Prevention (Adapted from Douglas, 1992).

Design Levels
1. Identify the material to be manufactured
2. Specify the input/output structure of the flowsheet
3. Design the recycle structure of the flowsheet
4. Specify the separation system
4a. General structure: phase splits
4b. Vapor recovery system
4c. Liquid recovery system
4d. Solid recovery system
5. Process integration
5a. Integrate process heating and cooling demands
5b. Identify process waste recycling and water reuse opportunities

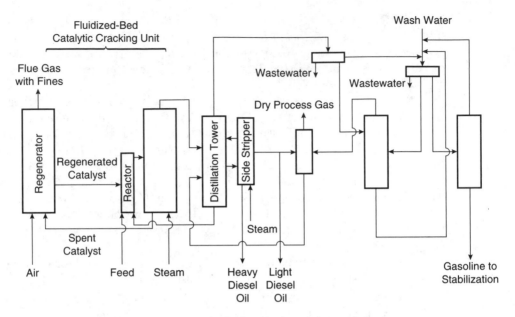

Figure 10.1-1   Process flow diagram of a fluidized-bed catalytic cracking unit.

contained in the recycled steam would be returned directly to the process. Catalyst regeneration consumes more than 11,000 lb/hr of steam. It may be possible to satisfy this duty with "dirty steam" as well, since the hydrocarbon content would be incinerated with the coke in the regenerator.

2) Used wash water is collected at several points and then purged from the process. If it could be recovered and recycled instead, or if recycled water from other sources could be used for washing in place of fresh water, fresh water usage and wastewater generation could both be reduced by about 10,500 lb/hr.

Three options were identified for separation systems (level 4):

1) Replace heating done by direct contacting with steam by heating with reboilers.
2) Place additional oil-water separators downstream of existing condensate collection points and recover hydrocarbons.
3) Improve gas-solid separation downstream of the regenerator to eliminate loss of catalyst fines. This might simply require better cyclone and/or ductwork design, or electrostatic precipitation.

These first four levels of the design hierarchy lead us to the types of process improvements described in Chapter 9—improvements in the reactor and improvements in the separation system. As Table 10.2-1 notes, the next step in the design process is to identify opportunities for process integration. This is the main topic of this chapter and the next several sections describe methods for process energy integration and methods for identifying process waste recycling and reuse opportunities.

10.2   PROCESS ENERGY INTEGRATION

Process streams frequently need to be heated, often to achieve the correct conditions for a desired reaction or to achieve separation of materials. Heating of process streams is generally done in furnaces or by contacting process streams with steam or other energy carrier fluids. No matter whether it is done in a furnace or using steam generated in a boiler, heating a process stream generally requires combustion of fuels, adding expense and environmental impacts to a process.

Process streams also frequently need to be cooled. This cooling is frequently done with cooling water which is circulated throughout the process. Cooling towers are used to keep cooling water temperatures at steady state values, but operating these cooling towers consumes energy and causes the loss of water through evaporation, adding expense and environmental impacts to a process.

The idea behind process heat or process energy integration is to use the heat from streams that need to be cooled for heating streams that need their temperature raised. Heat transfer between streams in a process prevents pollution by reducing the need for fuels and for cooling tower operation. Process heat integration is generally done using an analysis referred to as heat exchange network (HEN) synthesis. In heat exchange network synthesis, all of the heating and cooling requirements for a process are systematically examined to determine the extent to which streams that need to have their temperature raised can be heated by streams that need to be cooled. Heat integration is discussed in engineering undergraduate courses and detailed descriptions of HEN synthesis are available in most modern textbooks of chemical process design (e.g., Douglas, 1988). A simple example is described here in order to refresh readers on the basic concepts.

Figure 10.2-1 shows a heat balance diagram for a stream that needs to be heated from 50°C to 200°C and a stream that needs to be cooled from 200°C to

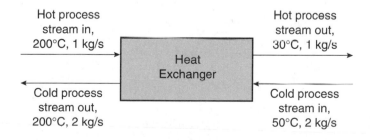

Figure 10.2-1 Heat balance diagram for a hot process stream and a cold process stream.

30°C. For the sake of simplicity, both streams in this example have a heat capacity of 1 kJ/(kg-°C). The stream that needs to be cooled has a flow rate of 1 kg/s and the stream that needs to be heated has a flow rate of 2 kg/s. Heating and cooling utilities could be applied to the two streams separately, as shown in Figure 10.2-2. However, the requirements for heating and cooling utilities are less if heat exchange between the two streams occurs.

There are two fundamental thermodynamic constraints to heat transfer. One is that the quantity of heat absorbed by the cold stream (the stream that needs to be heated) is equal to the quantity of heat lost by the hot stream (the stream that needs to be cooled). The other constraint is that heat flows from higher temperature streams to lower temperature streams.

One particularly useful way to graphically depict the streams to be heated and cooled in a flowsheet is called a "pinch" diagram. This diagram can be used to determine the extent to which heat transfer is possible and also to determine which hot streams should be paired with which cold streams. Heat transferred to and from the streams is on the y-axis in a pinch diagram and temperature is on the x-axis. Hot streams are represented as lines that slope downward and to the left, while cold streams are represented by lines that slope upwards and to the right. The hot and cold streams of Figure 10.2-1 are depicted in the pinch diagram of Figure 10.2-3.

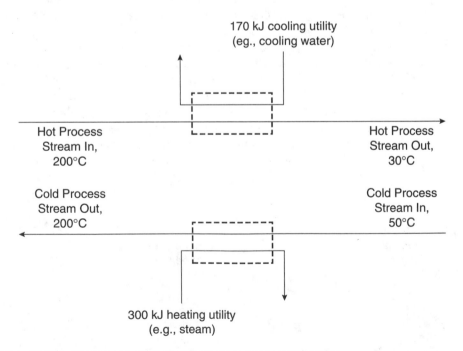

Figure 10.2-2 Heating and cooling requirements before heat integration.

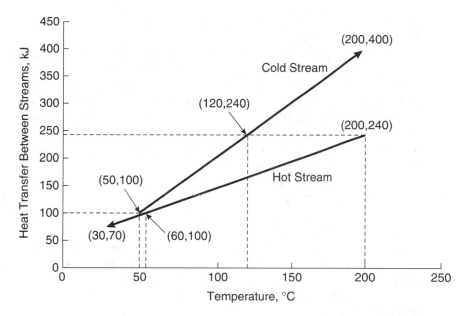

Figure 10.2-3   Hot stream and cold stream load line diagram for heat exchange network synthesis.

Note that the vector representing the cold stream (that needs to be heated) begins at a temperature of 50°C. It ends at a temperature of 200°C and at an enthalpy 300 kW higher than it started. Similarly, the hot stream (that needs to be cooled) begins at a temperature of 200°C and ends at a temperature of 30°C, and an enthalpy 170 kW lower than it started.

The enthalpy units in a pinch diagram, such as Figure 10.2-3, are relative. Because of this, both hot and cold stream lines are free to move vertically. However, because of the thermodynamic constraints, heat transfer between the streams can take place only in regions where the hot stream lies to the right of the cold stream (i.e., where the hot stream is at a higher temperature than the cold stream—in Figure 10.2-3, the region between the dashed lines that intersect the energy axis at 100 and 240 kW). The maximum theoretical heat transfer between the streams occurs when the two streams touch but do not cross. This point is called the thermal pinch. This theoretical maximum is not possible in practice because an infinitely large heat exchanger would be required to effect heat exchange at the point where the temperature difference between the streams goes to zero.

As the lines move apart, the region where the hot stream lies to the right of the cold stream becomes smaller and the amount of heat transferred between the streams decreases. As a result, the utility requirements for heating and cooling the streams to their target temperatures increase. In contrast, the heat exchanger for transferring heat from the hot stream to the cold stream gets smaller as the lines move apart. Thus, there is an optimum temperature driving force where total annualized costs (operating, or utility, costs plus capital costs of the heat exchanger) are minimized.

If the optimum temperature difference is 10°C, then the pinch diagram of Figure 10.2-3 shows the optimum heat transfer between the two streams to be 240 kJ − 100 kJ = 140 kJ. The diagram also shows that, under these conditions, the cold stream is heated from 50°C to 120°C in the exchanger and the hot stream is cooled from 200°C to 60°C. A diagram of the heat exchange network is given in Figure 10.2-4. Comparison of Figure 10.2-4 with Figure 10.2-2 shows that heat integration results in a substantial decrease in the utilities needed to heat and cool the two streams to their target temperatures.

This simple example of process heat integration illustrates a basic concept—that exchange of energy between process streams that need to be heated and process streams that need to be cooled (process energy integration) can reduce overall energy demand for a process. More complete treatments of HEN synthesis are available in standard chemical process design texts and this chapter will not describe these methods in detail. Rather the focus of the remainder of this chapter will be on a topic analogous to HEN synthesis, but one that is not normally addressed in chemical process designs—process mass integration and mass exchange network synthesis.

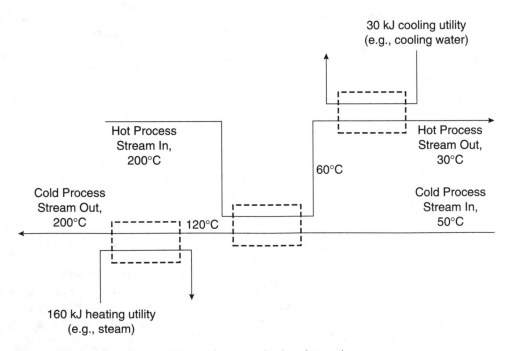

Figure 10.2-4   Heating and cooling requirements after heat integration.

10.3   PROCESS MASS INTEGRATION

Just as heat integration is the use of heat that would otherwise be wasted, mass integration is the use of materials that would otherwise be wasted. Three tools for determining process configurations that result in mass integration are described in this section. The first tool, source-sink mapping, is the most visual and intuitive of the three. Next, a strategy for determining optimum mixing, segregation, and recycle strategies is described. Finally, mass exchange network synthesis, which is the mass integration analogue for heat exchange network synthesis, is described.

10.3.1   Source-Sink Mapping

Source-sink mapping is used to determine whether waste streams can be used as feedstocks. It is the first quantitative tool for mass integration discussed in this chapter because it is one of the simplest and most visual tools for identifying candidate streams for mass integration.

The first step in creating a source-sink diagram is to identify the sources and sinks of the material for which integration is desired. For example, if water integration is desired, then wastewater streams (the "sources" of water) are identified. The processes that require water (the "sinks" of water) must also be identified. The flow rates of the sources and sinks must be known, keeping in mind that many sinks can accept a range of flow rates. Contaminants that are present in the source streams and that pose a potential problem for the sinks must be identified, and the tolerance of each sink for these contaminants must be known. Some processes require very pure feed, in which case using waste streams that contain the feed material as well as some contaminants is infeasible. However, many processes can make use of material that contains impurities, and some sinks have extremely liberal tolerances. The final piece of information that must be known before constructing a source-sink diagram is the concentration of contaminants that were identified as being potentially significant problems for the sinks.

Once all of these parameters are known, the source-sink diagram can be drawn. If only one contaminant is a concern, the diagram is two-dimensional, with source and sink flow rates plotted on the y-axis and contaminant concentration plotted on the x-axis. Each sink is represented by an area corresponding to its upper and lower limits of tolerance for flow rate and contamination, and each source is represented by a point.

As an example of the construction of a source-sink diagram, consider the sources and sinks described in Table 10.3-1. The material for which integration is sought (assume, for the moment that this is water) is available at the flow rate specified and the contaminant of concern is X. Figure 10.3-1 shows a source-sink diagram for the streams described in this table. Sources A, B, and C are shown as points in Figure 10.3-1 (because the flow rate and contaminant concentrations are point values), while sinks 1 and 2 are shown as shaded areas (because the flow rate needed and acceptable contaminant concentrations are ranges of values).

Table 10.3-1   Example Stream Data for Source-Sink Diagram.

Sources			Sinks				
				flow rate, kg/s		concentration of X, ppm	
Label	flow rate, kg/s	concentration of X, ppm	Label	max	min	max	min
A	3.0	7	1	4.8	4.0	5	0
B	5.0	15	2	2.5	2.1	1	0
C	1.0	4					

An initial examination of Figure 10.3-1 indicates that stream C could be used to partially satisfy the water demand for stream 1, since stream C's contaminant concentration falls within the range allowed for stream 1. No other direct reuse opportunities are available; however, stream A has a concentration that is not too far above the maximum allowable contaminant concentration for stream 1. Would it be possible to blend streams A and C and satisfy the contaminant constraint for stream 1?

Source streams whose concentration of contaminants is too high for feeding to any sinks (such as A) can be combined with low-concentration sources (such as C) to lower their concentration. In Figure 10.3-1, a point representing a combination of sources A and C is depicted. The flow rate of the combined streams is sim-

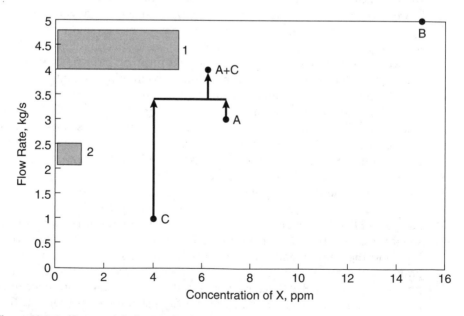

Figure 10.3-1   Source-sink diagram for the streams of Table 10.3-1.

ply the sum of the flow rates of the individual streams, and the concentration of compound X in the combined stream is the weighted average of the concentration in streams A and C,

$$\frac{3.0\,\frac{kg}{s} \times 7\,\text{ppm} + 1.0\,\frac{kg}{s} \times 4\,\text{ppm}}{3.0\,\frac{kg}{s} + 1.0\,\frac{kg}{s}} = 6.25\,\text{ppm.} \qquad \text{(Eq. 10-1)}$$

Note that this point representing the combined stream A-C has a flow rate within the acceptable range for sink 1, but its concentration of contaminant X is too high to allow the combined stream to be used directly in sink 1. In other words, streams A and C cannot be combined and used as the sole feedstock for sink 1.

To lower the concentration to within acceptable limits, uncontaminated material (fresh water with 0 concentration of X) must be used in addition to sources A and C. The most uncontaminated material that can be added to the stream to lower the concentration of X (while still using all of streams A and C to minimize water treatment requirements) is 0.8 kg/s, because more than that will create a stream with a larger flow rate than the upper bound allowed by sink 1. If 0.8 kg/s of uncontaminated material is added to streams A and C, the concentration of X in the combined stream will be

$$\frac{3.0\,\frac{kg}{s} \times 7\,\text{ppm} + 1.0\,\frac{kg}{s} \times 4\,\text{ppm} + 0.8\,\frac{kg}{s} \times 0\,\text{ppm}}{3.0\,\frac{kg}{s} + 1.0\,\frac{kg}{s} + 0.8\,\frac{kg}{s}} = 5.2\,\text{ppm.} \qquad \text{(Eq. 10-2)}$$

This concentration is still higher than the limit allowed by sink 1. To lower the concentration further without exceeding the limit on flow rate, only a portion of source A can be used. For example, if 2.8 kg/s of source A, all of source C, and 1.0 kg/s of uncontaminated material were combined, the resulting stream could be fed to sink 1. This is shown graphically in Figure 10.3-2.

The following example shows an application of source-sink mapping for a process that manufactures acrylonitrile.

Example 10.3-1   Source-Sink Mapping for Acrylonitrile Production.

Background

A simplified flowsheet for the production of acrylonitrile is given in Figure 10.3-3 (El-Halwagi, 1997). The chemistry of this process was described in Chapter 8. Oxygen, ammonia, and propylene are reacted to form a gaseous stream containing the product, ammonia, and water. This stream is sent to a condenser where most of the water is removed in a stream that is sent to treatment.

The gaseous stream from the condenser is then sent to a scrubber where it is washed into a liquid stream that is fed to the decanter. Off-gases from the scrubber contain a negligible amount of water, acrylonitrile, and ammonia.

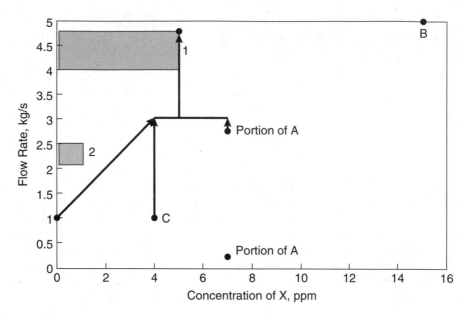

Figure 10.3-2   Source-sink diagram for Table 10.3-1, with sources and fresh feed combined to coordinate with a sink.

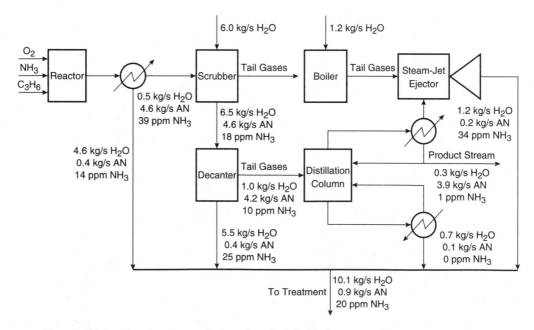

Figure 10.3-3   Flowsheet for production of acrylonitrile (AN).

The output of the scrubber is a mixture of water, acrylonitrile, and ammonia. This stream is sent to a decanter, where an aqueous layer and an organic layer (containing most of the acrylonitrile) form. All but 1 kg/s of the aqueous layer is sent to treatment, and the non-aqueous layer along with the remaining aqueous layer (1 kg/s) is sent to a distillation column in order to further purify it. A negligible amount of water is dissolved in the acrylonitrile layer, but 0.068 mass fraction acrylonitrile is found in the aqueous layer. Both the aqueous layer and the acrylonitrile layer from the decanter contain ammonia, with the concentration in the aqueous layer higher by a factor of 4.3. The entire non-aqueous layer and some of the aqueous layer from the decanter is sent to the distillation column for further purification of acrylonitrile. As long as the feed to the distillation column is between 15 and 25 ppm NH_3 and between 75% and 85% by mass acrylonitrile, the bottoms stream from the column does not change. This stream is sent to treatment.

The distillation column is operated at a vacuum, which is supplied by a steam-jet ejector. Some of the material in the distillation column is carried off in the ejectate stream, including 0.2 kg/s acrylonitrile. No appreciable quantity of water is pulled off in the ejectate, and the concentration of ammonia in the ejectate stream is a factor of 34 higher than the concentration of ammonia in the acrylonitrile product stream from the top of the distillation column. The ejectate stream is sent to treatment.

In this simplified example, integration of water and acrylonitrile is sought and the only contaminant of concern is ammonia. The liquid feed to the scrubber must be between 5.8 and 6.2 kg/s, and can consist of water or a mixture of water and acrylonitrile. The concentration of ammonia in the feed to the scrubber must not exceed 10 ppm. In contrast, the feed to the boiler cannot contain any ammonia or acrylonitrile and is required to be 1.2 kg/s.

Problem Statement
a. How many sources of water are there in the process? How many sinks?
b. Draw a source-sink diagram for the process, with total flow rate (water plus acrylonitrile) on the y-axis and ammonia concentration on the x-axis.
c. What wastewater streams can be used as feed for the scrubber?
d. What wastewater streams can be in the boiler?
e. If the amount of acrylonitrile fed to the scrubber is to be maximized, what will be the flows of wastewater from each wastewater stream to the scrubber?
f. Draw a diagram that shows the new flowsheet. If these flows are put into place, what is the flow rate of water and acrylonitrile and the concentration of ammonia in all of the streams downstream of the scrubber?
g. How does the product stream differ from the original configuration?
h. How does the quantity of fresh water feed differ?
i. How does the total stream sent to wastewater treatment differ?

Solution:
a. The sources of wastewater and waste acrylonitrile are the condenser, the decanter, the distillation column, and the ejectate. Sinks are the scrubber and the boiler.
b. Figure 10.3-4 shows a source-sink diagram for this process. Note that none of the wastewater streams can be used as feed for the boiler because they all contain acrylonitrile or ammonia or both.

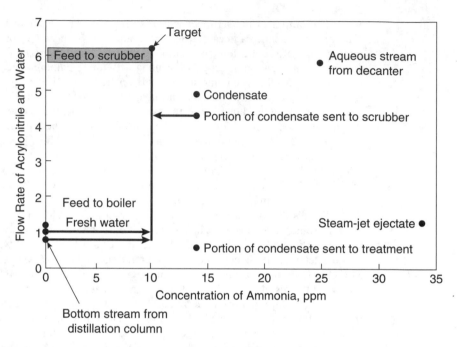

Figure 10.3-4   Source-sink diagram for acrylonitrile process (AN = acrylonitrile).

c. Examination of the source-sink diagram reveals that the bottom stream from the distillation column and the wastewater stream from the condenser are likely candidates for feed to the scrubber. In fact, because the bottom stream from the distillation column contains no ammonia and has the highest mass fraction of acrylonitrile of any waste stream, that entire stream should be fed to the scrubber. The maximum amount of acrylonitrile in the wastewater streams from the condenser is returned to the process when the scrubber is fed its maximum amount of feed (6.2 kg/s) at the maximum concentration of ammonia (10 ppm). If x is the flow rate of the condensate stream sent to the scrubber and y is the flow rate of fresh water to the scrubber, then balances for water and ammonia flow give:

$$0.8 \, \frac{\text{kg}}{\text{s}} + x + y = 6.2 \, \frac{\text{kg}}{\text{s}}, \qquad \text{(Eq. 10-3)}$$

$$\frac{0.8 \, \dfrac{\text{kg}}{\text{s}} \times 0 + x \times 14 \text{ ppm} + y \times 0}{0.8 \, \dfrac{\text{kg}}{\text{s}} + x + y} = 10 \text{ ppm.} \qquad \text{(Eq. 10-4)}$$

Solving for x and y gives x = 4.4 kg/s and y = 1.0 kg/s. In other words, 4.4 kg/s of the condenser wastewater stream (consisting of 4.0 kg/s water and 0.4 kg/s acrylonitrile) and all of the bottom stream from the distillation column along with 1 kg/s of fresh water can be used as scrubber feed. This is shown in the source-sink diagram of Figure 10.3-4.

To find the characteristics of each stream downstream of the scrubber, which are now different from those shown in Figure 10.3-3 (because the feed to the scrubber is different), do a mass balance of water, acrylonitrile, and ammonia over the scrubber. A diagram for the mass balance is shown in Figure 10.3-5.

The flow rate of water leaving the scrubber is

$$0.5\,\frac{kg}{s} + 1.0\,\frac{kg}{s} + 4.0\,\frac{kg}{s} + 0.7\,\frac{kg}{s} = 6.2\,\frac{kg}{s}, \tag{Eq. 10-5}$$

and the flow rate of acrylonitrile leaving the scrubber is

$$4.6\,\frac{kg}{s} + 0.35\,\frac{kg}{s} + 0.1\,\frac{kg}{s} = 5.1\,\frac{kg}{s}. \tag{Eq. 10-6}$$

The concentration of ammonia in the stream leaving the scrubber is found using the equation

$$\frac{39\,\text{ppm} \times 5.1\,\frac{kg}{s} + 0\,\text{ppm} \times 0.8\,\frac{kg}{s} + 0\,\text{ppm} \times 1.0\,\frac{kg}{s} + 14\,\text{ppm} \times 4.4\,\frac{kg}{s}}{5.1\,kg/s + 0.8\,\frac{kg}{s} + 1.0\,\frac{kg}{s} + 4.4\,\frac{kg}{s}} = 23\,\text{ppm}.$$

$$\tag{Eq. 10-7}$$

Next, mass balances are performed around the decanter, pictured in Figure 10.3-6, to find its outlet stream characteristics. Keep in mind that in the decanter, the stream from the scrubber forms two layers: an aqueous layer containing some acrylonitrile and ammonia, and an organic layer containing a negligible amount of water

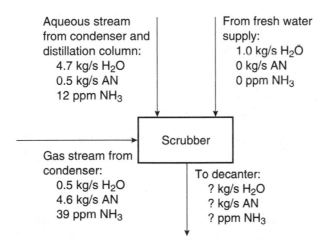

Aqueous stream from condenser and distillation column:
  4.7 kg/s H_2O
  0.5 kg/s AN
  12 ppm NH_3

From fresh water supply:
  1.0 kg/s H_2O
  0 kg/s AN
  0 ppm NH_3

Scrubber

Gas stream from condenser:
  0.5 kg/s H_2O
  4.6 kg/s AN
  39 ppm NH_3

To decanter:
  ? kg/s H_2O
  ? kg/s AN
  ? ppm NH_3

Figure 10.3-5 Mass balance diagram for scrubber in acrylonitrile process (AN = acrylonitrile).

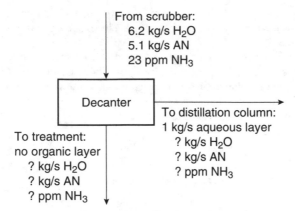

Figure 10.3-6   Mass balance diagram for decanter in acrylonitrile process (AN = acrylonitrile).

and some ammonia. All of the organic layer and 1 kg/s of the aqueous layer go to the distillation column and the remaining portion of the aqueous layer goes to wastewater treatment.

The relationship for acrylonitrile in the aqueous layer can be used to find the flow rate of water to the distillation column, as follows:

$$1\,\frac{\text{kg}}{\text{s}}\,\text{aqueous layer} = \frac{0.0678\ \text{kg acrylonitrile}}{\text{kg aqueous layer}} \times 1\,\frac{\text{kg}}{\text{s}}\,\text{aqueous layer} + z, \quad \text{(Eq. 10-8)}$$

where z is the flow rate of the water in the aqueous layer sent to the distillation column. Solving gives

$$z = 0.9\,\frac{\text{kg}}{\text{s}}\,\text{H}_2\text{O} \qquad\qquad \text{(Eq. 10-9)}$$

A water mass balance around the decanter gives the flow rate of water in the stream sent to treatment:

$$6.2\,\frac{\text{kg}}{\text{s}} - 0.9\,\frac{\text{kg}}{\text{s}} = 5.3\,\frac{\text{kg}}{\text{s}} \qquad\qquad \text{(Eq. 10-10)}$$

The flow rate of acrylonitrile in the stream sent to treatment is given by

$$\left(\frac{5.3\ \text{kg H}_2\text{O}}{\text{s}}\right)\left(\frac{0.0678\ \text{kg acrylonitrile}}{\text{kg H}_2\text{O}}\right) = \frac{0.4\ \text{kg acrylonitrile}}{\text{s}}. \qquad \text{(Eq. 10-11)}$$

From a mass balance for acrylonitrile, one can find the acrylonitrile being sent to the distillation column:

$$5.1\,\frac{\text{kg}}{\text{s}} - 0.4\,\frac{\text{kg}}{\text{s}} = 4.7\,\frac{\text{kg}}{\text{s}} \qquad\qquad \text{(Eq. 10-12)}$$

The concentration of ammonia in the streams leaving the decanter depends on the flow rates of the aqueous layer and organic layer, rather than on the flow rates of water and acrylonitrile leaving the decanter. Therefore, the next step is to find the flow rates of the aqueous layer and the organic layer. The total flow rate of the aqueous layer is equal to the total flow rate sent to treatment plus 1 kg/s that is sent to the distillation column, or

$$1.0\,\frac{kg}{s} + 0.4\,\frac{kg}{s} + 5.3\,\frac{kg}{s} = 6.7\,\frac{kg}{s} \qquad \text{(Eq. 10-13)}$$

and the total flow rate of the organic layer is 4.7 kg/s. Now, if x is the concentration in ppm of ammonia in the aqueous layer and y is the concentration in ppm of ammonia in the organic layer, then a mass balance for ammonia around the decanter gives

$$23\ \text{ppm} \times \left(5.1\,\frac{kg}{s} + 6.2\,\frac{kg}{s}\right) = \left(6.7\,\frac{kg}{s}\right)x + \left(4.6\,\frac{kg}{s}\right)y. \qquad \text{(Eq. 10-14)}$$

It is given that

$$x = 4.34\,y. \qquad \text{(Eq. 10-15)}$$

and solving for x and y gives $y = 8$ ppm and $x = 33$ ppm. The overall concentration of ammonia in the stream sent to the distillation column is

$$\frac{33\ \text{ppm} \times 1.0\,\dfrac{kg}{s} + 4.6\,\dfrac{kg}{s} \times 8\ \text{ppm}}{5.6\,\dfrac{kg}{s}} = 12\ \text{ppm} \qquad \text{(Eq. 10-16)}$$

Now, mass balances can be performed around the distillation column of Figure 10.3-7 in order to find the characteristics of the product stream and the steam-jet ejectate.

The bottoms stream is the same as before and there is no water carried away in the steam-jet ejectate, so a mass balance for water gives the water in the product stream, as follows:

$$0.9\,\frac{kg}{s} - 0.7\,\frac{kg}{s} = 0.2\,\frac{kg}{s}. \qquad \text{(Eq. 10-17)}$$

The amount of acrylonitrile in the steam-jet ejectate stream is given, and again, the bottoms from the distillation column do not change, so the amount of acrylonitrile in the product stream is given by

$$4.7\,\frac{kg}{s} - 0.2\,\frac{kg}{s} - 0.1\,\frac{kg}{s} = 4.4\,\frac{kg}{s}. \qquad \text{(Eq. 10-18)}$$

Next, a mass balance for ammonia, plus the relationship for ammonia concentrations in the product stream and the steam-jet ejectate, are used to find the concentration of ammonia in those streams. If x is the concentration of ammonia in ppm in the product stream and y is the concentration of ammonia in ppm in the steam-jet ejectate, then

$$12\ \text{ppm} \times 5.6\,\frac{kg}{s} = 0 \times 0.8\,\frac{kg}{s} + \left(4.6\,\frac{kg}{s}\right)x + \left(1.2\,\frac{kg}{s}\right)y \qquad \text{(Eq. 10-19)}$$

$$y = 34x. \qquad \text{(Eq. 10-20)}$$

Solving gives $x = 2$ ppm ammonia and $y = 50$ ppm ammonia.

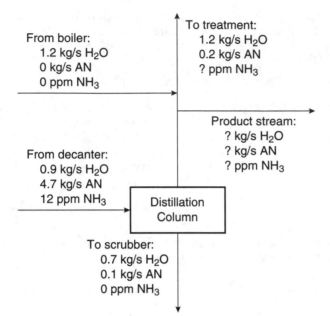

From boiler:
    1.2 kg/s H_2O
    0 kg/s AN
    0 ppm NH_3

To treatment:
    1.2 kg/s H_2O
    0.2 kg/s AN
    ? ppm NH_3

Product stream:
    ? kg/s H_2O
    ? kg/s AN
    ? ppm NH_3

From decanter:
    0.9 kg/s H_2O
    4.7 kg/s AN
    12 ppm NH_3

Distillation
Column

To scrubber:
    0.7 kg/s H_2O
    0.1 kg/s AN
    0 ppm NH_3

Figure 10.3-7 Mass balance diagram for the distillation column in the acrylonitrile process (AN = acrylonitrile).

The final flowsheet is given in Figure 10.3-8, and the differences between the outputs (wastewater and product stream) of the original configuration and the configuration with wastewater reuse are given in Table 10.3-2. The requirement for fresh water feed is 30% of the original process, and the flow rate of material sent to treatment is 60% of the original process. The mass fraction of acrylonitrile in the stream sent to treatment is lower than before by 15%, but the ammonia concentration in the stream sent to treatment is nearly a factor of two larger. Also, the concentration of ammonia in the product stream is twice what it was before wastewater reuse. More important than any of these changes, however, may be the increase in the production of acrylonitrile from 3.9 kg/s to 4.4 kg/s. With a market value of $0.60/kg and 350 days per year of production, this increase in product is worth $9,000,000 a year.

Table 10.3-2 Outputs from the Production of Acrylonitrile Before and After Wastewater Reuse.

Output Characteristic	Before Wastewater Reuse	After Wastewater Reuse
Fresh water feed required	71 kg/s	21 kg/s
Acrylonitrile in product stream	31 kg/s	4.4 kg/s
Flow rate to treatment	13.1 kg/s	7.7 kg/s
Mass fraction acrylonitrile in stream sent to treatment	0.092	0.078
Concentration of ammonia in stream sent to treatment	20 ppm	35 ppm
Concentration of ammonia in product stream	1 ppm	2 ppm

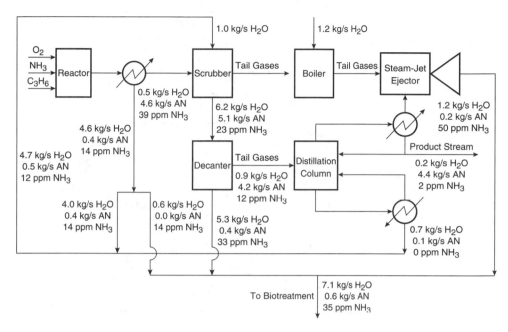

Figure 10.3-8   Flowsheet for acrylonitrile (AN) after identifying opportunity for water integration using source-sink mapping.

10.3.2 Optimizing Strategies for Segregation, Mixing, and Recycle of Streams

The source sink mapping described in the acrylonitrile example was a relatively simple example with just a few sources and sinks. As the processes to be analyzed become more complex and the number of sources and sinks increase, mathematical optimization techniques, coupled with process simulation packages, are generally employed to identify opportunities for recycle, segregation, and mixing of streams. The linear and non-linear mathematical programming techniques employed in these optimizations are beyond the scope of this text; however, the following example of source-sink matching for a chloroethane facility begins to illustrate the potential complexity of the problems that can be encountered.

Example 10.3-2   Source-Sink Mapping for Chloroethane Production.

Figure 10.3-9 is a flow diagram of a process for producing chloroethane. In this process, ethanol and hydrogen chloride are reacted in the presence of a catalyst to create chloroethane. The reaction is written as follows:

$$C_2H_5OH \rightarrow C_2H_5Cl + H_2O$$

Chloroethanol is also created in the reactor as an unwanted byproduct.

Two phases leave the reactor: an aqueous phase that is sent to wastewater treatment, and a gaseous phase that contains the product. The gaseous phase leaving the

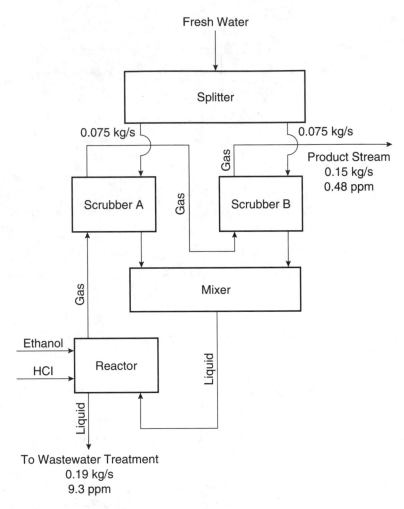

Figure 10.3-9 Process flow sheet for original configuration of chloroethane process. Concentrations of chloroethanol and total flow rates are given for the stream sent to wastewater treatment as well as the product stream. Flow rates of fresh water to scrubbers are also given.

reactor contains unreacted ethanol and hydrogen chloride as well as chloroethane and chloroethanol and is sent through two scrubbers in sequence in order to purify it enough for finishing steps and eventual sale. The aqueous streams from the two scrubbers are mixed and recycled back to the reactor because a fraction of the chloroethanol they contain can be converted into chloroethane (the desired product) via a reduction reaction whose rate is directly proportional to the concentration of chloroethanol in the aqueous stream entering the reactor. This conversion of chloroethanol to chloroethane is the key to minimizing the overall generation of chloroethanol in the process. Configuring the process streams so that the concentration of chloroethanol in the aqueous stream fed to the reactor results in as much destruction of chloroethanol as possible results in the minimum generation of this pollutant.

The goal of the analysis presented in this example is to determine how streams should be segregated, mixed, and recycled so that the net generation of chloroethanol is minimized. To provide a visual aid for the analysis, the process flowsheet can be redrawn as shown in Figure 10.3-10. This figure shows that each aqueous stream from each process unit can be split and sent to either wastewater treatment, sent back to the process unit from which it originated, or sent to one of the other process units. Mixers are located just prior to each process unit, and separators are located on the stream exiting each process unit. In the solution, the quantity of each stream that is sent to each option is determined. Note that the gaseous stream from the reactor must proceed through scrubber A, then through scrubber B.

Note that the process units have been assigned numbers. The reactor is unit number one, scrubber A is unit number two, scrubber B is unit number three, wastewater treatment is unit number four, and the source supplying fresh water is unit number five. The concentration of chloroethanol in ppm in the aqueous streams are given as x, and the concentration of chloroethanol in ppm in the gaseous streams is called y. Aqueous stream flow rates in kg/s are given as L and gaseous stream flow rates in kg/s are given as G. A subscript (one through five) denotes the unit number the aqueous stream is entering or leaving, and a superscript (in or out) indicates whether the stream is entering or leaving the unit. Thus, for example, the notations

$$L_1^{out} \text{ and } y_1^{out} \hspace{4cm} \text{(Eq. 10-21)}$$

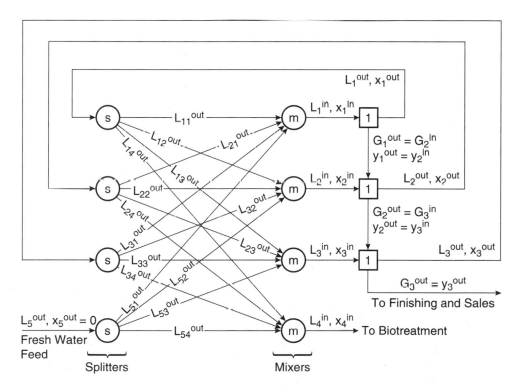

Figure 10.3-10   Diagram of chloroethane process showing every possibility for segregation, mixing, and recycle.

indicate the flow rate of the aqueous stream leaving the reactor and the concentration of chloroethanol in the gaseous stream leaving the reactor, respectively. In addition, the sixteen streams leaving the splitters are given special names. For example,

$$L_{11}{}^{out} \text{ and } L_{12}{}^{out} \qquad\qquad\qquad \text{(Eq. 10-22)}$$

are the portion of the aqueous stream leaving the reactor that is sent back to the reactor in kg/s and the portion of the aqueous stream leaving the reactor that is sent to scrubber A in kg/s, respectively.

At this point, a series of equations describing the mass balances and unit operations can be developed, involving dozens of equations and unknown variables. A list of the model variables is given in Table 10.3-3.

Table 10.3-3   Model variables with values for the original configuration of the chloroethane process.

Variable	Value from Figure 10.3-11
$L_{11}{}^{out}$	0 kg/s
$L_{12}{}^{out}$	0 kg/s
$L_{13}{}^{out}$	0 kg/s
$L_{14}{}^{out}$	0.19 kg/s
$L_{1}{}^{out}$	0.19 kg/s
$L_{21}{}^{out}$	? kg/s
$L_{22}{}^{out}$	0 kg/s
$L_{23}{}^{out}$	0 kg/s
$L_{24}{}^{out}$	0 kg/s
L_{2}	? kg/s
$L_{31}{}^{out}$	? kg/s
$L_{32}{}^{out}$	0 kg/s
$L_{33}{}^{out}$	0 kg/s
$L_{34}{}^{out}$	0 kg/s
L_{3}	? kg/s
$L_{51}{}^{out}$	0 kg/s
$L_{52}{}^{out}$	0.075 kg/s
$L_{53}{}^{out}$	0.075 kg/s
$L_{54}{}^{out}$	0 kg/s
$L_{5}{}^{out}$	? kg/s
$L_{1}{}^{in}$	? kg/s
$L_{4}{}^{in}$	0.19 kg/s
$x_{1}{}^{out}$	9.3 ppm
$x_{2}{}^{out}$	? ppm
$x_{3}{}^{out}$	? ppm
$x_{5}{}^{out}$	0 ppm
$x_{1}{}^{in}$	? ppm
$x_{2}{}^{in}$	0 ppm
$x_{3}{}^{in}$	0 ppm
$x_{4}{}^{in}$	9.3 ppm
G	0.15 kg/s
$y_{1}{}^{out}$	? ppm
$y_{2}{}^{out}$	? ppm
$y_{3}{}^{out}$	0.48 ppm

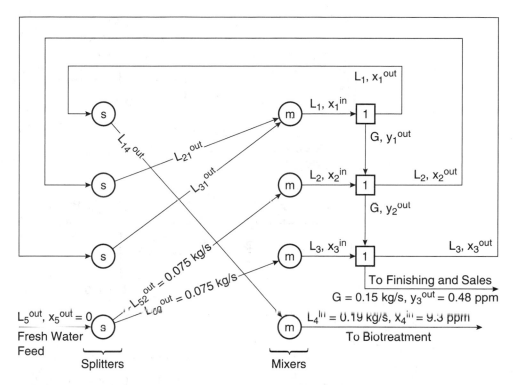

Figure 10.3-11 Diagram of original chloroethane process in the format of Figure 10.3-10.

The optimal mixing and recycle rates, where optimal is defined as the process that generates the least amount of chloroethane, can be determined using linear programming methods. The details are described by El-Halwagi (1997). The original process is shown in Figure 10.3-11 and the optimized process is shown in Figure 10.3-12 and Table 10.3-4.

Table 10.3-4 compares chloroethanol generation, wastewater flow rates, chloroethanol load to the wastewater treatment unit, and fresh water input between the optimized and the original process. Fresh water usage in the proposed process is 38% of that in the original process, and the load to the wastewater treatment unit is decreased from 0.19 kg/s to 0.10 kg/s. Chloroethanol loading to the wastewater treatment unit is decreased from 1.8 mg/s to 0.68 mg/s. Finally, the rate of generation of chloroethanol in the reactor is reduced from 1.9 mg/s to 0.76 mg/s.

Table 10.3-4 Chloroethane Process Characteristics Before and After Optimizing the Segregation, Mixing, and Recycle Strategy.

Parameter	Before Optimization	After Optimization
Volume to wastewater treatment, kg/s	0.19	0.10
Concentration of chloroethanol in wastewater stream, ppm	9.3	6.8
Chloroethanol load to wastewater treatment, mg/s	1.8	0.68
Net chloroethanol generation in the reactor, mg/s	1.9	0.76
Freshwater usage, kg/s	0.15	0.057

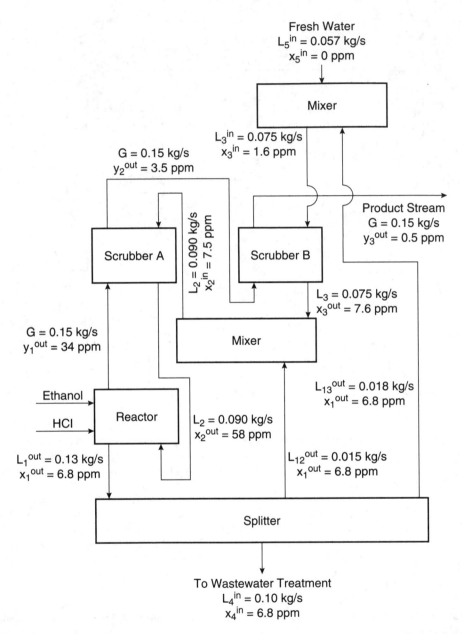

Figure 10.3-12   Chloroethane process with mixing, segregation, and recycle strategy that minimizes the production of chloroethanol.

10.3.3   Mass Exchange Network Synthesis

One of the more rigorous flowsheeting tools for mass integration is mass exchange network (MEN) synthesis. MEN synthesis is analogous to heat exchange network (HEN) synthesis, which was discussed earlier in this chapter. However, while the goal of HEN synthesis is energy efficiency, the goal of MEN synthesis is mass efficiency.

Unlike source-sink mapping and optimizing segregation, mixing, and recycle, MENs do not achieve mass integration through re-routing of process streams. Instead, they involve direct exchange of mass between streams. MEN synthesis was originally developed by Manousiouthakis and coworkers at UCLA (see, for example, El-Halwagi and Manousiouthakis, 1989), and is used to systematically generate a network of mass exchangers whose purpose is to preferentially transfer compounds that are pollutants in the streams in which they are found to streams in which they have a positive value. MEN synthesis can be used for any countercurrent, direct-contact mass-transfer operation, such as absorption, desorption, or leaching.

The case of phenol at petroleum refineries provides an example of a potential use of an MEN for pollution prevention. In these refineries, phenol is a pollutant in the water effluent of catalytic cracking units, desalter wash water, and spent sweetening waters. In other refinery streams, however, phenol can be a valuable additive. An MEN might therefore be used to transfer phenol to the streams where its presence is desirable, thus preventing phenol pollution in refinery wastewaters.

Remember from the earlier section in this chapter on heat integration that the limits to heat transfer are determined by energy balances and a positive driving force. Similarly, mass transfer is limited by mass balance constraints and equilibrium constraints. The constraints are (1) the total mass transferred by the rich stream (the stream from which a material is to be removed) must be equal to that received by the lean stream (the stream receiving the material) and (2) mass transfer is possible only if a positive driving force exists for all rich stream/lean stream matches. The means of incorporating the constraints into MEN synthesis are discussed in more detail below.

Because it has a positive value in some streams and cannot be referred to as a pollutant, the compound whose transfer is desired will be identified as the "solute" in the discussion that follows. A mass balance on the solute to be transferred from stream i to stream j of Figure 10.3-13 results in the equation

$$R_i(y_i^{in} - y_i^{out}) = L_j(x_j^{out} - x_j^{in}), \text{(Eq. 10-23)}$$

where R_i is the flow rate of rich stream i, L_j is the flow rate of lean stream j, y_i is the mass fraction of the solute in rich stream i, and x_j is the mass fraction of the solute in lean stream j. Note that rich streams are streams in which the solute concentration is higher than desired, and lean streams are streams in which the concentration of the solute is lower than desired. For the analysis in this chapter, the flow rates of

Figure 10.3-13   Mass balance diagram for mass exchange network synthesis

the streams are assumed to be constant. This is a good approximation if the concentration of solute in the streams is low and little transfer of material other than solute occurs.

Equilibrium between a rich and a lean stream can be represented by an equation of the form

$$y_i = m_j x_j^* + b_j, \qquad \text{(Eq. 10-24)}$$

where x_j^* is the mass fraction of the solute in stream j that is in equilibrium with the mass fraction y_i in stream i and m and b are constants. This linear type of equilibrium relationship is found in familiar expressions such as Raoult's Law, Henry's Law, and octanol/water partition coefficients. The constants of Equation 10-24 are thermodynamic properties and may be obtained through experimental data. The positive driving force constraint for mass transfer is satisfied when x_j is larger than x_j^*.

The tools of MEN synthesis are composition interval diagrams and load lines. A composition interval diagram (CID) depicts the lean and rich streams that are under consideration. CIDs for the rich and lean streams of Table 10.3-5 are shown in Figure 10.3-14. To construct this diagram, an arrow is drawn for each stream with its tail at the entering mass fraction and its head at the exiting mass fraction. The following example provides further illustration of the process of mapping lean and rich streams onto a CID.

Table 10.3-5   Stream Data for Two Rich Streams and One Lean Stream.

	Rich Stream				Lean Stream		
Stream	Flow Rate, kg/s	y^{in}	y^{out}	Stream	Flow Rate, kg/s	x^{in}	x^{out}
R_1	5	0.10	0.03	L	15	0.00	0.05
R_2	10	0.07	0.03				

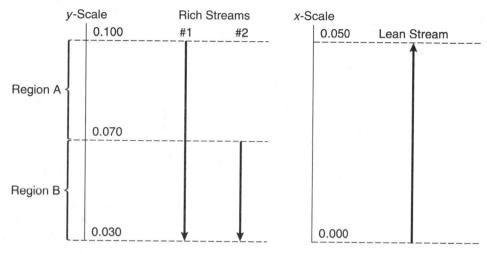

Figure 10.3-14   Composition interval diagram for the streams of Table 10.3-5.

Example 10.3-3   Constructing a Composition Interval Diagram

Construct a CID for the rich streams of Table 10.3-6. With the aid of this CID, calculate the mass transferred out of the rich streams in units of kg/s within each region of the CID. The mass transferred from the rich streams within each region is equal to $(y^{out}-y^{in})* \_R_i$, where y^{out} and y^{in} are the exiting and entering rich stream mass fractions, respectively, and $\_R_i$ is the sum of the rich stream flow rates in the region. Note that mass transfer is negative for the rich streams because they are losing mass.

Solution: The rich streams are mapped from Table 10.3-6 to generate the CID shown in Figure 10.3-15. The mass transferred in each region is:

$$\text{region } 1 = (y^{out} - y^{in}) \times R_i = (0.08 - 0.10)5 \text{ kg/s} = -0.10 \text{ kg/s}$$

$$\text{region } 2 = (0.07 - 0.08)(5 + 5) \text{ kg/s} = -0.10 \text{ kg/s}$$

$$\text{region } 3 = (0.03 - 0.07)(5 + 10 + 5) \text{ kg/s} = -0.80 \text{ kg/s}$$

$$\text{region } 4 = (0.01 - 0.03)5 \text{ kg/s} = -0.10 \text{ kg/s}$$

Table 10.3-6   Stream Data for Three Rich Streams and One Lean Stream.

	Rich Stream				Lean Stream		
Stream	Flow Rate, kg/s	y^{in}	y^{out}	Stream	Flow Rate, kg/s	x^{in}	x^{out}
R_1	5	0.10	0.03	L	15	0.00	0.14
R_2	10	0.07	0.03				
R_3	5	0.08	0.01				

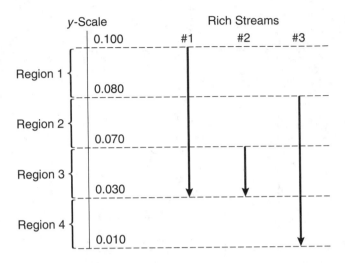

Figure 10.3-15 Composition interval diagram for the rich streams of Table 10.3-6 for Example 10.3-3.

In Figure 10.3-14, the compositions of the rich and lean streams are on separate axes. These axes can be combined by applying the equilibrium relationship. If the equilibrium relationship in the region of interest for the species considered in this problem is given by

$$y = 0.67x^*,$$

then a mass fraction of $y = 0.1$ in the rich stream is in equilibrium with a mass fraction of $x^* = 0.15$ in the lean stream. By converting the lean stream compositions of Figure 10.3-14 to the rich stream compositions with which they are in equilibrium, and vice versa, a combined CID with shared axes as shown in Figure 10.3-16 can be constructed.

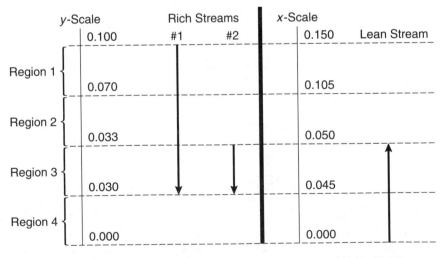

Figure 10.3-16 Combined composition interval diagram for the streams of Table 10.3-5.

Example 10.3-4   Constructing a Combined Composition Interval Diagram

Construct a CID for the three rich streams and the lean stream of Table 10.3-6. The equilibrium relationship in the region of interest is

$$y = 0.67x^*.$$

Solution: The combined CID is pictured in Figure 10.3-17.

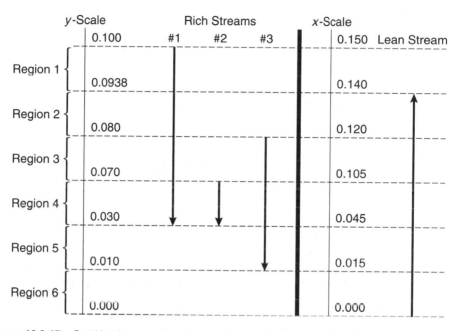

Figure 10.3-17   Combined composition interval diagram for the streams of Table 10.3-6 for Example 10.3-4.

Load lines depict the flow rate of solute transferred as a function of stream composition. These load lines are drawn on a set of axes that parallel the axes used for HEN synthesis (see Figure 10.2-3). The difference is that instead of heat transfer on the y-axis, mass transfer is used, and instead of temperature on the x-axis, concentration of the solute is used.

Constructing load lines for single streams is best explained through an example. Take R_1 of Table 10.3-5. At its inlet, this stream has not begun exchange of solute, so one endpoint of its load line is $y^{in} = 0.10$, mass exchanged = 0 kg/s. At its outlet, this stream has exchanged 5 kg/s × (0.03 − 0.10) or −0.35 kg/s, so the coordinates of the other endpoint are (0.03, −0.35 kg/s). The amount of mass transferred is negative at this endpoint because mass is being transferred out of the stream. This load line is shown in Figure 10.3-18. Note that there is an arrow pointing down and to the left to indicate the direction of transfer. In the following example, you are asked to construct the load line for the remaining stream of Table 10.3-5.

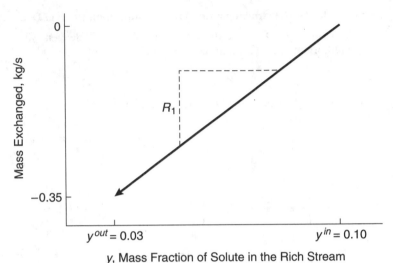

Figure 10.3-18   Load line for rich stream 1 of Table 10.3-5.

Example 10.3-5   Constructing Load Lines

On separate sets of axes, construct load lines for R_2 and L of Table 10.3-5. Remember that mass exchanged is equal to the flow rate multiplied by mass fraction out less mass fraction in.

Solution: One end point of the rich stream is at y_2^{in}, where no mass has been exchanged. Therefore, at kg/s = 0, $y = 0.07$. The other end point is (0.03, 10 kg/s × (0.03 − 0.07) = −0.4 kg/s). The end points of the lean stream load line are 0 kg/s, $x = 0$ and

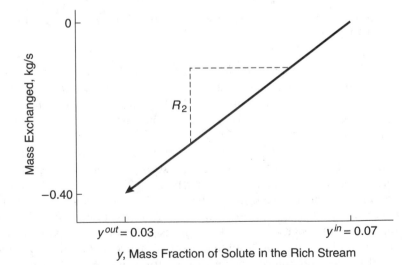

Figure 10.3-19   Load line for rich stream 2 of Table 10.3-5 for Example 10.3-5.

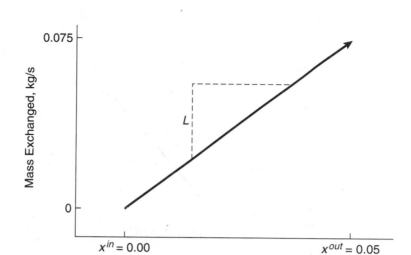

Figure 10.3-20   Load line for the lean stream of Table 10.3-5 for Example 10.3-5.

0.75 kg/s, $x = 0.05$. Plotting these points on the given axes, the graphic relationship between mass transferred (kg/s) and mass fraction shown in Figures 10.3-19 and 10.3-20 is obtained. Note that the slope of the load line for the rich stream is $(0.4-0)/(0.07-0.03)$ or 10 kg/s, which is the same as its flow rate. The same is true of the lean stream load line.

The load lines of Example 10.3-5 are for a single rich and a single lean stream. When there is more than one rich stream or more than one lean stream to consider, a line representative of the multiple streams must be constructed. This line, called a composite load line, is the sum of the individual load lines and is developed with the aid of a CID.

As an illustration, the composite load line for the rich streams of Table 10.3-5 is plotted in Figure 10.3-21. This composite line is the sum of the load lines of Figures 10.3-18 and 10.3-14. It consists of two segments, corresponding to the regions shown in the CID of Figure 10.19. In region A are the rich streams with mass fractions less than 0.1 and greater than 0.07 ($0.1 \le y \le 0.07$). Only R_1 falls into this category, so the total flow rate of the rich streams in this composition range is 5 kg/s. The starting point for the load line is $y = 0.1$ and 0.0 kg/s transferred. Recall that the mass transferred in each CID region is equal to the mass fraction exiting the region minus the mass fraction entering the region, multiplied by the sum of the flow rates in the region. Therefore, at $y = 0.07$, 5 kg/s $\times$ $(0.07-0.1)$ or -0.15 kg/s have been transferred, and the end point of the load line in this region is $(0.07, -0.15$ kg/s$)$. As before, the slope of the load line equals the mass flow rate of the stream. In region B are the rich streams with mole fractions less than 0.07 and greater than 0.03. Both rich streams fall into this region, and the endpoint of this

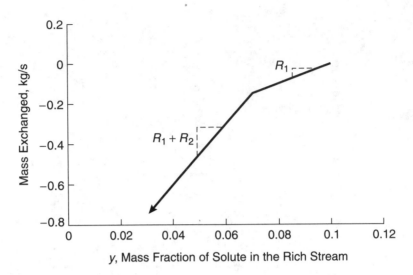

Figure 10.3-21   Composite load line for the rich streams of Table 10.3-5.

segment of the load line is y = 0.03, mass transferred = -0.15 kg/s + 15 kg/s $\times$ $(0.03 - 0.07) = -0.75$ kg/s. When the load line is plotted, it has a slope equal to the sum of the flow rates of all streams in this region. The starting point for this segment of the load line is the termination point of the previous segment.

Example 10.3-6   Constructing a Composite Load Line

Plot the composite load line for the rich streams of Table 10.3-6. Based on the total amount of mass transferred from the rich streams, calculate the minimum flow rate required for the lean stream. In the example that follows, the validity of using this simple overall mass balance technique for determining the minimum lean stream flow rate is examined.

Solution: The end points for the segments of the composite load line for the rich streams are: (0.1,0), (0.08,-0.10 kg/s), (0.07,-0.20 kg/s), (0.03,-1.0) and (0.01,-1.1). See Figure 10.3-22. Note that to plot a continuous load line, the cumulative mass transferred is used; the mass transferred in each interval is added to the mass transferred in all previous intervals. If the mass transferred by rich streams equals the mass gained by the lean stream, and (from before) total mass transferred from the rich streams is 1.10 kg/s, then from Table 10.3-6, mass gained by the lean stream is equal to $L(0.14-0.0)$, where L is the flow rate of the lean stream. Thus,

$$L(0.14 - 0.0) = 1.10$$

or

$$L = 7.86 \text{ kg/s.}$$

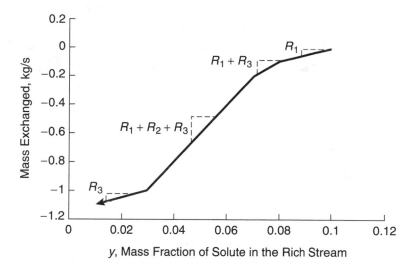

Figure 10.3-22   Composite load line for the rich streams of Table 10.3-6 for Example 10.3-6.

The next step in constructing load line diagrams is to plot the lean and rich streams on the same axes. As with combined CIDs, load line diagrams can be combined by making use of the equilibrium relationship, which for this specific numerical example is assumed to be

$$y = 0.67x^*$$

in the region of interest.

A combined figure can be made by following several different conventions, each of which gives the same final results. The convention used here for constructing the combined figure for the streams of Table 10.3-5 is to first plot the load line of the lean stream as shown in Figure 10.3-23. The rich stream composite load line is added to the figure after converting the rich stream mass fractions into the lean stream mass fractions with which they are in equilibrium. These conversions were made in order to construct the CID of Figure 10.3-16. The rich stream composite load line is free to move vertically; its placement determines which lean and rich streams contact each other. The rich stream load line has this freedom to move vertically because the values for mass exchanged on the y-axis are not absolute: they are useful only in relative terms, i.e., in terms of the differences in mass transferred between points. Therefore, the composite rich stream load line begins at x^* (the x-axis) = lean stream mass fraction with which the rich stream is in equilibrium = 0.10/0.67 = 0.15 and continues downward and to the left with a slope of $0.67R_1$. The next point falls at $x = 0.07/0.67 = 0.10$ where the slope changes to $0.67(R_1 + R_2)$. The load line ends where $x = 0.03/0.67 = 0.045$. The load lines for the lean and rich streams of Table 10.3-5 are plotted together in Figure 10.3-23. As stated before, the composite rich stream load line could have been located in any number of different vertical positions.

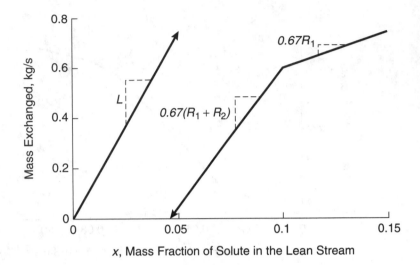

Figure 10.3-23 Combined load line diagram for the streams of Table 10.3-5.

In Figure 10.3-23, the lean stream load line falls to the left of the composite rich stream load line at every point. This indicates that the desired mass exchange is thermodynamically feasible (i.e., the equilibrium constraint is satisfied), and that the transfer could be accomplished using exchangers of finite size with no need for mass exchange into or out of any other streams. In Figure 10.3-24, the lean stream load line lies to the left of the rich stream load line, except at a point, called the pinch point, where the lines meet. Mass exchange in a case such as this is thermodynamically feasible, but would require an infinitely large mass exchanger (an infinite number of trays or stages, for example). Therefore, there is a practical requirement that conditions be manipulated so that a positive horizontal ε exists

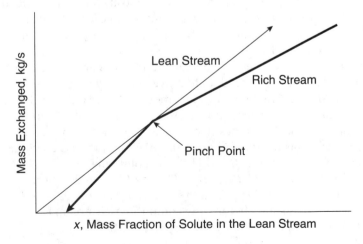

Figure 10.3-24 Load lines form a pinch.

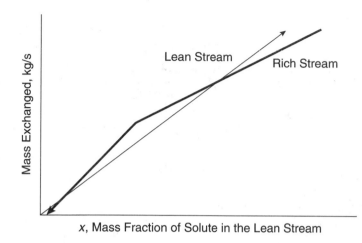

Figure 10.3-25 Mass transfer is thermodynamically infeasible where the rich stream load line lies to the left of the lean stream load line.

between the load lines. This ε is the driving force for mass transfer. If at any point the lean stream load line lies to the right of the rich stream load line, as shown in Figure 10.3-25, mass exchange in the desired direction is not thermodynamically feasible. In fact, if streams with such characteristics are contacted, mass exchange from the lean stream to the rich stream occurs. This infeasible situation could be made feasible by moving the rich stream load line down.

Example 10.3-7   Plotting Rich and Lean Stream Load Lines on a Single Set of Axes

Using the CID generated in Example 10.3-5, plot the load lines of the rich and lean streams described in Table 10.3-6 on the same diagram so that a pinch exists. Assign the lean stream a flow rate of 8 kg/s. Can the specified target concentrations be achieved solely through contact between the lean and rich streams? What does this indicate about the solution in Example 10.3-6 for the minimum lean stream flow rate required? Is there a lean stream flow rate at which all the desired mass exchange can occur solely through contact between the streams? If yes, what is that flow rate?

Solution: The easiest way to add the composite rich stream load line to the lean stream diagram is to determine where the pinch point occurs, plot it, and work outward. From examination of Figure 10.3-22, it appears that the point where $x = 0.07/0.67 = 0.104$ is the pinch point. The value for mass exchanged at this point is found using the equation for the lean stream load line, which is

$$\text{mass exchanged} = (8 \text{ kg/s})x.$$

Therefore, the first point to plot for the composite rich stream load line is (0.104, 8(0.104)kg/s = 0.832 kg/s). See Figure 10.3-26 for the remaining points on the composite rich stream load line. Target concentrations cannot be achieved solely through contact between the streams because the target concentration of the rich stream is lower than its concentration after contact with the lean stream (bottom left of Figure 10.3-26) and because the target concentration of the lean stream is higher than its concentration after contact with the rich stream (top right of Figure 10.3-26).

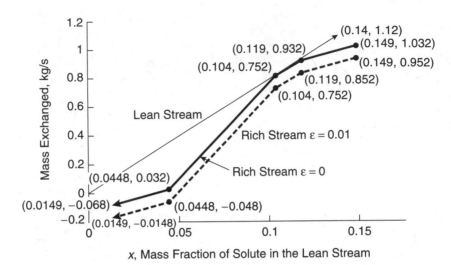

Figure 10.3-26   Combined load line diagram for the streams of Table 10.3-6 for Example 10.3-7.

The lean stream flow rate here (8 kg/s) is greater than the minimum lean stream flow rate calculated in Example 10.3-6, but it is insufficient to effect all the necessary mass exchange. There is no lean stream flow rate at which all the desired mass exchange can be achieved solely through contact between the streams.

Diagrams that combine the rich and lean stream load lines also show the amount of excess mass transfer capacity available from the lean stream and the amount of excess mass transfer capacity available from the rich stream. These regions were deliberately, if unrealistically, omitted in Figures 10.3-23, 10.3-24, and 10.3-25 for the sake of illustrating thermodynamic feasibility. In Figure 10.3-27, three regions labeled I, II, and III are identified. In region I, the lean stream has the capacity to exchange more mass and become richer, but there is a "shortage" of rich stream. The lean stream must be brought up to its specified concentration in a manner other than through mass exchange with the rich stream. For instance, the solute may be added to it. In region II, mass exchange can occur through contact between the rich and lean streams. In region III, the rich stream is capable of mass exchange, but there is a "shortage" of lean stream. An external lean stream mass separating agent is required to achieve the rich stream's target concentration. For example, an adsorbent such as activated carbon might be used to take up the excess solute, which is a pollutant in the rich stream. For the lean and rich streams depicted in Figure 10.3-27, the least amount of solute and external lean stream or mass separating agent is required when the rich stream load line is manipulated to form a pinch point ($\varepsilon = 0$). As ε increases, operating costs (the cost of solute and the cost of the mass separating agent) increase and capital costs (the cost of the network) decrease. As ε decreases, operating costs decrease and capital costs

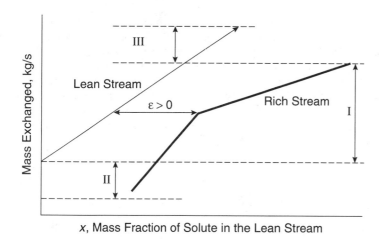

Figure 10.3-27 The three regions of a combined load line diagram.

x, Mass Fraction of Solute in the Lean Stream

increase. As with HEN synthesis, it is possible to find the ε at which total annualized costs are minimized.

MEN synthesis is powerful because of its graphic determination of the pinch point and its ability to show whether mass exchange is thermodynamically feasible. The culmination of all the previous steps is the synthesis of a mass exchange network. Combined load line diagrams show where the lean and rich streams contact each other in a mass exchange network. For example, inspection of Figure 10.3-26 reveals that rich stream 1 contacts the lean stream as the lean stream exits the mass exchange network. Then, where the lean stream mass fraction is $0.60/L_1 = 0.04$, the lean stream is split and one third contacts rich stream 1 while the remaining two thirds contacts rich stream 2. The lean stream enters this mass exchange network at the same point where both rich streams exit.

Example 10.3-8   Pairing Streams in a Mass Exchange Network (adapted from Allen and Rosselot, Pollution Prevention for Chemical Processes © 1997. This material is used by permission of John Wiley & Sons, Inc.)

Describe how the streams contact each other in the optimal mass exchange network of Figure 10.3-26. (This is the network where the value of _ is equal to 0.01.) To begin, rich stream 1 contacts the lean stream when the lean stream exits the mass exchange network. Further down the mass exchange network, the lean stream is split into two parts, with one third contacting rich stream 1 and two thirds contacting rich stream 2. Complete this description for the mass exchange network, and give the lean stream and rich stream mass fractions at which the splits/junctions occur.

Solution: As stated in the problem statement, rich stream 1 contacts the lean stream where the lean stream exits the mass exchange network. At this point, the mass fraction of solute in the lean stream is 0.952 kg/s / 8 kg/s = 0.119 (found from the equation of the lean stream load line) and the rich stream mass fraction is 0.149 × 0.67 = 0.10 (read directly from the composition interval diagram). Where the lean stream mass fraction is $0.852/8 = 0.107$ and the rich stream mass fraction is $0.119 \times 0.67 = 0.08$, the lean stream is split, with one half contacting rich stream 1 and the other half contacting rich stream 3. The lean stream is split three ways when its mass fraction is $0.752/8 = 0.094$ and the rich stream mass fraction is $0.104 \times 0.67 = 0.07$, with one fourth contacting rich stream 1, one half contacting rich stream 2, and the remaining fourth contacting rich stream 3. To find the mass fraction of solute in the lean stream when the rich streams exit the mass exchange network, the equation of the corresponding portion of the composite rich stream load line must first be found. It is known that the slope of this line is

$$m = 0.67(R_1 + R_2 + R_3) = 13.4 \text{ kg/s}.$$

The intercept is found by using one of the known points as follows:

$$0.752 \text{ kg/s} = 13.4 \text{ kg/s}(.07/.67) + b.$$

Therefore, $b = -0.648$ kg/s. The mass fraction at which this line crosses the x-axis can be found from

$$0 = (13.4 \text{ kg/s})x - 0.648 \text{ kg/s}.$$

This means that the rich streams exit the mass exchange network when the mass fraction of solute in them is $0.67 \times 0.648/13.4 = 0.0324$. The lean stream mass fraction at this point is 0. See Figure 10.3-28 for a flowsheet of this network. Stream matching around the pinch point is particularly complex (see Douglas, 1988 and El-Halwagi, 1997). MEN synthesis is described in more detail in El-Halwagi, 1997.

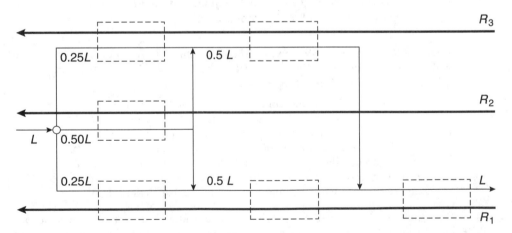

Figure 10.3-28   Mass exchange network showing stream matching for the streams of Table 10.3-6. Dashed lines indicate mass exchange units for Example 10.3-8.

10.4   CASE STUDY OF A PROCESS FLOWSHEET

In this section, the process flowsheet for a generic crude oil processing unit at a pe-troleum refinery is described, along with pollution prevention techniques that were developed for the unit (American Petroleum Institute, 1993). These pollution pre-vention techniques demonstrate the usefulness of both qualitative and quantitative flowsheeting tools and illustrate the complexity and integration found in processes in the chemical processing and refining industries. In this case study, proposed pol-lution prevention techniques, including both heat and mass integration, result in substantial environmental improvements at a cost savings.

One of the largest processing units in a petroleum refinery is the facility that separates crude oil into volatility fractions suitable for further processing (the crude unit). The hypothetical crude unit of this case study processes 175,000 barrels of light Arabian crude oil a day. It consists of a desalter (which removes salt and other contaminants from crude oil), an atmospheric distillation tower (so-called be-cause it operates at atmospheric pressure), and a vacuum distillation column (which operates at lower-than-atmospheric pressure in order to allow the column to separate low volatility materials at acceptable temperatures), as shown in Figure 10.4-1. Crude oil and water are the primary feed materials and several output streams are produced, including crude tower overhead (fuel gas and unstabilized gasoline), a light naphtha fraction, a kerosine fraction, a heavy distillate fuel frac-tion, an atmospheric gas oil (heavy) fraction, light vacuum gas oil, heavy vacuum gas oil, and vacuum residue.

While the crude unit is one of the largest processing units at a petroleum re-finery, it is important to remember that at a typical refinery, the streams from the crude unit are sent to many other processing units, many of which are reactors whose purpose is convert large hydrocarbon molecules into more saleable prod-ucts. These reactors include fluidized-bed catalytic crackers, hydroprocessers, and cokers. Other reactors create compounds with a higher octane rating, or combine small hydrocarbon compounds to create larger ones. Still other downstream pro-cessing units are used to purify and blend the refinery process streams. Finished re-finery products include gasoline, jet fuel, diesel fuel, fuel oil, waxes, asphalt, and petrochemical feedstocks. The boundaries of this case study, however, include only the crude unit and its input and output streams. The first major process unit, shown in Figure 10.4-2, is the desalter. Desalting removes salts in the crude that would cause corrosion of the process equipment. It also removes metals and suspended solids that would foul catalysts in downstream processing units.

In the first step of the desalting process, the crude oil is mixed with partially treated wastewater recycled from the refinery. Next, the oil is heated using a series of heat exchangers in preparation for desalting. In the desalter, which has two stages, hot oil and hot recycled water create a dispersed mixture of oil and water. The water extracts additional salts from the oil, and the salt-rich water (brine) is separated from the oil using an electric field. The brine from the desalter's second

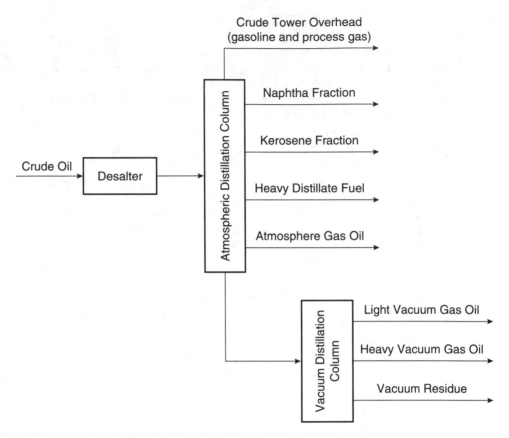

Figure 10.4-1   A simplified schematic of the petroleum refining crude unit.

stage is used as the wash water for the first stage, and brine from the first stage is sent to wastewater treatment after being cooled by heat exchange with desalter feed water and cooling water. The desalted crude is then sent to another series of heat exchangers, which elevate its temperature to 424°F.

After desalting and preheating, the crude oil is sent to the atmospheric pressure distillation process unit, shown in Figure 10.4-3. Fuel-fired heaters supply energy to the tower. Note that several streams are withdrawn from the column. Two of the side streams removed from the tower are sent to secondary strippers, where they are contacted with steam, and the products from these separation operations are sent to further processing operations at the refinery. The overhead product from the atmospheric distillation tower is cooled by heat exchange with the incoming crude oil and with cooling water. This overhead stream is collected in a drum and the fuel gases in the vapor phase are withdrawn, compressed, and sent to further processing operations at the refinery. Part of the gasoline in the condensed phase is used as a reflux for the atmospheric pressure distillation tower and the

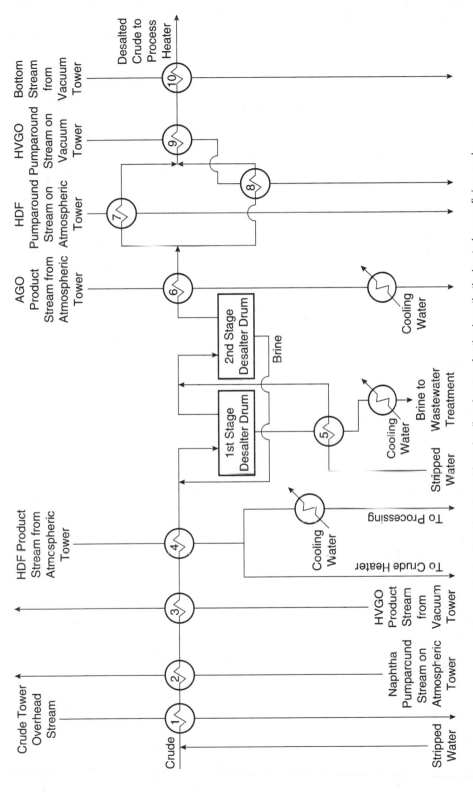

Figure 10.4-2 The desalter and crude oil preheaters for the hypothetical petroleum refining crude unit. Heat exchangers are numbered for cross-reference with Figures 10.4-3 and 10.4-4.

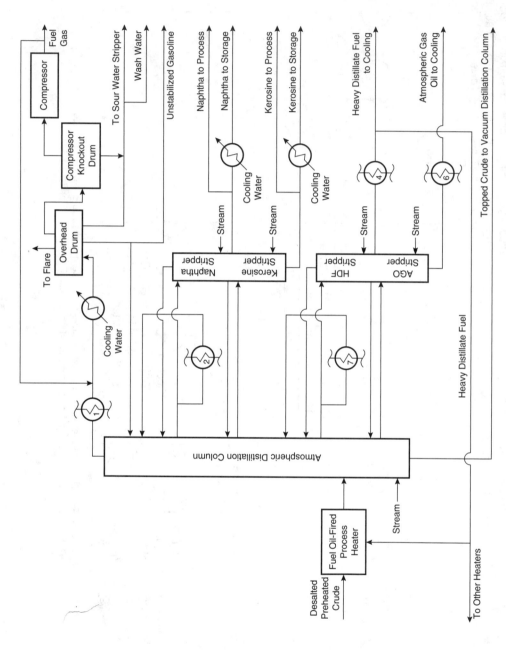

Figure 10.4-3 The atmospheric distillation tower for the hypothetical petroleum refining crude unit. Heat exchangers are numbered for cross-reference with Figure 10.4-2.

remainder is sent to the refinery. The bottom stream from the distillation column is sent to the next major process unit, the vacuum distillation tower.

In the vacuum distillation unit, shown in Figure 10.4-4, the bottoms from the atmospheric distillation unit are further fractionated into vacuum tower overhead, light vacuum gas oil, heavy vacuum gas oil, and vacuum tower bottoms. The energy for the tower is provided by a fuel-fired heater. Most of the product streams are sent directly to downstream processing at the refinery, with the exception of the overhead stream. The overhead stream is contacted with steam, then cooled and sent to a collection vessel called the overhead drum, where the oil and water phases are separated. The water stream from the overhead drum (which is contaminated with ammonia, hydrogen sulfide, and other components) is sent to the refinery's sour water stripper, which is a wastewater treatment process that recovers ammonia and hydrogen sulfide. The oil recovered from the overhead drum is sent to storage.

It is clear from this simplified process description that the crude unit in a refinery is a complex process that generates a variety of gaseous, liquid, and solid wastes. These wastes and emissions are described in Table 10.4-1.

Qualitative techniques for identifying pollution prevention opportunities were used to develop most of the proposed pollution prevention options for the crude unit. These options are described below:

1. Reboil with hot oil rather than steam to avoid oil/water contacting operations. Two additional side strippers must be added to the tower when this is done because the product specifications cannot be met with the existing side strippers when hot oil is used instead of steam for reboiling.
2. Add a liquid ring vacuum pump to the vacuum tower in order to reduce the pressure in the vacuum tower, which results in lower allowable operating temperatures, which in turn results in reduced cracking and fouling of the furnace tubes in the furnace, so that production of sour water is reduced.
3. Replace burners with new generation low-NOx burners and retrofit for flue-gas recirculation in order to reduce NOx emissions.
4. Reduce fugitive emissions by implementing a stringent inspection and maintenance program for piping components, using leakless valves when replacing small valves wherever it is economical to do so, using graphite or teflon packing and seals when repairing valves, specifying double seals when replacing pumps, installing rupture disks on pressure relief valves and venting them to a flare, modifying the compressor, blind-flanging all the vents and drains, eliminating flanges where possible, and making all the sampling systems closed-loop.
5. Segregate mildly-contaminated wastewater and treat it so that it can be reused.

In addition to the above strategies, pinch analysis to reduce external energy requirements showed that air emissions could be reduced substantially by increasing the surface area of the existing preheaters by 8% by adding three additional preheaters. The capital cost of these preheaters was estimated to be $2,268,000, while the

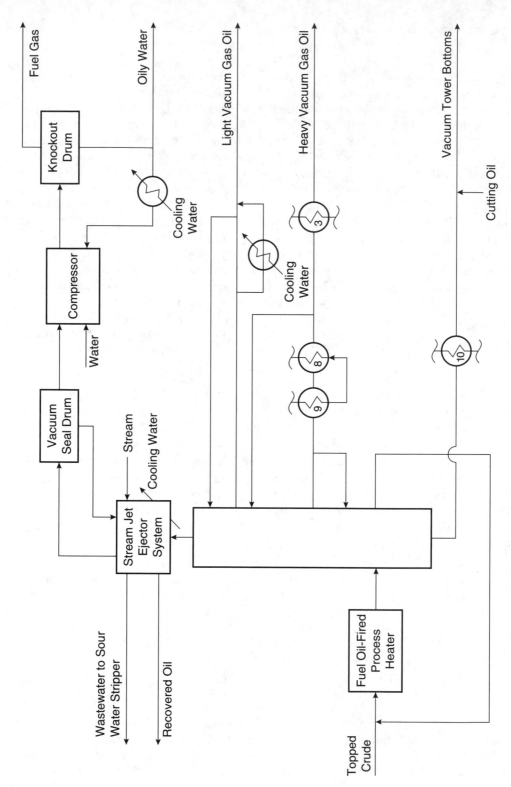

Figure 10.4-4 The vacuum distillation tower for the hypothetical petroleum refining crude unit. Heat exchangers are numbered for cross-reference with Figure 10.4-2.

Table 10.4-1   Major Waste and Emission Streams from a Petroleum Refining Crude Unit.

Type of Emission or Waste	Origin	Constituents
Stack air emissions	combustion of fossil fuels in boilers (for steam generation) and heaters (for heating the feed to the two distillation columns)	volatile organic compounds, carbon monoxide, particulate matter, sulfur dioxide, and nitrogen oxides
Fugitive air emissions	small leaks through packing on valves, pump seals, compressor seals, flanges, etc.; emissions from sewers; emissions from sampling operations; process fluid leaks into cooling water	volatile organic compounds whose composition depends on crude oil composition, but would include benzene, toluene, ethylbenzene, xylene, and cyclohexane
Cooling tower blowdown	cooling tower cools cooling water via evaporation; water evaporates more readily than the salts in the water; blowdown stream is necessary to prevent salts from getting too concentrated	salts
Condensate from steam drains		crude oil and other chemicals
Water plant sludge	sludge from purifying boiler and cooling tower water	salts
Oily wastewater	desalter wash water, sample drains, instrument drains, vessel drains	solids, phenols, chemical oxygen demand, ammonia, sulfides, crude oil components, heavy metals
Oily wastewater treatment sludges	treatment of oily wastewater generates a stream of slop oil, emulsified oil and water, and oily sludge	solids, crude oil components, heavy metals

annual savings in fuel costs were estimated to be $1,692,000. Therefore, the additional heat exchangers were projected to have a payback period of one and a third years.

Table 10.4-2 shows estimated wastes and emissions for the hypothetical crude processing unit with and without the pollution prevention measures. The measures were projected to decrease emissions of nitrogen oxides to air by 60%, and volatile organic compound emissions to air were projected to decrease by 93%. The quantity of oil and grease in wastewater was nearly halved, total suspended solids in wastewater were decreased by 32%, and sulfides in wastewater were decreased by 19%. Production of hazardous solid waste was decreased by over 90%, and a non-hazardous waste stream was generated.

Table 10.4-2 Estimates of Crude Unit Emissions and Waste Generation Before and After Pollution Prevention Alternatives are Implemented.

Emission or Waste	Without Additional Pollution Prevention Measures	With Additional Pollution Prevention Measures
Air emissions, tons/yr		
Nitrogen oxides	420	170
Carbon monoxide	180	170
Volatile organic compounds	180	12
Suspended particulate matter	23	21
Sulfur dioxide	3.3	3.0
Wastewater		
Oil and grease, gal/day	230	120
Total suspended solids, lb/day	11,000	7,500
Biological oxygen demand, lb/day	1,200	
Chemical oxygen demand, lb/day	4,600	4,600
Ammonia	570	570
Sulfides	160	130
Phenol	200	200
Hazardous waste, tons/day	6.3	0.5
Nonhazardous wastes, tons/day	none (mixed with hazardous)	3.7

SUMMARY

Even the simplest of chemical processes generally consist of a number of process units whose characteristics can influence overall waste generation. In this chapter, it was shown that pollution prevention can sometimes be achieved by examining the mass and energy integration of process units. Qualitative and quantitative tools for analyzing flowsheets are described. Case studies, in both the text of the chapter and in the accompanying problems, reveal both the power and the complexity of process integration.

REFERENCES

Allen, D.T. and Rosselot, K. S., *Pollution Prevention for Chemical Processes*, Wiley-Interscience, New York, 1997.

American Petroleum Institute (API), *Environmental Design Considerations for Petroleum Refining Crude Processing Units*, Publication 311, Washington, DC, 1993.

Crowl, D. A. and Louvar, J. F., *Chemical Process Safety: Fundamentals with Applications*, Prentice-Hall, Englewood Cliffs, NJ, 1990.

DuPont (1993), "DuPont Chambers Works Waste Minimization Project: Cambers Works—Deepwater, New Jersey," Report to the U.S. Government in satisfaction of consent decree in *U.S.v. DuPont,* Docket No.91cv768(JFG), May, 1993.

Douglas, J.M., *Conceptual Design of Chemical Processes*, McGraw-Hill, New York, 1988.

Douglas, J.M., "Process Synthesis for Waste Minimization," *Industrial and Engineering Chemistry Research,* 31, 238-243, 1992.

El-Halwagi, M. and Manousiouthakis, V., "Synthesis of Mass Exchange Networks", *AIChE Journal,* 35, 1233-1244 (1989).

El-Halwagi, M., *Pollution Prevention Through Process Integration: Systematic Design Tools*, Academic Press, San Diego, CA, 1997.

Rossiter, A. P. and Klee, H., "Hierarchical Process Review for Waste Minimization," *Waste Minimization Through Process Design*, A. P. Rossiter, ed., McGraw-Hill, New York, 1995.

PROBLEMS

1. Determine the amount of energy savings that could be achieved by putting in place a heat exchange network in the hot water system for a typical house.

   a) Begin by identifying the hot water streams from which heat can be extracted (e.g., dishwasher effluent, shower effluent). For each of these streams, estimate a water exit temperature and a daily flow rate.

   b) Assume that these hot streams will be contacted with the cold supply water entering the hot water heater and estimate the amount of energy that could be extracted from the hot streams. Determine the annual energy savings if the home uses an electric hot water heater and electricity costs $0.06/kwh. Make reasonable assumptions about the efficiency of the hot water heater (fraction of the electricity that goes into heated water used by the homeowner).

   c) The cost of an installed, non-contact, single tube, shell and tube exchanger for this application is approximately $500. Assume that the hot water exit lines already pass near the water heater so that little additional plumbing is required. Determine the time required to repay the installation cost using money saved in energy costs.

2. The equilibrium between the rich stream and the lean stream listed below is:

$$y = 1.5 \, x^{\,0.8}$$

Rich stream 1:	$y_{in} = 0.45$	$y_{out} = 0.1$	Flowrate = 3 kg/s
Lean stream:	$x_{in} = 0.45$	$x_{out} = 0.1$	Flowrate = 5 kg/s

   A second rich stream is available. The equilibrium relationship between the second rich stream and the lean stream is:

$$y = 1.5 \, x$$

   The second rich stream has inlet and outlet concentrations listed below.

Rich stream 2:	$y_{in} = 0.3$	$y_{out} = 0.1$	Flowrate = 3 kg/s

   Plot the rich streams and lean streams on a single set of axes (mass exchanged versus lean stream mass fraction) and determine the pinch point.

3. *Recovery of Benzene from the Aqueous Wastes of a Styrene Manufacturing Process* (adapted from El-Halwagi, 1997). Styrene is manufactured by dehydrogenating ethylbenzene over an oxide catalyst at 600 to 650°C. Steam is generally co-injected with the ethylbenzene into the reactor, shown in Problem 3, Figure 1. Styrene and hydrogen gas are the primary reaction products. Byproducts include benzene, ethane, toluene and methane. The products and byproducts from the reactor are cooled to approximately ambient temperature. Light products such as hydrogen, methane and

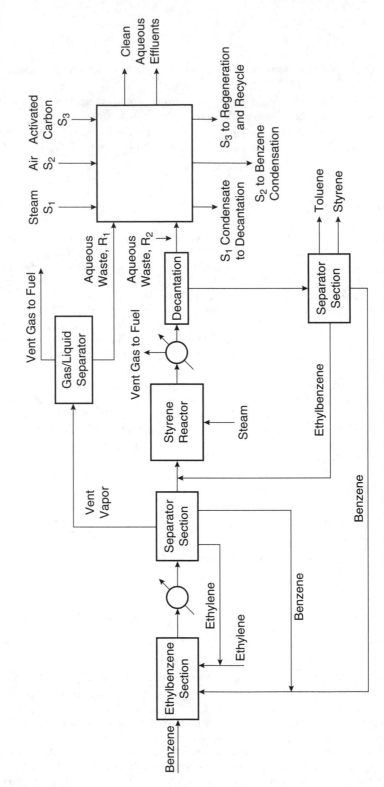

Problem 3, Figure 1. Flow diagram for styrene manufacturing facility

ethane which do not condense when cooled to ambient temperatures are vented. The condensed materials are sent to a decanter where an organic layer and an aqueous layer form. The organic layer containing benzene, toluene, and unreacted ethylbenzene is recycled. The aqueous portion of the wastewater stream (R_2) generated in the decanter is saturated with benzene and must be treated.

Another wastewater stream (R_1) is generated earlier in the process, during the manufacture of ethylbenzene. The ethylbenzene fed to the styrene manufacturing operation is generally produced on site from ethylene and benzene and produces a condensate wastewater stream that is saturated with benzene.

Thus, there are two sources of benzene saturated wastewater (approximately 1770 ppm or 0.00177 kg benzene/kg water) in the process: R_1 (1000 kg/hr) and R_2 (69,500 kg/hr). The concentrations of benzene must be reduced to at most 57 parts per billion in the wastewater.

Design a recycle and reuse system for the benzene using steam stripping and adsorption onto activated carbon. The system is shown in Problem 3, Figure 2. In the design:

- A steam stripping unit is used which removes benzene from the wastewater and recycles it. This unit produces a wastewater stream saturated with benzene (70 kg/h; see Problem 3, Figure 2) that must be sent to the recovery system.
- Spent activated carbon is regenerated with steam. The steam is recovered by condensation , producing a wastewater stream saturated with benzene (37 kg/h; see Problem 3, Figure 2). This wastewater is sent to the benzene recovery system.

a) Draw a composition interval diagram for the benzene rich streams.

b) Data on the lean stream compositions are given below. The target compositions are based on a variety of factors. For example, the supply composition of benzene in activated carbon corresponds to the residual benzene remaining on the carbon after regeneration. Using these data, draw the composition interval diagram for the lean streams.

Lean Stream	Supply composition (kg benzene/kg benzene free stream)	Target Composition (kg benzene/kg benzene free stream)
Steam	0	1.620
Activated Carbon	0.00003	0.200

c) Using the equilibrium relationships shown below, draw a pinch diagram for this system.

$$y_i = m_i \times x_i$$

where y_i and x_i are the mass ratios (kg benzene/kg benzene free stream) in the rich and lean streams. Assume m_i for steam is 0.001 and m_i for activated carbon is 0.00071.

d) Determine the location of the pinch point assuming no minimum driving force for mass transfer and calculate the amount of benzene recovered. What is the value of the recovered benzene if it can be sold for $0.20/kg? Repeat your calculation for a minimum mass transfer driving force of 0.001.

e) Compare your results to the flow rates shown in Problem 3, Figure 2 and discuss any differences.

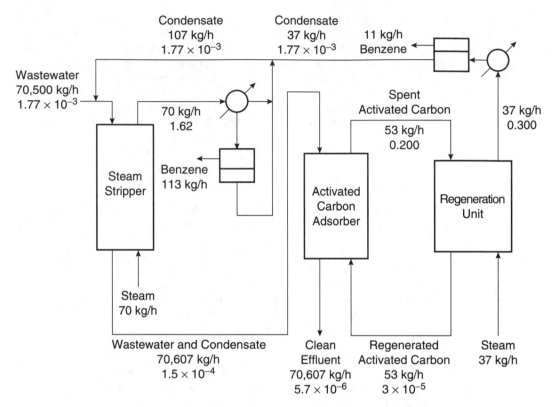

Problem 3, Figure 2.   Mass exchange network for the styrene facility. All flow rates are in kg of benzene free stream per hour; numbers without units represent mass ratios, in kg of benzene per kg of benzene free stream.

4. Etching of copper, using an ammoniacal solution, is an important operation in the manufacture of printed circuit boards. During etching, the concentration of copper in the ammoniacal solution increases. Etching is most efficiently carried out for copper concentrations between 10 and 13 weight% in the solution, while etching efficiency almost vanishes at higher concentrations (15–17 weight%). In order to maintain the etching efficiency, copper must be continuously removed from the spent ammoniacal solution through solvent extraction. The regenerated ammoniacal etchant can then be recycled to the etching line.

   The etched printed circuit boards are washed with water to dilute the concentration of the contaminants on the board surface to an acceptable level. The extraction of copper from the effluent rinse water is essential for both environmental and economic reasons, since decontaminated water is returned to the rinse vessel.

   A flowsheet of the etching process is shown in Problem 4, Figure 1.

   Design a mass exchange network that will recover copper from the rinse water and spent ammoniacal etchant. The characteristics of the rich streams are:

Rich stream 1:	$y_{in} = 0.13$	$y_{out} = 0.1$	Flowrate = 0.25 kg/s
Rich stream 2:	$y_{in} = 0.06$	$y_{out} = 0.02$	Flowrate = 0.02 kg/s

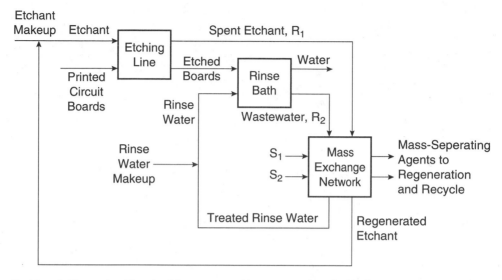

Problem 4, Figure 1. Flowsheet for copper etching process described in Problem 4.

Two lean streams are available. One lean stream for extracting copper is an aliphatic hydroxyoxime (S_1). It can be loaded to a copper mass fraction of 0.07. When regenerated, it has a copper mass fraction of 0.03. The equilibrium relationship between this lean stream and the two rich streams is given by:

$$y = 0.734\,x + 0.001$$

where y is the rich stream copper mass fraction and x is the lean stream copper mass fraction.

A second lean stream for extracting copper is an aromatic hydroxyomine (S_2). It can be loaded to a copper mass fraction of 0.2. When regenerated, it has a copper mass fraction of 0.001. The equilibrium relationship between this lean stream and the two rich streams is given by:

$$y = 1.5\,x + 0.1$$

where y is the rich stream copper mass fraction and x is the lean stream copper mass fraction.

a) Draw a composition interval diagram for the rich streams and the lean streams. Express all concentrations in units of the rich stream mass fractions.

b) Determine the maximum amount of copper that can be removed using S_1. What flow rate of S_1 does this correspond to?

c) Determine the maximum amount of copper that can be removed using S_2. What flow rate of S_2 does this correspond to?

d) Determine the maximum amount of copper that can be removed using both S_1 and S_2.

e) If the cost of the aromatic stream (S_2) is half the cost of the aliphatic stream (S_1), identify the optimal ratio of the two lean streams that still recovers the maximum amount of copper.

CHAPTER 11

Evaluating the Environmental Performance of a Flowsheet

by
David R. Shonnard

11.1   INTRODUCTION

After the process flowsheet has been established and energy and mass efficiency measures have been applied (Chapter 10), it is appropriate for a detailed environmental impact evaluation to be performed. The end result of the impact evaluation will be a set of environmental metrics (indexes) which represent the major environmental impacts or risks of the entire process. A number of indexes are needed to account for potential damage to human health and to several important environmental media. The indexes can be used in several important engineering applications during process design, including the ranking of technologies, the optimizing of in-process waste recycle/recovery processes, and the evaluation of the modes of reactor operation.

In quantitative risk assessment, it is shown that impacts are a function of dose, that dose is a function of concentration, and that concentration is a function of emission rate. Therefore, emissions from a process design flowsheet are the primary piece of information required for impact assessment. The concentrations in the relevant compartments of the environment (air, water, soil) are dependent upon the emissions and the location and chemical and physical properties of the pollutants. A suitable fate and transport model can transform the emissions into environmental concentrations. Finally, information regarding toxicity or inherent impact is required to convert the concentration-dependent doses into probabilities of harm (risk). Based on this understanding of risk assessment, the steps for environmental impact assessment are grouped into three categories, a) estimates of the rates of release for all chemicals in the process, b) calculation of environmental fate and transport and environmental concentrations, and c) the accounting for multiple measures of risk using toxicology and inherent environmental impact information.

Ideally, one would prefer to conduct a *quantitative risk assessment* when comparing the environmental performance of chemical process designs. Although this approach is preferred when the source and receptor are well defined and localized, it is not well suited for industrial releases that often affect not only local, but also regional and global environments. Also, the computing resources needed to perform a quantitative risk assessment for all release sources and receptors would tax the abilities of even the largest chemical manufacturer. A more achievable approach is to abandon quantitative risk assessment in preference to the assessment of *potential* environmental and health risks. The establishment of the potential impacts of chemical releases is sufficient for comparing the environmental risks of chemical process designs. In this chapter, material is presented which establishes methodologies for assessing the potential for environmental impacts of chemical processes and their designs.

In this development, we will utilize the concept of impact *benchmarking*. First introduced for the assessment of global warming and ozone-depletion potentials of refrigerants in the early 1990s, benchmarking takes the ratio of the environmental impact of a chemical's release to the impact of the identical release of a well-studied compound. A value greater than 1 for this dimensionless quantity indicates that the chemical has a greater potential for environmental impact than the benchmark compound. The product of the benchmarked environmental impact potential with the process emission rate results in the equivalent emission of the benchmark compound in terms of environmental impact. In this text, we adopt the benchmarking concept when assessing the environmental and toxicological impact potentials of releases from chemical processes. (Heijungs et. al. 1992)

Section 11.2 is a description of a multimedia compartment model approach for determining fate and transport of chemical releases into the environment. This model predicts the long-time and large-spatial scale distribution of chemicals using multiple compartments as the physical structure for the environment. Section 11.3 is a presentation of a Tier 3 environmental assessment (Tier 1 and Tier 2 assessments are discussed in Chapter 8) consistent with the goal of efficiently comparing chemical process designs.

11.2 ESTIMATION OF ENVIRONMENTAL FATES OF EMISSIONS AND WASTES

After a chemical is released into a compartment of the environment, either to the air, the surface water, or the soil, there are several transport and reaction processes that affect the ultimate concentrations in each of these compartments. There is also more than one modeling approach for calculating these concentrations.

Two important issues arise when choosing the type of environmental fate and transport model—accuracy and ease of use. Accuracy depends on how rigorously the model incorporates environmental processes into its description of mass transport and reaction. Ease of use relates to the data requirements and computational demands which the model places on the environmental assessment. One very common modeling approach used in environmental applications focuses on transport and fate processes occurring on only one compartment. The familiar atmospheric dispersion models for predicting air concentrations downwind from stationary sources (stacks) are examples of these single compartment models (Seinfeld and Pandis, 1997; Thibodeaux, 1996). Other very important examples are groundwater dispersion models that are used to predict concentration profiles in contaminated plumes downgrading from subsurface pollution sources (Fetter, 1993). The advantages of these models are that they require little chemical and environmental-specific data and provide relatively accurate results using modest computer resources. The disadvantage is that they provide concentration information in only one compartment, a severe limitation when multiple environmental impacts are under consideration. Several single compartment models can be linked together, thus providing multi-compartmental (multimedia) insights into environmental fate and transport. Unfortunately, the computational requirements are very large, making practical implementation difficult for routine chemical process evaluation.

The second major modeling approach is to use multimedia compartment models (MCMs), which predict chemical concentrations in several environmental compartments simultaneously (Mackay, 2001; McKone, 1994; Cohen et al., 1990). The advantages of MCMs are that they require modest data input, they are relatively simple and computationally efficient, and they account for several intermediate transport mechanisms and degradation. Limitations include a general lack of experimental data that can be used to verify their accuracy, and the general belief that they can provide only order-of-magnitude estimates of environmental concentrations (Mackay and Paterson, 1991). Keeping these limitations in mind, we present and demonstrate with example problems the Level III multimedia fugacity model of Mackay (2001). This model predicts the steady-state concentrations of a chemical in four environmental compartments (air (1), surface water (2), soil (3), and sediment (4)) in response to a constant emission into an environmental region of defined volume. The numbering convention used above for the compartments will be followed in the remainder of the text.

The surface area selected for the model is 10^5 km^2, about the area of Ohio, Greece, or England. The fraction of the area covered by water and by soil is a region-specific value, but typical values of 10% water and 90% land are often used (Mackay, 2001). The surface area of sediment is the same as the water compartment surface area. The atmosphere height is set at 1000 m, which is the typical height affected by pollutants emitted at the earth's surface. The depth of the water compartment is 20 m and those of the soil and sediment layers are assumed to be 10 cm and 1 cm, respectively. The atmosphere compartment contains a condensed (aerosol) phase having a volume fraction in air of 2×10^{-11} or about 30 μg/m^3. Though the aerosol phase is present in only a small amount, a large fraction of a chemical in the air may be associated with aerosols. The water compartment contains suspended sediments of volume fraction 5×10^{-6} (5 mg/L) and organic carbon content of 20%. Fish are included at a volume fraction of 10^{-6} and are assumed to contain 5% lipid into which hydrophobic chemical can partition. The soil compartment is assumed to contain 20% by volume of air, 30% water, and the remainder solids. The organic carbon content of soils is 2%. All of these parameters could be modified, if appropriate, but these values are reasonable and will be used in the examples presented in this chapter.

Figure 11.2-1 is a schematic diagram of the chemical processes occurring in the model domain which can affect the concentrations in each of the four compartments. Chemical may directly enter compartments by emissions (E_i) (moles/hr) and advective inputs ($G_{Ai}C_{Bi}$) (moles/hr). There is transfer of chemical between compartments by diffusive and non-diffusive processes characterized by intermediate transfer values (D_{ij}) (moles/(Pa•hr)). Chemical may enter or exit compartments by advective (bulk flow) mechanisms having a transfer value D_{Ai}, and chemical may disappear by reaction within each compartment having a loss value of D_{Ri}.

MCMs use the concept of fugacity in describing mass transfer and reaction processes. Fugacity is a thermodynamic property of a chemical and is defined as the "escaping tendency" of the chemical from a given environmental phase (air, water, soil organic matter, etc.). Thus for example, a volatile organic compound having a low water solubility will exhibit a large escaping tendency from the water phase (large water fugacity). Partitioning of a chemical between environmental phases can be described by the equilibrium criterion of equal fugacity f (Pa) in all phases. The fugacity is equal to partial pressure in the dilute limit typical of most environmental concentrations. Another feature of fugacity is that it is generally proportional to concentration, $C = fZ$, where Z is termed the fugacity capacity (Pa•m^3/mole) and C is the concentration (mole/m^3).

It is illustrative to develop expressions for relating fugacity to concentration in different environmental phases. We analyze seven phases: air, water, solids (soil, sediment, and suspended sediment), fish, and aerosol. Following that, we present several prominent intermediate transport mechanisms and reaction expressions for inclusion in the model. Finally, the model equations are presented and solved for the fugacity and ultimately the molar concentrations in each compartment.

11.2.1 Fugacity and Fugacity Capacity

Air Phase (1). The fugacity in the air phase is rigorously defined as

$$f = y \, \phi \, P_T \approx P$$

where y is the mole fraction of the chemical in the air phase, ϕ is the fugacity coefficient (dimensionless) which accounts for non-ideal behavior, and P_T is the total pressure (Pa). At the relatively low pressures (1 atm) encountered in the environment, $\phi = 1$, making the fugacity equal to the partial pressure (P) of the chemical in air. The concentration is related to partial pressure (and fugacity) by the ideal gas law,

$$C_1 = n/V = P/(RT) = f/(RT) = f Z_1$$

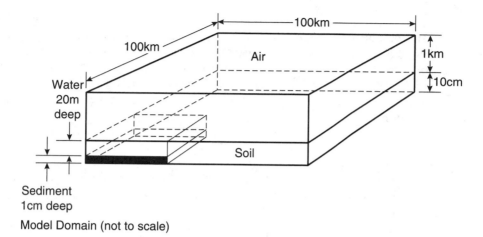

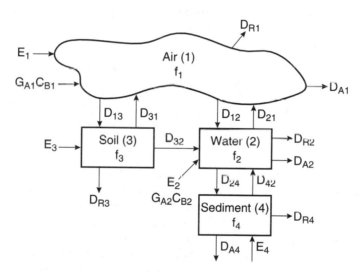

Figure 11.2-1 Schematic diagram of fugacity level III model domain and the intermedia transport mechanisms.

where n is moles of the chemical in a volume V (m³), R is the gas constant (8.31 Pa•m³/[mole•K]), and T is the absolute temperature (K). The air phase fugacity capacity Z_1 is $1/RT$ and has a value of 4.04×10^{-4} moles/(m³•Pa) at 25°C. The fugacity capacity is shown to be independent of chemical, being a constant value at a constant system temperature.

Water Phase (2). In aqueous solution, the fugacity of a chemical is

$$f = x\,\gamma\,P^s$$

where x is the mole fraction, γ is the activity coefficient in the Raoult's Law convention, and P^s is the saturation vapor pressure of pure liquid chemical at the system temperature (Pa). It can be shown that in most situations involving dilute concentrations typical of environmental problems, the activity coefficient is a constant (not varying with concentration (x)). Thus, there is a linear relationship between f and concentration in aqueous solution, C_2 (moles/m³). Rearranging the equation above results in the desired relationship,

$$C_2 = x/v_w = f/(v_w \gamma P^s) = f/H = f Z_2$$

where v_w is the molar volume of solution (water, 1.8×10^{-5} m³/mole), H is the Henry's Law constant for the chemical (Pa$\bullet$m³/mole), and Z_2 is the water fugacity capacity for each chemical, which is the inverse of the chemical's Henry's Law constant.

Soil Phase (3). Chemicals associated with the soil or sediment phases are almost always sorbed into the natural organic matter in the soil and are in equilibrium with the water phase concentration. A linear relationship has been observed between the sorbed concentration (C_s, moles/kg soil or sediment) and the aqueous concentration (C_2, moles/L solution).

$$C_s = K_d C_2$$

where K_d is the equilibrium distribution coefficient (L solution /kg solids) and the slope of the linear sorption isotherm. Because natural organic matter is composed mainly of carbon, the distribution coefficient is related to the fraction of organic carbon in the soil or sediment by

$$K_{oc} = K_d / \phi_3$$

where K_{oc} is the organic carbon-based distribution coefficient (L/kg) and ϕ_3 is the mass fraction of organic carbon in the soil phase (g organic carbon/g soil solids). The octanol-water partition coefficient has been shown to correlate very well with K_{oc} with the equality $K_{oc} = 0.41 K_{ow}$ being a good approximation for most chemicals (for other estimation methods, see Chapter 5). We find it convenient to relate the concentration per volume of the sorbed phase (C_3 for soil solids) to the fugacity by multiplying C_s by the phase density (ρ_3, kg solid/ m³ solid) and by relating C_2 to partial pressure (fugacity) through the Henry's Law constant.
Thus:

$$\rho_3 C_s = [1/H] K_{oc} \phi_3 \rho_3 f/1000 = Z_3 f$$

where the factor of 1000 is used to convert L to m³.

Similar expressions for fugacity capacities (Z) are obtained for the other environmental phases which make up the four environmental compartments. A summary of these equations is given in Table 11.2-1. Included in Table 11.2-1 are the fugacity capacities for each of the environmental compartments—the air, surface water, soils, and sediments—which are summations of individual phase Z values weighted by their respective volume fractions in the compartment.

Table 11.2-1   Fugacity Capacity (Z) Values for the Various Phases and Compartments in the Environment.

Environmental Phases		Phase Densities (kg/m³)
Air phase	$Z_1 = 1/(R\,T)$	1.2
Water phase	$Z_2 = 1/H$	1,000
Soil phase	$Z_3 = [1/H]\,K_{oc}\,\phi_3\,\rho_3/1000$	2,400
Sediment phase	$Z_4 = [1/H]\,K_{oc}\,\phi_4\,\rho_4/1000$	2,400
Suspended sediment phase	$Z_5 = [1/H]\,K_{oc}\,\phi_5\,\rho_5/1000$	2,400
Fish phase	$Z_6 = [1/H]\,0.048\,\rho_6\,K_{ow}$	1,000
Aerosol phase	$Z_7 = [1/(R\,T)]\,6\times10^6/P^S_L$	

where

R = gas constant (8.314 Pa• m³/[mole•K])
T = absolute temperature (K)
H = Henry's Law constant (Pa•m³/mole)
K_{oc} = organic-carbon partition coefficient ($= 0.41\,K_{ow}$)
K_{ow} = octanol-water partition coefficient
ρ_i = phase density for phase i (kg/m³)
ϕ_i = mass fraction of organic carbon in phase i (g/g)

Environmental Compartments

Air compartment (1)	$Z_{C1} = Z_1 + 2\times10^{-11}\,Z_7$	(approximately 30 µg/m³ aerosols)
Water compartment (2)	$Z_{C2} = Z_2 + 5\times10^{-6}\,Z_5 + 10^{-6}\,Z_6$	(5 ppm solids, 1 ppm fish by vol.)
Soil compartment (3)	$Z_{C3} = 0.2\,Z_1 + 0.3\,Z_2 + 0.5\,Z_3$	(20% air, 30% water, 50% solids)
Sediment compartment (4)	$Z_{C4} = 0.8\,Z_2 + 0.2\,Z_4$	(80% water, 20% solids)

Note: for solid aerosols $P^S_L = P^S_S / \exp\{6.79(1-T_M/T)\}$ where T_M is the melting point (K).
Adapted from Mackay et al. (1992).

11.2.2   Intermedia Transport

Chemicals move between environmental compartments by diffusive and non-diffusive processes. Diffusive processes, such as volatilization from water to air or soil to air, can proceed in more than one direction, depending on the sign of the fugacity difference between compartments. The diffusive rate of transfer N_{ij} (moles/h) from a compartment i to compartment j is defined by;

$$N_{ij} = D_{ij}(f_i) \text{ (moles/h)}$$

where D_{ij} (mole/(Pa•hr)) is an intermedia transport parameter for diffusion from compartment i to j and f_i is the fugacity in compartment i, serving to drive the chemical into adjoining compartments. A comparable expression exists for the molar rate of transfer from compartment j to i. The difference in these molar rates is the net rate of transfer between these compartments due to a disequilibrium in the compartmental fugacities. In parallel to diffusive transport is non-diffusive (one-way) transport between compartments, such as rain washout and wet/dry deposition of atmospheric particles to soil and water, and sediment deposition and resuspension. This transport can be described by

$$N = GC = GZf = Df \,(\text{moles/h})$$

where G (m³/h) is a volumetric flow rate of the transported material (rainwater, suspended sediment, etc.) and C (moles/m³) is its phase concentration. Figure 11.2-1 illustrates all of the intermediate diffusive and non-diffusive transport mechanisms within the model domain. In the following discussion, each intermediate transport parameter will be derived.

Air/Water Transport (D_{12} and D_{21})

Three processes are included in air-to-water transport: diffusion (absorption), washout by rain, and wet/dry deposition of aerosols. The conventional two-film approach is taken for absorption from air to water compartments through the atmosphere/water interface using air-side ($k_A = 5$ m/h) and water-side ($k_W = 0.05$ m/h) mass transfer coefficients. For the sake of organization, we rename the mass transfer coefficients as $k_A = u_1$ and $k_W = u_2$ and follow this convention for the remaining derivations. The intermediate transport parameter for absorption is

$$D_{VW} = 1/(1/(u_1 A_W Z_1) + 1/(u_2 A_W Z_2))$$

where A_W is the interfacial area between the atmosphere and the surface water. For rain washout, a rainfall rate u_3 of 0.876 m/yr (10^{-4} m/h) is assumed. The D value for rain washout is

$$D_{RW} = u_3 A_W Z_2$$

For wet and dry deposition of aerosols, the deposition velocity u_4 is taken to be the sum of these parallel transport mechanisms (6×10^{-10} m/h) and therefore the D value becomes

$$D_{QW} = u_4 A_W Z_7$$

Since these mechanisms operate in parallel, we can define a cumulative D value for the air-to-water transfer (D_{12}) as

$$D_{12} = D_{VW} + D_{RW} + D_{QW}$$

The water-to-air transport is just the reverse of the absorption mechanism described above and therefore, the D value for water-to-air transport (D_{21}) is

$$D_{21} = D_{VW}$$

Air-Soil Transport (D_{13} and D_{31})

For air-to-soil transport, identical treatments of rain washout (D_{RS}) and wet/dry deposition (D_{QS}) are taken as in the air-to-water transport case. The only difference is that the correct area term is the air/soil interface area, A_S. For diffusion from air to soil, the chemical must traverse a thin mass-transfer resistance film at the atmosphere/soil interface before diffusing through the soil air phase or the soil water phase, both of which have resistances of their own. The value of this mass transfer coefficient at the soil surface u_5 is the same as the air-side mass transfer coefficient for the atmosphere-water interface u_1 (5 m/h). Diffusion through the soil air or water phases is hampered by the presence of the soil solids, and as a result, the molecular diffusion coefficients of the chemical in either air or water decreases substantially. The Millington-Quirk relationship is employed as outlined in the work by Jury et al. (1983, 1984) to decrease the diffusion coefficients by a factor of about 20. Thus, the effective air diffusion coefficient becomes 0.05 * 0.02 m^2/h = 10^{-3} m^2/h and the effective water diffusion coefficient becomes 0.05 * 2×10^{-6} m^2/h = 10^{-7} m^2/h. The effective diffusion coefficients divided by the path length for diffusion in soil (half the soil depth, 0.05 m) yields the mass-transfer coefficients for diffusion in the soil water $u_6 = 2×10^{-6}$ m/h and soil air $u_7 = 0.02$ m/h. Downward flow of water in the soil pores is likely to result in a water transport velocity of about 10^{-5} m/h. Thus, u_6 is taken to be a larger value in order to account for this, 10^{-5} m/h. The soil diffusion processes in the air and water occur in parallel but are in series with the air film at the soil surface. The final equation for air-to-soil diffusion D value is

$$D_{VS} = 1/(1/D_S + 1/(D_{SW} + D_{SA}))$$

where

$$D_S = u_5 A_S Z_1$$
$$D_{SW} = u_6 A_S Z_2$$
$$D_{SA} = u_7 A_S Z_1$$

The total D value for all air-to-soil processes is given by

$$D_{13} = D_{VS} + D_{QS} + D_{RS}$$

For soil-to-air diffusion transport, the D_{31} value is equal to D_{VS}.

Water-Sediment Transport (D_{24} and D_{42})

Diffusion from the water column to the sediment is characterized by a mass-transfer coefficient u_8 or 10^{-4} m/h, which is the molecular diffusivity in water (2×10^{-6} m^2/h) divided by the path length (0.02 m). Ignored are the processes of bioturbation and shallow water current-induced turbulence which would increase u_8. The D value is $u_8 A_W Z_2$. Deposition of suspended sediment is assumed to occur at a rate of 5000 m^3/h over an area $A_W = 10^{10}$ m^2. Thus the suspended sediment deposition velocity u_9 is 5000 m^3/h /A_W = 5×10^{-7} m/h. The water to sediment D value is thus

$$D_{24} = u_8 A_W Z_2 + u_9 A_W Z_5$$

where Z_5 is the Z value for the suspended sediment.

Sediment to water is treated similarly to D_{24}. Resuspension is assumed to occur at a rate which is 40% that of deposition. Therefore, the resuspension velocity u_{10} is 2×10^{-7} m/h and the D value for sediment to water transfer is

$$D_{42} = u_8 A_W Z_2 + u_{10} A_W Z_4$$

Soil to Water (D_{32})

Soil-to-water transfer occurs by surface run-off. The rate of water run-off is assumed to occur at 50% the rate of rainfall. The run-off water velocity u_{11} is then $0.5\, u_3 = 5 \times 10^{-5}$ m/h. The solids contained in the run-off water are assumed to be at a volumetric concentration of 200 ppm in the water. The run-off solids velocity u_{12} is $200 \times 10^{-6}\, u_{11}$. The D value is

$$D_{32} = u_{11} A_S Z_2 + u_{12} A_S Z_3$$

where A_S is the soil surface area and Z_2 is the Z value for the water and Z_3 is the Z value for the soil solids. A summary of the intermediate transport parameters is shown in Table 11.2-2.

An additional non-diffusive transport mechanism which removes chemical from the sediment is burial. The D value (D_{A4}) is equal to

$$D_{A4} = u_B A_W Z_4,$$

where u_B is the sediment burial rate (2×10^{-7} m/h).

Advective Transport

Chemical may directly enter into compartments by emissions and advective inputs from outside the model region. The total rate of inputs for each compartment i is

$$I_i = E_i + G_{Ai} C_{Bi}$$

where E_i (moles/h) is the emission rate, G_{Ai} (m^3/h) is the advective flow rate, and C_{Bi} (moles/ m^3) is the background concentration external to compartment i. Chem-

Table 11.2-2 Intermedia Diffusive and Non-Diffusive Mass Transfer Coefficients (meters/hr).

u_1	air-side mass-transfer coefficient	5
u_2	water-side mass-transfer coefficient	0.05
u_3	rainfall rate	10^{-4}
u_4	wet/dry aerosol deposition velocity	6×10^{-10}
u_5	soil air phase diffusion mass-transfer coefficient	0.02
u_6	soil water phase diffusion mass-transfer coefficient	10^{-5}
u_7	soil air surface mass-transfer coefficient	5
u_8	sediment water diffusion mass-transfer coefficient	10^{-4}
u_9	suspended sediment deposition velocity	5×10^{-7}
u_{10}	sediment resuspension velocity	2×10^{-7}
u_{11}	soil water run-off velocity	5×10^{-5}
u_{12}	soil suspended solids run-off velocity	10^{-8}

icals may also exit the model domain from compartments by advective (bulk flow) processes having transfer values (D_{Ai})

$$D_{Ai} = G_{Ai}Z_{Ci}$$

where Z_{Ci} is the compartment i fugacity capacity (Table 11.2-1).

11.2.3   Reaction Loss Processes

Reaction processes occurring in the environment include biodegradation, photolysis, hydrolysis, and oxidation. A good approximation for reaction processes in the dilute limit commonly found in the environment is to express them as first order with rate constant k_R (hr^{-1}). The rate of reaction loss for a chemical in a compartment N_{Ri} (moles/hr) is

$$N_{Ri} = k_{Ri} V_i C_i = k_{Ri} V_i Z_{Ci}f = D_{Ri}f$$

V_i is the compartment volume, C_i is the molar concentration of the chemical, and i refers to a specific compartment. Rate constants are compound-specific and have been tabulated for several compounds in the form of a reaction half-life $t_{1/2}$, defined as the time required for the concentration to be reduced by one half of the initial by reaction (Mackay et al., 1992; also see Chapter 5). Tabulated half lives for compounds might represent the combined reaction mechanisms listed above, which can occur simultaneously in a given compartment. The relationship between $t_{1/2}$ and k_R for a first order reaction is

$$k_R = -\ln(0.5) / t_{1/2}$$

A summary of the D values for intermedia transport, advection, and reaction is shown in Table 11.2-3.

Table 11.2-3 D Values in the Mackay Level III Model (Adapted from Mackay and Paterson, 1991).

Compartment	Process	Individual D	Total D
air (1) water (2)	diffusion	$D_{VW} - 1/(1/(u_1A_WZ_1) + 1/(u_2A_WZ_2))$	
	rain washout	$D_{RW} = u_3A_WZ_2$	$D_{12} = D_{VW} + D_{RW} + D_{QW}$
	wet/dry deposition	$D_{QW} = u_4A_WZ_7$	$D_{21} = D_{VW}$
air (1)-soil (3)	diffusion	$D_{VS} = 1/(1/(u_5A_SZ_1) +$	
		$1/((u_6A_SZ_2) + (u_7A_SZ_1)))$	
	rain washout	$D_{RS} = u_3A_SZ_2$	$D_{13} = D_{VS} + D_{QS} + D_{RS}$
	wet/dry deposition	$D_{QW} = u_4A_SZ_7$	$D_{31} = D_{VS}$
water (2)-sediment (4)	diffusion	$u_8A_WZ_2$	$D_{24} = u_8A_WZ_2 + u_9A_WZ_5$
	deposition	$u_9A_WZ_5$	
sediment (4)-water (2)	diffusion	$u_8A_WZ_2$	$D_{42} = u_8A_WZ_2 + u_{10}A_WZ_4$
	resuspension	$u_{10}A_WZ_4$	
soil (3)-water (2)	water runoff	$u_{11}A_SZ_2$	$D_{32} = u_{11}A_SZ_2 + u_{12}A_SZ_3$
	soil runoff	$u_{12}A_SZ_3$	$D_{23} = 0$
advection (bulk flow)	emissions and		
	bulk flow in	$I_i = E_i + G_{Ai}C_{Bi}$	for compartment i
	bulk flow out	$D_{Ai} = G_{Ai}Z_{Ci}$	
reaction		$D_{Ri} = k_{Ri} V_i Z_{Ci}$	for compartment i

Table 11.2-4   Mole Balance Equations for the Mackay Level III Fugacity Model.

Air	$I_1 + f_2 D_{21} + f_3 D_{31} = f_1 D_{T1}$
Water	$I_2 + f_1 D_{12} + f_3 D_{32} + f_4 D_{42} = f_2 D_{T2}$
Soil	$I_3 + f_1 D_{13} = f_3 D_{T3}$
Sediment	$I_4 + f_2 D_{24} = f_4 D_{T4}$

where the lefthand side is the sum of all gains and the righthand side is the sum of all losses, $I_i = E_i + G_{Ai} C_{Bi}$, I_4 usually being zero. The D values on the right hand side are:

$$D_{T1} = D_{R1} + D_{A1} + D_{12} + D_{13}$$
$$D_{T2} = D_{R2} + D_{A2} + D_{21} + D_{24}$$
$$D_{T3} = D_{R3} + D_{A3} + D_{31} + D_{32}$$
$$D_{T4} = D_{R4} + D_{A4} + D_{42}$$

The solution for the unknown fugacities in each compartment is:

$$f_2 = (I_2 + J_1 J_4 / J_3 + I_3 D_{32} / D_{T3} + I_4 D_{42} / D_{T4}) / (D_{T2} - J_2 J_4 / J_3 - D_{24} D_{42} / D_{T4})$$
$$f_1 = (J_1 + f_2 J_2) / J_3$$
$$f_3 = (I_3 + f_1 D_{13}) / D_{T3}$$
$$f_4 = (I_4 + f_2 D_{42}) / D_{T4}$$

where
$$J_1 = I_1 / D_{T1} + I_3 D_{31} / (D_{T3} D_{T1})$$
$$J_2 = D_{21} / D_{T1}$$
$$J_3 = 1 - D_{31} D_{13} / (D_{T1} D_{T3})$$
$$J_4 = D_{12} + D_{32} D_{13} / D_{T3}$$

11.2.4   Balance Equations

As indicated in Figure 11.2-1, there must be a balance between the rates of input from all emissions/bulk flow and intermedia transport and the rates of output from intermedia transport, advection, and reaction loss processes within each compartment at steady-state. We write mole balance equations for each compartment as summarized in Table 11.2-4.

The fugacity calculations outlined in the previous pages are obviously very complex. Routine hand calculations of environmental fugacities using this model are prohibitively time consuming. Fortunately, spreadsheet programs are available for carrying out these calculations (Mackay et al., 1992, Volume 4). Using these programs and equipped with a relatively small number of chemical-specific input partitioning and reaction parameters, environmental fate calculations can be quickly performed as shown in the following example problem.

Example 11.2-1   Multimedia Concentrations of Benzene, Ethanol, and Pentachlorophenol.

Benzene, ethanol, and pentachlorophenol (PCP) are examples of organic pollutants with very different environmental properties, as shown in Table 11.2-5. Benzene and ethanol are volatile (high vapor pressures) and have comparatively short reaction half-lives. Pentachlorophenol has a long reaction half live, low volatility and water solubility, and strong sorptive properties (high K_{ow}). Benzene is the most reactive in air and ethanol is the most reactive in water, soil, and sediment.

Table 11.2-5   Data Entry Values for the Mackay Level III Spreadsheet

Environmental Property	Unit	Location in Spreadsheet	Benzene	Ethanol	PCP
Molecular weight	g/mole	C6	78.11	46.07	266.34
Melting point	°C	C7	5.53	−115	174
Dissociation constant	log pK$_a$	C8			4.74
Solubility in water	g/m^3	C11	1.78×10^2	6.78×10^5	14
Vapor pressure	Pa	C12	1.27×10^4	7.80×10^3	4.15×10^{-3}
Octanol-water coefficient	log K$_{ow}$	C13	2.13	−0.31	5.05
Half-life in air	hr	C33	1.7×10	5.5×10	5.50×10^2
Half-life in water	hr	C34	1.7×10^2	5.5×10	5.50×10^2
Half-life in soil	hr	C35	5.5×10^2	5.5×10	1.7×10^3
Half-life in sediment	hr	C36	1.7×10^3	1.7×10^2	5.50×10^3

Use the Mackay level III spreadsheet to determine the amounts of each chemical (moles) and their percentages in the four environmental compartments at steady-state for three distinct emissions scenarios:

a) 1000 kg/hr emitted into the air only.
b) 1000 kg/hr emitted into the water only.
c) 1000 kg/hr emitted into the soil only.

Discuss the compartmental distributions and the total amount of each chemical in the model domain in light of the environmental property data presented above.

Solution: After entering the environmental properties for each chemical in the tabulated spreadsheet locations, one can have the spreadsheet recalculate the resulting environmental fugacities, molar concentrations, and finally molar amounts in each compartment. For emission into air, locations F276–F279 contain the amounts in the four compartments; air, water, soil, and sediment. Locations G276–G279 contain the corresponding percentages in these compartments. Similar results are contained in rows 286–289 for emission into water and in rows 296–299 for emission into soil. Table 11.2-5 highlights these results for all three emission scenarios and for each of the three chemicals.

Discussion of Results: There are several key items to summarize from Table 11.2-6, all of which will help us interpret how the model performs. First, the majority of the chemical can be found in the compartment into which the chemical was emitted. The percentages in each compartment relay this information. The only exception is for PCP when emitted into the air. The chemical has such a low vapor pressure (4.15×10^{-3} Pa) that rain washout and wet/dry deposition effectively remove it from the atmosphere, leading to accumulation in the soil. Second, the total amount of the chemical in each compartment of the environment increases with increasing reaction half-life, as shown by the relatively large amounts of PCP compared to benzene and ethanol. Note that PCP has relatively large values of reaction half-life in each compartment compared to the other two chemicals.

Further use of the Mackay level III model will occur in Section 11.3, where environmental concentrations will be incorporated into health risk assessment of chemical process designs.

Table 11.2-6 Environment Compartment Molar Amounts and Percentages for Benzene, Ethanol, and PCP (pentachlorophenol).

Chemical	Amounts (moles)				Total	Percentages (%)			
(emission scenario)	air	water	soil	sediment	(moles, kg)	air	water	soil	sediment
Benzene (a)	2.52×10^5	7.46×10^1	4.15×10^1	2.67×10^{-1}	2.52×10^5, 1.97×10^4	99.95	0.03	0.02	0.00
Benzene (b)	8.29×10^4	1.69×10^6	1.36×10^1	6.05×10^3	1.78×10^6, 1.39×10^5	4.66	95.00	0.00	0.34
Benzene (c)	2.49×10^5	1.94×10^3	1.17×10^5	6.94×10^0	3.68×10^5, 2.87×10^4	67.63	0.53	31.84	0.00
Ethanol (a)	9.20×10^5	3.81×10^4	3.23×10^4	2.89×10^1	9.90×10^5, 4.56×10^4	92.89	3.84	3.27	0.00
Ethanol (b)	3.52×10^3	1.59×10^6	1.24×10^2	1.21×10^3	1.59×10^6, 7.32×10^4	0.22	99.70	0.01	0.07
Ethanol (c)	1.58×10^4	9.64×10^4	1.59×10^6	7.30×10^1	1.70×10^6, 7.83×10^4	0.93	5.65	93.42	0.00
PCP (a)	1.98×10^4	1.99×10^5	7.56×10^6	7.80×10^3	7.79×10^6, 2.07×10^6	0.25	2.55	97.09	0.11
PCP (b)	1.00×10^0	1.66×10^6	4.77×10^2	6.51×10^4	1.72×10^6, 4.58×10^5	0.00	96.19	0.03	3.78
PCP (c)	2.61×10^1	4.95×10^4	8.93×10^6	1.94×10^3	8.99×10^6, 2.39×10^6	0.00	0.55	99.43	0.02

11.3   TIER 3 METRICS FOR ENVIRONMENTAL RISK EVALUATION OF PROCESS DESIGNS

In this section, we learn how to combine emissions estimation, environmental fate and transport information, and environmental impact data to obtain an assessment of the potential risks posed by releases from chemical process designs. This methodology will be applied in a systematic manner for the quantitative evaluation of a completed flowsheet for a chemical process design. We use the multimedia compartment model described in Section 11.2 to calculate environmental concentrations that are used by several of the indices. Although no single methodology has gained universal acceptance, several useful methodologies for indexing environmental and health impacts of chemicals have recently appeared in the literature. Many of the indexing methods include metrics for abiotic as well as biotic impacts. In the abiotic category, global warming, stratospheric ozone depletion, acidification, eutrification, and smog formation are often included. In the biotic category, human health and plant, animal, and other organism health are impacts of concern. For issues of environmental and economic sustainability, resource-depletion indexes reflect long-term needs for raw materials use. A review of several of these methodologies would indicate that many environmental metrics (indexes) have been constructed by employing separate parameters for the inherent impact potential (IIP) and exposure potential (EP) of an emitted chemical. The index is normally expressed as a product of inherent impact and exposure, following risk assessment guidelines (NRC, 1983; Heijungs et al., 1992; SETAC, 1993), although summation-based indexes have also been used (Davis et al., 1994; Mallick et al., 1996).

In this text, we define and use nine environmental and health-related indexes for chemical process impacts, as shown in Table 11.3-1. These impacts affect local, regional, and global environmental issues. Global warming and stratospheric ozone depletion are problems with potentially global implications for a large proportion of the earth's population. Smog formation and acid deposition are regional problems that can affect areas in size ranging from large urban basins up to a significant fraction of a continent. Issues of toxicity and carcinogenicity are often of highest concern at the local scale in the vicinity of the point of release.

The general form of a dimensionless environmental risk index is defined as;

$$(Dimensionless\ Risk\ Index)_i = \frac{[(EP)(IIP)]_i}{[(EP)(IIP)]_B} \qquad \text{(Eq. 11-1)}$$

Table 11.3-1   Environmental Impact Index Categories for Process Flowsheet Evaluation.

Abiotic Indexes	Health-Related Indexes	Ecotoxicity Indexes
Global warming	Inhalation toxicity	Fish Aquatic Toxicity
Stratospheric ozone depletion	Ingestion toxicity	
Acid deposition	Inhalation carcinogenicity	
Smog formation	Ingestion carcinogenicity	

where B stands for the benchmark compound and i the chemical of interest. To estimate the index I for a particular impact category due to all of the chemicals released from a process, we must sum the contributions for each chemical weighed by their emission rate.

$$I = \sum_i (Dimensionless\ Risk\ Index)_i \times m_i \qquad \text{(Eq. 11-2)}$$

The following is a brief summary of environmental and health indexes which have been used to compare impacts of chemicals, processes, or products.

11.3.1 Global Warming

A common index for global warming is the global warming potential (GWP), which is the cummulative infrared energy capture from the release of 1 kg of a greenhouse gas relative to that from 1 kg of carbon dioxide (IPCC, 1991):

$$GWP_i = \frac{\int_0^n a_i C_i dt}{\int_0^n a_{CO_2} C_{CO_2} dt} \qquad \text{(Eq. 11-3)}$$

where a_i is the predicted radiative forcing of gas i (Wm^{-2}) (which is a function of the chemical's infrared absorbance properties and C_i), C_i is its predicted concentration in the atmosphere (ppm), and n is the number of years over which the integration is performed, for example, 100 years. The concentration is a function of time (t), primarily due to loss within the troposphere by chemical reaction with hydroxyl radicals. For carbon dioxide, n = 120 years. Several authors have developed models to calculate GWP and as a result, some variation in GWP predictions have appeared (Fisher, 1990a; Derwent, 1990; Lashof & Ahuja, 1990; Rotmans, 1990). A list of "best estimates" for GWPs has been assembled from these model predictions by a panel of experts convened under the Intergovernment Panel on Climate Change (IPCC, 1991 and 1996) and have appeared on separate lists (Heijungs et al., 1992; Goedkoop, 1995).

In Appendix D, Table D-1 is a list of global warming potentials for several important greenhouse gases. The global warming potential for each chemical is influenced mostly by the chemical's tropospheric residence time and the strength of its infrared radiation absorbence (band intensities). All of these gases are extremely volatile, do not dissolve in water, and do not adsorb to soils and sediments. Therefore, they will persist in the atmosphere after being released from sources. The product of the GWP and the mass emission rate of a greenhouse chemical results in the equivalent emission of carbon dioxide, the benchmark compound. The global warming index for the entire chemical process is the sum of the emissions-weighted GWPs for each chemical,

$$I_{GW} = \sum_i (GWP_i \times m_i) \qquad\qquad \text{(Eq. 11-4)}$$

where m_i is the mass emission rate of chemical i from the entire process (kg/hr). This step will provide the equivalent process emissions of greenhouse chemicals in the form of the benchmark compound, CO_2.

The global warming index as calculated above accounts for direct effects of the chemical, but most chemicals of interest are so short-lived in the atmosphere (due to the action of hydroxyl radicals in the troposphere) that they disappear (become converted to CO_2) long before any significant direct effect can be felt. However, organic chemicals *of fossil fuel origin* will have an indirect global warming effect because of the carbon dioxide released upon oxidation within the atmosphere and within other compartments of the environment. In order to account for this indirect effect for organic compounds with atmospheric reaction residence times *less than 1/2 year,* an indirect GWP is defined (Shonnard and Hiew, 2000) as

$$GWP_i(indirect) = N_C \frac{MW_{CO_2}}{MW_i} \qquad\qquad \text{(Eq. 11-5)}$$

where N_C is the number of carbon atoms in the chemical i and the molecular weights MW convert from a molar to a mass basis for GWP, as originally defined. Organic chemicals whose origin is in renewable biomass (plant materials) have no global warming impact because the CO_2 released upon environmental oxidation of these compounds is, in principle, recycled into biomass within the natural carbon cycle.

Example 11.3-1   Global Warming Index for Air Emissions of 1,1,1-Trichloroethane from a Production Process

1,1,1-Trichloroethane (TCA) is used as an industrial solvent for metal cleaning, as a reaction intermediate, and for other important uses (US EPA, 1979-1991). A major processing route for TCA is by hydrochlorination of vinyl chloride in the presence of an $FeCl_3$ catalyst to produce 1,1-dichloroethane, followed by chlorination of this intermediate. Sources for air emissions include distillation condenser vents, storage tanks, handling and transfer operations, fugitive sources, and secondary emissions from wastewater treatment. We wish to estimate the global warming impact of the air emissions from this process, including direct impacts to the environment (from 1,1,1-TCA) and indirect impacts from energy usage (CO_2 and NOx release) in the analysis. Data below show the major chemicals that impact global warming when emitted from the process.

Determine the global warming index for the process and the percentage contribution for each chemical.

Data: Air Emissions (15,500 kg 1,1,1-TCA/hr)

Chemical	m_i (kg/hr)	GWP_i
TCA	10	100
CO_2	7,760	1
N_2O	.14	310

TCA emissions were estimated using data for trichloroethylene (US EPA, 1979-1991).

CO_2 and N_2O emission rates were estimated from a life cycle assessment of ethylene production (Allen and Rosselot, 1997; Boustead, 1993).

Solution: Using Equation 11.3-4, the process global warming index is

$$I_{GW} = (10 \text{ kg/hr})(100) + (7,760 \text{ kg/hr})((1) + (.14 \text{ kg/hr})(310)$$

$$= 1,000 + 7,760 + 43.4$$

$$= 8,803.4 \text{ kg/hr}$$

The percent of the process I_{GW} for each chemical is;

1,1,1-TCA: (1,000/8,803.4)×100 =	11.4%
CO_2: (7,760/8,803.4)×100 =	88.1%
N_2O: (43.4/8,803.4)×100 =	0.5%

Discussion: This case study demonstrates that the majority of the global warming impact from the production of 1,1,1-TCA is from the energy requirement of the process and not from the emission of the chemical with the highest global warming potential. This analysis assumes that a fossil fuel was used to satisfy the energy requirements of the process. If renewable resources were used (biomass-based fuels), the impact of CO_2 on global warming would be significantly reduced. Finally, the majority of the global warming impact of 1,1,1-TCA could very well be felt during the use stage of its life cycle, not the production stage. A complete life cycle assessment (see Chapter 13) of 1,1,1-TCA is necessary to demonstrate this.

11.3.2   Ozone Depletion

The ozone depletion potential (ODP) of a chemical is the predicted time- and height-integrated change $\delta[O_3]$ in stratospheric ozone caused by the release of a specific quantity of the chemical relative to that caused by the same quantity of a benchmark compound, trichlorofluoromethane (CFC-11, CCl_3F) (Fisher et al., 1990b).

$$ODP_i = \frac{\delta[O_3]_i}{\delta[O_3]_{CFC-11}} \qquad \text{(Eq. 11-6)}$$

Model calculations for ODP have been carried out using one- and two-dimensional photochemical models. A list of ODPs for a small number of chemicals has been assembled by a committee of experts (WMO, 1990b and 1992b) and have appeared on separate lists (Heijungs et al., 1992; Goedkoop, 1995). The product of the ODP and the mass emission rate of a chemical i results in the equivalent impact of an emission of CFC-11. Appendix D, Table D-2 hows a list of ozone depletion potential values for important industrial compounds. Data on the tropospheric reaction lifetimes (τ), stratospheric atomic oxygen reaction rate constant (k), and number of chlorines in each molecule (X) are also listed. Notice that the brominated compounds in this table have much larger ODPs than the chlorinated species. Also, it is thought that fluorine does not contribute to ozone depletion

(Ravishankara, et al., 1994). Like the global warming chemicals in Appendix D Table D-1, the chemicals in Appendix D, Table D-2 will exist almost exclusively in the atmosphere after being emitted by sources. The ozone depletion index for an entire chemical process is the sum of all contributions from emitted chemicals multiplied by their emission rates. The equivalent emission of CFC-11 for the entire process is then;

$$I_{OD} = \sum_i (ODP_i \times m_i) \qquad \text{(Eq. 11-7)}$$

11.3.3   Acid Rain

The potential for acidification for any compound is related to the number of moles of H^+ created per number of moles of the compound emitted. The balanced chemical equation can provide this relationship;

$$X \quad + \quad \bullet\bullet\bullet\bullet\bullet \quad \rightarrow \quad \alpha H^+ \quad + \quad \bullet\bullet\bullet\bullet\bullet \qquad \text{(Eq. 11-8)}$$

where X is the emitted chemical substance that initiates acidification and α (moles H^+/mole X) is a molar stoichiometric coefficient. Acidification is normally expressed on a mass basis and therefore the H^+ created per mass of substance emitted (η_i, moles H^+/kg i) is:

$$\eta_i = \frac{\alpha_i}{MW_i} \qquad \text{(Eq. 11-9)}$$

where MW_i is the molecular weight of the emitted substance (moles i/kg i). As before, we can introduce a benchmark compound (SO_2) and express the acid rain potential (ARP_i) of any emitted acid-forming chemical relative to it (Heijungs et al., 1992):

$$ARP_i = \frac{\eta_i}{\eta_{SO_2}} \qquad \text{(Eq. 11-10)}$$

The number of acidifying compounds emitted by industrial sources is limited to a rather small number of combustion byproducts and other precursor or acidic species emitted directly onto the environment. Appendix D, Table D-3 lists the acid rain potentials for several common industrial pollutants. The total acidification potential of an entire chemical process is defined similarly to I_{GW} and I_{OD}.

$$I_{AR} = \sum_i (ARP_i \times m_i) \qquad \text{(Eq. 11-11)}$$

11.3.4   Smog Formation

The most important process for ozone formation in the lower atmosphere is photo-dissociation of NO_2:

$$NO_2 + hv \longrightarrow O(^3P) + NO$$

$$O(^3P) + O_2 + M \longrightarrow O_3 + M$$

$$O_3 + NO \longrightarrow NO_2 + O_2$$

where M is nitrogen or molecular oxygen. This cycle results in O_3 concentration being in a photostationary state dictated by the NO_2 photolysis rate and ratio of $[NO_2]/[NO]$. The role of VOCs in smog formation is to form radicals which convert NO to NO_2 without causing O_3 destruction, thereby increasing the ratio $[NO_2]/[NO]$, and increasing O_3.

$$VOC + \bullet OH \longrightarrow \bullet RO_2 + \text{other oxidation products}$$

$$\bullet RO_2 + NO \longrightarrow NO_2 + \text{radicals}$$

$$\text{radicals} \longrightarrow \bullet OH + \text{other oxidation products}$$

The tendency of individual VOCs to influence O_3 levels depends upon its hydroxyl radical ($\bullet OH$) rate constant and elements of its reaction mechanism, including radical initiation, radical termination, and reactions which remove NOx. Simplified smog formation potential indexes have been proposed based only on VOC hydroxyl radical rate constants, but these have not correlated well with model predictions of photochemical smog formation (Allen, et al., 1992; Japar, et al., 1991).

Incremental reactivity (IR) has been proposed as a method for evaluating smog formation potential for individual organic compounds. It is defined as the change in moles of ozone formed as a result of emission into an air shed of one mole (on a carbon atom basis) of the VOC (Carter and Atkinson, 1989). Several computer models have been developed to evaluate incremental reactivity (Bufalini and Dodge, 1983; Carter and Atkinson, 1989; Carter, 1994; Chang and Rudy, 1990; Dodge, 1984). In general, predicted VOC incremental reactivities are greatest when NOx levels are high relative to reactive organic gases (ROG) and lowest (or even negative) when NOx is relatively low. Therefore, the ratio ROG/NOx is an important model parameter. Lists of incremental smog formation reactivities for many VOCs have been compiled (Carter, 1994; Heijungs et al., 1992). An estimation methodology has also been developed which circumvents the need for computer model predictions, though the practical use of this method is limited due to lack of detailed smog reaction mechanisms for a large number of compounds (Carter and Atkinson, 1989). Although several reactivity scales are possible, the most relevant for comparing VOCs is the maximum incremental reactivity (MIR), which occurs under high NOx conditions when the highest ozone formation occurs (Carter, 1994).

The smog formation potential (SFP) is based on the maximum incremental reactivity scale of Carter (Carter, 1994):

$$SFP_i = \frac{MIR_i}{MIR_{ROG}} \qquad \text{(Eq. 11-12)}$$

where MIR_{ROG} is the average value for background reactive organic gases, the benchmark compound for this index. This normalized and dimensionless index is similar to the one proposed by the Netherlands Agency for the Environment (Heijungs et al., 1992). Appendix D, Table D-4 contains a listing of calculated MIR values for many common volatile organic compounds found in fuels, paints, and solvents. Most of the chemicals in Appendix D, Table D-4 are volatile and will maintain a presence in the atmosphere after release into the air, with the exception of the higher molecular weight organics. The total smog formation potential is the sum of the MIRs and emission rates for each smog-forming chemical in the process. The process equivalent emission of ROG is;

$$I_{SF} = \sum_i (SFP_i \times m_i) \qquad \text{(Eq. 11-13)}$$

Example 11.3-2   Solvent Recovery from a Gaseous Waste Stream: Effect of Process Operation on Indexes for Global Warming, Smog Formation, and Acidification

A gaseous waste stream is generated within a plastic film processing operation from a drying step. The stream (12,000 scfm) is currently being vented to the atmosphere and it contains 0.5% (vol.) of total VOCs having equal mass percentages of toluene and ethyl acetate with the balance being nitrogen. Figure 11.3-1 is a process flow diagram of an absorption technology configuration to recovery and recycle the VOCs back to

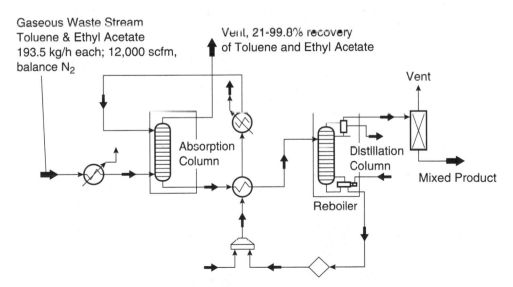

Figure 11.3-1   Schematic diagram (from HYSYS, Hyprotech, Calgary, Canada) of a solvent recovery and recycle process using absorption into heavy oil (n-tetradecane) followed by distillation.

the film process (Sangwichien, 1998). Since the waste stream may already meet environmental regulations for smog formation and human toxicity, the key issue is how much of the VOCs to recover and how much savings on solvent costs can be realized. In this problem, we do not deal with the economic issues, but rather show that when considering environmental impacts, there are trade-offs for several impacts depending on the percent recovery of the VOCs.

The gaseous waste stream enters the absorption column where the VOCs (toluene and ethyl acetate) transfer from the gas phase to the absorption oil (tetradecane). The effectiveness of this transfer depends largely on the oil flow rate, as the percent recovery of VOCs increases with increasing oil flow rate. The VOCs are separated from the absorption oil in the distillation column and the oil is then recycled back to the absorption column after cooling. The VOCs are recovered as a mixed product from the condenser of the distillation column and stored in a tank for re-use in the plastic film process. The main emission sources are the absorption column, the vent on the distillation column, the vent on the storage tank (not shown), utility related pollutants, and fugitive sources.

Solution: Table 11.3-2 shows the effect of absorber oil flow rate on the emissions from the solvent recovery process. A commercial process simulator (HYSYS) was used to generate mass and energy balances and to calculate the VOC emission rates from the absorber unit. Within the Environmental Fate and Risk Assessment Tool (EFRAT,[©] refer to Appendix F for a list of software resources.) EPA emission factors and correlations were used to calculate VOC emission rates from the distillation column, storage tank, and fugitive sources. CO_2, CO, TOC, NOx, and SOx emission rates were also calculated within EFRAT based on the energy requirements of the process and an assumed fuel type (fuel oil no. 4). Figure 11.3-2 shows the recovery of toluene and ethyl acetate as a function of absorption oil flow rate in the process. As the absorber oil flow rate is increased, the emissions of toluene and ethyl acetate from the absorber unit decrease, reflecting an increased percent recovery from the gaseous

Table 11.3-2   Air Emission Rates of Chemicals From the Solvent Recovery Process of Figure 11.3-1 (Adapted from Hiew, 1998).

Absorber Oil Flow Rate (kgmol/hr)	Emission Rate (kg/hr)							
	Toluene	Ethyl Acetate	CO_2	CO	TOC	NOx	SOx	n-C14
0	193.55	193.55	0	0.0	0.0	0.0	0.0	0.0
10	119.87	185.87	37	0.013	0.001	0.05	0.41	4.28
20	53.11	178.37	74	0.027	0.001	0.11	0.81	4.83
50	0.97	160.4	183	0.066	0.003	0.26	1.99	4.67
100	0.02	128.07	360	0.129	0.007	0.52	3.39	4.23
200	0.02	59.95	714	0.257	0.013	1.03	7.82	4.13
300	0.02	12.87	1,067	0.385	0.019	1.54	11.69	4.06
400	0.03	1.70	1,420	0.512	0.026	2.05	15.56	4.05
500	0.03	0.27	1,773	0.639	0.032	2.56	19.42	4.04

Adapted from Hiew (1998), using EFRAT[©] and HYSIS[©]. See Appendix F for a list of software resources.

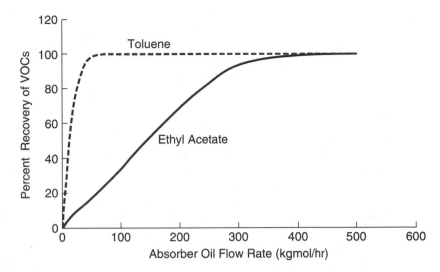

Figure 11.3-2   VOC recovery efficiency for the solvent recovery process of Figure 11.3-1. Adapted from Hiew (1998) and Shonnard and Hiew (2000).

waste stream. Most of the toluene (99.5%) is recovered at a flow rate of only 50 kg-moles/hr. To recover a significant percentage of ethyl acetate requires a much larger oil flow rate. Toluene is recovered more quickly with oil flow rate compared to ethyl acetate because the oil is more selective towards toluene. Emissions of the utility-related pollutants (CO_2, CO, TOC, NOx, and SOx) increase in proportion to the oil flow rate. The emissions of the absorption oil (n C14) remains relatively constant with oil flow rate.

Relative risk indexes for global warming, smog formation, and acidification have been calculated for the solvent recovery process at each flow rate. These values were generated by applying Equations 11-4, 11-13 and 11-11, respectively,

$$I_{GW} = \sum_i (GWP_i \times m_i)$$

$$I_{SF} = \sum_i (SFP_i \times m_i)$$

$$I_{AR} = \sum_i (ARP_i \times m_i)$$

using the emission rates in Table 11.3-2 and the impact potential values for each chemical (Appendix D, Tables D-1, D-3, and D-4). For the smog formation potential (SFP=MIR) of ethyl acetate, the average MIR of the ethers (1.13) and ketones (0.87) listed in Appendix D, Table D-4 were used as an approximation. As an example calculation, the smog formation index of the process will be determined at an absorption oil flow rate of 50 kgmole/hr.

		$SFP_i \bullet m_i$
Toluene:	(0.87)(0.97 kg/hr)	0.84 kg/hr
Ethyl Acetate:	(0.32)(160.4 kg/hr)	51.33 kg/hr
Tetradecane:	(0.1)(4.67 kg/hr)	0.47 kg/hr
Total:		**52.64 kg/hr**

Shown in Figures 11.3-3 through 11.3-5 are the relative impact indexes for the solvent recovery process of Figure 11.3-1. We observe in Figure 11.3-3 that the global warming index is minimized by operating the process at approximately 50 kgmole/hr. An explanation for this behavior follows next. At an oil flow rate of 0 kgmole/hr, all of the VOCs are emitted directly to the air, resulting in an elevated global warming impact after the organics are oxidized to CO_2. Nearly a 40% reduction in the global warming index is realized by operating the process at an absorption oil flow rate of 50 kgmole/hr. However, above 50 kgmole/hr, the process utilities increase at a substantial rate compared to the rate of additional recovery of the VOCs, driving the index higher. Therefore, the optimum flow rate is approximately 50 kgmole/hr for global warming. As shown in Figure 11.3-4, the acid rain index for the process increases in nearly direct proportion to the absorption oil flow rate. This behavior occurs because the only acidifying species emitted from the process are from the process utility requirements (SOx and NOx), which increase in proportion to the absorption oil flow rate. The optimum flow rate for acidification would be at 0 kgmole/hr for the absorption oil flow rate. The smog formation index (Figure 11.3-5) shows a significant decrease in the index with absorption oil flow rate up to 50 kgmole/hr (recovery of toluene) and a gradual decrease from 50 to 500 kgmole/hr (recovery of ethyl acetate). The flow rate for minimizing the smog formation index is therefore about 500 kgmole/hr.

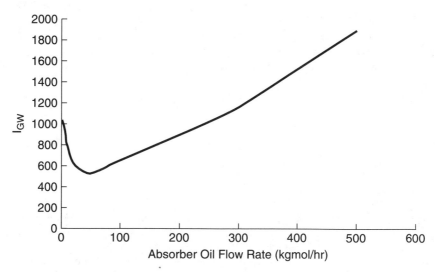

Figure 11.3-3   The global warming index for the solvent recovery process of Figure 11.3-1. Adapted from Hiew (1998) and Shonnard and Hiew (2000).

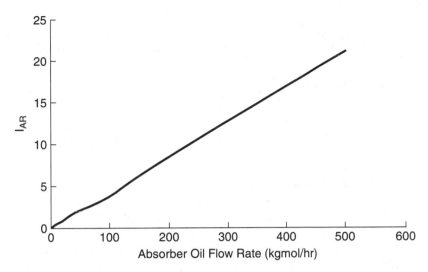

Figure 11.3-4 The acid rain index for the solvent recovery process of Figure 11.3-1. Adapted from Hiew (1998) and Shonnard and Hiew (2000).

Discussion: These indexes demonstrate the complexities in evaluating chemical processes using multiple indexes of environmental performance. It is not possible to identify a single absorption oil flow rate that simultaneously minimizes all three indexes. However, we can see that significant reductions in the global warming (42%) and the smog formation (82%) indexes are realized at an oil flow rate of 50 kgmole/hr, with only a relatively modest increase in the acid rain index. This observation suggests

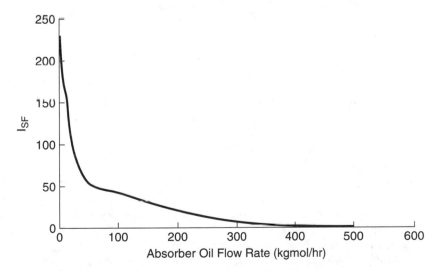

Figure 11.3-5 The smog formation index for the solvent recovery process of Figure 11.3-1. Adapted from Hiew (1998) and Shonnard and Hiew (2000).

that a decision to operate the process at 50 kgmole/hr is a good compromise. In reality, the decision to operate the process at any given flow rate will only be made after economic and safety considerations have been taken into account.

11.3.5 Toxicity

As explained in Chapter 2, chemical toxicity to humans and ecosystems is a function of dose and response. The dose is dependent on a complex series of steps involving the manner of release, environmental fate and transport of chemicals, and uptake mechanisms. The final two steps dictate the extent of exposure. Key questions which affect the administered dose include: Where are the emissions released to—the air, water, or soil? Are the chemicals altered by environmental reactions or are the chemicals persistent? How are the chemicals taken up by the body? Through breathing contaminated air? Drinking contaminated water? By direct contact with and transfer through the skin? The effective dose is dependent on processes occurring in the body including absorption, distribution, storage, transformation, and elimination. The response by the target organ in the body is a very complex function of chemical structure and modes of action and is the purview of the field of toxicology.

Clearly, the complexity of toxicology precludes an exact determination of all adverse effects to human and ecosystem health from the release of a chemical. From an engineering point of view, an exact assessment may not be necessary. Similar to the potential impact indexes presented for global warming, stratospheric ozone depletion, smog formation, and acidification, we develop and use *toxicity potentials* for non-carcinogenic and carcinogenic health effects for ingestion and inhalation routes of exposure. Both inhalation and ingestion are thought to be the dominant routes of exposure for human contact with toxic chemicals in the environment.

Non-Carcinogenic Toxicity

Non-carcinogenic toxicity in humans is thought to be controlled by a threshold exposure, such that doses below a threshold value do not manifest a toxic response whereas doses above this level do. A key parameter for each chemical is therefore its reference dose (RfD (mg/kg/d) or reference concentration (RfC (mg/m^3)) for ingestion and inhalation exposure, respectively. Exposures to concentrations in the water or air which result in doses or concentrations above these reference levels is believed to cause adverse effects. Lists of RfD and RfC data are available in electronic or paper copy form (US EPA, 1997; US EPA, 1994). Because RfDs and RfCs are not available for all chemicals, we use lethal doses (LD_{50}) and concentrations (LC_{50}) as additional toxicological parameters for health assessments. Lists of LD_{50} and LC_{50} are tabulated in additional sources (NTP, 1997). Threshold Limit Values (TLVs), Permissible Exposure Limits (PELs), and Recommended Exposure Limits (RELs) are additional toxicity properties that, like RfD and RfC, are based on low-dose studies.

For the purpose of an approximate assessment of risk, concentrations in the air or water will be calculated using the multimedia compartment model shown in Section 11.2. The toxicity potential for ingestion route exposure is defined in this text as:

$$INGTP_i = \frac{[(C_{i,w})(2\ L\ /\ d)\ /\ (70\ kg)]\ /\ (RfD_i)}{[(C_{Toluene,w})(2\ L\ /\ d)\ /\ (70\ kg)]\ /\ (RfD_{Toluene})} \qquad \text{(Eq. 11-14)}$$

$C_{i,w}$ and $C_{Toluene,w}$ are the steady-state concentrations of the chemical and the benchmark compound (toluene) in the water compartment after release of 1000 kg/hr of each into the water compartment, as predicted by the multimedia compartment model of Section 11.2. The factor of 2 L/d and 70 kg are the standard ingestion rate and body weight used for risk assessment (Pratt, et al., 1993). The product of the concentration and the ingestion rate divided by the body weight provides the exposure dose. This exposure dose is divided by the reference dose to determine whether this dose poses a toxicological risk. The ratio of these risks for the chemical and the benchmark compound results in the ingestion toxicology potential for chemical i.

The toxicity potential for inhalation exposure is defined similarly as:

$$INHTP_i = \frac{C_{i,a}\ /\ RfC_i}{C_{Toluene,a}\ /\ RfC_{Toluene}} \qquad \text{(Eq. 11-15)}$$

where $C_{i,a}$ and $C_{Toluene,a}$ are the concentrations of chemical i and of the benchmark compound (toluene) in the air compartment of the environment after release of 1000 kg/hr of each into the air compartment, as predicted by the multimedia compartment model. The doses are not shown in the equation because the inhalation rate (20 m^3/d) and body weights (70 kg) cancel out. In this equation, the ratio of the risks for inhalation exposure is the potential for inhalation toxicity relative to the benchmark compound.

In order to determine a non-carcinogenic toxicity index for the entire process, we must multiply each chemical's toxicity potential with its emission rate from the process and sum these for all chemicals released.

$$I_{ING} = \sum_i (INGTP_i \times m_i) \qquad \text{(Eq. 11-16)}$$

Similarly for inhalation route toxicity;

$$I_{INH} = \sum_i (INHTP_i \times m_i) \qquad \text{(Eq. 11-17)}$$

Carcinogenic Toxicity

In a similar method as outlined for non-carcinogenic toxicity, we develop two indexes for cancer-related risk, based on predicted concentrations of chemicals in the air and water from a release of 1000 kg/hr. The concentrations are converted to doses using standard factors and then the risk for the chemical and a benchmark compound, benzene, is calculated. The carcinogenic potential for a chemical is determined by taking the ratio of the chemical's risk to that for the benchmark compound. The ingestion route carcinogenic potential for a chemical is:

$$INGCP_i = \frac{(C_{i,w})(SF_i)}{(C_{Benzene,w})(SF_{Benzene})} \qquad \text{(Eq. 11-18)}$$

where SF $(mg/kg/d)^{-1}$, the cancer potency slope factor, is the slope of the excess cancer versus administered dose data. The dose-response data is normally taken using animal experiments and extrapolated to low doses. The higher the value of SF, the higher is the carcinogenic potency of a chemical. Lists of SF values for many chemicals can be found in the following references (US EPA, 1997; US EPA, 1994). Because SFs are not yet available for all chemicals of interest, weight of evidence (WOE) classifications have been tabulated for many industrial chemicals by consideration of evidence by a panel of experts. The definitions of each weight of evidence classification is shown in Table 11.3-3 along with a numerical hazard value (HV). The value of HV can be used in Equations 11-18 and 11-19 in the absence of SF data. Data for WOE can be found in the following sources (NIHS, 1997; OSHA, 1997; IRIS, 1997).

A similar definition for the inhalation carcinogenic potential for a chemical is:

$$INHCP_i = \frac{(C_{i,a})(SF_i)}{(C_{Benzene,a})(SF_{Benzene})} \qquad \text{(Eq. 11-19)}$$

The carcinogenic toxicity index for the entire process is again a summation for each carcinogen. For ingestion, it is:

$$I_{CING} = \sum_i (INGCP_i \times m_i) \qquad \text{(Eq. 11-20)}$$

and for inhalation,

$$I_{CINH} = \sum_i (INHCP_i \times m_i) \qquad \text{(Eq. 11-21)}$$

Example 11.3-3   Toxicity Evaluation of the Solvent Recovery Process in Figure 11.3-1

Toxicity evaluation of the toluene and ethyl acetate recovery and recycle process design is conducted in a fashion similar to the previous example problem. We are concerned with three compounds in this analysis, toluene, ethyl acetate, and hexane (a surrogate for products of incomplete combustion in utility consumption). There are no carcinogenic compounds present in the design, therefore the two carcinogenic indexes will be ignored. This example illustrates how LD_{50}/LC_{50} can be used interchangeably with RfDs and RfCs when data gaps occur.

Data and Results: The emission rates of these compounds appeared in the previous example problem and are used again here. The concentrations of these chemicals and

Table 11.3-3   Weight of Evidence (WOE) Classifications (IRIS, 1997; Davis et al., 1994).

Group	Definition	HV
A	Human carcinogen. This classification is used only when there is sufficient evidence from epidemiologic studies to support a causal association between exposure to the agent and cancer.	5
B	Probable human carcinogen. This group is divided into two subgroups, B1 and B2. Subgroup B1 is usually used when there is limited WOE of human carcinogenicity based on epidemiologic studies. Group B2 is used when there is sufficient WOE of carcinogenicity based on animal studies, but inadequate evidence or no data from epidemiologic studies.	B1=4 B2=3.5
C	Possible human carcinogen. This classification is used when there is limited evidence of carcinogenicity in animals in the absence of human data.	1.5
D	Not classifiable as to human carcinogenicity. This classification is generally used when there is inadequate human and animal evidence of carcinogenicity or when no data are available.	0
E	Evidence of non-carcinogenicity for human. This classification is used when agents show no evidence of carcinogenicity in at least two adequate animal tests in different species or in both adequate epidemiologic and animal studies.	0

the benchmark compound (toluene again) are calculated using the Mackay model of Section 11.2. The input data and the resulting concentrations are provided in the following table. The calculations were conducted using a standard emission of 1000 kg/hr of each compound into the air compartment when evaluating both ingestion toxicity and inhalation toxicity. This approach was adopted rather than using the actual emission rates of each compound, because only the ratios of concentrations are needed in the index calculation, and the concentration ratios are not a function of emission rate using the Mackay model.

Chemical	Molecular Weight	Melting point (°C)	Fugacity ratio	Vapor pressure @25°C (Pa)	Solubility (g/m^3)	Log K_{OW}
Toluene	92.13	−95.0	1.0	3800	550	2.70
Ethyl acetate	88.11	−82.0	1.0	12000	80800	0.70
Hexane	86.17	−95.3	1.0	20000	10	4.00

Chemical	Half life (hr)				Concentration (g/m^3)	
	Air	Water	Soil	Sediment	Air	Water
Toluene	17	550	1700	5500	1.97×10^{-7}	4.00×10^{-7}
Ethyl acetate	55	55	170	550	4.36×10^{-7}	5.00×10^{-6}
Hexane	17	550	1700	5500	1.97×10^{-7}	1.50×10^{-9}

The toxicological properties (RfDs, RfCs) are incomplete for the three chemicals in this design. We are forced to use LD$_{50}$ and LC$_{50}$ data when gaps occur. The following table summarizes the toxicology data and calculated ingestion and inhalation toxicity potentials using the air and water concentrations in the table above and the toxicity Equations 11-14 and 11-15.

	Inhalation RfC (mg/m^3)	Oral RfD (mg/kg/day)	LC$_{50}$ (ppm)	LD$_{50}$ (mg/kg)	Toxicity Potentials INHTP	INGTP
Toluene	0.4	0.2	4000	5000	1.0	1.0
Ethyl acetate		0.9	3200		2.8	2.8
Hexane	0.2			28700	2.0	6.5×10^{-4}

Figures 11.3-6 and 11.3-7 show the change in process inhalation and ingestion toxicity index with absorption oil flow rate using the emission rate data tabulated in the previous example problem and concentrations calculated by the Mackay model.

Discussion: The inhalation toxicity is reduced with increasing absorption oil flow rate due to the removal of both toluene and ethyl acetate, and to a much lesser extent by hexane (TOC). The inhalation and ingestion index behavior is nearly identical since the inherent toxicity potentials, INHTP and INGTP, are virtually the same (as shown above). These toxicity indexes can be reduced by 39% by operating the process at 50 kgmoles/hr absorber flow rate. Keep in mind that interchanging RfCs with LC$_{50}$s will introduce additional uncertainties in the evaluation.

SUMMARY

This chapter has outlined a systematic methodology for evaluating environmental and health-based impacts for chemical process designs. Multiple impact indexes are included for process evaluation because of the complexity of pollutant interactions

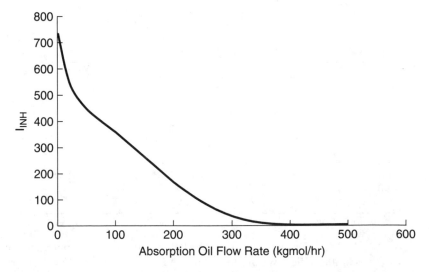

Figure 11.3-6   Inhalation toxicity index for the solvent recovery and recycle process.

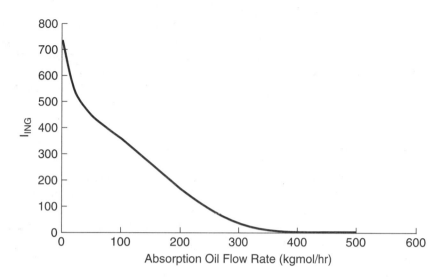

Figure 11.3-7   Ingestion toxicity index for the solvent recovery and recycle process.

with the environment and with human health. The methodology includes pollutant release or emission estimation (Chapter 8), environmental fate and transport of pollutants, and relative risk assessment using the benchmarking concept. The methodology was applied for the evaluation of air emissions from a process flow-sheet utilizing a commercial chemical process simulator to generate the material and energy balances for the process. From this analysis, we were able to assess the environmental performance of the process as one of the process parameters, the absorber oil flow rate, was varied over a wide range. This exercise provided insights into how energy consumption within the process drives up certain environmental impacts and how the recovery and recycle of VOCs drives down others. The trends for impact with respect to absorber oil flow rate are clear, but making a decision based on these trends is not straightforward. Since trade-offs do occur among these impact indexes, questions such as which impacts are most important need to be addressed. Nonetheless, the environmental information provided will allow for more sound process design decisions.

It is important to stress that the methodology is general, and can accommodate releases to the water and the soil as well as to the air, since the multimedia compartment model can predict environmental concentrations for all of these release mechanisms. Another issue to consider is the uncertainties involved in the assessment of environmental risk. There are many sources of uncertainty, particularly for emission estimation and environmental fate calculations. The magnitude of these uncertainties may be quite large, depending upon the emission estimation method and on chemical-specific environmental properties. It is important to understand the magnitude of these uncertainties in order to decide whether significant differences actually occur when comparing the environmental impacts of process operating conditions or of various process technologies. Uncertainty analysis for environmental impact assessment is an active research area in chemical engineering and in environmental science and engineering. The topic is beyond the scope of this introductory textbook, but methods of evaluating uncertainty are available, and may include Monte Carlo simulation and propagation of error analysis.

REFERENCES

Allen, D.T., Bakshani, N., and Rosselot, K.S., "Pollution Prevention: Homework & Design Problems for Engineering Curricula," American Institute of Chemical Engineers, New York, NY, 155 pages, 1992.

Allen, D.T. and Rosselot, K.S. "Pollution Prevention for Chemical Processes," 1st ed., John Wiley & Sons, New York, NY, 434 pages, 1997.

Bufalini, J. J. and Dodge, M. C. "Ozone-forming potential of light saturated hydrocarbons," *Environmental Science and Technology* 1983, 17, 308.

Boustead, I., "Ecoprofiles of the European Plastics Industry, Report 1-4," PQMI, European Centre for Plastics in the Environment, Brussels, May, 1993.

Carter, W. P. and Atkinson, R. "A computer modeling study of incremental hydrocarbon reactivity," *Environmental Science and Technology* 1989, 23, 864-880.

Carter, W. P. L., "Development of ozone reactivity scales for volatile organic compounds," *Air & Waste* 1994, 44, 881-899.

Chang, T. Y. and Rudy, S. J., "Ozone-forming potential of organic emissions from alternative-fueled vehicles," *Atmospheric Environment* 1990, 24A, 2421.

Cohen, Y., Tsia, W., Chetty, S. L. and Mayer, G. J., "Dynamic partitioning of organic chemicals in regional environments: A multimedia screening-level modeling approach," *Environ. Sci. Technol.* 1990, 24, 1549-1558.

Davis, G.A., Kincaid, L., Swanson, M., Schultz, T., Bartmess, J., Griffith, B., and Jones, S., "Chemical Hazard Evaluation for Management Strategies: A Method for Ranking and Scoring Chemicals by Potential Human Health and Environmental Impacts," United States Environmental Protection Agency, EPA/600/R-94/177, September 1994.

Derwent, R.G., "Trace Gases and Their Relative Contribution to the Greenhouse Effect," Atomic Energy Research Establishment, Report AERE-R13716, Harwell, Oxon, 1990.

Dodge, M. C., "Combined effects of organic reactivity and NMHC/NOx ratio on photochemical oxidant formation—a modeling study," *Atmospheric Environment* 1984, 18, 1857.

Fetter, C.W., *Contaminant Hydrogeology,* Prentice-Hall, Inc., pg. 458, 1993.

Fisher, D.A., Hales, C.H., Wang, W., Ko, M.K.W., and Sze, N.D., "Model calculations of the relative effects of CFCs and their replacements on global warming," Nature, vol. 344, 513-516, 1990a.

Fisher, D.A., Hales, C.H., Filkin, D.L., Ko, M.K.W., Sze, N.D., Connell, P.S., Wuebbles, D.J., Isaksen, I.S.A., and Stordal, F., "Model calculations of the relative effects of CFCs and their replacements on stratospheric ozone," Nature, Vol. 344, 508-512, 1990b.

Goedkoop, M., "The Eco-indicator 95, Final Report," Netherlands Agency for Energy and the Environment (NOVEM) and the National Institute of Public Health and Environmental Protection (RIVM), NOH report 9523, 1995

Heijungs, R., Guinée, J.B., Huppes, G., Lankreijer, R.M., Udo de Haes, H.A., Sleeswijk, A. Wegener, "Environmental Life Cycle Assessment of Products. Guide and Backgrounds," NOH Report Numbers 9266 and 9267, Netherlands Agency for Energy and the Environment (Novem), 1992.

Hiew, D.S., "Development of the Environmental Fate and Risk Assessment Tool (EFRAT) and Application to Solvent Recovery from a Gaseous Waste Stream," Masters Thesis, Department of Chemical Engineering, Michigan Technological University, 1998.

IPCC, "Radiative Forcing of Climate, Climate Change—The IPCC Scientific Assessment: 1990 (WMO)," Cambridge University Press, 45 - 68, 1991.

IPCC, "Climate Change 1994—Radiative Forcing of Climate Change and an Evaluation of the IPCC IS92 Emission Scenarios," Intergovernmental Panel on Climate Change, J.T. Houghton et al. (ed.), Cambridge University Press, 1994.

IPCC, "Climate Change 1995—The Science of Climate Change," Intergovernmental Panel on Climate Change, J.T. Houghton et al. (ed.), Cambridge University Press, 1996.

IRIS, Integrated Risk Information System (IRIS), http://www.epa.gov/ngispgm3/iris/Substance_List.html, US Environmental Protection Agency, 7/15/97.

Japar, S. M., Wallington, T. J., Rudy, S. J., and Chang, T. Y., "Ozone-forming potential of a series of oxygenated organic compounds," *Environmental Science and Technology* 1991, 25, 415-420.

Lashof, D.A. and Ahuja, D.R., "Relative contributions of greenhouse gas emissions to global warming," Nature, Vol. 344, 529-531, 1990.

Mackay, D. *Multimedia Environmental Models, The Fugacity Approach* Second Edition, CRC Press 2001; pg. 272.

Mackay, D, and Paterson, S., "Evaluating the multimedia fate of organic chemicals: A level III fugacity model," Environmental Science and Technology, V. 25 (3), pg 427-436, 1991.

Mackay, D., Shiu, W., and Ma, K., *Illustrated Handbook of Physical-Chemical Properties and Environmental Fate for Organic Chemicals*, 1st edition, Vol. 1-4, Lewis Publishers, Chelsea, MI, 1992.

Mallick, S.K., Cabezas, H., Bare, J.C., and Sikdar, S.K., "A pollution reduction methodology for chemical process simulators," Industrial & Engineering Chemistry Research, Vol. 35, (11), 4128-, 1996.

McKone, T. E., 1994, "CALTOX™, A Multimedia Total Exposure Model for Hazardous Waste Sites," Version 1.5, IBM/PC, Microsoft Excel Spreadsheet Program, National Technical Information Service, Springfield, VA, Order No. PB95-100467.

NRC (National Research Council), "Risk Assessment in the Federal Government: Managing the Process," Committee on Institutional Means for Assessment of Risks to Public Health, National Academy Press, Washington, D.C. (1983).

NTP, National Toxicology Program, Chemical Health & Safety Data, http://ntp-db.niehs.nih.gov/Main_pages/Chem-HS.HTML, 7/15/97.

OSHA, Occupational Safety & Health Administration, http://www.osha-slc.gov/ChemSamp_toc/ChemSamp_toc_by_chn.html, Chemical Sampling Information, Table of Contents by Chemical Name, U.S. Department of Labor, 7/15/97.

Pouchert, C.J., "The Aldrich Library of FT-IR Spectra, Vapor Phase," 1st edition, Vol. 2, pg. 165-229, 1989.

Pratt, G.C., Gerbec, P.R., Livingston, S.K., Oliaei, F., Bollweg, G.L., Paterson, S., and Mackay, D., "An indexing system for comparing toxic air pollutants based upon their potential environmental impacts," Chemosphere, Vol. 27 (8), 1359-1379, 1993.

Ravishankara, A.R., Turnipseed, A.A., Jensen, N.R., Barone, S., Mills, M., Howard, C.J., and Solomon, S., "Do hydrofluorocarbons destroy stratospheric ozone?," Science, Vol. 263, 71-75, 1994.

Rotmans, J.; *IMAGE*, an Integrated Model to Assess the Greenhouse Effect," Maastrich, pp 205-224, 1990.

Sangwichien, C., "Modeling and Evaluating Solvent Recovery Technologies Applied to Industrial Pollution Prevention," Masters Thesis, Department of Chemical Engineering, Michigan Technological University, 1998.

Seinfeld, J.H. and Pandis, S., "Atmospheric Chemistry and Physics: Air Pollution to Climate Change," John Wiley & Sons, New York, NY, pg 1326, 1997.

SETAC, Society for Environmental Toxicology and Chemistry, "Guidelines for Life-Cycle Assessment: Code of Practice," Brussels, Belgium, 1993.

Shonnard, D.R. and Hiew, D.S., "Comparative Environmental Assessments of VOC Recovery and Recycle Design Alternatives for a Gaseous Waste Stream," *Environmental Science and Technology*, 34(24), 5222-5228, 2000.

Thibodeaux, L.J., *Environmental Chemodynamics,* Second Edition, John Wiley & Sons, New York, NY, pg 593, 1996.

US EPA, Health Effects Assessment Summary Tables (HEAST), 1994.

US EPA, Integrated Risk Information System (IRIS), 1997.

US EPA, *Locating and Estimating Air Emissions From Sources,* U.S. Environmental Protection Agency, Air CHIEF, Version 8.0, Office of Air Quality Planning and Standards, EPA 454/C-00-003 December 2000.

WMO, "Halocarbon Ozone Depletion and Global Warming Potential, Scientific Assessment of Stratospheric Ozone: 1989," World Meteorological Organization, Report number 20, Vol. 1, 1990a.

WMO, "Executive Summary: Supporting Evidence and Other Results, Scientific Assessment of Stratospheric Ozone: 1989," World Meteorological Organization, Report number 20, Vol. 1, 1990b.

WMO, "Radiative Forcing of Climate, Scientific Assessment of Global Warming: 1991," World Meteorological Organization, Report number 25, Vol. 1, 1992a.

WMO, "Ozone Depletion and Chlorine Loading Potential, Scientific Assessment of Ozone Depletion: 1991," Global Ozone Research and Monitoring Project, World Meteorological Organization, Report number 25, Vol. 1, 1992b.

PROBLEMS

1. **Ethanol as a Substitute Octane-Boosting Additive to Automobile Fuels.** In response to requirements of the 1990 Clean Air Act Amendments, automobile fuels sold in some urban areas must contain 10% ethanol. The reasons for adding ethanol are 1)

to reduce emissions of carbon monoxide from tail pipes, and 2) to boost the octane rating of the resulting fuel blend. The maximum incremental reactivity (MIR) values for ethanol and other potential octane boosters are provided below.

Ethanol   1.34
Toluene   2.70
Xylenes   7.10
Base Fuel   1.5

Calculate by what percentage the ozone-producing potential of an ethanol fuel blend (10% ethanol, 90% base fuel) is reduced compared to a fuel blend containing 10% toluene and 90% base fuel and another blend containing 10% xylenes and 90% base fuel, respectively. Use the provided MIR values in Appendix D, Table D-4 for this calculation and assume that the MIR of the blend is a summation of each components MIR weighted by its fraction in the blend.

(For a comprehensive discussion of the use of ethanol as a fuel blending component, see National Research Council, "Ozone-forming Potential of Reformulated Gasoline," National Academy Press, Washington, D.C., 1999)

2. **Hydroxyl Radical and Reaction with Ethene.** The second-order rate constant for the reaction of ethene with OH• is approximately $k = 2.7 \times 10^{-13}$ (cm^3 / (molecule • sec)). The expression for the rate of reaction of ethene is

$$Rate - k\,[OH\bullet]\,[Ethene]$$

If [OH•] is maintained at 10^{-12} moles per liter in a test chamber, and the initial ethene concentration is 100 ppm, estimate:
   a) the pseudo first-order rate constant ($k\,[OH\bullet]$) for ethene under these conditions.
   b) the amount of ethene remaining after 1 hour
   Hint: convert [OH•] from moles per liter of air to molecules per cm^3 first before answering part a.

3. **Global Warming Parameters.** The most important chemical parameters which affect the global warming potential of a greenhouse gas are the reaction lifetime (τ) and the infrared absorbence band strength (BI). Demonstrate the importance of these parameters by performing the following tasks.
   a) From Appendix D, Table D-1, group chemicals having the highest values of both τ and BI and determine whether these have the highest values of GWP.
   b) Determine the sensitivity of GWP to τ and BI. Assume that GWP is a continuous function of τ and BI. We can write a change in GWP (dGWP) as functions of changes in τ (dτ) and BI (dBI) as follows;

$$d(GWP) = \left(\frac{\partial GWP}{\partial \tau}\right)_{BI} \bullet d\tau + \left(\frac{\partial GWP}{\partial BI}\right)_{\tau} \bullet dBI$$

where $(\partial GWP/\partial \tau)_{BI}$ is the partial derivative of GWP with respect to τ holding BI constant. The other partial derivative is similarly defined. Estimate the partial derivatives for GWP using the data listed in Appendix D, Table D-1.
   c) Estimate each term on the right hand side of the equation above by using dτ = 10 years and dBI as 400 atm^{-1} cm^{-2}. Which term contributes the most to changes in GWP (d(GWP))?

4. **Incremental Reactivities of Volatile Organic Compounds.** Appendix D, Table D-4 shows maximum incremental reactivities (MIR) for many volatile organic compounds found in urban areas. Use entries from the first column of data in this table to answer the following questions.

   a) How do the averages for each of the classes of compounds compare to the average for the entire list (Base ROG Mixture)?

   b) If you were in charge of reformulating a fuel for automobiles, what class of compounds would you retain for a more detailed consideration as a substitute fuel component if one of the main considerations is reducing smog formation potential? Based on this analysis, why has ethanol been used as a substitute? What other considerations (environmental and non-environmental) have to be taken into account in choosing a compound to add to make a reformulated gasoline?

5. **Environmental Risk Index Calculations for Solvent Recovery and Recycle.** Calculate and then compare the environmental and human health indexes for the solvent recovery process shown in Example 11.3-2 for absorber oil flow rates of 0 and 100 kgmoles/hr. Use the emission rate data shown in Table 11.3-2 of Example 11.3-2 and use hexane as a surrogate (representative compound) for TOC. In your analysis, calculate the indirect global warming potential for each VOC. Assume that Fuel Oil #4 is being used to satisfy the energy demand of the processes and that this fuel contains 1% sulfur. Use the RfD, RfC, LD_{50}, and LC_{50} values and employ the multimedia concentrations of each component in the air and water as shown in Example 11.3-3.

6. **Carbon Dioxide Emission Factors (Chapter 8).** Confirm the CO_2 emission factors listed in the text (Tables 8.3-5 and 8.3-6) for fuel oil and natural gas. As an approximation, assume that fuel oil is composed entirely of n-decane ($C_{10}H_{22}$) and that natural gas is 100% methane (CH_4). Use the ideal gas law and standard conditions of 0 °C and 1 atmosphere pressure for the natural gas calculation. The specific gravity of n-decane is 0.73. The reaction stoichiometries for the combustion reactions are:

$$C_{10}H_{22} + 15\,1/2\,O_2 \rightarrow 10\,CO_2 + 11\,H_2O$$

$$CH_4 + 2\,O_2 \rightarrow CO_2 + 2\,H_2O$$

CHAPTER 12

Environmental Cost Accounting

by
Kirsten Sinclair Rosselot and David T. Allen

12.1 INTRODUCTION

Costs associated with poor environmental performance can be devastating. Waste disposal fees, permitting costs, and liability costs can all be substantial. Wasted raw material, wasted energy and reduced manufacturing throughput are also consequences of wastes and emissions. Corporate image and relationships with workers and communities can suffer if environmental performance is substandard. But, how can these costs be quantified?

This chapter will review the tools available for estimating environmental costs. Many of these tools are still in their developmental stages, and practices therefore vary widely from company to company. In general, however, traditional accounting practices have acted as a barrier to implementation of green engineering projects because they hide the costs of poor environmental performance. Many companies are now giving more consideration to all significant sources of environmental costs. The principle is that if costs are properly accounted for, business management practices that foster good economic performance will also foster superior environmental performance.

The relationships between economic and environmental performance are examined in a number of steps. First, in Section 12.2, a few key terms are defined to simplify and clarify the presentation of material. In Section 12.3, the magnitude and types of environmental costs typically encountered by companies are reviewed. Then, in Section 12.4, a framework for assessing environmental costs is described. Finally, Sections 12.5 through 12.8 describe specific methods for evaluating environmental costs.

Prerequisites to fully understanding the benefits of environmental accounting practices are an understanding of the time value of money and some familiarity with present value, payback period, internal rate of return, and other financial evaluation calculations. These concepts are covered in textbooks on engineering economics (see, for example, Valle-Riestra, 1983). Also, it is assumed that the reader understands how to evaluate potential environmental impacts associated with products and processes (see Chapter 11).

12.2 DEFINITIONS

Environmental accounting is still in its infancy and terminology is therefore continually evolving. Precise definitions of terms are often elusive. Nevertheless, to keep the discussion presented in this chapter clear, it is useful to define a number of terms as they will be used in this text. Many of these definitions are drawn from an introduction to environmental accounting prepared by the US Environmental Protection Agency (US EPA, 1995).

Internal costs, or private costs, are costs that are borne by a facility. Costs for materials and labor are examples of internal costs. **External costs,** or societal costs, on the other hand, are the costs to society of the facility's activities. The cost associated with a loss of fishable waters due to pollutants discharged by a facility to a stream is an example of an external cost. Often, environmental fees, regulations, and requirements act to internalize what would have otherwise been an external cost, so that a facility that produces waste must pay to reduce its quantity or toxicity or pay a premium for its disposal. This chapter focuses primarily on internal costs.

There are two types of accounting that are pursued at most large facilities: management accounting and financial accounting. **Management accounting** is the collection of information that helps a firm to make internal decisions. This information is not usually disclosed to the public, and each firm has its own style and accounting requirements. Green accounting (accounting that promotes environmentally sound practices) refers to managerial accounting practices. **Financial accounting,** in contrast, is the information collected for reporting to stockholders, the Securities and Exchange Commission (which oversees trade and investment practices of companies in the United States), and banks. Financial accounting practices tend to be fairly uniform across companies and are governed by generally accepted accounting principles.

A typical management accounting system for a manufacturer would include categories for direct materials and labor (costs that are clearly and exclusively associated with a product or service), manufacturing overhead, sales, general and administrative overhead, and research and development. Environmental expenses can be hidden in any or all of these categories, but are charged most often as **overhead.** Overhead costs, as opposed to costs of direct materials and labor for production, are often referred to as **indirect costs** and consist of any costs that the accounting system either pools facility-wide and does not allocate among activities, or that are allocated on the basis of a formula. Overhead generally includes indirect materials and labor, capital depreciation, rent, property taxes, insurance, supplies, utilities, and repair and maintenance. It can also include labor costs ranging from supervisor salaries to janitorial services. Often, even the direct environmental costs that could be assigned to a particular process, product, or activity, such as waste disposal, are lumped together facility-wide. This is often done because of practices such as using a single waste disposal company to manage all of a facility's waste. Other environmental costs, such as the costs of filling out forms for reporting waste management practices, are also hidden in the overhead category. Because environmental costs are not traditionally allocated to the activity that is generating wastes, some of the benefits of green engineering projects are masked.

Full-cost accounting is a type of managerial accounting that is considered to be "green." In full-cost accounting, as many costs as possible are allocated to product, product lines, processes, services, or activities. Full-cost accounting is not a strictly environmental activity. For instance, it is pursued because it is useful in determining the profitability of processes and products and in setting prices. Even though full-cost accounting does not focus particularly on environmental costs, it promotes improved environmental performance because the costs of producing waste for individual processes or products are revealed, providing management with a better idea of the true costs associated with the generation of wastes and emissions.

Activity-based costing is similar to full-cost accounting except that the costs are allocated to specific measures of activity. For example, in activity-based costing, the cost of generating a particular kind of waste per pound of production might be measured. Another example would be determining the cost of chemical inputs per item for painting.

Capital is the wherewithal a facility has to produce goods or to bring in income. **Capital budgeting,** sometimes called investment analysis or financial evaluation, is supported by information from accounting activities. Each firm has its own capital budgeting process for making decisions about how capital will be spent. How a firm employs the standard evaluation measures such as rate of return, payback period, or net present value, to analyze potential products is individualized. In addition, each firm has self-defined hurdles for determining which projects are worthwhile. For example, for firms using rate of return to evaluate projects, the minimum internal rate of return required to fund a project varies from one company to the next, as do techniques for estimating future interest and inflation rates.

Total cost assessment, discussed in more detail later, is a capital budgeting procedure that requires a comprehensive analysis of savings and costs (especially environmental costs and savings) beyond the capital and operating costs that are conventionally considered in capital budgeting. **Life-cycle costing** is another type of capital budgeting in which the costs of a project from its conception (e.g., the research and development phase) to its retirement (e.g., salvage value) are assessed. Note that in this chapter, life-cycle costing is assumed to include only internal costs and is not to be confused with life-cycle assessment, which is the assessment of the environmental impacts of a product, process, or activity from raw material extraction to final disposal (see Chapter 13). Life-cycle costing affects decisions about capital expenditure because replacement and closure costs (also called back-end or exit costs) are often hidden, as are up-front costs like research and development.

These are the primary terms that will be used in this chapter. It is useful to keep in mind that precise definitions remain in flux, and vary from organization to organization, so the terminology used in this chapter is not universal. Nevertheless, it is generally recognized that in environmental accounting, words like "full" (e.g., full-cost accounting), "total" (e.g., total cost assessment), and "life-cycle" (e.g., life-cycle costing) are used to indicate that not all costs are captured in traditional accounting and capital budgeting practices.

Green Accounting Practices and Other Quality Management Paradigms

Readers familiar with various quality management paradigms might have noticed that green accounting and capital budgeting practices are frequently compatible with general strategies for improving business management. These various strategies tend to work together to form a general philosophy of quality improvement, and companies that are accustomed to tracking and improving the productivity of labor and capital are just now realizing that it benefits them to do the same for energy and resources.

In quality management, many companies seek external certification of their management systems through the International Standards Organization (ISO) or similar organizations at a national level.

ISO, the International Standards Organization, is an independent standard-setting body with representatives from all industrialized nations. The initial focus of standard-setting organizations such as ISO was to improve the quality and uniformity of internationally-traded merchandise. More recently, ISO standards have been set for management practices. ISO 9000 is a business management standard for quality systems and certification is fairly common now. To be certified, companies must show that they have the required quality management systems in place. ISO has also developed standards for environmental management. ISO 14000 environmental management standards are similar to ISO 9000 quality management standards except that they focus on environmental management, of which total cost accounting is a component. Some of the standards for ISO 14000 are still under development. Note that ISO 14000 certification is based on whether or not a company has systems in place for managing environmental responsibilities, not on environmental performance.

12.3 MAGNITUDES OF ENVIRONMENTAL COSTS

The definitions in the previous section made clear that not all environmental costs are captured in traditional accounting and capital budgeting practices. Nevertheless, some measures of environmental costs are available, providing a rough indication of the magnitude of environmental costs and the variation of those costs among industry sectors.

Among the easiest environmental costs to track are the costs associated with treating emissions and disposing of wastes. Direct costs of pollution abatement are

Table 12.3-1   Pollution Abatement Expenditures by U.S. Manufacturing Industries (data reported by US Congress, Office of Technology Assessment, 1994; original data collected by U.S. Census Bureau).

Industry sector	Pollution control expenditures (as a % of sales)	Pollution control expenditures (as a % of value added)	Pollution control capital expenditures (as a % of total capital expenditures)
Petroleum	2.25%	15.42%	25.7%
Primary metals	1.68%	4.79%	11.6%
Paper	1.87%	4.13%	13.8%
(pulp mills)	(5.70%)	(12.39%)	(17.2%)
Chemical manufacturing	1.88%	3.54%	13.4%
Stone products	0.93%	1.77%	7.2%
Lumber	0.63%	1.67%	11.1%
Leather products	0.65%	1.37%	16.2%
Fabricated materials	0.65%	1.34%	4.6%
Food	0.42%	1.11%	5.3%
Rubber	0.49%	0.98%	2.0%
Textile	0.38%	0.93%	3.3%
Electric products	0.49%	0.91%	2.9%
Transportation	0.33%	0.80%	3.0%
Furniture	0.38%	0.73%	3.4%
Machinery	0.25%	0.57%	1.9%

tracked by the U.S. Census Bureau, and have been increasing steadily. Expenditures in 1972 totaled $52 billion (in 1990 dollars) and have been projected to grow to approximately $140 billion (1990 dollars), or 2.0-2.2% of Gross National Product, in the year 2000 (for a review and analysis of these data, see US Congress, Office of Technology Assessment, 1994).

These expenditures are not distributed uniformly among industry sectors. As shown in Table 12.3-1, sectors such as petroleum refining and chemical manufacturing spend much higher fractions of their net sales and capital expenditures on pollution abatement than other industrial sectors. Therefore, in these industrial

Table 12.3-2   Summary of Environmental Costs at the Amoco Yorktown Refinery (Heller, et al., 1995).

Cost category	Percentage of annual non-crude operating costs
Waste treatment	4.9%
Maintenance	3.3%
Product requirements	2.7%
Depreciation	2.5%
Administration, compliance	2.4%
Sulfur recovery	1.1%
Waste disposal	0.7%
Fees, fines, penalties	0.2%
Total costs	21.9%

Table 12.3-3 Summary of Environmental Costs at the DuPont LaPorte Chemical Manufacturing Facility (Shields, et al., 1995).

Cost category	Percentage of manufacturing costs
Taxes, fees, training, legal	4.0
Depreciation	3.2
Operations	2.6
Contract waste disposal	2.4
Utilities	2.3
Salaries	1.8
Maintenance	1.6
Engineering services	1.1
Total	19.1%

sectors, minimizing costs by preventing wastes and emissions will be far more strategic an issue than in other sectors.

Pollution abatement costs reported by individual companies both reflect these general trends and provide more detail about the magnitude and the distribution of environmental expenditures. For example, Tables 12.3-2 and 12.3-3 show the distribution of environmental costs reported by the Amoco Yorktown refinery and DuPont's LaPorte chemical manufacturing facility (Heller, et al., 1995; Shields, et al., 1995). In the case of the Amoco refinery, only about a quarter of the quantified environmental costs are associated with waste treatment and disposal—the costs summarized in Table 12.3-2. Costs associated with removing sulfur from fuels, meeting other environmentally-based fuel requirements and maintaining environmental equipment were greater than the costs associated with waste treatment and disposal. This indicates that the magnitude of environmental costs is substantially greater than that reported in Table 12.3-2—and that these costs may be hard to identify.

Table 12.3-3 shows that the profile of environmental costs at a DuPont chemical manufacturing facility exhibits many of the same characteristics. Waste treatment and disposal costs are less than a quarter of the annual, quantifiable, environmental costs.

Taken together, Tables 12.3-1 through 12.3-3 demonstrate that environmental costs are substantial, but that quantifying these costs is challenging. The next several sections of this chapter present a framework and procedures for estimating these costs.

12.4 A FRAMEWORK FOR EVALUATING ENVIRONMENTAL COSTS

Engineering projects are generally not undertaken unless they are financially justifiable. Projects designed to improve environmental performance beyond regulatory requirements usually must compete financially with all other projects under consideration at a facility. Fortunately, improved environmental performance is frequently profitable. Since the potential profitability of environmental projects is difficult to assess, it is common to neglect many of the financial benefits of improved environmental performance when projects are analyzed. That is why a better understanding of methods for estimating environmental costs and benefits serves to promote green engineering.

In this section, the types and magnitudes of costs associated with emissions and waste generation are described and categorized. Five categories, or tiers, of costs will be considered, following the framework recommended in the Total Cost Assessment Methodology developed by the American Institute of Chemical Engineers' Center for Waste Reduction Technologies (AIChE CWRT, 2000; see Appendix F). The tiers are:

- Tier I—Costs normally captured by engineering economic evaluations.
- Tier II—Administrative and regulatory environmental costs not normally assigned to individual projects.
- Tier III—Liability costs.
- Tier IV—Costs and benefits, internal to a company, associated with improved environmental performance.
- Tier V—Costs and benefits, external to a company, associated with improved environmental performance.

Tier I costs are the types of costs quantified in traditional economic analyses. Specific examples are provided in Table 12.4-1. As discussed in Sections 12.1 through 12.3, traditional accounting systems that focus on Tier I costs fail to capture some types of environmental costs. Examples of some of the costs that are frequently overlooked by traditional methods are listed in Table 12.4-2.

The costs listed in Table 12.4-2 are generally charged to overhead and therefore may be "hidden" when project costs are evaluated. These will be referred to as Tier II or hidden costs. Note that these costs are actually borne by facilities regardless of whether facilities choose to quantify them or assign them to project or product lines.

A less tangible set of costs are those designated as Tier III—liability costs. An accounting definition of liability is a "probable future sacrifice of economic benefits arising from present obligations to transfer assets or provide services in the future"

Table 12.4-1   Costs that are traditionally evaluated during financial analyses of projects.

- Capital equipment
- Materials
- Labor
- Supplies
- Utilities
- Structures
- Salvage value

Table 12.4-2   Environmental costs that are often charged to overhead.

- Off-site waste management charges
- Waste treatment equipment
- Waste treatment operating expenses
- Filing for permits
- Taking samples
- Filling out sample reporting forms
- Conducting waste and emission inventories
- Filling out hazardous waste manifests
- Inspecting hazardous waste storage areas and keeping logs
- Making and updating emergency response plans
- Sampling stormwater
- Making chemical usage reports (some states)
- Reporting on pollution prevention plans and activities (some states)

(Financial Accounting Standards Board Concept Statement No. 6, Paragraph 35 (1985); Institute of Management Accountants Statement No. 2A Management Accounting Glossary (1990)). Liability costs could include:

- Compliance obligations
- Remediation obligations
- Fines and penalties
- Obligations to compensate private parties for personal injury, property damage, and economic loss
- Punitive damages
- Natural resource damages

A final set of costs are designated as Tier IV or Tier V, which can be referred to as image or relationship costs (AIChE CWRT, 2000). These costs arise in relationships with customers, investors, insurers, suppliers, lenders, employees, regulators, and communities. They are perhaps the most difficult to quantify.

Thus, a basic framework for estimating costs and benefits associated with environmental activities consists of 5 tiers, beginning with the most tangible costs and extending to the least quantifiable costs. The remaining sections of this chapter will focus on methods for estimating Tier II, III, IV, and V costs. Tier I costs, by definition, are captured effectively by conventional accounting methods and are described in detail in texts on engineering economics (see, for example, Valle-Riestra, 1983). The description of Tier II costs in Section 12.5 focuses on methods for quantifying reporting, notification, and compliance costs. These are costs that are certain, yet are often difficult to separate from general overhead expenditures.

Estimating Tier III, IV, and V costs poses different challenges. These costs are often due to unplanned events, such as incidents that result in civil fines, remediation costs, or other charges. While these events are not planned, they do occur, and it is prudent to estimate the expected value of these costs. Arriving at an expected value for Tier III, IV, and V costs will involve estimating three distinct parameters:

1. The probability that an event will occur.
2. The costs associated with the event.
3. When the event will occur.

For example, if the goal is to estimate the expected value of a civil fine or penalty (a Tier III cost), the likelihood that a fine will be assessed and the likely magnitude of that fine must be calculated. If the probability of a fine being assessed is 0.1 (1 chance in 10) per year and the likely magnitude of the fine is $10,000, the expected annual cost due to fines would be $1000. For events that will occur in future years, such as costs of complying with anticipated future regulations, knowledge of when the event will occur is critical to determining the present value of the expected costs. These estimation methods are described in Sections 12.6 through 12.8.

12.5   HIDDEN ENVIRONMENTAL COSTS

Table 12.4-2 described a number of emission and waste management charges that are frequently viewed as overhead costs, and therefore can be overlooked by traditional accounting systems. These charges can be grouped into a number of broad categories, specifically waste treatment costs, regulatory compliance costs, and hidden capacity costs.

Waste treatment costs are the most straightforward to estimate. They are frequently hidden because many facilities charge the capital and operating costs of centralized air and water treatment facilities to overhead, rather than to specific processes. Specific treatment costs will vary from facility to facility and will depend strongly on the types of pollutants being treated. However, order-of-magnitude estimates of treatment costs can be estimated using values suggested by Douglas and co-workers (Schultz, 1998), as shown in Table 12.5-1.

Table 12.5-1 Order-of-Magnitude Estimates of Treatment Costs Developed by Douglas and Co-workers (Schultz, 1998).

Treatment technology	Operating cost ($/lb)	Capital cost ($/lb)
Air treatment	1.5×10^{-4}	1.0×10^{-3}
Water treatment		
Water flow	7.4×10^{-5}	7.4×10^{-4}
Organic loading	0.25	0.74
Incineration		
Organics/water	0.32	NA
Organic solids	0.80	NA
Landfill	0.12	NA
Deep well	0.30	NA

Example 12.5-1 (Adapted from Schultz, 1998)

A preliminary process design for a process to produce Bis (2-Hydroxyethyl) Tereph-thalate (BHET) from oxygen, ammonia, xylene and ethylene glycol results in the following estimates of raw material requirements and waste generation:

Raw Materials per mole of BHET (Molecular weight (MW) = 254)
  1 mole para-xylene (MW= 106; cost=$0.40/lb)
  2 moles ammonia (MW= 17; cost=$0.065/lb)
  2 moles ethylene glycol (MW= 62; cost=$0.176/lb)
  3+ moles oxygen (derived from air—no material acquisition cost)

Wastes generated per pound of product
  3.17 pounds of gaseous effluent to be treated
  0.39 pound of water to be treated
  0.01 pound of organic solid waste to be incinerated

Provide a preliminary estimate of waste treatment costs and compare these to raw material costs per pound of product.

Solution: The costs of raw materials per pound-mole of product are:

$$106*\$0.40 + 2*17*\$0.065 + 2*62*\$0.176 = \$66.4 \text{ per 254 lb product}$$
$$= \$0.26 \text{ per pound product}$$

The waste disposal operating costs are:

$$3.17*\$0.00015 + 0.39*\$0.000074 + 0.01*\$0.80 = \$0.0085 \text{ per pound}$$

The costs total about 3% of raw material costs (reasonably consistent with the data presented in Section 12.3) and are dominated by the costs of incineration.

A second major category of hidden environmental costs are the personnel costs associated with meeting environmental regulations. These costs are difficult to account for because environmental reporting and recordkeeping is frequently performed by corporate staff who divide their time between many different processes and facilities. Nevertheless, it is possible to estimate the time required to meet notification, reporting, manifesting, and other administrative tasks associated with environmental record keeping. Appendix E contains worksheets that can be used for this purpose.

Finally, two major costs associated with waste generation that are frequently overlooked are lost raw materials and lost capacity. As an example, consider a process that converts raw material A into product P and waste W. If the yield for the process is increased from 90% to 95%, waste generation and therefore waste disposal costs are cut in half. Not as obvious, however, is the fact that for a given quantity of raw material, the yield of product has increased by 5.5% (5% increase in yield/90% base yield). Further, the same processing equipment (reactors) are able to increase production, and the costs for separating product from wastes may decrease dramatically. Savings due to increased production capacities and increased use of raw materials can often be more substantial than avoided treatment costs.

Example 12.5-2

A chemical manufacturing facility buys raw material for $0.50 per pound and produces 90 million pounds per year of product, which is sold for $0.75 per pound. The process is typically run at 90% selectivity and the raw material that is not converted into product is disposed of at a cost of $0.80 per pound (by incineration). A process improvement allows the process to be run at 95% selectivity, allowing the facility to produce 95 million pounds per year of product. What is the net revenue of the facility (product sales − raw material costs − waste disposal costs) before and after the change? How much of the increased net revenue is due to increased sales of product and how much is due to decreased waste disposal costs?

Solution: The net revenue before the change is:

$$(90 \text{ million pounds} * \$0.75/\text{pound} - 100 \text{ million pounds raw material} * \$0.50/\text{pound}$$
$$-10 \text{ million pounds waste} * \$0.80) = \$9.5 \text{ million/year}$$

The net revenue after the change is:

$$(95 \text{ million pounds} * \$0.75/\text{pound} - 100 \text{ million pounds raw material} * \$0.50/\text{pound}$$
$$- 5 \text{ million pounds waste} * \$0.80) = \$17.25 \text{ million/year}$$

Of the difference ($7.75 million), about half ($3.75 million) is due to increased product sales and the remainder is due to decreased disposal cost. Note that the disposal cost assumed in this example is very high and thus represents a likely upper bound on these costs. It should also be noted that the cost of capital depreciation per pound of product is reduced after the change.

12.6   LIABILITY COSTS

Tier III (liability) costs include future compliance costs and compliance obligations, potential civil and criminal fines and penalties, potential remedial costs of contamination, potential compensation and punitive damages, potential judgements for natural resource damage, Potentially Responsible Party (PRP) liabilities for off-site contamination, and potential industrial process risk. Estimation methodologies for each of these costs have been developed through the AIChE's Center for Waste Reduction Technologies (AIChE CWRT, 2000) using the data described in Table 12.6-1. A summary of most of the available estimation methodologies has been assembled by the US EPA (US EPA, 1996).

It is beyond the scope of this chapter to describe the methodologies for estimating all of these costs. (See Appendix F for sources of additional information.) Instead, since the procedures for estimating the costs in each of the categories are similar, our focus will be on the procedures. The procedures will be illustrated by considering the cost categories of civil and criminal fines and penalties, and Potentially Responsible Party liabilities for off-site contamination.

Table 12.6-1   Sources of Data Used in AIChE CWRT Total Cost Assessment Methodology (AIChE CWRT, 2000).

Type III cost	Data sources
Compliance obligations	EPA's Basis and Purpose Documents (BPDs), Back ground Information Documents (BIDs), and Economic Impact Analysis (EIA) prepared by the US EPA for proposed National Emission Standards for Hazardous Air Pollutants (NESHAPs)
Civil and criminal fines and penalties	EPA's Integrated Data for Enforcement Analysis (IDEA) database
Remedial costs of contamination	Federal Remediation Technologies Roundtable website (case studies for 141 remedial full-scale and demonstration projects); data on the types of contaminants, remedial technologies, and overall project costs
Compensation and punitive damages	Compilation of individually reported compensation amounts for toxic torts from published literature
Natural resource damage	Compilation of individually reported natural resource damage amounts from published literature
Potentially Responsible Party liabilities for off-site contamination	EPA CERCLIS* database
Industrial process risk	EPA ARIP** database
	Production downtime (company-specific, e.g., daily cost of production downtime)

*Comprehensive Environmental Response, Compensation, and Liability Information System
**Accident Release Information Program

In each case, arriving at an expected value of the liability cost will involve estimating three parameters:

1. The probability that an event will occur.
2. The costs associated with the event.
3. When the event will occur.

Consider first the cost category of civil and criminal fines and penalties. Even the best-run manufacturing facilities have occasional violations of environmental statutes. These might be violations due to inadequate reporting or notification (often called paperwork violations), or violations due to process upsets. Most companies keep historical records of these violations, and these can be used to estimate the probability of future fines and penalties. In estimating the probability of a fine or penalty, it should be recognized that not all process units are equally likely to be fined. Factors influencing the probability of a fine or penalty include (AIChECWRT, 2000):

- The extent that spill control measures will be in place.
- The history and reputation of the plant or company.
- The local culture and visibility of the operation to non-governmental organizations.

- How well the administrative requirements of monitoring, recording, and recordkeeping will be maintained.
- The toxicity of the potential contaminants.
- The chance for a large release.

Because probabilities of fines and penalties can vary widely from company to company, this section will assume that these probabilities are known, either through company data or through estimates based on information assembled by the CWRT (AIChE CWRT, 2000).

Estimated magnitudes of fines and penalties vary by statute, as shown in Table 12.6-2. They also vary greatly in magnitude within a given governing statute. For example, most civil fines under the Safe Drinking Water Act are under $20,000, but the largest fines can be as high as $2,500,000. This skewed distribution results in large differences between average and median values for fines and penalties.

Example 12.6-1 illustrates how probabilities of occurrence can be combined with estimated costs to lead to an expected value of civil fine and penalties.

Table 12.6-2   Summary of Penalty Data Assembled for the Total Cost Accounting Methodology of the AIChE CWRT (AIChE CWRT, 2000).

	Administrative Fines				Civil Judicial Fines			
Statute	Number of Cases	Average	Median	Maximum	Number of Cases	Average	Median	Maximum
CAA	486	$21,000	$10,000	$300,000	157	$486,000	$150,000	$11,000,000
CWA	767	$19,000	$10,000	$150,000	111	$669,000	$201,000	$14,040,000
EPCRA	885	$18,000	$7,000	$210,000	3	$31,000	$13,000	$74,000
FIFRA	456	$12,000	$3,000	$876,000	6	$8,000	$2,000	$39,000
RCRA	904	$31,000	$1,000	$1,020,000	44	$795,000	$163,000	$8,000,000
SDWA	160	$7,000	$3,000	$125,000	18	$247,000	$20,000	$2,500,000
TSCA	662	$65,000	$14,000	$4,000,000	7	$50,000	$33,000	$142,000

CAA: Clean Air Act
CWA: Clean Water Act
EPCRA: Emergency Planning and Community Right to Know Act
FIFRA: Federal Insecticide, Fungicide and Rodenticide Act
RCRA: Resource Conservation and Recovery Act
SDWA: Safe Drinking Water Act
TSCA: Toxic Substances Control Act

Example 12.6-1

A manufacturing facility operates under an air permit and generates an industrial hazardous waste. The facility has a good record of compliance with air regulations (1 violation in the past 5 years due to a release during an emergency shutdown) and has had two violations under RCRA during the past five years—both due to improper completion of hazardous waste manifest reports. Estimate the annual costs due to civil and administrative fines and penalties.

Solution: Based on the historical data, the probability of an air release resulting in a fine is 0.2/year. If these releases are due to an emergency shutdown and the emergency release is properly reported, an administrative fine, rather than a civil fine, might be anticipated. The expected value of this cost could be calculated using either the average or median value of administrative fines under the Clean Air Act.

Expected annual cost of clean air act fines based on median fine = 0.2 * (10,000) = $2,000

Expected annual cost of clean air act fines based on average fine = 0.2 * (21,000) = $4,000

In contrast, if the violation resulted in a civil fine the expected costs would be:

Expected annual cost of clean air act fines based on median fine = 0.2 * (150,000)

$$= \$30,000$$

Expected annual cost of clean air act fines based on average fine = 0.2 * (486,000)

$$= \$100,000$$

Again, based on historical data, the annual probability of a violation of RCRA is 0.4. Assuming that a paperwork violation would result in an administrative fine, the expected cost would be:

Expected annual cost of RCRA fines based on median fine = 0.4 * (1,000) = $400

Expected annual cost of RCRA fines based on avereage fine = 0.4 *(31,000) = $12,000

The range of costs calculated in this example point out that fines and penalties can either be relatively minor costs or they can be major costs. The range of values highlights the importance of collecting company-specific data in estimating likely fines and penalties.

Consider next another category of Tier III costs, Potentially Responsible Party liabilities for off-site contamination. These costs arise when a facility is identified as responsible for site contamination, and therefore must bear the cost of remediating the site. The probability of a remediation cost occurring is, of course, strongly dependent on the practices used in managing wastes and emissions. Company-specific data should be used whenever possible in estimating these probabilities. Again, this section will assume that these probabilities are known, either through company data or through estimates based on information assembled by the CWRT (AIChE CWRT, 2000).

The magnitude of remediation liabilities can be large, as shown in Table 12.6-3, and can depend on a number of factors, including

- the number of responsible parties at the site
- the volume of waste disposed at the site relative to other parties
- the toxicity of the contaminants
- future use of the site

Again, the costs can vary greatly in magnitude and the skewed nature of the cost distribution results in large differences between average and median values for

Table 12.6-3   Typical Remediation Costs (AIChE CWRT, 2000).

	Average	Low	Median	High
Soil/Sediment Remediation Cost	$20,861,000	$114,000	$2,602,000	$192,395,000
Ground Water Remediation Cost	$8,366,000	$246,000	$2,820,000	$53,847,000

fines and penalties. Example 12.6-2 illustrates how these data might be used to arrive at expected values for remediation costs.

Example 12.6-2

A manufacturing facility generates an industrial hazardous waste and sends that waste to a landfill. In order to anticipate future remediation costs, the company collects data on the number of remediation actions with which operators of similar disposal sites have been associated. The data indicate that on average, none of the similar sites have required remediation after 5 years of operation, 10% of similar disposal sites have required groundwater remediation after 10 years of operation and 30% of similar sites have required groundwater remediation after 15 years of operation.

The landfill that the facility uses has been in operation for five years and is used in roughly equal amounts by 5 manufacturing facilities. Estimate the expected remediation costs over the next 10 years.

Solution: Based on the historical data, the probability of groundwater remediation being required in the next year is 0.02, based on a linear interpolation of probability of remediation. The expected value of the groundwater remediation cost in the first year is:

Expected first year cost of groundwater remediation based on median cost

$$= 0.02 * (2,820,000) = \$60,000$$

If the costs are shared equally between 6 potentially responsible parties (5 generators of waste and the operator of the landfill), the expected cost in the first year is $10,000.

The probability of groundwater remediation increases between year 1 and year 2 by 2% (from a cumulative probability of 2% to a cumulative probability of 4%). Therefore the expected additional cost of failure occurring in year 2 is the same as in year 1—$10,000. The remediation costs are likely to escalate in year 2 relative to year 1, but if the cost is then converted back to a present value, the present value of the remediation cost can be assumed to be the same as the current remediation cost. Thus, the present value of the year 2 remediation cost is approximately $10,000. The expected present values of remediation costs is the same in years 3–5.

In year 6, the incremental probability of remediation costs increases from 10% to 14% (again assuming a linear interpolation of remediation probability). The expected present value of the cost in years 6–10 would be $20,000.

Thus, the approximate present value of the remediation costs in years 1–10 is $150,000 ($10,000 per year in years 1–5 and $20,000 per year in years 6–10).

Example 12.6-2 illustrates again the importance of having relevant data on the probability of the occurrence of environmental costs. The costs could be relatively modest, but might also range into hundreds of millions of dollars.

12.7 INTERNAL INTANGIBLE COSTS

Even more difficult to estimate than liability costs are a set of environmental costs and benefits that are referred to as intangibles. This section briefly reviews types of intangible costs experienced directly by companies (internal intangible costs) and suggests sources of data for estimating these costs. Section 12.8 describes intangible costs borne by individuals and organizations external to companies.

Major categories of internal intangible costs are listed in Table 12.7-1, along with sources of data relevant to estimating these costs. These data sources are described at length in the AIChE CWRT's Total Cost Accounting methodology.

Brief definitions of each of these cost categories are provided below:

Staff (productivity, morale, turnover, union negotiating time): Poor environmental performance, particularly as reflected in workplace conditions, may lead to increased rates of illness, lower productivity, and more staff turnover.

Market share: Limited anecdotal evidence exists relating negative environmental incidents to loss in market share; other evidence points to the positive influence of "green" handbooks and other environmental ratings.

License to operate: This is not the direct costs associated with obtaining legally required permits; rather, these are costs associated with issues such as delays in receiving permits.

Investor relationships: Relationships with investors can be, at least in part, reflected in stock price.

Lender relationships: Relationships with lenders can be, at least in part, reflected in bond ratings.

Community and regulator relationships: Relationships with regulators and the community are related to license to operate.

Table 12.7-1 Sources of Data on Internal Intangible Costs (AIChE CWRT, 2000).

Type IV Cost	Data Sources
Staff (productivity, morale, turnover, union negotiating time)	Published literature on costs of injuries in specific industries; published literature on costs to employers of mortality and illness.
Market share	Published literature on market values of environmental reputation; published literature on loss of market share after environmental incidents; published literature on market share effects of negative news reports.
License to operate	Historical data on permitting.
Investor relationships	Published literature on the effects on share value of environmental reputation; published literature on decreases in stock prices following environmental incidents; published literature on the effect of negative news reports on share price.
Lender relationships	Data on the effect of environmental incidents on credit ratings.
Community relationships	Costs and benefits of public relations programs.
Regulator relationships	Costs of new regulations.

It is beyond the scope of this chapter to describe the methodologies for estimating all of these costs. Developing cost estimates is made difficult by the variability and uncertainty in much of the data. As an example of this variability and uncertainty, consider the problem of estimating the response of stock prices to environmental reputation. Of the literature reviewed by the AIChE CWRT(2000), some found positive associations between a positive environmental reputation and higher stock prices. Other studies found relationships between negative environmental news or performance and lower stock prices. These studies used widely ranging measures of environmental performance, from emissions reported through the Toxic Release Inventory, to the number of oil spills and whether companies sign on to a set of corporate environmental principles. Thus, it is difficult to design a cost estimation methodology that can employ this broad range of data. Further complicating the situation is the fact that some studies found little to no relationship between stock price and environmental performance.

Nevertheless, these internal intangible costs are widely regarded as real, albeit extremely difficult to quantify.

12.8   EXTERNAL INTANGIBLE COSTS

External intangible costs are costs borne broadly by communities due to emissions, wastes, resource depletion, and habitat destruction. Examples of these external impacts, which are sometimes referred to as externalities, along with sources of data that can be used to estimate their costs, are listed in Table 12.8-1.

It is beyond the scope of this chapter to describe the methodologies for estimating all of these costs. Instead, since the procedures for estimating the costs in

Table 12.8-1   Sources of Data on External Intangible Costs (AIChE CWRT, 2000).

Type V Cost	Data Sources
Pollutant discharges to air	Costs per ton of greenhouse gas emitted; costs per case of disease or mortality; published literature on the social costs of global warming.
Pollutant discharges to surface water	Cost of lost fishing habitat and fisheries resources, using published literature; cost of market transfers of water for environmental protection.
Pollutant discharges to ground water/deep well	Costs of fresh water use; costs to desalinate.
Pollutant discharges to land	Published literature on willingness-to-pay scales, related to recreational land use or conservation of land; costs and benefits of preserving undeveloped land.
Natural habitat impacts	Published data on the costs of restoring wetlands, habitats or species; violation of societal benefits of wetlands; published literature on willingness-to-pay scales, related to preservation of natural habitat.

each of the categories have similar features, our focus will be on the procedures. The procedures will be illustrated by considering the cost categories of air pollutant discharges.

A number of studies have recently appeared attempting to determine actual health-related costs associated with air pollutants, especially ozone and particulate matter. These studies attempt to quantify direct health costs and lost work time associated with air pollutant morbidity and attempt to account for air pollutant mortality by valuing lost earning power and other factors. Typically, when such estimates are done for large urban areas such as Los Angeles and Houston, the costs associated with concentrations of ozone and particulate matter in excess of National Ambient Air Quality Standards are billions of dollars per year. Attributing these externalities to individual emission sources is possible, but would be region-specific and has rarely been done. In contrast, most valuations of externalities rely on surveys that assess the public's willingness to pay for avoidance of the impacts. The results of willingness-to-pay surveys and other measures of external costs vary widely. The AIChE CWRT (2000), for example, quotes values of external costs ranging from $0.22 to $19 per ton of CO emissions. Particulate matter externalities range from $600 to $26,000 per ton. Similar ranges are reported for Hg, sulfur dioxide, oxides of nitrogen, and a variety of hazardous air pollutants.

REFERENCES

American Institute of Chemical Engineers Center for Waste Reduction Technologies (AIChE CWRT), "Total Cost Assessment Methodology," AIChE, New York (ISBN 0-8169-0807-9), 2000.

Hall, J.V., Winer, A.M., Kleinman, M.T., Lurmann, F.W., Brajer, V., and Colome, S.D. (1992), "Valuing the health benefits of clean air," *Science,* 255, 812–817.

Heller, M., Shields, P.D., and Beloff, B. (1995) "Environmental Accounting Case Study: Amoco Yorktown Refinery," Green Ledgers: Case Studies in Corporate Environmental Accounting, D. Ditz, J. Ranganathan, and D. Banks, eds., World Resources Institute (ISBN 1-56973-032-6), Washington, D.C.

Kennedy, M., *Total Cost Assessment for Environmental Engineers and Managers*, John Wiley & Sons, New York, 1997.

Lurmann, F.W., Hall, J.V., Kleinman, M., Chinkin, L.R., Brajer, V., Meacher, D., Mummery, F., Arndt, R.L., Funk, T.H., Alcorn, S.H., and Kumar, N., "Assessment of the Health Benefits of Improving Air Quality in Houston, Texas," Final report by Sonoma Technologies to the City of Houston (STI-998460-1875-FR), November, 1999.

Schultz, M.A., "A Hierarchical Decision Procedure for the Conceptual Design of Pollution Prevention Alternatives for Chemical Processes," Ph.D. Thesis, University of Massachusetts, 1998.

Shields, P., Heller, M., Kite, D., and Beloff, B. (1995) "Environmental Accounting Case Study: DuPont," Green Ledgers: Case Studies in Corporate Environmental Accounting, D. Ditz, J. Ranganathan, and D. Banks, eds., World Resources Institute (ISBN 1-56973-032-6), Washington, D.C.

United States Congress, Office of Technology Assessment (OTA), "Industry, Technology and the Environment: Competitive Challenges and Business Opportunities," OTA-ITE-586 (Washington, D.C.: US Government Printing Office, January 1994).

United States Environmental Protection Agency (US EPA), "Pollution Prevention Benefits Manual," Office of Policy Planning and Evaluation and Office of Solid Waste, 1989.

United States Environmental Protection Agency (US EPA), "A Primer for Financial Analysis of Pollution Prevention Projects," EPA 600R-93-059, April 1993.

United States Environmental Protection Agency (US EPA), "Valuing Potential Environmental Liabilities for Managerial Decision-Making: A Review of Available Techniques," EPA 742R-96-003, December, 1996.

United States Environmental Protection Agency (US EPA), "An Introduction to Environmental Accounting as a Business Management Tool: Key Concepts and Terms," EPA 742-R-95-001, June, 1995.

United States Environmental Protection Agency (US EPA), "Searching for the Profit in Pollution Prevention: Case Studies in the Corporate Evaluation of Environmental Opportunities," EPA 742-R-98-005, April 1998.

Valle-Riestra, J. F., *Project Evaluation in the Chemical Process Industries*, McGraw-Hill, New York, 1983.

PROBLEMS

1. A preliminary process design for a process to produce cyclohexanone ($0.73/lb) and cyclohexanol ($0.83/lb) from cyclohexane ($0.166/lb) and oxygen (from air) results in the following estimates of raw material requirements and waste generation (see Chapter 8):

   Raw Materials per mole of cyclohexanone/cyclohexanol
   Avg. Molecular weight (MW = 99)
   1.1 mole cyclohexane (MW = 84)
   2 moles oxygen (derived from air—no material acquisition cost)

   Wastes generated per pound of product
   0.060 pound of organics in the gaseous effluent to be treated
   0.2 pound of organic aqueous wastes to sent to water treatment

   Provide a preliminary estimate of waste treatment costs and compare these to raw material costs per pound of product (Calculate the organic loading in the liquid waste and the total quantity of air to be treated by mass balance).

2. Select a process documented in the AP-42 documents at www.epa.gov/chief/ and estimate the costs of waste treatment per pound of product.

3. A chemical manufacturing facility buys raw material for $0.60 per pound and produces 90 million pounds per year of product, which is sold for $0.75 per pound. The process is typically run at 90% selectivity and the raw material that is not converted into product is disposed of at a cost of $0.80 per pound (by incineration). A process improvement allows the process to be run at 98% selectivity, allowing the facility to produce 98 million pounds per year of product. What is the net revenue of the facility (product sales − raw material costs − waste disposal costs) before and after the

change? How much of the increased net revenue is due to increased sales of product and how much is due to decreased waste disposal costs?

4. Lurmann, et al. (1999) have estimated the costs associated with ozone and fine particulate matter concentrations above the National Ambient Air Quality Standards (NAAQSs) in Houston. They estimated that the economic impacts of early mortality and morbidity associated with elevated fine particulate matter concentrations (above the NAAQS) are approximately $3 billion/year. Hall, et al. (1992) performed a similar assessment for Los Angeles. In the Houston study, Lurmann, et al examined the exposures and health costs associated with a variety of emission scenarios. One set of calculations demonstrated that a decrease of approximately 300 tons/day of fine particulate matter emissions resulted in a 7 million person-day decrease in exposure to particulate matter concentrations above the proposed NAAQS for fine particulate matter, 17 fewer early deaths per year, and 24 fewer cases of chronic bronchitis per year. Using estimated costs of $300,000 per case of chronic bronchitis and $6,000,000 per early death, estimate the social cost per ton of fine particulate matter emitted. How does this compare to the range of values quoted by the AIChE CWRT? Review the procedures for estimating costs (see Hall, et al., 1992) and comment on the uncertainties associated with the methodology.

5. Browse the website of the World Business Council for Sustainable Development (www.wbcsd.ch) and identify a case study of a company improving business performance through eco-efficiency. Write a one page summary of the case study.

PART III

Moving Beyond the Plant Boundary

OVERVIEW

Part II presented tools for evaluating and improving the environmental performance of chemical processes, but the analysis ended at the boundary of the flowsheet. While it is appropriate for chemical engineers to focus on the process flowsheet, where the chemical process design engineer has the most control, it is also important to recognize that chemical manufacturing processes are linked to both suppliers and customers. Customers will be concerned about the environmental performance of chemical products that they use, and so process design engineers must increasingly become stewards for their products. In addition, chemical processes are linked to their suppliers, so engineers must be aware of the linkages between their processes and other chemical processes and other industrial sectors. The group of chapters listed below addresses these issues of product stewardship and industrial networks.

1. Chapter 13, covering life-cycle assessment, presents emerging product stewardship tools.
2. Chapter 14, covering industrial ecology, presents tools for analyzing mass and energy flows within networks of industrial processes.

More specifically, Chapter 13 presents a methodology for tracking the flows of energy, materials, and waste streams required in the manufacture, use, and disposal of products. This methodology, called life-cycle assessment, is used to identify opportunities for improving the environmental performance of products throughout their life cycle, from raw material extraction to waste disposal. Chapter 13 covers the basic principles of life-cycle assessment, and demonstrates the applications and limitations of the methodology through a series of case studies. Chapter 14 examines the flows of energy, materials, and wastes among industrial sectors. These studies attempt to identify potential exchanges of material and energy between industrial processes that create the type of symbiotic relationships often found in ecosystems—hence the name industrial ecology. Chapter 14 illustrates the principles of industrial ecology through a number of case studies.

CHAPTER 13

Life-Cycle Concepts, Product Stewardship and Green Engineering

by
Kirsten Rosselot and David T. Allen

13.1 INTRODUCTION TO PRODUCT LIFE CYCLE CONCEPTS

Products, services, and processes all have a life cycle. For products, the life cycle begins when raw materials are extracted or harvested. Raw materials then go through a number of manufacturing steps until the product is delivered to a customer. The product is used, then disposed of or recycled. These product life-cycle stages are illustrated in Figure 13.1-1, along the horizontal axis. As shown in the figure, energy is consumed and wastes and emissions are generated in all of these life cycle stages.

Processes also have a life cycle. The life cycle begins with planning, research and development. The products and processes are then designed and constructed. A process will have an active lifetime, then will be decommissioned and, if necessary, remediation and restoration may occur. Figure 13.1-1, along its vertical axis, illustrates the main elements of this process life cycle. Again, energy consumption, wastes, and emissions are associated with each step in the life cycle.

Traditionally, product and process designers have been concerned primarily with product life-cycle stages from raw material extraction up to manufacturing. That focus is changing. Increasingly, chemical product designers must consider how their products will be recycled. They must consider how their customers use their products. Process designers must avoid contamination of the sites at which their processes are located. Simply stated, design engineers must become stewards for their products and processes throughout their life cycles. These increased responsibilities for products and processes throughout their life cycles have been recognized by a number of professional organizations. Table 13.1-1 describes a Code of Product Stewardship developed by the Chemical Manufacturers' Association (now named the American Chemistry Council).

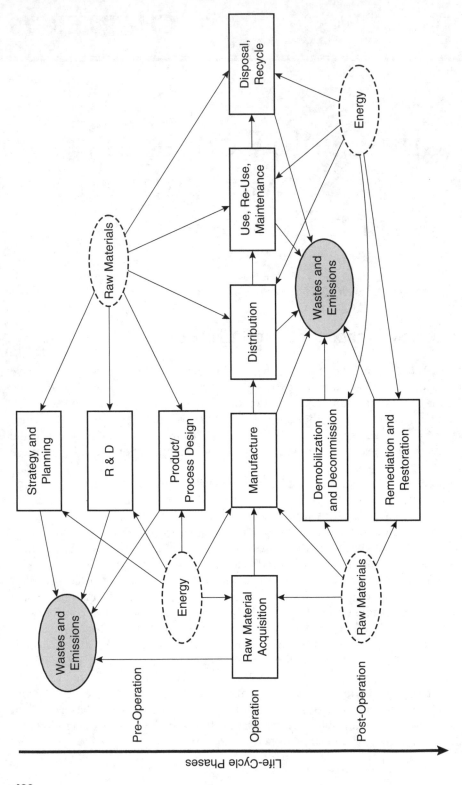

Figure 13.1-1 Product life cycles include raw material extraction, material processing, use, and disposal steps, and are illustrated along the horizontal axis. Process life cycles include planning, research, design, operation, and decommissionning steps and are shown along the vertical axis. In both product and process life cycles, energy and materials are used at each stage of the life cycle and emissions and wastes are created.

Table 13.1-1   The Chemical Manufacturers' Association (American Chemistry Council) Product Stewardship Code.

The purpose of the Product Stewardship Code of Management Practices is to make health, safety and environmental protection an integral part of designing, manufacturing, marketing, distributing, using, recycling and disposing of our products. The Code provides guidance as well as a means to measure continuous improvement in the practice of product stewardship.

The scope of the Code covers all stages of a product's life. Successful implementation is a shared responsibility. Everyone involved with the product has responsibilities to address society's interest in a healthy environment and in products that can be used safely. All employers are responsible for providing a safe workplace, and all who use and handle products must follow safe and environmentally sound practices.

The Code recognizes that each company must exercise independent judgment and discretion to successfully apply the Code to its products, customers and business.

Relationship to Guiding Principles

Implementation of the Code promotes achievement of several of the Responsible Care Guiding Principles:

- To make health, safety and environmental considerations a priority in our planning for all existing and new products and processes;
- To develop and produce chemicals that can be manufactured, transported, used and disposed of safely;
- To extend knowledge by conducting or supporting research on the health, safety and environmental effects of our products, processes and waste materials;
- To counsel customers on the safe use, transportation and disposal of chemical products;
- To report promptly to officials, employees, customers and the public, information on chemical-related health or environmental hazards and to recommend protective measures;
- To promote the principles and practices of Responsible Care by sharing experiences and offering assistance to others who produce, handle, use, transport or dispose of chemicals.

Effective product and process stewardship requires designs that optimize performance throughout the entire life cycle. This chapter provides an introduction to tools available for assessing the environmental performance of products and processes throughout their life cycle. The primary focus is on product life cycles, but similar concepts and tools could be applied to process life cycles. Sections 13.2 and 13.3 present quantitative tools used in product life cycle assessments (LCAs). Section 13.4 presents more qualitative tools. Section 13.5 describes a number of applications for these tools and Section 13.6 summarizes the main points of the chapter.

13.2   LIFE-CYCLE ASSESSMENT

Life-cycle studies range from highly detailed and quantitative assessments that characterize, and sometimes assess, the environmental impacts of energy use, raw material use, wastes, and emissions over all life stages, to assessments that qualitatively identify and prioritize the types of impacts that might occur over a life cycle. As shown in this chapter, different levels of detail and effort are appropriate for the different ways the life-cycle information is used. In this section, the steps involved in conducting detailed, highly quantitative life-cycle assessments are described.

13.2.1 Definitions and Methodology

There is some variability in life-cycle assessment terminology, but the most widely accepted terminology has been codified by international groups convened by the Society for Environmental Toxicology and Chemistry (SETAC) (see, for example, Consoli, et al., 1993). Familiarity with the terminology of life-cycle assessment makes communication of results easier and aids in understanding the concepts presented later in this chapter. To begin, a **life-cycle assessment** (LCA) is the most complete and detailed form of a life-cycle study. A life-cycle assessment consists of four major steps.

Step 1: The first step in an LCA is to determine the scope and boundaries of the assessment. In this step, the reasons for conducting the LCA are identified; the product, process, or service to be studied is defined; a functional unit for that product is chosen; and choices regarding system boundaries, including temporal and spatial boundaries, are made. But what is a functional unit, and what do we mean by system boundaries? Let's look first at the system boundaries.

The **system boundaries** are simply the limits placed on data collection for the study. The importance of system boundaries can be illustrated by a simple example. Consider the problem of choosing between incandescent light bulbs and fluorescent lamps in lighting a room. During the 1990s the US EPA began its Green Lights program, which promoted replacing incandescent bulbs with fluorescent lamps. The motivation was the energy savings provided by fluorescent bulbs. Like any product, however, a fluorescent bulb is not completely environmentally benign, and a concern arose during the Green Lights program about the use of mercury in fluorescent bulbs. Fluorescent bulbs provide light by causing mercury, in glass tubes, to fluoresce. When the bulbs reach the end of their useful life, the mercury in the tubes might be released to the environment. This environmental concern (mercury release during product disposal) is far less significant for incandescent bulbs. Or is it? What if we changed our system boundary? Instead of just looking at product disposal, as shown in the first part of Figure 13.2-1, what if the entire product life cycle were considered, as shown in the second part of Figure 13.2-1? In a comparison of the incandescent and fluorescent lighting systems, if the system boundary is selected to include electric power generation as well as disposal, the analysis changes. Although mercury is a trace contaminant in coal, the burning of coal is the greatest contributor of Hg releases to the atmosphere. Since an incandescent bulb requires more energy to operate, the use of an incandescent bulb results in the release of more mercury to the atmosphere than the use of a fluorescent bulb. Over the lifetime of the bulbs, more mercury can be released to the

Lighting Life-Cycle System

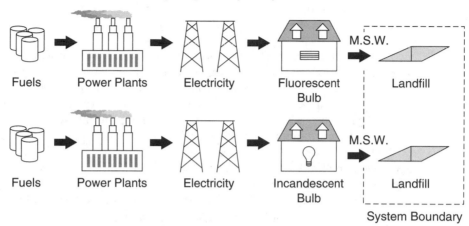

Lighting Life-Cycle System

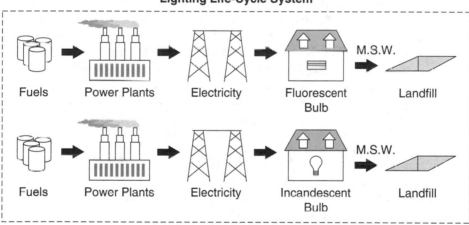

Figure 13.2-1   The importance of system boundaries in life-cycle assessment is illustrated by the case of lighting systems. As noted in the text, fluorescent bulbs contain mercury and if these bulbs are sent directly to municipal solid waste landfills, mercury might be released into the environment. Use of incandescent bulbs would result in a smaller amount of mercury in the municipal solid waste stream. Thus, an analysis focusing on just municipal solid waste disposal would conclude that fluorescent bulbs release more mercury to the environment than incandescent bulbs. If a larger system is considered, however, the conclusion changes. Mercury is a trace contaminant in coal, and when coal is burned to generate electricity, some mercury is released to the atmosphere. Since an incandescent bulb requires more energy to operate, the use of an incandescent bulb can result in the release of more mercury to the atmosphere than the use of a fluorescent bulb. Over the lifetime of the bulbs, more mercury can be released to the environment due to energy use than due to disposal of fluorescent bulbs.

environment due to the burning of coal than due to the disposal of fluorescent bulbs. Thus, the simple issue of determining which bulb, over its life cycle, results in the release of more mercury depends strongly on how the boundaries of the system are chosen.

As this simple example illustrates, the choice of system boundaries can influence the outcome of a life-cycle assessment. A narrowly defined system requires less data collection and analysis, but may ignore critical features of a system. On the other hand, in a practical sense it is impossible to quantify all impacts for a process or product system. In our simple example, should we also assess the impacts of mining the metals, and making the glass used in the bulbs we are analyzing? In general, we would not need to consider these issues if the impacts are negligible, compared to the impacts associated with operations over the life of the equipment. On the other hand, for specific issues, such as mercury release, some of these ancillary processes could be important contributors. What is included in the system and what is left out is generally based on engineering judgement and a desire to capture any parts of the system that may account for 1% or more of the energy use, raw material use, wastes or emissions.

Another critical part of defining the scope of a life-cycle assessment is to specify the **functional unit.** The choice of functional unit is especially important when life-cycle assessments are conducted to compare products. This is because functional units are necessary for determining equivalence between the choices. For example, if paper and plastic grocery sacks are to be compared in an LCA, it would not be appropriate to compare one paper sack to one plastic sack. Instead the products should be compared based on the volume of groceries they can carry. Because fewer groceries are generally placed in plastic sacks than in paper sacks, some LCAs have assumed a functional equivalence of two plastic grocery sacks to one paper sack. Differing product lifetimes must also be evaluated carefully when using life-cycle studies to compare products. For example, a cloth grocery sack may be able to hold only as many groceries as a plastic sack, but will have a much longer use lifetime that must be accounted for in performing the LCA. As shown in the problems at the end of this chapter, the choice of functional unit is not always straightforward and can have a profound impact on the results of a study.

Step 2: The second step in a life-cycle assessment is to inventory the inputs, such as raw materials and energy, and the outputs, such as products, byproducts, wastes, and emissions, that occur and are used during the life cycle. This step, shown conceptually in Figure 13.2-2, is called a **life-cycle inventory,** and is often the most time consuming and data intensive portion of a life-cycle assessment. Examples of life-cycle inventories and more detail concerning the structure of a life-cycle inventory are provided in the next section.

Step 3: The output from a life-cycle inventory is an extensive compilation of specific materials used and emitted. Converting these inventory elements into an assessment of environmental performance requires that the emissions and material use be transformed into estimates of environmental impacts. Thus, the third step in a life-cycle assessment is to assess the environmental impacts of the inputs and out-

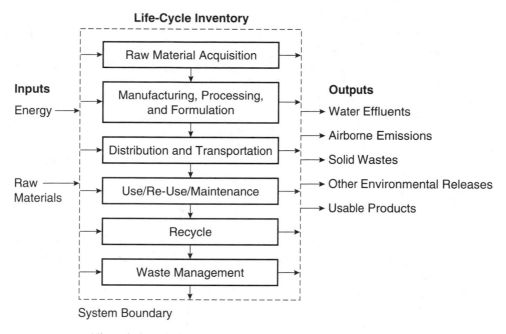

Figure 13.2-2 Life-cycle inventories account for material use, energy use, wastes, emissions, and co-products over all of the stages of a product's life cycle.

puts compiled in the inventory. This step is called a **life-cycle impact assessment.** This topic will be discussed in detail in Section 13.2.3.

Step 4: The fourth step in a life-cycle assessment is to interpret the results of the impact assessment, suggesting improvements whenever possible. When life-cycle assessments are conducted to compare products, for example, this step might consist of recommending the most environmentally desirable product. Alternatively, if a single product were analyzed, specific design modifications that could improve environmental performance might be suggested. This step is called an **improvement analysis** or an **interpretation step.**

13.2.2 Life-Cycle Inventories

A life-cycle inventory is a set of data and material and energy flow calculations that quantifies the inputs and outputs of a product life cycle. Some of the values that are sought during the inventory process are objective quantities derived using tools such as material and energy balances. As is shown later in this section, other values are more subjective and depend on choices and assumptions made during the assessment.

Before describing in detail the data elements associated with a life-cycle inventory, take a moment to review the stages of a product life cycle. The first stage in a product life cycle, as shown along the horizontal axis of Figure 13.1-1, is raw material acquisition. Examples of raw material acquisition are timber harvesting, crude oil extraction, and mining of iron ore. After raw material acquisition is the material manufacture stage, where raw materials are processed into the basic materials of product manufacture. Felled trees are processed into lumber and paper, for example. Crude oil is processed into fuels, solvents, and the building blocks of plastics. These materials move to the product manufacture stage where they are used to make the final product. In this stage, paper and plastic may be made into cups, steel turned into car bodies, or solvents and pigments turned into paints. The next stage of the life-cycle stage is use. Some products, such as automobiles, generate significant emissions and wastes during use, while other products, such as grocery sacks, have negligible material and energy flows associated with the use of the product. The final life-cycle stage consists of disposal or recycling.

Recycling can occur in several ways. A product might be reused, which is what happens when a ceramic cup is washed and reused instead of being thrown away. The product could be re-manufactured, where the materials it contains are used to make another product. A newspaper, for example, might be made into another newspaper or might be shredded and used for animal bedding. Finally, products might be recycled to more basic materials, through processes such as plastics depolymerization or automobile disassembly which yield commodity materials such as monomers and steel.

Tracking material flows, over all of the stages of a life cycle, is required for a comprehensive life-cycle inventory. Even for a simple product made from a single raw material in one or two manufacturing steps, the data collection effort can be substantial. Table 13.2-1 shows a summary of an inventory of the inputs and outputs associated with the production of one kilogram of a relatively simple product: ethylene. Consider each element in the table.

The first set of data in the table are energy requirements. These are the hydrocarbon fuels and electric power sources used in extracting the raw materials for ethylene production and for running the ethylene manufacturing process (an energy-intensive operation). The next set of data elements are referred to as feedstock energy. The main raw materials of ethylene production (oil and gas) are also fuels. The energy content of this feedstock for ethylene production is reported in units of energy rather than mass (a common practice among life-cycle study practitioners) so that it can be combined with the energy that was required in the production process.

A second set of entries in the table describes non-fuel raw material use. These include iron ore, limestone, water, bauxite, sodium chloride, clay, and ferromanganese. As shown in this table, these data are often aggregated over the life cycle and reported as aggregate quantities. Thus, water use would include water used in oil field production as well as steam used in the ethylene cracker. Some of the entries may seem obscure, but only serve to point out the complex nature of product life cycles. For example, the limestone use is due in part to acid gas scrubbing in various parts of the product life cycle.

A final set of inventory elements are the wastes and emissions. Some subjectivity is introduced here in deciding which materials to report. For example, some life cycle inventories do not report the release of carbon dioxide, a global warming gas, or

Table 13.2-1   Life-Cycle Inventory Data for the Production of 1 kg of Ethylene (Boustead, 1993).

Category	Input or Output	Unit Average
Energy content fuels, MJ	Coal	0.94
	Oil	1.8
	Gas	6.1
	Hydroelectric	0.12
	Nuclear	0.32
	Other	<0.01
	Total	9.2
Feedstock, MJ	Coal	<0.01
	Oil	31
	Gas	29
	Total	60
Total Fuel + Feedstock		69
Raw Materials, mg	Iron ore	200
	Limestone	100
	Water	1,900,000
	Bauxite	300
	Sodium chloride	5,400
	Clay	20
	Ferromanganese	<1
Air emissions, mg	Dust	1,000
	Carbon monoxide	600
	Carbon dioxide	530,000
	Sulfur oxides	4,000
	Nitrogen oxides	6,000
	Hydrogen sulfide	10
	Hydrogen chloride	20
	Hydrocarbons	7,000
	Other organics	1
	Metals	1
Water emissions, mg	Chemical oxygen demand	200
	Biological oxygen demand	40
	Acid, as $H+$	60
	Metals	300
	Chloride ions	50
	Dissolved organics	20
	Suspended solids	200
	Oil	200
	Phenol	1
	Dissolved solids	500
	Other nitrogen	10
Solid waste, mg	Industrial waste	1,400
	Mineral waste	8,000
	Slags and ash	3,000
	Nontoxic chemicals	400
	Toxic chemicals	1

the use of water. Neglecting these inventory elements implies that they are not important. More subtle subjectivity can arise in defining exactly what is and what is not a waste. Consider the example of a paper plant that debarks wood. The wood that is not used in the pulp making operation is commonly burned for energy recovery within the pulping operation. Some life-cycle practitioners may count this material as a waste that is subsequently used as a fuel. Other life-cycle practitioners might regard the material as an internal process stream. The environment does not recognize a difference between these two material accounting methods, but a life-cycle inventory applying one type of material accounting would appear to predict larger quantities of solid waste that a life-cycle inventory that employed different material accounting practices.

Take a moment to review the entries in Table 13.2-1 in order to obtain an idea of the level of effort necessary to inventory the inputs and outputs.

Table 13.2-1 provides life-cycle inventory data for a single material: ethylene. A complex product such as a computer would have a very complicated life-cycle framework. Computers are made up of many products (semiconductors, casing, display, etc.) that are themselves made from diverse materials, some of which require sophisticated manufacturing technologies. Life-cycle inputs and outputs for each of these sub-products would need to be inventoried in a life-cycle assessment of a computer (see, for example the life-cycle inventory of a computer workstation performed by the Microelectronics and Computer Technology Corporation (MCC, 1993)).

Other complexities in life-cycle inventories arise when processes have **co-products.** To illustrate the concept of co-product allocation, consider the allocation of inputs and outputs for the processes shown in Figure 13.2-3. The left-hand side of this figure shows a process where one input results in two products and one type of emission. If a life-cycle inventory is being performed on one of the two products, then the input and emissions must be allocated between the two products. Part of the life cycle of ethylene can be used as an example. Ethylene is made, in part, from a petroleum liquid referred to as naphtha. Naphtha is produced in petroleum refineries, which have crude oil as their primary input. The refinery produces a variety of products, including gases, gasoline, other fuels, asphalt, and the naphtha used to make ethylene. Data on emissions and crude oil usage are generally available for the refinery as a whole, and the fraction of the refinery's crude oil use and emissions due to naphtha production must generally be assigned using an allocation procedure. One commonly used allocation procedure is based on mass of products. As shown in Figure 13.2-3, the input and emissions attributed to each of the products can be allocated based on the mass of the co-products. In the naphtha/refinery example, the crude oil usage and emissions might be assigned in the following way:

crude oil use assigned to
naphtha production = (crude oil use for entire refinery/
total mass of products produced
by refinery) * mass of naphtha
produced by refinery

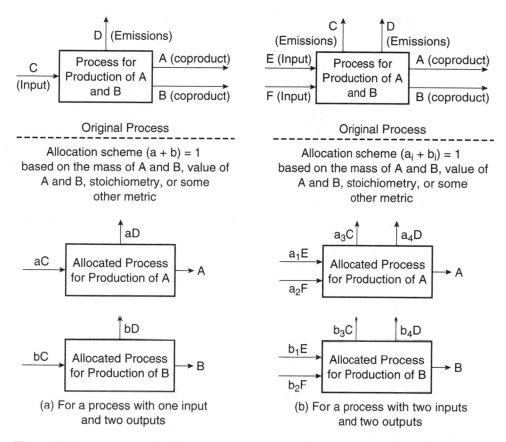

Figure 13.2-3 Allocating material use, energy use, and emissions among multiple products that are manufactured in the same processes can be difficult.

In most life-cycle inventories, allocation of material use, energy use, and emissions among co-products is based on mass. Sometimes, however, the co-product is a byproduct that would not be produced solely for its own merit, and allocation based on value might be more appropriate. As a graphic example, consider raising cattle, which produces beef and manure product streams. Clearly, the cattle rancher is in business to produce beef, not manure. Yet, if inputs and emissions to the cattle ranching were allocated based on the mass of the two products, most of the inputs and emissions would be assigned to the manure (other, less graphic examples could have been chosen, such as a pharmaceutical production process that creates a recyclable solvent by product). Clearly, in some situations, allocation based on mass is not appropriate. In such cases, inputs and emissions are sometimes allocated based on the value of the products generated.

Returning to the naphtha/refinery example, an alternative co-product allocation would be:

crude oil use assigned to
naphtha production = (crude oil use for entire refinery/
total value of products produced
by refinery) * value of naphtha
produced by refinery

Issues of co-product allocation can be complex, even when single inputs exist. The situation can become more complex when the number of inputs and emissions increases. The right-hand side of Figure 13.2-3 shows a process where multiple inputs result in multiple products. Properly allocating inputs in this situation requires great understanding of the process. If there are multiple inputs and some are converted solely into one co-product, any allocation of those inputs to the other co-products would be misleading.

Another area of life-cycle inventories where subjective decisions are made is in allocation of inputs and outputs for products that are recycled or that are made from recycled goods. Some life-cycle practitioners treat products made from recycled materials as if they had no raw material requirements, while others allocate a portion of the raw material requirements from the original product to the product made from recycled materials. Consider the example of a synthetic garment made out of the polyethylene terephthalate (PET) recovered from recycled milk bottles. In performing a life-cycle inventory, it is clear that the total amount of raw materials required for the combined milk bottle/garment system has been reduced by recycling the PET. But, if a life-cycle inventory were to be done on one of these products, how would the raw materials be allocated? Would it be appropriate, for example, to assume that a garment made completely out of recycled PET required the use of no raw materials? Or, would it be more appropriate to assume that some fraction (say 50%) of the raw materials required to produce the milk bottles should be assigned to the life-cycle inventory of the garment? There are no correct answers to these questions, and different life-cycle practitioners make different assumptions. Sometimes these assumptions, which do not appear explicitly in tables of results such as Table 13.2-1, can have a significant effect on the inventory data.

Perhaps the most important uncertainty in life-cycle inventories, however, is due to the quality of data available on the processes being inventoried and the level of aggregation of the data. Overall data quality issues, such as whether data are direct measured values or are based on engineering estimation methods, are fairly straightforward to identify and deal with. Data aggregation issues can be more subtle. Consider two examples of cases where data aggregation has impacted the findings of a life-cycle inventory. A first example is provided by the comparison of electric vehicles to gasoline-powered vehicles in the Los Angeles area. Table 13.2-2 and Figure 13.2-4 provide data summaries from two different life-cycle inventories that compared electric vehicles to gasoline-powered vehicles. The data shown in Table 13.2-2 indicate that driving an electric vehicle results in far less emission of

Table 13.2-2 Comparison of Electric Vehicle and Gasoline Powered Vehicle Emissions Based on Electricity Generation within Southern California (Electric Power Research Institute, 1994).

Emission type	Emissions of gasoline-powered, Ultra Low Emission Vehicles (ULEVs) in California (g/mi)	Emissions of Electric Vehicles based on an average of power plants within air quality districts in California (g/mi)
Reactive organic gases	0.191	0.003
Nitrogen oxides	0.319	0.011
Carbon monoxide	1.089	0.024
Particulate matter	0.018	0.004

reactive organic gases, nitrogen oxides, carbon monoxide and particulate matter, than driving an ultra-low emission vehicle (ULEV) an equivalent distance. In contrast, the data from Figure 13.2-4 indicates that an electric vehicle emits more of certain pollutants (such as NO_2 and sulfur oxides) than a gasoline-powered vehicle (based on EPA emission data). Why is there such a dramatic difference? The answer is related to **data aggregation.** The data reported in Figure 13.2-4 were based on a life-cycle inventory that used nationally averaged emissions data for electric power generation. In contrast, the data reported in Table 13.2-2 were based on a life-cycle inventory that used emissions data for electric power generation in Southern California. Since emissions from power generation in Southern California are much lower than the average for the rest of the United States, and because a large

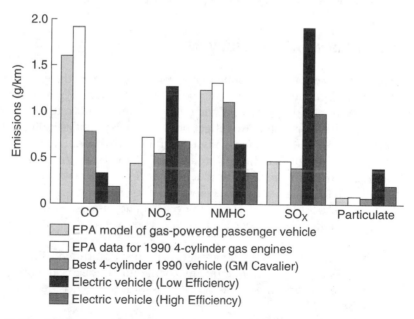

Figure 13.2-4 Comparison of air pollutant emissions for gasoline- and electricity-powered motor vehicles. (Graedel and Allenby, 1995)

fraction of the total emissions associated with an electric vehicle are due to the power generation required to fuel the vehicle, the studies lead to very different results. It can be presumed that both studies are technically correct, but they present a different picture of the relative benefits of electric vehicles.

A second example of the importance of the level of data aggregation emphasizes the difference between well-operated and poorly operated facilities. Epstein (1995) tabulated the emissions and transfers of wastes associated with 166 refineries in the United States. She found that the total emissions and waste transfers reported by refineries, through the Toxic Release Inventory, averaged 7 pounds per barrel of refinery capacity. The 10% of the refineries with the highest reported emissions and waste transfers released more than 18 pounds per barrel, while the 10% of the refineries with the lowest reported emissions and transfers released less than 0.3 pounds per barrel. Thus, a life cycle inventory that used emissions data from a refinery with low emissions might lead to very different conclusions than a life cycle inventory based on data from a refinery that has high emissions.

In summary, this section has described the basic elements of a life-cycle inventory. In performing the inventory, a number of assumptions are made concerning functional units, system boundaries, co-product allocation, data aggregation methods, and other parameters. These assumptions, as illustrated in the simple examples in the text, and as illustrated by a number of the problems at the end of the chapter, can have a significant impact on the findings of a life-cycle inventory. It is prudent to define and explore these assumptions and uncertainties before arriving at conclusions based on life-cycle inventory data.

13.3 LIFE-CYCLE IMPACT ASSESSMENTS

Life-cycle inventories do not by themselves characterize the environmental performance of a product, process, or service. This is because overall quantities of wastes and emissions, and raw material and energy requirements must be considered in conjunction with their potency of effect on the environment. Simply stated, a pound of lead emitted to the atmosphere has a different environmental impact than a pound of iron emitted to surface waters. To develop an overall characterization of the environmental performance of a product or process, throughout its life cycle, requires that life-cycle inventory data be converted into estimates of environmental impact.

The process of producing life-cycle impact assessments is generally divided into three major steps (Fava, et al., 1992). They are:

- **Classification,** where inputs and outputs determined during the inventory process are classified into environmental impact categories; for example, methane, carbon dioxide, and CFCs would be classified as global warming gases.
- **Characterization,** where the potency of effect of the inputs and outputs on their environmental impact categories is determined; for example, the rela-

tive greenhouse warming potentials (see Chapter 11) of methane, carbon dioxide, and CFCs would be identified in this step.

- **Valuation,** where the relative importance of each environmental impact category is assessed, so that a single index indicating environmental performance can be calculated.

Note that the classification and characterization steps are generally based on scientific data or models. The data may be incomplete or uncertain, but the process of classification and characterization is generally objective. In contrast, the valuation step is inherently subjective, and depends on the value society places on various environmental impact categories.

Each of the three steps is discussed in more detail below.

13.3.1   Classification

As a first step in life-cycle impact assessment, inputs and outputs that were the subject of the inventory are classified into environmental impact categories. Examples of environmental impact categories are given in Table 13.3-1. Note that some impact categories might apply to very local phenomena (for example, aquatic toxicity to organisms found only in certain ecosystems), while other impact categories are global (for example, stratospheric ozone depletion and global warming).

As an example of classification, consider the list of air emissions inventoried for a study that examined polyethylene, shown in Table 13.3-2. Nitrogen oxides emissions would be classified as photochemical smog precursors, global warming gases, and acid precipitation and acid deposition precursors. Carbon monoxide emissions, on the other hand, would be classified as a smog precursor.

13.3.2   Characterization

The second step of impact assessment, characterization, quantifies impact for each inventory item by integrating the inventory amount with the potential to cause an impact; i.e., potency factor. For example, if the impact category is global warming,

Table 13.3-1   Examples of Environmental Impact Categories.

Global warming
Stratospheric ozone depletion
Photochemical smog formation
Human carcinogenicity
Atmospheric acidification
Aquatic toxicity
Terrestrial toxicity
Habitat destruction
Depletion of nonrenewable resources
Eutrophication

Table 13.3-2   Selected Air Emissions from the Production
of One Kilogram of Polyethylene

	kg emissions per kg of polyethylene
Nitrogen oxides	0.0012
Sulfur dioxide	0.009
Carbon monoxide	0.0009

then relative global warming potentials can be used to weight the relative impact of emissions of different global warming gases. In Chapter 11, relative global warming potentials were described. Other weighting factors were presented for smog formation potential, atmospheric acidification potential, and other categories. Once these **potency factors** are established, the inventory values for inputs and outputs are combined with the potency factors to arrive at **impact scores.**

The process of calculating impact scores was described in detail in Chapter 11. In this chapter, that discussion is not repeated. Rather, the emphasis in this chapter is on the new issues that arise when applying these impact scoring methods to life-cycle data, and on the range of variation in impact scoring systems that have been employed around the world.

Consider first the new issues that arise when impact scoring systems are applied to life-cycle data. In Chapter 11, impact scoring systems were applied to processes. For this type of application, the location of the emissions can be specified and the time at which the emissions occur can be specified. In contrast, for life-cycle assessment data, the spatial location of the emissions may not be known and the temporal distribution of the emissions may be uncertain. For example, in a life-cycle assessment for an automobile, emissions, energy use, and material use may be distributed all over the world since automotive components are manufactured all over the world and users may operate vehicles all over the world. The energy use, material use, and emissions would also be distributed over a product lifetime that may last for more than a decade. In general, life-cycle impact assessments do not account for this spatial and temporal distribution of energy use, material use, and emissions. Energy use, material use, and emissions are summed over the life cycle and the weighting or potency factors are then applied to these summed inventory elements. Does it make sense to sum the emissions of (for example) carbon dioxide from activities all over the world over a period of more than a decade, as would be done in a life-cycle impact assessment of an automobile? The answer to that question, of course, depends on project boundaries, and the spatial scales and time scales over which the impact occurs. It may be appropriate, for example, to sum worldwide emissions of global warming gases in a life-cycle study. It may be inappropriate to do the same summation for a type of impact that depends strongly on local conditions. Compounds that contribute to acid rain, for example, may not be an environmental concern in areas where the soil is well-buffered and acid rain is not a problem. Similarly, the release of nitrates in one area might cause eutrification, while the release of phosphates might be the cause of eutroification in another area.

A summary of the concerns associated with spatial and temporal averaging of emissions is given in Table 13.3-3. Some recent life-cycle impact assessment methods have attempted to account for spatial and temporal variability of potency factors, but this remains a relatively underdeveloped area of life-cycle impact assessment. Most life-cycle impact assessments continue to assume that inventory data can be summed over the entire life cycle without accounting for spatial and temporal distributions.

Table 13.3-3   Impact Categories Frequently Considered in Life-Cycle Assessments. Listed in Table 13.3-1, these range from local to global in their spatial extent and operate over time scales ranging from hours to decades. These spatial and temporal characteristics of impacts should be compared to the spatial and temporal resolution of data collected in life-cycle studies (adapted from Owens, 1997).

Impact categories	Spatial scale	Temporal scale
Global warming	global	decades/centuries
Stratospheric ozone depletion	global	decades
Photochemical smog formation	regional/local	hours/day
Human carcinogenicity	local	hours (acute)–decades (chronic)
Atmospheric acidification	continental/regional	years
Aquatic toxicity	regional	years
Terrestrial toxicity	local	hours (acute)–decades (chronic)
Habitat destruction	regional/local	years/decades
Depletion of nonrenewable resources	global	decades/centuries
Eutrophication	regional/local	years

Example 13.3-1   Impact assessment scores for the manufacture of polyethylene.

An inventory of the manufacture of a one kilogram of polyethylene showed that air emissions of carbon dioxide, carbon monoxide, nitrogen oxides, and sulfur oxides were 1.3 kg, 0.0009 kg, 0.012 kg, and 0.009 kg, respectively (Boustead, 1993). In a typical impact assessment scheme, emissions are multiplied by potency factors to arrive at impact scores. If carbon dioxide is assigned a characterization score of one for global warming, calculate the total global warming score for the polyethylene. Assume that the other emissions have no global warming potential. Assume that carbon monoxide, nitrogen oxides, and sulfur dioxide have been assigned characterization scores of 0.012, 0.78, and 1.2 for human toxicity, respectively. Calculate the impact scores for each compound and the overall score for each impact category. Also calculate the total human toxicity score. Remember that, while it would be incorrect to add the emissions of the four compounds together, their impact scores can be combined. Is it correct to add the overall global-warming score to the overall human toxicity score to arrive at a single impact score?

Solution: The global warming impact assessment score is

$$1.3 \times 1 = 1.3$$

The human toxicity score for carbon dioxide is zero, while for carbon monoxide it is

$$0.0009 * 0.012 = 0.000011.$$

For nitrogen oxides it is

$$0.012 * 0.78 = 0.0094.$$

And for sulfur dioxide it is

$$0.009 * 1.2 = 0.011.$$

The overall score for this group of chemicals for global warming is 1.3 and the overall score for human toxicity is

$$0 + 0.000011 + 0.0094 + 0.011 = 0.020.$$

Adding the global warming score for this set of compounds to the human toxicity score would be inappropriate.

Another issue that arises in impact assessment is the choice of potency factors. In Chapter 11, a single set of potency factors is presented. In practice, there are numerous impact scoring systems available. Many of these characterization schemes have been developed by life-cycle researchers (e.g., Guinée et al., 1996; Fava et al., 1993). Also, a number of schemes for weighting releases to the environment have been developed for reasons other than life-cycle assessment, and these can be adopted for life-cycle characterization (e.g., Wright, et al., 1997; Pratt et al., 1996; US EPA, 1997). At times, different life-cycle impact systems will lead to different results. As an example, consider an inventory of the releases of organochlorine compounds to the Great Lakes Basin, which was performed by Rosselot and Allen (1999). Three different potency factor schemes for human and ecological toxicity impact categories were used to rank the inventory data. The results of the rankings are shown in Table 13.3-4. Ideally, each of the potency factor schemes would result in the same rank ordering of chemicals; however, the data show that the different potency schemes lead to different rank ordering of some of the compounds. The potency schemes agree in their rankings of trichloroethylene, 1,2-dichloroethane, and PCBs, but disagree in their rankings of dichloromethane, endosulfan, and hexachlorobutadiene. Note that not all of the characterization systems listed in the table were created for the purpose of conducting life-cycle impact assessments. Instead, some of them were developed in order to rank emissions. Also note that ranking these compounds by mass of release (the order in which they are listed in the table) would give very different results than ranking them by potency of effect for any of the characterization schemes. Thus, while not all potency factors lead to identical results, ignoring the concept of potency and considering only the mass of emissions may place too great an emphasis on relatively benign compounds that are emitted in large amounts.

Why would different potency schemes lead to different results? The answer is simple. The methods are often based on different criteria. Some commonly used potency factors (Swiss Federal Ministry for the Environment (BUWAL),

Table 13.3-4   Rankings of 1993 Releases of Chlorinated Organic Compounds in the Great Lakes Basin for Several Potency Factor Schemes. Compounds are listed in descending order of quantity released.

Compound	EPA Human Risk[1]	Dutch Human Tox.[2]	Dutch Aquatic Tox.[2]	EPA Eco. Risk[1]	MPCA Tox. Score[3]	Dutch Terr. Tox.[2]
Tetrachloroethylene	8	4	5	10	6	3
Dichloromethane	13	5	8	10	2	9
Trichloroethylene	8	12	12	10	10	10
1, 1, 1-Trichloroethane	8	1	6	10	7	4
Chloroform	6	8	4	6	1	6
1, 2-Dichloroethane	8	6	7	8	4	7
PCBs	1	2	2	1	5	1
Endosulfan	2	7	1	2	11	2
Carbon tertrachloride	3	10	10	2	3	11
Vinyl chloride	6	9	13	6	9	13
Chlorobenzene	8	14	14	8	13	14
Benzyl chloride	13	15	15	10	14	15
Hexachlorobutadiene	3	13	11	4	12	8
2, 4-Dichlorophenol	13	11	9	10	15	12
2, 3, 7, 8-TCDD	3	3	3	4	8	5

[1]US EPA: Waste Minimization Prioritization Tool, used to rank pollutants.
[2]Dutch: Guinée et al., 1996, considers environmental fate and transport, developed specifically for life-cycle assessment.
[3]MPCA: Pratt et al., 1993, Minnesota Pollution Control Agency system for ranking air pollutants. May be based on human or animal effects.

Postlethwaite and de Oude, 1996), are based on data from environmental regulations. In these systems, each emission is characterized based on the volume of air or water that would be required to dilute the emission to its legally acceptable limit. For example, if air quality regulations allowed 1 part per billion by volume of a compound in ambient air, then one billion moles of air (22.4 billion liters of air at standard temperature and pressure) would be required to dilute one mole of the compound to the allowable standard. This volume per unit mass or mole of emission is called the critical dilution volume and can vary across political boundaries. Other potency factor systems are based on relative risk, but establishing relative risks requires assumptions about the type of environment that the emissions are released to. These assumptions may differ in the various impact assessment schemes.

13.3.3   Valuation

The final step in life-cycle impact assessment, valuation, consists of weighting the results of the characterization step so that the environmental impact categories of highest importance receive more attention than the impact categories of least concern. There is no generally accepted method for aggregating values obtained from the evaluations of different impact categories to obtain a single environmental impact score. Some of the approaches that have been employed are listed in Table 13.3-5. Some methods assign valuations of high, medium, or low to the impact categories based on the extent and irreversibility of effect, so that stratospheric ozone depletion might receive a high rating and water usage might receive a low rating.

Table 13.3-5   Strategies for Valuing Life-cycle Impacts (Christiansen, 1997).

Life-cycle impact assessment approach	Description
Critical volumes	Emissions are weighted based on legal limits and are aggregated within each environmental medium (air, water, soil).
Environmental Priority System (Steen and Ryding, 1992)	Characterization and valuation steps combined using a single weighting factor for each inventory element (see example below). Valuation based on willingness-to-pay surveys.
Ecological scarcities	Characterization and valuation steps combined using a single weighting factor for each inventory element. Valuation based on flows of emissions and resources relative to the ability of the environment to assimilate the flows or the extent of resources available.
Distance to target method	Valuation based on target values for emission flows set in the Dutch national environmental plan.

Valuation schemes based on the "footprint" of the inputs and outputs have been suggested. In these schemes, characterization would be conducted so that the air, water, land, and other resources required to absorb the inputs and outputs are quantified. These quantities could then be normalized according to the amount of each resource available, on either a local or global basis, and added within resource category. The resource with the highest combined normalized value is the one that is being most adversely impacted. In fact, it would be possible to arrive at a single value that represented the total fraction of the earth's resources required to buffer the inputs and outputs over the life cycle being studied.

Data on the public's willingness to pay for various environmental health categories have also been used in developing valuation schemes. However, there is very little data of actual scenarios where people paid a premium based solely on environmental preferability, and most willingness-to-pay information is based on surveys.

The following example illustrates the use of the Environmental Priority Strategy (EPS) system, developed in Sweden, which combines characterization and valuation into single values. Impact categories for this system include biodiversity, human health, ecological health, resources, and aesthetics. Environmental indices are assigned to compounds by considering six factors:

- Scope: the general impression of the environmental impact.
- Distribution: the extent of the affected area.

- Frequency or intensity: the regularity and intensity of the problem in the affected area.
- Durability: the permanence of the effect.
- Contribution: the significance of one kilogram of the emission of that substance in relation to the total effect.
- Remediability: the relative cost to reduce the emission by one kilogram.

Data from willingness-to-pay studies were used in developing the indices. Note that with this system, impacts are aggregated, and environmental value judgements and priorities are built into the indices.

Example 13.3   Selected Environmental Indices from the Environmental Priority Strategies System

In the EPS system, environmental indices are multiplied by the appropriate quantity of raw materials used or emissions released to arrive at Environmental Load Units (ELUs), which can then be added together to arrive at an overall ELU for the subject of the life-cycle study. Table 13.3-6 gives selected environmental weighting factors from the EPS system. Calculate the environmental load units due to air emissions from one kilogram of ethylene production. Emissions are 0.53 kg, 0.006 kg, 0.0006 kg, and 0.004 kg of carbon dioxide, nitrogen oxides, carbon monoxide, and sulfur oxides, respectively (Boustead, 1993).

Table 13.3-6   Selected Environmental Indices from the EPS System (in Environmental Load Units Per Kilogram) (Steen and Ryding, 1992).

Raw Materials		Air Emissions		Water Emissions	
Cobalt	76	Carbon monoxide	0.27	Nitrogen	0.1
Iron	0.09	Carbon dioxide	0.09	Phosphorous	0.3
Rhodium	1,800,000	Nitrogen oxides	0.22		
		Sulfur oxides	0.10		

Solution: Total ELUs due to air emissions are

0.53 kg CO_2 H 0.09 ELU/kg CO_2
0.006 kg NO_x H 0.22 ELU/kg NO_x
0.0006 kg CO H 0.27 ELU/kg CO
0.004 kg SO_x H 0.10 ELU/kg SO_x
= 0.05 ELU.

Note that if quantities of raw materials or water emissions were given, the ELUs for these inputs would be added to the ELUs for the air emissions.

Valuation occurs implicitly in every life-cycle study, because the attributes chosen for inventorying, such as air emissions and energy usage, reflect the values of the practitioners and the organization funding the study. Also, the choice of impact categories to be evaluated in the classification and characterization steps implicitly includes valuation. For example, odor is not typically included as an impact category, implicitly suggesting that it is of minimal importance relative to impacts such as ecotoxicity and human toxicity.

While there is no widely accepted procedure for aggregating impact scores across different impact categories, aggregation within impact categories takes place widely. It would be impractical, for example, to have a separate impact category for every biological species. Some impact assessment schemes have separate impact factors for aquatic and terrestrial life but within those broad categories, the response of different species to the same dose of a compound is very different.

Because valuation is subjective, many practitioners stop at the characterization step. If a life-cycle study was conducted to compare two products and the impact scores for each impact category were higher for one product than the other, valuation is not needed to determine which product is environmentally superior. This rarely happens, however; typically products being compared and design alternatives for a single product have some positive features and some less desirable features (recall the examples from Chapter 11). Each alternative has an environmental footprint with unique characteristics, meaning that any design choice typically means tradeoffs between categories of impacts.

13.3.4   Interpretation of Life-Cycle Data and Practical Limits to Life-Cycle Assessments

While the process of a life-cycle assessment might seem simple enough in principle, in practice it is subject to a number of practical limitations. In performing the inventory, system boundaries must be chosen so that completion of the inventory is possible, given the resources that are available. Even if sufficient resources are available, the time required to perform a comprehensive life-cycle inventory may be limiting. Then, even if the necessary time and resources are available, life-cycle data are subject to uncertainty for the reasons cited earlier in this section.

The limitations of life-cycle inventories are then carried forward into the impact assessment stage of life-cycle studies, and the impact assessment methodologies add their own uncertainties. For example, potency factors are not available for all compounds in all impact categories. Issues of temporal and spatial aggregation, as described in this section arise. Finally, valuation adds an element of subjectivity into the analyses.

This is not to say that life-cycle assessments are without value. Rather, despite the uncertainties involved, these assessments provide invaluable information for decision-making and product stewardship. They allow environmental issues to be evaluated strategically, throughout the entire product life cycle. The challenge is to take advantage of these valuable features of life-cycle assessments while bearing in mind the difficulties and uncertainties.

The next section describes methods for managing the uncertainties and effort required for life-cycle assessments. Once these tools are described, application and interpretation of life-cycle information will be examined.

13.4   STREAMLINED LIFE-CYCLE ASSESSMENTS

The use of life-cycle studies falls along a spectrum from a complete spatial and temporal assessment of all the inputs and outputs due to the entire life cycle (which may never be accomplished in practice, both because of a lack of information and because it would require a tremendous amount of effort and expense) to an informal consideration of the environmental stresses that occur over a product or process life cycle. This spectrum is illustrated in Figure 13.4-1. The further a study falls to the right on the spectrum, the more expensive and time-consuming the study will be. In this chapter, an analysis that includes an inventory of all inputs and outputs and all life-cycle stages (including an assessment of which ones are significant enough to be included in the inventory), an impact assessment, and an improvement analysis will be called a life-cycle assessment and a study that falls to the left in the spectrum of complexity will be said to involve the use of **life-cycle concepts.** Studies in between the two extremes will be called **streamlined life-cycle assessments.** Streamlined life-cycle assessments are conducted in order to find the most important life-cycle stages or type of inputs and outputs for more detailed study. Also, they can be used to identify where the most significant environmental issues occur.

13.4.1   Streamlined Data Gathering for Inventories and Characterization

The importance of product stewardship and growing awareness of the importance of product and process life cycles, coupled with a growing frustration with the complexity and data intensity of traditional life-cycle assessments has led to a new type of product life-cycle evaluation, often referred to as a streamlined life-cycle assessment. There are many ways that a life-cycle assessment can be streamlined. A study might build extensively on previously completed life-cycle assessments. A life-cycle assessment for polyethylene, for example, might rely on the data presented in the previous section on ethylene and focus on extending the supply chain through the polymerization of ethylene into polyethylene. Similarly, data collected

Figure 13.4-1 Life-cycle studies fall along a spectrum of difficulty and complexity, beginning with the use of life-cycle concepts and ending with complete life-cycle assessments.

Life-Cycle Thinking   Life-Cycle Assessment

in previous studies may indicate that certain impact categories or life-cycle inventory categories can be safely neglected without a meaningful effect on the results of the study.

Other approaches for making life-cycle studies easier to accomplish include omission of product components or materials. The omission can be based on whether the components or materials contribute significantly to the product's overall environmental impacts. Some practitioners routinely exclude any component that accounts for less than 1% of the total product weight. This could result in inadequate study results, because some small components, such as semiconductor devices in computers, can have large environmental impacts relative to their weight (see Box 1). There are other ways to decide whether a component or material should be included or omitted in a life-cycle study, such as its economic value, which in turn reflects resource scarcity and ease of manufacturing and is at least loosely tied to environmental importance. Energy use (which is sometimes relatively simple to find data for) or toxicity might also be considered.

Environmental impact categories are sometimes neglected in streamlined life-cycle studies. Similarly, a selected set of inputs or outputs might be chosen for inventorying. Some products are known to have heavy impacts due to process wastes but require little energy, making an inventory of energy requirements less necessary than an inventory of gaseous, liquid, and solid residues.

Another possible shortcut to completing a life-cycle study would be to leave out life-cycle stages. Short-lived products, such as single-use packaging, usually have environmental impacts that are dominated by raw material acquisition and materials manufacture and disposal. In contrast, the use phase dominates for long-lived products that require resources during use. For example, in a streamlined life-cycle assessment of electric vehicle batteries, Steele and Allen (1998) considered only the recycling and disposal life-cycle stages.

13.4.2 Qualitative Techniques for Inventories and Characterization

One of the more common techniques employed in streamlined life-cycle studies involves conducting qualitative rather than quantitative analyses. For example, instead of quantifying the number of units of energy required to produce a product, the energy usage could be characterized as high, medium, or low. Qualitative evaluations can be enormously helpful in reducing the time and resources necessary for providing life-cycle information, because detailed inventory information is not necessary. However, there is a risk of failing to assess different life-cycle stages and products in a comparable manner. For example, energy usage during manufacture of a car may be high compared to manufacture of many products, but compared to the tens of tons of fuel required in the use stage of a typical car during its lifetime, an evaluation of high for energy use during manufacture would be inappropriate. Qualitative approaches for streamlined life-cycle assessments have been developed by a number of researchers.

Box 1   A streamlined life-cycle study of a computer workstation

A computer workstation is a complex product involving an enormous range of materials and components. Conducting a full life-cycle assessment on a product of this complexity would be extremely difficult, yet life-cycle data can prove extremely useful in identifying areas for environmental improvement. A streamlined life-cycle assessment for a workstation was performed by an industry team coordinated by the Microelectronics and Computer Technology Corporation (MCC, 1993). The workstation components considered in the MCC study included the cathode ray tube (display), plastic housings, semiconductors, and printed wiring boards. A streamlined life-cycle assessment was able to identify, for a variety of life-cycle inventory categories, which workstation components were of primary concern. For example, product disposal was dominated by issues related to cathode ray tubes. Hazardous waste generation was dominated by semiconductor manufacturing. Energy use was dominated by the consumer use stage of the life cycle. Somewhat surprisingly, semiconductor manufacturing was identified as a significant factor in material use. Results are summarized in Figure 13.4-2. The study was used to guide research and technology development for the microcomputer industry.

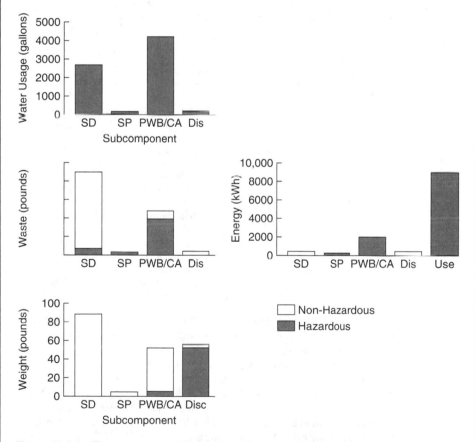

Figure 13.4-2   Energy use, material use, water use, and waste generation in computer workstation life cycles. Note that different subcomponents of the workstation (SD = semiconductor device, SP = semiconductor packaging, PWB/CA = printed wiring board and computer assembly, DIS = display) dominate different inventory categories. (MCC, 1993)

Example 13.4-1   Computer Display Project: LCDs vs CRTs

The U.S. EPA's Design for the Environment (DfE) Computer Display Project, in partnership with the electronics industry, has evaluated the life-cycle environmental impacts of liquid crystal display (LCD) and cathode ray tube (CRT) technologies that can be used in desktop computer monitors, by combining life-cycle assessment (LCA) and streamlined Cleaner Technologies Substitutes Assessment (CTSA) aproaches (U.S. EPA, 1995). The LCA examined environmental impacts through the collection of inventory data (i.e., material inputs/outputs) for all life-cycle stages, from raw materials acquisition through end-of-life. The material inventory was collected from display and component manufacturers and suppliers in the U.S., Japan, Korea, and Taiwan (U.S. EPA, 1998).

By integrating the inventory amounts with the potential to cause an impact, the impacts of the CRT and LCD technologies were assessed for the following major impact categories: atmospheric resources, water quality, natural resource use, and human health/ecological impacts (see table below).

		Relative Impacts	
Impact Category	Impact Subcategory	*CRT's*	*LCD's*
Atmospheric Resources	local air quality	Higher	Lower
	regional air quality	Higher	Lower
	global impacts	Higher	Lower
Water Quality	surface water	Higher	Lower
	ground water	Higher	Lower
Natural Resource Use	energy consumption	Higher	Lower
	material/resource use	Higher	Lower
	landfill space	Higher	Lower
Human Health Impacts	occupational (multiple pathways)	Higher	Lower
	general public (multiple pathways)	Higher	Lower
	aesthetic (odor)	Higher	Lower
Ecological Impacts	aquatic toxicity	Lower	Higher
	terrestrial toxicity	Higher	Lower

As presented in table above, preliminary results indicate that when aggregating data for all life-cycle stages, CRTs have greater potential impacts for most categories, except aquatic toxicity.

The study indicates numerous opportunities to improve both CRT and LCD designs (e.g., by reducing or eliminating mercury use in LCDs or lead in CRTs) and to reduce energy consumption during LCD and CRT manufacturing. Manufacturers can also evaluate substitutes or implement better management practices for particular problematic materials, such as sulphur hexaflouride, which is used in LCD manufacturing and contributes to global warming.

Some life-cycle practitioners evaluate both the quantity of the inputs and outputs and their impacts in a single qualitative process. An example of such a system is the Environmentally Responsible Product Assessment, developed at Bell Laboratories/Lucent Technologies, which relies on the use of expert evaluations of extensive checklists, surveys, and other information (Graedel, 1998). Scores from 0 to 4 (with 4 indicating environmental preferability) are assigned to the life stages and inventory categories listed in Table 13.4-1. The table shows that there are five life-cycle stages and five inventory categories to assess, for a total of 25 assessments per product. After all the scores are assigned, they are added together to arrive at an overall score. The maximum value for this overall score, therefore, is 4×25, or 100.

As an example, the scoring guidelines and protocols for just one of the 25 elements in the Environmentally Responsible Product Assessment matrix is given in Table 13.4-2. These guidelines and protocols are for the materials choice category of the premanufacture life-cycle stage. Parallel systems for assessing services, processes, facilities, and infrastructures have also been developed at Bell Laboratories/Lucent Technologies (Graedel, 1998).

The results of the Environmentally Responsible Product Assessment can be plotted on a chart such as the one shown in Figure 13.4-3. A product that is relatively environmentally benign would have all the points of the chart clustered around the center, and an environmentally damaging product would have points that fell towards the outside circumference.

In these qualitative, streamlined life-cycle analyses, functional units and allocation methods are not explicitly considered. However, use of virgin materials is penalized, so credit is given for using recycled materials. Scores developed by different individuals tend to fall within 15% of each other, which is an indication of the uncertainty in the results. Evaluations tend to be based on comparisons to a standard and focus on whether or not best practices are being followed. Therefore, this scheme might be useful in improving already-designed products rather than a product that is in the early design phases; however, it is not as useful for comparing completely different means to fulfilling a need. For example, they may be helpful in identifying whether aqueous or chlorinated organic solvents are environmentally preferable, but not for comparing a process change that makes cleaning unnecessary compared to the use of aqueous or chlorinated solvents.

Streamlined life-cycle assessment methods could be devised to produce results with an absolute basis. For example, inputs and outputs could be assigned qualitative

Table 13.4-1   Life Stages and Inventory Categories Evaluated in the Environmentally Responsible Product Assessment Matrix (Graedel, 1998).

Inventory Category/Life Stage	Premanufacture	Product manufacture	Product delivery	Product use	Refurbishment, recycling, disposal
Materials choice					
Energy use					
Solid residues					
Liquid residues					
Gaseous residues					

Table 13.4-2 Environmentally Responsible Product Matrix Scoring Guidelines for the Materials Choice Inventory Category During the Premanufacture Life-Cycle Stage (Graedel, 1998).

Score	Condition
0	For the case where supplier components/subsystems are used: No/little information is known about the chemical content in supplied products and components.
	For the case where materials are acquired from suppliers: A scarce material is used where a reasonable alternative is available. (Scarce materials are defined as antimony, beryllium, boron, cobalt, chromium, gold, mercury, platinum, iridium, osmium, palladium, rhodium, rubidium, silver, thorium, and uranium.)
4	No virgin material is used in incoming components or materials.
1, 2, or 3	Is the product designed to minimize the use of scarce materials (as defined above)?
	Is the product designed to utilize recycled materials or components wherever possible?

inventory scores that correspond to a specific functional unit and an absolute value. Product inventory matrices with types of inputs and outputs for rows and life-cycle stages for columns would be filled out with evaluations of high, medium, low, and none. For illustrative purposes, consider the life-cycle inputs and outputs of one kilogram of glass. One might assign a score of low to compounds whose air emissions are believed to be greater than zero but less than 0.001 kilogram, a score of medium if air

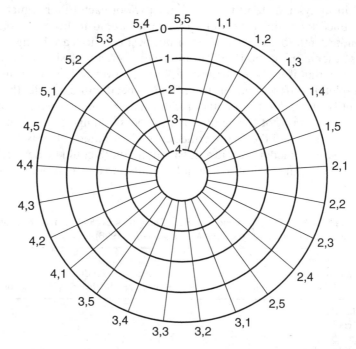

Figure 13.4-3 The target plot for the Environmentally Responsible Product Assessment matrix. Each radial axis represents one of the 25 life-cycle stage/inventory category combinations from Table 13.4-1 (Graedel, 1998).

emissions of a compound are believed to be between 0.001 kilogram and 0.1 kilogram, and a score of high if air emissions of a compound are believed to be greater than 0.1 kilogram. While it might be difficult (or even impossible) to inventory the inputs and outputs of a life cycle to within one or two significant digits, less effort is required to arrive at estimates accurate to within an order of magnitude.

Potency factor matrices with evaluations of high, medium, low, and none for each of the inputs and outputs from the product inventory matrix could also be devised. The rows of these matrices would be types of inputs and outputs and the columns would be environmental impact categories. Numerical scores could then be assigned to the qualitative evaluations. For example, low could be assigned a score of one; medium, a score of two; and high, a score of three. The inventory scores could then be added to the potency scores to arrive at impact assessment scores. This streamlined life-cycle assessment technique would require a large number of relatively simple evaluations. Complex products made from many separate components would be more amenable to this type of streamlined assessment because scores with an absolute basis could be weighted and summed over the components involved. Also, the potency factor scores would need to be developed only once for each type of input and output.

Many such streamlined assessments could be devised and it is beyond the scope of this chapter to review all of the methods that have appeared. Nevertheless, it is useful to keep in mind the features of a well-designed, streamlined study. A good streamlined life-cycle assessment has a goal, an inventory, an impact assessment, and an improvement analysis, just as a comprehensive life-cycle assessment does. All of the relevant life stages are evaluated, if only to say they are being omitted and why, and all of the relevant inventory categories are evaluated (again, perhaps only to say why they are being omitted). In a streamlined life-cycle assessment, evaluations for inventory categories and impact assessments may be qualitative instead of quantitative.

13.4.3   Pitfalls, Advantages, and Guidance

Streamlined life-cycle assessments and life-cycle concepts play a particularly important role in green engineering—more so than comprehensive life-cycle assessments. This is because of the nature of the design cycle of a product or process. There is a rule of thumb that 80% of the environmental costs of a product are determined at the design phase (Graedel, 1998). Modifications made to the product at later stages can therefore have only modest effects. Thus it is in the early design phase that life-cycle studies for improving the environmental performance of a product are most useful. However, in the design phase, materials have not been selected, facilities have not been built, packaging has not been determined, so a comprehensive, quantitative life-cycle assessment is impossible at the time when it would be most useful. Instead, preferable materials and processes can be identified through the use of an abbreviated life-cycle study early in the design cycle where it is most effective. This is discussed further in the later sections on design of products and processes.

13.5 USES OF LIFE-CYCLE STUDIES

According to a survey of organizations actively involved in life-cycle studies, the most important goal of life-cycle studies is to minimize the magnitude of pollution (Ryding, 1994). Other goals include conserving non-renewable resources, including energy; ensuring that every effort is being made to conserve ecological systems, especially in areas subject to a critical balance of supplies; developing alternatives to maximize the recycling and reuse of materials and waste; and applying the most appropriate pollution prevention or abatement techniques. As discussed in this section, life-cycle studies have been applied in many ways in both the public and private sectors for uses such as developing, improving, and comparing products.

13.5.1 Product Comparison

The most widely publicized life-cycle studies are those that have been conducted for the purpose of comparing products. A life-cycle study comparing cloth and disposable diapering systems, another study comparing plastic and paper cups, and another one comparing polystyrene clamshells and paper wrappings for sandwiches are examples of studies that received a great deal of attention from the press (see problems at the end of the chapter). Product comparison studies are often sponsored by organizations that have a vested interest in the results, and because of the open-ended nature of life-cycle studies, there is always room for criticism of the assumptions that were made and the data that were gathered in the course of the study. Because the results of these high-profile product comparison life-cycle studies have generated a great deal of controversy and debate, they have created skepticism over the value of life-cycle studies. This has diverted attention away from some of the less controversial applications, such as studies conducted in order to improve products.

13.5.2 Strategic Planning

One of the most important uses for manufacturers of life-cycle studies is to provide guidance in long-term strategic planning concerning trends in product design and materials (Ryding, 1994). By their nature, life-cycle studies include environmental impacts whose costs are external to business (e.g., habitat destruction) as well as internal (e.g., the cost of waste generation). Assessing these external costs is key to strategic environmental planning, as regulations tend to internalize what are currently external costs of doing business.

13.5.3 Public Sector Uses

Life-cycle studies are also used in the public sector. Policymakers report that the most important uses of life-cycle studies are 1) helping to develop long-term policy regarding overall material use, resource conservation, and reduction of environmental impacts and risks posed by materials and processes throughout the product life cycle, 2) evaluating resource effects associated with source reduction and alternative waste management techniques, and 3) providing information to the public about the resource characteristics of products or materials (Ryding, 1994).

Some of the most visible of the applications of life-cycle studies are environmental or ecolabeling initiatives. Examples of ecolabels from around the world are given in Figure 13.5-1. Besides environmental labeling programs, public sector uses of life-cycle studies include making procurement decisions and developing regulations.

Figure 13.5-1   Ecolabels from Around the World.

For example, the United States Environmental Protection Agency used life-cycle information when making a decision about regulation of industrial laundries whose effluent was a problem because of the oily shop rags they laundered. The concern was that tighter regulations may have made the costs of industrial laundering so expensive that a shift from cloth shop rags to disposable shop rags would occur. Would this be a benefit to the environment? Life-cycle concepts provide some insights. A summary of this case and other the uses of life-cycle studies in public policy initiatives has been assembled by Allen, et al. (1997).

13.5.4   Product Design and Improvement

Product comparisons have received the most attention from the press, but in a survey, manufacturers state that the most important uses of life-cycle studies are 1) to identify processes, ingredients, and systems that are major contributors to environmental impacts, and 2) to compare different options within a particular process with the objective of minimizing environmental impacts (Ryding, 1994).

Manufacturers have more potential for influencing the environmental impacts of products than any other "owners" of life-cycle stages. This is because they can exert some influence over the environmental characteristics of the supplies they use, because manufacturing processes account for a large portion of the wastes generated in the United States, and because manufacturers determine to some extent the use and disposal impacts of the products they make.

Choosing Suppliers

As stated before, manufacturers have some potential to influence the environmental characteristics of the companies from whom they purchase supplies. This is illustrated by the efforts of Scott Paper Company in their procurement of pulp for paper products (Fava and Consoli, 1996). Scott decided to use a life-cycle approach to environmental impacts as opposed to its traditional focus on environmental concerns only at plants that it owns when it found that the issues of major concern were not in the life-cycle stages directly controlled by the company.

Scott's first step was to require its pulp suppliers in Europe to provide detailed information about their emissions, energy use, manufacturing processes, and forestry practices. In their impact assessment, Scott ranked the environmental impact categories by consulting with opinion leaders. They found that there was considerable variation in performance among suppliers, and that the suppliers that were ranked worst in one environmental impact category tended to be worst in other categories as well. The poorest performing suppliers were shown the potential for improvement, and if they did not choose to proceed, Scott no longer used them as a supplier. As a result of this program, Scott changed about 10% of its pulp supply base. Scott publicized their efforts and its products were seen as environmentally preferable by consumers and environmental advocacy groups.

Improving Existing Products

The results of a life-cycle study conducted for the purpose of product improvement are shown in Table 13.5-1. This table shows the results of an inventory of the energy required to produce one kilogram of polyethylene. The majority of fuel required to make polyethylene is in the organic matter that instead of being burned for energy is converted to polyethylene. In fact, the values in the column titled "Feedstock Energy" are about 75% of the total energy requirements. This inventory showed that the focus of efforts to reduce the life-cycle energy requirements of polyethylene are best spent on reducing the mass of polyethylene in products (i.e., to make them as light as possible).

Another life-cycle study conducted for the purpose of product improvement (Franklin Associates, Ltd., 1993) showed that the energy requirements of the use stage of a polyester blouse are 82% of the total energy requirements over the life cycle. Furthermore, the greatest potential for reducing the energy requirements over the life cycle consisted of switching to a cold water wash and line dry instead of a warm water wash and drying in a clothes dryer. Such a switch would reduce the energy requirements of the use stage by 90%. Thus, one of the greatest environmental improvements that could be made in a garment is to make it cold water washable.

In another life-cycle study for product improvement of clothing, it was shown that the means of transportation used in delivering a garment to a customer can have a profound impact on the garment's life-cycle energy requirements (Hopkins et al., 1994). This study showed that in the case where next-day shipping is used, transportation and distribution energy requirements can be 28% of manufacturing life-cycle energy requirements. Transportation and distribution of products generally contribute negligibly to the energy requirements of a product and are frequently neglected in life-cycle studies. Prior to this study, the garment manufacturer was unaware that their choice of delivery mode could contribute significantly to the energy required over the life cycle of their products.

In yet another life-cycle study conducted for product improvement, the components of a computer workstation were assessed to reveal which were responsible for the majority of raw material usage, wastes, emissions, and energy consumption

Table 13.5-1   Average Gross Energy Required to Produce One Kilogram of Polyethylene (Boustead, 1993).

Fuel Type	Fuel Production and Energy, MJ	Delivered Energy, MJ	Feedstock Energy, MJ	Total Energy, MJ
Electricity	5.31	2.58	0.00	7.89
Oil Fuels	0.53	2.05	32.76	35.34
Other	0.47	8.54	33.59	42.60
Total	6.31	13.17	66.35	85.83

(Box 1; MCC, 1993). The components studied included semiconductors, semiconductor packaging, printed wiring boards and computer assemblies, and display monitors. One of the findings of the study was that the majority of energy usage over a workstation's life cycle occurs during the use phase of the display monitor. Therefore, to reduce the overall energy usage of a computer workstation, efforts are best directed at reducing the energy consumed by the monitor. Semiconductor manufacture was found to dominate hazardous waste generation and was also found to be a significant source of raw material usage. This is in spite of the fact that by weight, semiconductors are a very small portion of a workstation.

Another life-cycle study was conducted by a light switch maker in Europe as a result of a competitor gaining market share by claiming to manufacture a cadmium-free switch (Besnainou and Coulon, 1996). The life-cycle inventory showed that the cadmium contained in the contactor of the switch for both manufacturers was negligible compared to the cadmium used in plating operations during manufacture. In effect, only one of the manufacturers made switches that contained cadmium, but neither switch was truly "cadmium-free." Also, the life-cycle study revealed that the biggest environmental gains could be had by reducing the electricity consumed by the switch over its ten-year lifetime. This result is surprising because the electricity consumed is small per event and only becomes important when totaled over the life cycle.

Using Life-Cycle Concepts in Early Product Design Phases

Traditionally, performance, cost, cultural requirements, and legal requirements have set the boundaries for the design of products. Increasingly, environmental aspects are included with this core group of design criteria and life-cycle studies can be used to assess environmental performance. Optimizing environmental performance from the beginning of the design process has the possibility of the largest gains, but it is a moving target as markets, technologies, and scientific understanding of impacts change. However, as stated earlier, roughly 80% of the environmental costs of a product are determined at the design phase, and modifications made to the product at later stages may have only modest effects. Thus it is in the early design phase that life-cycle studies for improving the environmental performance of a product are most useful.

Motorola has developed a matrix for streamlined life-cycle assessment that is intended in part to specifically address early design (Graedel, 1998). The matrix is shown in Table 13.5-2. There are five life-cycle stages and three impact assessment categories (one of which is divided into two subcategories) in the matrix. Motorola intends to use this matrix in three succeedingly quantitative phases: the initial design concept phase, the detailed drawings phase, and the final product specifications phase. In the initial design phase, the matrix elements can be filled out by asking a series of yes and no questions. An overall score is computed by adding the yes answers, and changes in that score show progress in the product's environmental characteristics. This example is typical of emerging trends in product design.

Table 13.5-2   Motorola's Life-Cycle Matrix (Graedel, 1998).

Impact		Part Sourcing	Manufacturing	Transportation	Use	End of Life
Sustainability	Resource use					
	Energy use					
Human health						
Eco health						

13.5.5   Process Design

Industrial process changes should be given strategic thought because they are generally in place for decades and retrofits tend to be expensive and difficult. While the life-cycle stages of a process are different from those of a product (as shown in Figure 13.1-1), the types of inputs and outputs and impact categories are the same. Generally, process choices (including choices of feed materials for the process) are likely to have more impact over the life cycle of the process than production of the equipment itself.

Jacobs Engineering has developed a life-cycle matrix tool that has been applied to processes (as opposed to products) (Graedel, 1998). This matrix is shown in Table 13.5-3. This tool identifies five inventory categories and seven environmental impact categories at two spatial scales (shop level and global). A base case process is determined and elements in the matrix are assigned $+1$, 0, or -1, depending on whether the alternative is an improvement, equivalent, or worse than the base case. Note that not all life-cycle stages are explicitly identified in this scheme.

SUMMARY

Life-cycle studies are a uniquely useful tool for assessing the impact of human activities. These impacts can only be fully understood by assessing them over a life cycle, from raw material acquisition to manufacture, use, and final disposal. Life-cycle techniques have been adopted in industry and the public sector to serve a variety of purposes, including product comparison, strategic planning, environmental labeling, and product design and improvement.

Life-cycle assessments have four steps. The first is scoping, where boundaries are determined and strategies for data collection are chosen. The second step is an inventory of the inputs and outputs of each life-cycle stage. Next is an impact assessment, where the effects of the inputs and outputs are evaluated. The final step is an improvement analysis. Even for simple products, comprehensive life-cycle studies require a great deal of time and effort. Also, no matter how much care is taken in preparing a study, the results obtained have uncertainty.

Table 13.5-3 Jacobs Engineering Impact Analysis Matrix for Evaluating Alternative Processes.

Impacting Parameters

Risk Area	Shop Level						Global Level					
	Material Inputs	Energy Inputs	Atmospheric Emissions	Aqueous Wastes	Solid Wastes	Total	Material Inputs	Energy Inputs	Atmospheric Emissions	Aqueous Wastes	Solid Wastes	Total
Global warming												
Ozone-depleting resource utilitization												
Non-renewable resource utilization												
Air quality												
Water quality												
Land disposal												
Transportation effects												
Total												

Yet, life-cycle studies remain useful. Environmental concerns that are identified early in product or process development can be most effectively and economically resolved and life-cycle studies can be used as tools to aid in decision-making.

QUESTIONS FOR DISCUSSION

1. Life-cycle assessments have been performed comparing the environmental impacts of beverage delivery systems, lighting systems, grocery sacks, and numerous other products. In all of these studies, determining a functional unit is a critical part of performing the comparison. For each of the product comparisons listed below, suggest a functional unit, describe how an equivalency between products could be established using the functional unit, and estimate that equivalency factor.

    For example, if the goal were to compare paper and plastic grocery sacks, the functional unit would be providing packaging for a given market basket of groceries. Equivalency could be established by going to grocery stores, purchasing identical market baskets of goods and packaging the goods using both paper and plastic. Studies of this type have concluded that 2 plastic sacks are equivalent to one paper sack.

    a) Fluorescent bulbs vs. Incandescent bulbs
    b) Cloth vs. Disposable diapers
    c) 12 ounce aluminum cans vs. 16-ounce glass beverage containers

2. Tables 13.4-1, 13.5-2 and 13.5-3 each present frameworks for performing streamlined or semi-quantitative analyses of impacts throughout a life cycle. Compare and contrast the features of the tables, keeping in mind the types of products or processes that the frameworks are designed for. For a product of your own choosing, develop a matrix which could be used in performing a streamlined life-cycle assessment (see, for example, Steele and Allen, 1998).

REFERENCES

Allen, D.T., Bakshani, N., and Rosselot, K.S., "Homework and Design Problems for Engineering Curricula," American Institute of Chemical Engineers, New York, 155pp. (1992).

Allen, D.T., Consoli, F.J., Davis, G.A., Fava, J.A., and Warren, J.L., "Public Policy Applications of Life Cycle Assessment," SETAC, Pensacola, FL, 127 pp. (1997).

Besnainou, J. and Coulon, R, "Life-Cycle Assessment: A System Analysis," in *Environmental Life-Cycle Assessment,* M. A. Curran, ed., McGraw-Hill, New York, 1996.

Boustead, I., "Ecoprofiles of the European Plastics Industry, Reports 1-4," PWMI, European Centre for Plastics in the Environment, Brussels, May 1993.

Christiansen, K, ed., "Simplifying LCA: Just a Cut? Final Report from the SETAC-EUROPE LCA Screening and Streamlining Working Group," SETAC, Brussels, 1997.

Consoli, F., Allen, D., Boustead, I., Franklin, W., Jensen, A., de Oude, N., Parrish, R., Perriman, R., Postlethwaite, D., Quay, B., Seguin, J., Vigon, B., editors, "Guidelines for Life Cycle Assessment: A Code of Practice," SETAC Press, Pensacola, Florida, 1993.

Epstein, L.N., Greetham, S., and Karuba, A., "Ranking Refineries," Environmental Defense Fund, Washington, D.C., November, 1995.

Electric Power Research Institute, "EV Emission Benefits in California," Technical Brief, Customer Systems Division, RB2882, Palo Alto, Ca., May, 1994.

Fava, J and Consoli, F. "Application of Life-Cycle Assessment to Business Performance," in *Environmental Life Cycle Assessment,* M.A. Curran, ed. McGraw Hill, New York, 1996.

Franklin Associates, Ltd., "Resource and Environmental Profile Analysis of a Manufactured Apparel Product," Prairie Village, KS, June 1993.

Graedel, T.E. and Allenby, B. R. Industrial Ecology, Prentice-Hall, Inc. Englewood Cliffs, New Jersey, 1995.

Graedel, T. E., *Streamlined Life-Cycle Assessment,* Prentice-Hall, Inc., Englewood Cliffs, New Jersey, 1998.

Guinée, J., L. van Oers, D van de Meent, T. Vermiere, M. Rikken, "LCA Impact Assessment of Toxic Releases," Dutch Ministry of Housing, Spatial Planning and the Environment, The Hague, The Netherlands, May 1996.

Hopkins, L., Allen, D. T., and Brown, M., "Quantifying and Reducing Environmental Impacts Resulting from Transportation of a Manufactured Garment," *Pollution Prevention Review,* 4(4), 1994.

Microelectronics and Computer Technology Corporation (MCTC), "Life-Cycle Assessment of a Computer Workstation," Report no. HVE-059-94, Austin, TX, 1994.

Owens, J.W., "Life Cycle Assessment: Constraints on Moving from Inventory to Impact Assessment," *Journal of Industrial Ecology,* 1(1) 37–49 (1997).

Postlethwaite, D and de Oude, N. T., "European Perspective," *Environmental Life-Cycle Assessment,* M. A. Curran, ed., McGraw-Hill, New York, 1996.

Pratt G. C., Gerbec, P.E., Livingston, S. K., Oliaei, F., Gollweg, G. L., Paterson, S., and Mackay, D., "An Indexing System for Comparing Toxic Air Pollutants Based upon Their Potential Environmental Impacts," *Chemosphere,* 27(8), 1350–1379, 1993.

Rosselot, K.S. and Allen, D.T., "Chlorinated Organic Compounds in the Great Lakes Basin: Impact Assessment," submitted to *Journal of Industrial Ecology,* 1999.

Ryding, S., "International Experiences of Environmentally Sound Product Development Based on Life-Cycle Assessment," Swedish Waste Research Council, AFR Report 36, Stockholm, May 1994.

Steele, N. and Allen, D.T., "An Abridged Life-Cycle Assessment of Electric Vehicle Batteries," *Environmental Science and Technology,* Jan. 1, 1998, 40A-46A.

Steen, B. and Ryding, S., "The EPS Enviro-Accounting Method: An Application of Environmental Accounting Principles for Evaluation and Valuation of Environmental Impact in Product Design," Stockholm: Swedish Environmental Research Institute (IVL), 1992.

United States Environmental Protection Agency (US EPA), Waste Minimization Prioritization Tool, EPA530-R97-019, June 1997.

United States Environmental Protection Agency (US EPA), Design for the Environment Computer Display Project (fact sheet), EPA 744-F-98-010, 1998.

United States Environmental Protection Agency (US EPA), Cleaner Technologies Substitutes Assessment: A Methodology and Resource Guide, EPA 744-R-95-002, 1995.

Wright, M., Allen, D., Clift, R., and Sas, H., "Measuring Corporate Environmental Performance: The ICI Environmental Burden System," *Journal of Industrial Ecology,* 1 (4), 117–127 (1998).

PROBLEMS

1. (From Allen, et al., 1992.) At the supermarket checkstand, customers are asked to choose whether their purchases should be placed in unbleached paper grocery sacks or in polyethylene grocery sacks. Some consumers make their choice based on the perception of the relative environmental impacts of these two products. This problem will quantitatively examine life-cycle inventory data on the energy use and air emissions for these two products.

Life-cycle inventories for paper and polyethylene grocery sacks have resulted in the data given below, and these data will be used in comparing the two products. Assume that the functional unit to be used in this comparison is a defined volume of groceries to be transported, and that based on this functional unit, 2 plastic sacks are equivalent to one paper sack.

Air Emissions and Energy Requirements for Paper and Polyethylene Grocery Sacks (Allen, et al., 1992).

Life-cycle Stages	Paper sack air emissions (oz/sack)	Plastic sack air emissions (oz/sack)	Paper sack air energy req'd (Btu/sack)	Plastic sack air energy req'd (Btu/sack)
Materials manufacture plus product manufacture plus product use	0.0316	0.0146	905	464
Raw materials acquisition plus product disposal	0.0510	0.0045	724	185

Note: These data are based on past practices and may not be current.

a) Using the data in the table, determine the amount of energy required and the quantity of air pollutants released per plastic sack. Also determine the amount of energy required and the quantity of air pollutants released for the quantity of paper sacks capable of carrying the same volume of groceries as the plastic sack. Both the air emissions and the energy requirements are functions of the recycle rate, so perform your calculations at three recycle rates: 0%, 50% and 100% recycled. Note that a 50% recycle rate indicates that half of the sacks are disposed of and the other half are recycled after the product use stage of their life cycle.

b) Plot the energy requirements calculated in part a as a function of the recycle rate for both sacks. Do the same for the air emissions. Compare the energy requirements and air emissions of the sacks at different recycle rates.

c) Discuss the relative environmental impacts of the two products. Do the results allow for a comprehensive comparison?

d) The material and energy requirements of the plastic sacks are primarily derived from petroleum, a non-renewable resource. In contrast, the paper sacks rely on petroleum to only a limited extent and only for generating a small fraction of the manufacturing and transportation energy requirements. Compare the amount of petroleum required for the manufacture of two polyethylene sacks to the amount of energy necessary to provide 10% of the energy required in the manufacture of one paper sack. Assume 0% recycle and that 1.2 lb of petroleum is required to manufacture 1 lb of polyethylene. The higher heating value of petroleum is 20,000 BTU/lb.

e) In this problem, we have assumed that 2 plastic sacks are equivalent to one paper sack. Does the uncertainty in the equivalency between paper and plastic sacks affect any of your conclusions?

2. (From Allen, et al., 1992.) Disposable diapers, manufactured from paper and petroleum products, are one of the most convenient diapering systems available, while cloth diapers are often believed to be the most environmentally sound. The evidence is not so clear-cut, however. This problem will quantitatively examine the relative energy requirements and the rates of waste generation associated with diapering systems.

Three types of diapering systems are considered in this problem: home-laundered cloth diapers, commercially-laundered cloth diapers, and disposable diapers containing a super-absorbent gel. The results of life-cycle inventories for the three systems are given below.

Energy Requirements and Waste Inventory per 1000 Diapers (Allen, et al., 1992).

Impact	Disposable diapers	Commercially-laundered cloth diapers	Home-laundered cloth diapers
Energy requirements (million BTU)	3.4	2.1	3.8
Solid waste (cubic feet)	17	2.3	2.3
Atmospheric emissions (lb)	8.3	4.5	9.6
Waterborne wastes (lb)	1.5	5.8	6.1
Water requirements (gal)	1300	3400	2700

a) The authors of the report from which the data in the table are taken found that an average of 68 cloth diapers were used per week per baby. Disposable diaper usage is expected to be less because disposable diapers are changed less frequently and never require double or triple diapering. In order to compare the diapering systems, determine the number of disposable diapers required to match the performance of 68 cloth diapers, assuming:

15.8 billion disposable diapers are sold annually
3,787,000 babies are born each year
children wear diapers for the first 30 months
disposable diapers are used on 85% of children.

b) Complete the table given below. Remember to use the equivalency factor determined for cloth and disposable diapers determined in part a. Based on the assumptions you made in part a, how accurate are the entries in the table?

Ratio of Impact to Home-Laundered Impact.

Impact	Disposable diapers	Commercially-laundered cloth diapers	Home-laundered cloth diapers
Energy requirements (million BTU)	0.50	0.55	1.0
Solid waste (cubic feet)			1.0
Atmospheric emissions (lb)			1.0
Waterborne wastes (lb)			1.0
Water requirements (gal)			1.0

c) Using the data given below, determine the percentage of disposable diapers that would need to be recycled in order to make the solid waste landfill requirements equal for cloth and disposable systems.

Impact of Recycle Rate on Solid Waste
for Diapering Systems.

Percentage of diapers recycled	Solid waste per 1000 diapers (Cubic feet)
0	17
25	13
50	9.0
75	4.9
100	0.80

3. The University of Michigan has developed a case study analyzing the decision, made by McDonald's, to replace polystyrene clamshell containers with other container systems. The case study is described at: http://www.umich.edu/~nppcpub/resources/compendia/chem.e.html. Review the materials on the web site and write a one-page summary of the case.

4. Choose two similar products and perform streamlined life-cycle assessments for them using the methods described in this chapter and the methods described by Graedel (1998). Assess whether the streamlined approach is effective in comparing the products you have chosen.

CHAPTER 14

Industrial Ecology

by
David T. Allen

14.1 INTRODUCTION

The environmental performance of chemical processes is governed not only by the design of the process, but also by how the process integrates with other processes and material flows. Consider a classic example—the manufacture of vinyl chloride.

Billions of pounds of vinyl chloride are produced annually. Approximately half of this production occurs through the direct chlorination of ethylene. Ethylene reacts with molecular chlorine to produce ethylene dichloride (EDC). The EDC is then pyrolyzed, producing vinyl chloride and hydrochloric acid.

$$Cl_2 + H_2C=CH_2 \Rightarrow Cl\ H_2C\text{-}CH_2\ Cl$$

$$Cl\ H_2C\text{-}CH_2\ Cl \Rightarrow H_2C=CH\ Cl + HCl$$

In this synthesis route, one mole of hydrochloric acid is produced for every mole of vinyl chloride. Considered in isolation, this process might be considered wasteful. Half of the original chlorine winds up, not in the desired product, but in a waste acid. But the process is not operated in isolation. The waste hydrochloric acid from the direct chlorination of ethylene can be used as a raw material in the oxychlorination of ethylene. In this process, hydrochloric acid, ethylene, and oxygen are used to manufacture vinyl chloride.

$$HCl + H_2C=CH_2 + 0.5\ O_2 \Rightarrow H_2C=CHCl + H_2O$$

By operating both the oxychlorination pathway and the direct chlorination pathway, as shown in Figure 14.1-1, the waste hydrochloric acid can be used as a raw material and essentially all of the molecular chlorine originally reacted with ethylene is incorporated into vinyl chloride. The two processes operate synergistically and an efficient design for the manufacture of vinyl chloride involves both processes.

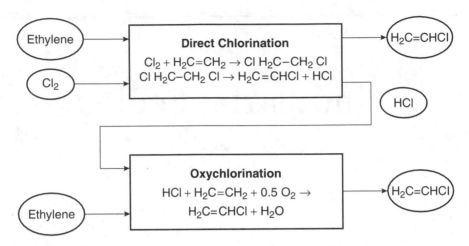

Figure 14.1-1 Byproduct hydrochloric acid from the direct chlorination of ethylene is used as a raw material in the oxychlorination process; by operating the two processes in tandem, chlorine is used efficiently.

Additional efficiencies in the use of chlorine can be obtained by expanding the number of processes included in the network. In the network involving direct chlorination and oxychlorination processes, both processes incorporate chlorine into the final product. Recently, more extensive chlorine networks have emerged linking several isocyanate producers into vinyl chloride manufacturing networks (McCoy, 1998). In isocyanate manufacturing, molecular chlorine is reacted with carbon monoxide to produce phosgene:

$$CO + Cl_2 \Rightarrow COCl_2$$

The phosgene is then reacted with an amine to produce an isocyanate and byproduct hydrochloric acid:

$$RNH_2 + COCl_2 \Rightarrow RNCO + 2\,HCl$$

The isocyanate is subsequently used in urethane production, and the hydrochloric acid is recycled. The key feature of the isocyanate process chemistry is that chlorine does not appear in the final product. Thus, chlorine can be processed through the system without being consumed. It may be transformed from molecular chlorine to hydrochloric acid, but the chlorine is still available for incorporation into final products, such as vinyl chloride, that contain chlorine. A chlorine-hydrogen chloride network incorporating both isocyanate and vinyl chloride has developed in the Gulf Coast of the United States. The network is shown in Figure 14.1-2. Molecular chlorine is manufactured by Pioneer and Vulcan Mitsui. The molecular chlorine is sent to both direct chlorination processes and to isocyanate manufacturing. The byproduct hydrochloric acid is sent to oxychlorination processes or calcium chloride manufacturing. The network has redundancy in

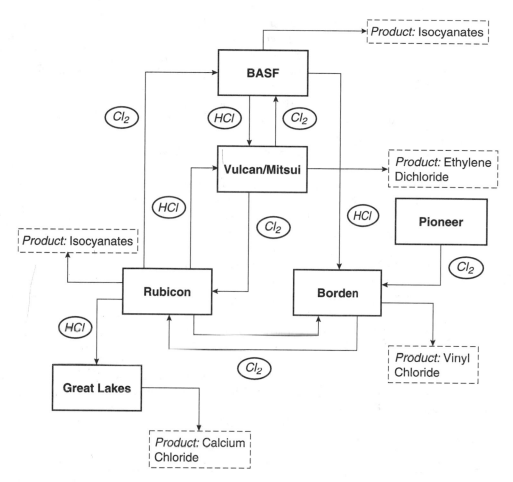

Figure 14.1-2 Chlorine flows in combined vinyl chloride and isocyanate manufacturing (McCoy, 1998).

chlorine flows, such that most processes could rely on either molecular chlorine or hydrogen chloride.

Consider the advantages of this network to the various companies (Francis, 2000):

- Vulcan/Mitsui effectively rents chlorine to BASF and Rubicon for their isocyanate manufacturing; the chlorine is then returned in the form of hydrochloric acid for ethylene dichloride/vinyl chloride manufacturing.
- BASF and Rubicon have guaranteed supplies of chlorine and guaranteed markets for their byproduct HCl.

Even more complex networks could, in principle be constructed. As shown in Table 14.1-1, chlorine is used in manufacturing a number of non-chlorinated products. Table 14.1-1 lists, for selected reaction pathways, the pounds of chlorinated intermediates used along the supply chain, per pound of finished product. This ranking provides one indication of the potential for networking these processes with processes for manufacturing chlorinated products (see Rudd, et al., 1981, or Chang, 1996).

An examination of individual processes, such as those listed in Table 14.1-1, can be useful in building process networks, but the individual process data do not reveal whether efficient use of chlorine is a major or a minor issue in chemical manufacturing. To determine the overall importance of these flows, it is useful to consider an overall chlorine balance for the chemical industry. The overall flows of chlorine into products and wastes, as well as the recycling of chlorine in the chemical manufacturing sector, is shown in Figure 14.1-3. The data indicate that roughly a third of the total chlorine eventually winds up in wastes. By employing the types of networks shown in Figures 14.1-1 and 14.1-2, the total consumption of chlorine could be reduced.

Identifying which processes could be most efficiently integrated is not simple and the design of the ideal network depends on available markets, what suppliers and markets for materials are nearby, and other factors. What is clear, however, is that the chemical process designers must understand not only their process, but also processes that could supply materials, and processes that could use their byproducts. And the analysis should not be limited to chemical manufacturing. Continuing with our example of waste hydrochloric acid and the manufacture of vinyl chloride, byproduct hydrochloric acid could be used in steel making or byproduct hydrochloric acid from semiconductor manufacturing might be used in manufacturing chemicals.

Table 14.1-1 Partial Listing of Non-Chlorinated Chemical Products That Utilize Chlorine in Their Manufacturing Processes (Chang, 1996).

Product	Synthesis pathway	Pounds of chlorinated intermediates per pound of product
Glycerine	Hydrolysis of epichlorohydrin	4.3
Epoxy resin	Epichlorohydrin via chlorohydrination of allyl chloride, followed by reaction of epichlorohydrin with bisphenol-A	2.3
Toluene diisocyanate	Phosgene reaction with toluenediamine	2.2
Aniline	Chlorobenzene via chlorination of benzene, followed by reaction of chlorobenzene with ammonia	2.2
Phenol	Chlorobenzene via chlorination of benzene, followed by dehydrochlorination of chlorobenzene	2.1
Methylene diphenylene diisocyanate	Phosgene reaction with aniline (also produced with chlorinated intermediates)	1.5
Propylene oxide	chlorohydration of propylene	1.46

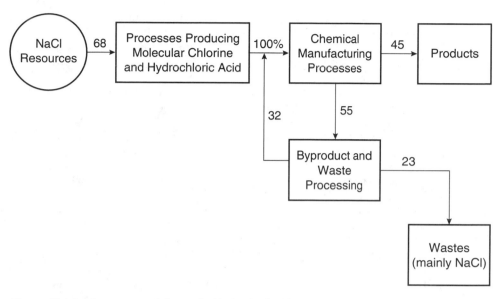

Figure 14.1-3 A summary of flows of chlorine in the European chemical manufacturing industry (Francis, 2000)

Finding productive uses for byproducts is a principle that has been used for decades in chemical manufacturing. What is relatively new, however, is the search for chemical byproduct uses in industries that extend far beyond chemical manufacturing. This chapter will examine both of these topics—the overall flows of raw materials, products and byproducts in chemical manufacturing industries—as well as the potential for combining material and energy flows in chemical manufacturing with material and energy flows in other industrial sectors. Variously called byproduct synergy, zero waste systems, or even industrial ecology, the goal of this design activity is to create industrial systems that are as mass-efficient as possible.

Section 14.2 provides an overview of material flows in chemical manufacturing and describes analysis methods that can be used to optimize flows of materials. Section 14.3 examines case studies of exchanges of materials and energy across industrial sectors and the emerging concept of eco-industrial parks. Finally, Section 14.4 briefly attempts to assess the potential benefits of byproduct synergies.

Box 1   What Is Industrial Ecology?

The phrase "Industrial Ecology" evokes powerful images and strong reactions, both positive and negative. To some, the phrase conjures images of industrial systems that mimic the mass conservation properties of natural ecosystems. Powerful analogies can be drawn between the evolution of natural ecosystems and the potential evolution of industrial systems. Billions of years ago, the Earth's life forms consumed the planet's stocks of materials and changed the composition of the atmosphere. Our natural ecosystems evolved slowly to the intricately balanced, mass-conserving networks that exist today. Can our industrial systems evolve in the same way, but much more quickly? These are interesting visions and thought-provoking concepts. But is Industrial Ecology merely a metaphor for these concepts? Is there any engineering substance to the emerging field of Industrial Ecology?

As demonstrated in this chapter, Industrial Ecology is much more than a metaphor and it is a field where engineers can make significant contributions. At the heart of Industrial Ecology is the knowledge of how to reuse or chemically modify and recycle wastes—making wastes into raw materials. Chemical engineers have practiced this art for decades. The history of the chemical manufacturing industries provides numerous examples of waste streams finding productive uses. Nonetheless, even though the chemical manufacturing industries now provide excellent case studies of Industrial Ecology in practice—tightly networked and mass-efficient processes—there is much left to be done. While the chemical manufacturing industries are internally integrated, there is relatively little integration between chemical manufacturing and other industrial sectors and between chemical manufacturers and their customers.

Engineers could take on design tasks such as managing the heat integration between a power plant and an oil refinery or integrating water use between semiconductor and commodity material manufacturing. The goal is to create even more intricately networked and efficient industrial processes—an industrial ecology. Not all of the tools needed to accomplish these goals are available yet, but this chapter begins to describe the basic concepts and suggests the types of tools that the next generation of process engineers will require.

14.2 MATERIAL FLOWS IN CHEMICAL MANUFACTURING

The chemical manufacturing industries are a complex network of interrelated processes. An individual process typically relies on other chemical manufacturing processes for raw materials and as markets for its products. Take the manufacture of styrene as an example. Styrene manufacturing relies on ethylene and benzene, manufactured in other processes, for raw materials. The styrene is not sold as a consumer product; rather, it is used as a raw material for polystyrene manufacturing. Additional complexity arises from the fact that most sequences of chemical manufacturing process are not unique. There are generally a variety of pathways available for manufacturing products.

As a relatively simple example of the multiple pathways available in chemical synthesis, again consider styrene. Styrene is produced from ethylene and benzene, but the source of the ethylene might be naptha, or refinery gases. Benzene might be produced by dehydrogenation of cyclohexane, dealkylation of toluene or separation from crude oil. These options provide multiple pathways from raw materials to styrene. Each route has raw material requirements, energy requirements, water usage, and rates of emissions and waste generation.

Selecting the most environmentally benign and most economical route is a difficult proposition. It is made even more difficult when the entire chemical supply chain is considered. For example, in methanol production, the methanol is produced using carbon monoxide. The carbon monoxide in turn may be produced through a partial oxidation of a material that is currently wasted by another process. On the other hand, to convert the carbon monoxide into methanol requires hydrogen, which is an energy-intensive material. Evaluations of the environmental features of producing a chemical product should examine the entire chemical raw material supply chain, but to realistically examine these supply chains requires comprehensive, integrated models of material flows in the chemical process industries.

Fortunately, such models have been developed. Rudd and co-workers have developed basic material and energy flow models of over 400 chemical processes associated with the production of more than 200 chemical products (Rudd, et al., 1981), describing a complex web of chemical manufacturing technologies.

An understanding of material flows in these networks can be used at a variety of levels. First, the material flow networks can be used simply to identify potential users and suppliers of materials, and to identify networks of processes that are strategically related. For example, for the types of networks considered in Section 14.1, it would be useful to have lists of processes that produce and consume hydrochloric acid. A partial list is given in Table 14.2-1.

Once consumers and producers of targeted chemicals are identified, material and energy flow models can be used to construct networks. The network that makes the most sense depends on the features that are to be optimized. Analyses have been performed to identify networks that minimize energy consumption (Sokic, et al., 1990a,b), the use of toxic intermediates (Yang, 1984; Fathi-Afshar and Yang, 1985), and chlorine use (Chang and Allen, 1997). Other analyses have considered the response of networks to perturbations in energy supplies (Fathi-Afshar, et al., 1981) and restrictions on the use of toxic substances (Fathi-Afshar and Rudd, 1981). Regardless of the application, however, the material flow model of the chemical manufacturing web provides the basic information necessary to identify and optimize networks of processes.

Table 14.2-1   Partial List of Processes That Produce or Consume Hydrochloric Acid. Such lists are useful in identifying potential material exchange networks.

Processes that consume hydrochloric acid	Processes that produce hydrochloric acid
Chlorobenzene via oxychlorination of benzene	Adiponitrile via chlorination of butadiene
Chloroprene via dimerization of acetylene	Benzoic acid via chlorination of toluene
Ethyl chloride via hydrochlorination of ethanol	Carbon tetrachloride via chlorination of methane
Glycerine via hydrolysis of epichlorohydrin	Chloroform via chlorination of methyl chloride
Methyl chloride via hydrochlorination of methanol	Ethyl chloride via chlorination of ethanol
Perchloroethylene via oxychlorination of ethylene dichloride	Methyl chloride via chlorination of methane
Trichloroethylene via oxychlorination of ethylene dichloride	Perchloroethylene via chlorination of ethylene dichloride
	Phenol via dehydrochlorination of chlorobenzene
	Trichloroethylene via chlorination of ethylene dichloride

Table 14.2-2   Processes for Reducing Chlorine Use in Chemical Manufacturing.

Process description
Chlorine via electrolysis of hydrogen chloride (Ker-Chlor process)
Chlorine via oxidation of hydrogen chloride ($CuCl_2$ catalyst)
Chlorine via oxidation of hydrogen chloride (HNO_3 catalyst)

Yet another use of comprehensive material flow models is in the evaluation of new technologies (Chang and Allen, 1997). Consider once again the case of chlorine use in chemical manufacturing. Rather than generating complex networks involving HCl and molecular chlorine, as described in Section 14.1, it might be preferable to use a chemistry that converts waste HCl into molecular chlorine. Several processes have been proposed and are listed in Table 14.2-2.

These processes will only be successful if they can compete with the reuse of by-product HCl, in the types of networks described in Section 14.1. Data on material and energy flows in the chemical manufacturing web can again be used to assess the competitiveness of new chemical pathways, such as the technologies listed in Table 14.2-2.

14.3   ECO-INDUSTRIAL PARKS

The examples of process networking described in Sections 14.1 and 14.2 dealt exclusively with chemical manufacturing. Yet, the types of material and energy flows found in chemical manufacturing (solvents, acids, water, energy, salts) are used in a wide variety of industrial sectors. It would therefore seem reasonable to consider designing industrial networks that involve a variety of industries.

One of the classic examples of this type of network is a group of facilities located at Kalundborg in Denmark. At Kalundborg, an oil refinery, a sulfuric acid plant, a pharmaceutical manufacturer, a coal burning power plant, a fish farm, and a gypsum board manufacturer form an industrial network, exchanging flows of energy and mass. As shown conceptually in Figure 14.3-1, the power plant and the refinery exchange steam, gas, and cooling water. Waste heat from the power plant is used in district residential heating and to warm greenhouses and a fish farm. Ash from coal combustion at the power plant is shipped to cement manufacturers. Calcium sulfate from the scrubbers at the power plant is sent to the gypsum board manufacturer. Treated process sludges from the pharmaceutical plant are sent to local farmers for use as fertilizer, and the refinery sends hot liquid sulfur from the desulfurization of crude oil to a sulfuric acid manufacturer (Ehrenfeld and Gertler, 1997).

A more detailed examination of the exchanges of material and energy at Kalundborg reveals a number of interesting features.

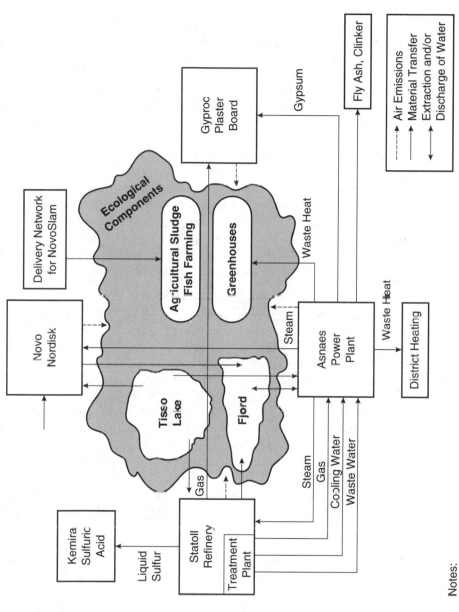

Notes:

(1) This figure is not drawn to scale, nor is it an accurate geographic depiction.

(2) Unused residuals resulting from all activities in the industrial ecopark are eventually released into the biosphere.

Figure 14.3-1 The industrial network at Kalundborg, Denmark. (Ehrenfeld and Gertler, 1997)

- The ecopark developed over a period of more than 30 years. Some material and energy exchanges have occurred for decades and the exchanges continue to grow in extent.

- The exchanges have the potential to be remarkably efficient. For example, the power plant is able to use some of the waste heat and steam produced through power generation by sending it to the refinery, greenhouses, the fish farm, and the district heating system. If markets were found for all of the waste steam, up to 90% of the heat from the plant's combustion of coal could be utilized. The only losses would be energy escaping with the stack gases. By contrast, typical coal-burning power plants in the United States use heat from combustion solely to generate electricity, at an efficiency of about 40%.

- Material and energy exchanges provide economic benefits to the participants. In some cases, such as the power plant's sale of calcium sulfate to the gypsum board manufacturer, the direct economic benefits do not fully cover the recovery costs. In these cases, the exchanges are driven by regulations, such as those requiring the scrubbing of power plant stack gases to remove SO_2. The exchanges simply lower the cost of compliance by making it unnecessary to landfill or otherwise dispose of the waste generated by the scrubbers. In other cases, such as the use of power plant waste heat in the refinery, the exchanges are self-supporting.

The central facilities in the Kalundborg Ecopark are the power plant and the oil refinery. Many of the exchanges either originate from or go to the power plant or the refinery. While using a power plant or a refinery as a central facility is a concept that could be successful in other locations, many other approaches are possible. Consider, for example, an eco-industrial park in North Texas where the central facility is a steel mill. This facility, shown conceptually in Figure 14.3-2, utilizes scrap cars as the primary feed material. The steel from the vehicles goes to an electric arc furnace, producing a variety of steel products. The furnace also produces a significant quantity of electric arc furnace (EAF) dust, which contains significant quantities of zinc, lead, and other metals. In the North Texas facility, the EAF dust is sent to a cement kiln where the trace metals (copper, sulfur, manganese, chromium, nickel, zinc, lead, and others) have value. Automobile Shredder Residue can be burned for energy recovery, or some of the plastics in the residue can be separated.

Another alternative for electric arc furnace dust, currently being explored in Europe, is as a feed for zinc and lead recovery operations. The recovered zinc can then be used in producing galvanized steel products and batteries can be used as an alternative source of zinc.

These two case studies illustrate the basic principles of ecoparks—integrating flows of energy and materials in diverse industrial operations, increasing mass and energy efficiency. The two cases examined in this section involved exchanges between facilities that are located adjacent to each other; however, co-location of facilities is not always necessary.

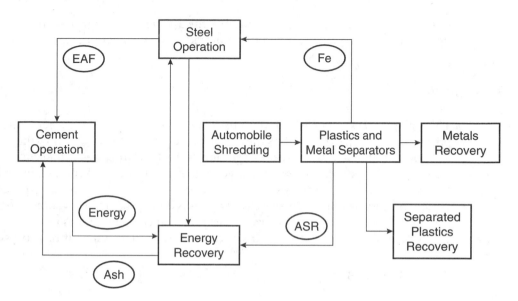

Figure 14.3-2   Material flows in an eco-industrial park in North Texas.

14.4   ASSESSING OPPORTUNITIES FOR WASTE EXCHANGES AND BYPRODUCT SYNERGIES

The previous sections have demonstrated anecdotally that productive uses can be found for selected waste streams. Are these anomalies, or are there large quantities of waste materials that can be productively used? This question is difficult to answer with certainty, but a few simple examples may illustrate the potential for finding new uses for waste.

One estimate of the potential for industrial exchanges of materials and energy can be drawn from a simple examination of energy flows in the United States. Approximately a third of the 80–100 quadrillion BTU of energy consumed annually in the United States is used for electric power generation. Of the energy used in electricity generation, roughly 2/3 is lost as waste heat. This means that roughly a quarter of all energy demand in the United States could be met through the utilization of lost heat. Combined heat and power systems are emerging throughout the country to take advantage of such opportunities, but much remains to be done.

A second example of the potential for conservation through material exchanges involves another ubiquitous material—water. Water is used in virtually all industrial processes and major opportunities exist for reuse since, in general, only a small amount of water is consumed; most water in industrial applications is used for cooling, heating, or processing of materials, not as a reactant. Further, different industrial processes and industrial sectors have widely varying demands for water quality. For example, waste water from a semiconductor manufacturing facility that requires ultrapure water may be suitable for a variety of other industrial applications. Thus, water exchanges and reuse provide a significant opportunity. An example of such opportunities is described by Keckler and Allen (1999).

SUMMARY

This chapter has emphasized that the environmental performance of chemical processes is governed not only by the design of the process, but also by how the process integrates with other processes. Integration with other processes can occur through exchanges of material, through exchanges of energy, and through common use of utilities, such as cooling and process waters. To design efficient and economical processes, designers must systematically search out markets for byproducts; they should consider using byproducts from other processes as raw materials; and, perhaps most significantly, they should not restrict their searches to chemical manufacturing processes.

REFERENCES

Chang D., and Allen, D.T., "Minimizing chlorine use: Assessing the trade-offs between cost and chlorine use in chemical manufacturing," *Journal of Industrial Ecology,* 1(2), 111–134 (1997).

Ehrenfeld, J. and Gertler, N., "Industrial Ecology in Practice, The evolution of interdependence at Kalundborg," *Journal of Industrial Ecology,* 1(1), 67–80 (1997).

Fathi-Afshar, S., and Rudd, D.F., "Impact of restrictions on toxic substances on the production of synthetic materials," Polymer-Plastics Technology and Engineering, 16, 99–118, 1981.

Fathi-Afshar, S., Maisel, D.S., Rudd, D.F., Trevino, A.A., and Yuan, W.W., "Advances in petrochemical technology assessment," Chemical Engineering Science, 36, 1487–1511, 1981.

Fathi-Afshar, S. and Yang, J., "Designing the optimal structure of the petrochemical industry for the minimum cost and least gross toxicity of chemical production," Chemical Engineering Science, 40, 781–797, 1985.

Francis, C.G. personal communication, 2000.

Keckler, S.E. and Allen, D.T., "Material Reuse Modeling: A Network Flow Programming Approach," *Journal of Industrial Ecology,* in press (1999).

M. McCoy, "Chlorine Links Gulf Coast Firms," Chemical and Engineering News, September 7, 1998 pp 17–20.

Rudd, D. F., Fathi-Afshar, S., Trevino, A.A., and Stadtherr, M.A., *Petrochemical Technology Assessment,* John Wiley & Sons, New York, 1981.

Sokic, Milorad, Cvetkovic, R., and Trifunovic, Z., "Thermodynamic aspects of the utilization of coal-based raw materials within the system of the petrochemical industry," Canadian Journal of Chemical Engineering, 65, 662–671, 1990a.

Sokic, Milorad, Zdravkovic, S., and Trifunovic, Z., "Multiobjective approach to the structuring of an efficient system for producing petrochemicals from alternative raw materials," Canadian Journal of Chemical Engineering, 68, 119–126, 1990b.

Yang, J., "Designing the optimal structure of the petrochemical industry for the minimum cost and the least gross toxicity of chemical production," M.S. Thesis, University of California, Los Angeles, 1984.

PROBLEMS

1. Identify processes that produce and consume the following chemicals. Report your results in a form similar to Table 14.2-1. Can you identify potential networks of processes that could exchange these materials?

   Ammonia

   Hydrogen

2. At the Kalundborg ecopark, waste heat in the form of steam is sent from the Asnæs Power Station to the Statoil refinery (140,000 tons/year), to the Novo Nordisk pharmaceutical manufacturing facility (215,000 tons/year), and to district heating (225,000 tons/year). The power plant is rated at 1,500 megawatts, and the steam has a recoverable heat of 1,000 BTU/lb. Each year the power plant burns approximately 4.5 million tons of coal rated at 10,000 BTU/lb.

   (a) Calculate the fraction of the energy from coal combustion that goes to electricity generation, to the refinery, to the pharmaceutical plant, and to district heating. What is the total rate of energy utilization?

   (b) Not all of these energy demands will operate on similar cycles. Project the daily and seasonal variations in demand and suggest ways for the power plant to meet these needs.

   (c) Calculate the quantity of residential heating oil consumption that is displaced by the use of steam. If oil costs $2.00 per gallon and each gallon has a heating value of approximately $1.5*10^5$ BTU, what is the value of this resource?

3. The case studies presented in Section 14.3 highlighted the opportunities for co-locating refineries and power plants, cement kilns and steel mills. Use the information available in the Toxic Release Inventory (www.epa.gov/Tri/) to identify locations of refineries, power plants, steel mills and cement plants in your state. Are any co-located? Can you suggest other industries in your state that might be able to exchange materials or energy?

Details of the Nine Prominent Federal Environmental Statutes

(Adapted from Lynch, 1995)

THE TOXIC SUBSTANCES CONTROL ACT (TSCA) OF 1976

Incidents in which highly toxic substances such as polychlorinated biphenyls (PCBs) began appearing in the environment and in food supplies prompted the federal government to create a program to assess the risks of chemicals before they are introduced into commerce. The Toxic Substances Control Act (TSCA) was enacted in October 11, 1976. TSCA empowers EPA to screen new chemicals or certain existing chemicals to ensure that their production and use do not pose "unreasonable risk" to human health and the environment. However, TSCA requires EPA to balance the economic and social benefits against the purported risks.

Information Gathering

TSCA requires EPA to gather information on all chemicals manufactured or processed for commercial purposes in the United States. The first version of the "TSCA Inventory" contained 55,000 chemicals, and if a chemical is not found on this list, it is considered to be a new chemical and is subject to the Premanufacture Notification requirements. To aid in the gathering of information on existing compounds, companies that manufacture, import, or process any chemical substance are required to submit a report detailing chemical and processing information. This information includes the chemical identity; name and molecular structure; categories of use; amounts manufactured or processed; byproducts from manufacture, processing, use, or disposal; environmental/health effects of chemical and byproducts; and exposure information. Companies must also keep records of any incidents involving the chemical that resulted in adverse health effects or environmental damage.

Existing Chemicals Testing

TSCA may require companies to conduct chemical testing and then submit more detailed data to EPA compared to the information gathering activities listed above. EPA can request this additional data for chemicals that reside on a separate list compiled from the TSCA Inventory by an Interagency Testing Committee. Chemicals that become listed are typically either produced in very high volumes or they may pose unreasonable risk or injury to health or the environment. The list can contain no more than 50 chemicals and EPA is required to recommend a test rule or remove the chemical from the list within one year of its listing. Once a test rule has been promulgated, a regulated entity (a chemical manufacturer) has 90 days from the initiation of the test rule to submit the data.

New Chemical Review

Chemical manufacturers, importers, and processors are required to notify EPA within 90 days of introducing a new chemical into commerce by submitting a Pre-manufacturing Notice (PMN). The PMN contains information on the identity of the chemical, categories of use, amounts intended to be manufactured, number of persons exposed to the chemical, the manner of disposal, and data on the chemical's effects on health and the environment. EPA can require a PMN to be submitted on any existing chemical that is being used in a significantly different manner than prior known usages. EPA has 90 day from the submission of the PMN to assess the risks of the new chemical or new usage of an existing chemical. If the risks are deemed to be unreasonable based on the information in the PMN and other data that are generally available, EPA is required to take steps to control such risks. These steps might include limiting the production or use of the chemical or ruling a complete ban of the chemical. If data contained in the PMN is insufficient such that EPA cannot make a determination of the risks, the production of that chemical may be banned until such data is made available.

Regulatory Controls and Enforcement

EPA has several options to control the risk of chemicals that have been deemed to pose unreasonable risk, ranging from banning the chemical (most burdensome), to limiting its production and use (less burdensome), to requiring warning labels at the point of sale (least burdensome). EPA is required to use the least burdensome regulatory control considering the chemical's societal and economic benefits. This does not mean that the least burdensome control is always used, but rather it requires EPA to consider the benefits before applying regulatory controls. EPA is authorized to conduct inspections of facilities for manufacturing, processing, storing, or transporting regulated chemicals and items eligible for inspection may include records, files, controls, and processing equipment. "Knowing or willful" violations of TSCA are punishable as crimes that carry up to 1 year imprisonment and up to $25,000 per day of violation.

THE FEDERAL INSECTICIDE, FUNGICIDE, AND RODENTICIDE ACT (FIFRA) OF 1972

The Federal Insecticide, Fungicide, and Rodenticide Act was originally enacted in 1947, but has been amended several times, most notably in 1972. Because all pesticides are toxic to plants and animals, they may pose an unacceptable risk to human health and the environment. FIFRA is a federal regulatory program whose purpose is to assess the risks of pesticides and to control their usage so that any exposure poses an acceptable level of risk.

Registration of Pesticides

Before any pesticide can be distributed or sold in the United Stated, it must be registered with the EPA. The decision by EPA to register a pesticide is based on the data submitted by the pesticide manufacturer in the registration application. The data in the registration application is difficult and expensive to develop, and must include the crop or insect to which it will be applied. In addition, the data must support the claim that it is effective against its intended target, that it allows adequate safety to those applying it and that it will not cause unreasonable harm to the environment. The use of the term "unreasonable harm" is equivalent to requiring the EPA to consider a pesticide's environmental, economic, and social benefits and costs. Pesticides are registered for either general or restricted use. EPA requires that restricted pesticides be applied by a certified applicator. A registration is valid for five years, upon which time it automatically expires unless a re-registration petition is received. FIFRA requires older pesticides that were never subject to the current registration requirements to be registered if their use is to continue. It is estimated that there are over 35,000 older pesticides that were never registered during their prior usage. EPA can cancel a pesticide's registration if it is found to present unreasonable risk to human health or the environment. Also, a registration may be revoked if the pesticide manufacturer does not pay EPA the registration maintenance fee.

Labeling

Labels must be placed on pesticide products that indicate approved uses and restrictions. The label must contain the pesticide's active ingredients, instructions on approved applications to crops or insects, and any limitations on when and where it can be used. It is a violation of FIFRA to use any pesticide in a manner that is not consistent with the information contained on the product label.

Enforcement

It is unlawful to sell or distribute any unregistered pesticide or any pesticide whose composition or usage is different from the information contained in its registration. It is also a violation if FIFRA record keeping, reporting, and inspection requirements are not met. The use of registered pesticides that were approved for restricted use only in any manner other than stated on the FIFRA registration also constitutes a violation. Finally, it is unlawful to submit false data and registration claims. The power to enforce FIFRA is given to the states; however, the state implementation and enforcement programs must be substantially equivalent to the federal program. Any violation of FIFRA is punishable by a civil fine of up to $5,000 while knowing violations of registration requirements may have criminal fines of up to $50,000 and 1 year imprisonment. Fraudulent data submissions may be punishable by up to $10,000 or up to 3 years imprisonment.

THE OCCUPATIONAL SAFETY AND HEALTH ACT (OSH ACT) OF 1970

The OSH Act was enacted in December 29, 1970 in order to ensure safe working conditions for men and women. The agency that oversees the implementation of the OSH Act is the Occupational Safety and Health Administration (OSHA). Each state is authorized to develop its own safety and health plan, but it may adopt the federal program and must meet all federal standards. All private facilities having more than 10 employees must comply with the OSH Act requirements, though certain employment sectors are exempt from the majority of the Act's regulatory provisions. For example, excluded are certain segments of the transportation industry which are covered by the Department of Transportation regulations, the mining industry which is regulated by the Mine Safety and Health Administration, and the atomic energy industry which must comply with the Nuclear Regulatory Commission standards.

Workplace Health and Safety Standards

The OSH Act requires OSHA to set workplace standards to ensure a safe and healthy work environment. These include health standards, which provide protection from harmful or toxic substances by limiting the amount to which a worker is exposed, and safety standards, which are designed to protect workers from physical hazards, such as faulty or potentially dangerous equipment. When establishing health standards, OSHA considers the short term (acute), long term (chronic), and carcinogenic health effects of a chemical or a chemical mixture. These standards take the form of maximum exposure concentrations for chemicals and requirements for labeling, use of protective equipment, and workplace monitoring.

Hazard Communication Standard

The OSH Act's Hazard Communication Standard requires that several standards be met by manufacturers or importers of chemicals and also for the subsequent users of them. These requirements include the development of hazard assessment data, the labeling of chemical substances, and the informing and training of employees in the safe use of chemicals. Chemical manufacturers and importers are required to assess both the physical and health hazards of the chemicals they make or use. This information must be assembled in a material safety data sheet (MSDS) in accordance with OSH Act standards and accompany any sale or transfer of the chemical. Chemical manufacturers and importers must also label chemicals according to OSH Act standards whenever a chemical leaves their control and must train their employees on the safe handling of chemicals in the workplace. Employers must keep a copy of the MSDS in the workplace for each chemical used. Employers must also develop a written hazards communication plan which outlines the implementation plan for informing and training employees on the safe handling of chemicals in the workplace. Employers that use manufactured chemicals must also label those containers according to OSH Act standards.

Record Keeping and Inspection Requirements

Employers must keep records of all steps taken to comply with OSH Act requirements, including the company's safety policies, hazard communication plan, and employee training programs. In addition, employers must keep records of all work-related injuries and deaths and report them periodically to OSHA. Employers must keep records of employee exposure to potentially toxic chemicals and keep them for 30 years. An OSHA Compliance Safety and Health Officer is authorized to enter all covered facilities as part of a general inspection schedule in order to review safety policies and records and to inspect manufacturing equipment. After inspection, a closing meeting is held between the inspector and company health and safety representative to discuss any potential OSH Act violations.

Enforcement

Based on the inspection, a citation may be issued for any OSH Act violations. These citations must be posted in a prominent location in the facilities for at least 3 days. *De Minimus* violations are not considered serious enough to threaten employee safety and health. Serious violations present a real potential for employee harm and may involve penalties of up to $7,000. Willful or repeated violations carry penalties of up to $70,000 per violation.

CLEAN AIR ACT (CAA) OF 1970

The Clean Air Act is actually an amendment of an earlier law (the 1955 Air Pollution Act had weak regulatory provisions) and has been amended eight times, most notably in 1977 and 1990. The CAA is intended to control the discharge of air pollution by establishing uniform ambient air quality standards that are in some instances health-based and in others, technology-based. Mobile and stationary sources of air pollution must comply with source-specific emission limits that are intended to meet these ambient air quality standards. In addition, the CAA addresses specific air pollution problems such as hazardous air pollutants, stratospheric ozone depletion, and acid rain. The 1990 amendments of the CAA revised the hazardous air pollutant regulatory program, instituted a market-based emissions trading system for sulfur dioxide, created strict tail-pipe emission standards for the most polluted urban areas, created a market for reformulated and alternative fuels, and instituted a comprehensive state-run operating permit program.

One of the most important steps in achieving the goals of the CAA was the establishment of the national ambient air quality standards (NAAQS). These are maximum allowable concentrations of specific chemicals monitored in the ambient, or background, air that meet or exceed health-based criteria. Table A.3-1 is a list of the primary and secondary NAAQS for the criteria pollutants, carbon monoxide, lead, nitrogen dioxide, tropospheric ozone, particulate matter, and sulfur dioxide. The NAAQS primary standards are human health-related while the secondary standards are intended to protect a broader range of environmental harm (soils, crops, vegetation, and wildlife); thus they are more restrictive than primary standards.

State Implementation Plan

The CAA requires states to develop individualized state implementation plans (SIPs) that outline how they intend to achieve national ambient air quality standards (NAAQS). The SIP-NAAQS system is an example of "cooperative federalism." The federal government ensures that the provisions of the CAA are implemented but states are responsible for controlling local sources of air pollution. Thus, the state regulatory agencies establish source-specific emission limits on mobile and stationary sources at a sufficient level to ensure compliance with federal quality standards. Under the CAA, EPA establishes the NAAQS, reviews state-authored SIPs to ensure that they will achieve the NAAQS, and may take over state programs if they fail to implement the SIP effectively.

New Source Performance Standards

The CAA allows emission limits to be set on new sources that are more restrictive than limits on existing sources. These standards are termed New Source Performance Standards (NSPS). The reasoning behind this standard is that controls can be incorporated more easily into new processes than they can be retrofitted into

Table A.3-1   Criteria Pollutants and the National Ambient Air Quality Standards.

Pollutant	Primary Standard (Human Health Related)		Secondary (Welfare Related)	
	Type of Average	*Concentration[a]*	*Type of Average*	*Concentration*
CO	8-hour[b]	9 ppm (10 mg/m^3)	No Secondary Standard	
	1-hour[b]	35 ppm (40 mg/m^3)	No Secondary Standard	
Pb	Maximum Quarterly Average	$1.5 \text{ }\mu\text{g/m}^3$	Same as Primary Standard	
NO_2	Annual Arithmetic Mean	0.053 ppm ($100 \text{ }\mu\text{g/m}^3$)	Same as Primary Standard	
O_3	1-hour[c]	0.12 ppm ($235 \text{ }\mu\text{g/m}^3$)	Same as Primary Standard	
	8-hour[d]	0.08 ppm ($157 \text{ }\mu\text{g/m}^3$)	Same as Primary Standard	
PM_{10}	Annual Arithmetic Mean	$50 \text{ }\mu\text{g/m}^3$	Same as Primary Standard	
	24-hour[e]	$150 \text{ }\mu\text{g/m}^3$	Same as Primary Standard	
$PM_{2.5}$	Annual Arithmetic Mean[f]	$15 \text{ }\mu\text{g/m}^3$	Same as Primary Standard	
	24-hour[g]	$65 \text{ }\mu\text{g/m}^3$	Same as Primary Standard	
SO_2	Annual Arithmetic Mean	0.03 ppm ($80 \text{ }\mu\text{g/m}^3$)	3-hour[b]	0.50 ppm ($1,300 \text{ }\mu\text{g/m}^3$)
	24-hour[b]	0.14 ppm ($365 \text{ }\mu\text{g/m}^3$)		

[a] Parenthetical value is an equivalent mass concentration.
[b] Not to be exceeded more than once per year.
[c] Not to be exceeded more than once per year on average.
[d] 3-year average of annual 4th highest concentration.
[e] The pre-existing form is exceedance-based. The revised form is the 99th percentile.
[f] Spatially averaged over designated monitors.
[g] The form is the 98th percentile.

Source: 40 Code of Federal Register (CFR) Part 50, revised standards issued July 18, 1997. Adapted from US EPA (1998).

existing processes. EPA established which categories of industrial sources can be subject to these standards, and the emission limits are set by considering the best available emission control technologies, other health and environmental impacts that may occur during the application of the control technology, and energy usage issues. Because the new source standards are uniformly established nationwide, they create a level playing field where companies are discouraged from locating in states that do not require these strict pollution controls.

A New Source Review program has been established by the CAA in order to review new processes and significant modifications to existing processes and to prevent significant deterioration of ambient air quality. Before construction can begin, the operator must obtain a permit and demonstrate 1) that the source will comply with ambient air quality standards, 2) that the source will utilize the best available control technology, 3) that their emissions will not cause a violation of the NAAQS in nearby areas, and 4) that new or modified sources must achieve offsets, that is reductions in emissions of the same pollutant, in a greater than 1:1 ratio.

Hazardous Air Pollutants

The CAA has identified 189 hazardous air pollutants (HAPs) that are subject to more stringent emission controls than the six criteria air pollutants. Any stationary source that emits 10 tons per year of any HAP or 25 tons per year of any combination of HAPs is subject to these CAA provisions. EPA is required to develop source-specific emission standards that require installation of technologies that will result in the maximum achievable degree of control (MACT). If an existing source can demonstrate that it has achieved or will achieve a reduction of 99 percent of hazardous air pollution emissions before enactment of the MACT standards, it may receive a six-year extension of its compliance deadline.

Enforcement

Civil penalties for violations of the Clean Air Act may involve fines of up to $25,000 per day of violation. Penalties for knowing violations of the CAA are up to $250,000 per day in fines and up to 5 years imprisonment. Corporations may be fined up to $500,000 per violation and repeat offenders may receive double fines. Knowing violations that involve releases of HAPs may trigger fines of up to $250,000 per day and up to 15 years imprisonment. Corporations may be fined $1,000,000.

THE CLEAN WATER ACT (CWA) OF 1972

The Clean Water Act (CWA) was first enacted in October 18, 1972 and is the first comprehensive federal program designed to reduce pollutant discharges into the nation's waterways ("zero discharge" goal). Another goal of the CWA is to make water bodies safe for swimming, fishing, and other forms of recreation ("swimmable" goal). This act has resulted in significant improvements in the quality of the nation's waterways since its enactment. The CWA defines a pollutant rather broadly, as "dredged spoil, solid waste, incinerator residue, sewage, garbage, sewage sludge, munitions, chemical wastes, biological materials, radioactive materials, heat, wrecked or discarded equipment, rock, sand, cellar dirt and industrial, municipal, and agricultural waste discharged into water" (CWA §502(14), 33 U.S.C.A. §1362). The CWA has two major components, the National Pollutant Discharge Elimination System (NPDES) permit program and the Publicly Owned Treatment Works (POTW) construction program.

Publicly Owned Treatment Works (POTW) Construction Program

This program originally provided grants to POTW so that they could upgrade their facilities from primary to secondary treatment. Primary treatment involves removing a portion of the suspended solids and organic matter using operations such as screening and sedimentation. Secondary treatment removes residual organic matter using microorganisms in large mixed basins. Federal grants, having no repayment obligations, were available for as much as 55% of the total project costs. The 1987 amendments converted the grant program into a revolving loan program in which municipalities can obtain low interest loans that must be repaid.

National Pollutant Discharge Elimination System (NPDES) Permit Program

The statute classifies water pollution sources as point sources and nonpoint sources. Point sources are any discrete conveyance (pipe or ditch) that introduces pollutants into a water body. Point sources are further divided into municipal (from POTWs) and industrial. An example of a nonpoint source is runoff from agricultural lands. Nonpoint sources are the last major source of uncontrolled pollution discharge into waterways. The National Pollutant Discharge Elimination System (NPDES) permit program requires any point source of pollution to obtain a permit. The NPDES permit program is another example of a cooperative federal-state regulatory program. The federal government established national standards (e.g., effluent guidelines), and the states are given flexibility in achieving these standards. NPDES permits contain effluent limits, either requiring the installation of specific pollutant treatment technologies or adherence to specified numerical discharge limits. In establishing the NPDES limits, the state regulatory agency considers the federal effluent guidelines and the desired water quality standards established by the state for the intended use of the waterway (drinking water source, recreation, agricultural, etc.).

Monitoring/Inspection Requirements

NPDES permit holders must monitor discharges, collect data, and keep records of the pollutant levels of their effluents. These records must be submitted to the agency that granted the NPDES permit to ensure that the point source is not exceeding the effluent discharge limits. The permitting agency is authorized to inspect the permit holder's records and collect effluent samples to verify compliance with the CWA.

Industrial Pretreatment Standards

Industrial sources that discharge into sewers that eventually enter POTWs are termed "indirect discharge" sources. These sources do not need to obtain a NPDES permit, but may have to apply for state or local permits and must comply with EPA pretreatment standards. Pretreatment standards reflect the best available control technology (BACT) and are designed to remove the most toxic pollutants and to minimize the "pass through" of these components into receiving waters from POTWs. Indirect dischargers can obtain removal credits if they can demonstrate that the POTW can effectively remove a particular pollutant down to acceptable levels.

Dredge and Fill Permits and Discharge of Oil or Hazardous Substances

A permit must be obtained from the United States Army Corp of Engineers before any discharge of dredge or fill materials into navigable waterways, including wetlands, occurs. The CWA also prohibits discharge of oil or hazardous substances into any navigable waters and provides mechanisms for the clean up of oil and hazardous substance spills. Any person in charge of a vessel or facility must notify the Coast Guard's National Response Center and also state officials whenever such a spill occurs above a certain quantity. Failure to do so may result in up to 5 years imprisonment.

Enforcement

Civil penalties may be as high as $25,000 per day for violations of the CWA provisions. Criminal violations for repeated negligent conduct may be as high as $50,000 per day and up to 2 years imprisonment. Repeated knowing violations can result in fines of up to $100,000 per day and 6 years imprisonment. Repeated knowing endangerment violations of the CWA can bring fines as high as $500,000 and 15 years imprisonment. Organizations can be fined as much as $1,000,000. Violations that involve false monitoring and reporting are subject to a $10,000 fine and up to 2 years imprisonment.

RESOURCE CONSERVATION AND RECOVERY ACT (RCRA) OF 1976

The Resource Conservation and Recovery Act was enacted to regulate the disposal of both non-hazardous and hazardous solid wastes to land, encourage recycling, and promote the development of alternative energy sources based on solid waste materials. In reality, RCRA also regulates any waste material that is disposed to land, including liquids, gases, and mixtures of liquids with solids and gases with solids. RCRA's Subtitle C provisions regarding the management and disposal of hazardous wastes have become the key provisions. RCRA was significantly amended by the Hazardous and Solid Waste Amendments (HSWA) in 1984. The provisions of the HSWA affect hazardous waste disposal facilities by restricting the disposal of hazardous waste and regulating underground storage tanks containing hazardous substances or petroleum. RCRA's Subtitle C establishes provisions that must be complied with by hazardous waste generators, transporters of hazardous waste, and facilities that treat, store, or dispose of hazardous waste. RCRA represents a "cradle-to-grave" regulatory system that manages hazardous waste throughout its life cycle in order to minimize the risks that these wastes pose to the environment and to human health.

Identification/Listing of Hazardous Waste

If wastes exhibit any of four hazardous characteristics (ignitibility, corrosivity, reactivity, or toxicity), they are considered to be hazardous. A material can also be designated as a hazardous waste if EPA lists it as such. Three hazardous waste lists have been compiled by EPA. The first list contains approximately 500 wastes from non-specific sources and includes specific chemicals. The second list of hazardous wastes is from specific industry sources, for example hazardous wastes from the petroleum refining industry. The third list includes wastes from commercial chemical products, which when discarded or spilled, must be managed as hazardous wastes. Specifically exempted from being hazardous wastes are household waste, agricultural wastes that are returned to the ground as fertilizer, and wastes from the extraction, beneficiation, and processing of ores and minerals, including coal.

Generator Requirements

EPA defines a generator as any facility that causes the generation of a waste that is listed as a hazardous waste under RCRA provisions. A generator of hazardous waste must obtain an EPA identification number within 90 days of the initial generation of the waste. RCRA requires generators to properly package hazardous waste for shipment off-site and to use approved labeling and shipping containers. Generators must maintain records of the quantity of hazardous waste generated and where the waste was sent for treatment, storage, or disposal, and must file this data in biennial reports to the EPA. Generators must prepare a Uniform Hazardous Waste Manifest, which is a shipping document that must accompany the waste at all times. A copy of the manifest is sent back to the generator by the treatment facility to ensure that the waste reached its proper destination.

Other Requirements

RCRA imposes requirements on transporters of hazardous waste as well as on facilities that treat, store, and dispose of hazardous wastes. Transporters are any persons that transport hazardous waste by air, rail, highway, or water from the point of generation to the final destination of treatment, storage, or disposal. The final destinations are termed treatment, storage, and disposal facilities (TSDFs) by EPA. Transporters must adhere to the Uniform Hazardous Waste Manifest system when shipping hazardous waste, which includes retaining copies of manifests for a period of three years. A facility that accepts hazardous waste for the purpose of changing the physical, chemical, or biological character of the waste and with the intent of rendering the waste nonhazardous, making the waste amenable for transport or recovery, or reducing the waste volume is defined as a treatment facility by RCRA. Storage facilities are intended for holding wastes for a short period of time until such time as the waste is shipped to a treatment or disposal facility elsewhere. A disposal facility is a location that is engineered to safely accept hazardous waste in various forms (drums, solids, etc.) for long term internment. These facilities must monitor the environment within and adjacent to the facility to assure that hazardous waste components are not leaving the site in concentrations that threaten the environment or human health. Generators who store hazardous waste on site for more than 90 days or who treat or dispose of hazardous waste themselves are considered TSDFs by RCRA.

Enforcement

Failure to comply with RCRA Subtitle C or EPA compliance orders carries a civil penalty of up to $25,000 per day of violation. Violations may result in the revocation of the RCRA permit. Criminal penalties for violations may be as high as $50,000 per day for each violation and/or 2 years imprisonment. Fines and jail time may double for repeat offenders. When a person violates RCRA and in the process knowingly endangers another individual, fines may reach $250,000 per day and up to 15 years imprisonment. Organizations may be fined as much as $1,000,000.

THE COMPREHENSIVE ENVIRONMENTAL RESPONSE, COMPENSATION, AND LIABILITY ACT (CERCLA) OF 1980

The contamination of Love Canal in upstate New York with industrial toxic materials and the subsequent evacuation of hundreds of families from the vicinity, alerted the federal government of the need to clean up this and other related sites. The Comprehensive Environmental Response, Compensation, and Liability Act (CERCLA) of 1980 began a process of identifying and cleaning up the many sites of uncontrolled hazardous waste disposal at abandoned sites, at industrial complexes, and at federal facilities. EPA is responsible for creating a list of the most hazardous sites of contamination, which is termed the National Priority List (NPL). As of 1994, there were 1,232 facilities, including 150 federal facilities, on the NPL and an additional 340 to 370 sites are expected to be added to the NPL before September 30, 1999. CERCLA established a $1.6 billion Hazardous Substance Trust Fund (Superfund) to initiate cleanup of the most contaminated sites. Superfund (the trust fund) allows for the cleanup of sites for which parties responsible for creating the contamination cannot be identified because of bad record keeping in the past, or are no longer able to pay, are bankrupt, or are no longer in business. The Superfund Amendments and Reauthorization Act (SARA) of 1986 increased the Superfund appropriation to $8.5 billion through December 31, 1991, extended and expanded the tax for Superfund, and stipulated a preference for remedial action to be cleanup rather than containment of hazardous waste. In addition, Superfund was extended to September 30, 1994 with an additional $5.1 billion. As of this printing, the Superfund program continues to operate via yearly US EPA budget appropriations, fund interest, and cost recoveries from PRPs (see below), though no new appropriations have been added to the trust fund since 1995. Under the CERCLA provisions, EPA can make two responses to sites of hazardous waste contamination. These are short-term emergency response to spills or other releases, and long-term remedial actions, which may actually occur long after the site is listed on the NPL, and which is designed to achieve a permanent state of cleanup.

Potentially Responsible Party (PRP) Liability

After a site is listed in the NPL, EPA identifies potentially responsible parties (PRPs) and notifies them of their potential CERCLA liability. If the cleanup is conducted by the EPA, the PRPs are responsible for paying their share of the cleanup costs. If the cleanup has not begun, PRPs can be ordered to complete the cleanup of the site. PRPs are 1) present or 2) past owners of hazardous waste disposal facilities, 3) generators of hazardous waste who arrange for treatment or disposal at any facility, and 4) transporters of hazardous waste to any disposal facility. Liability for PRPs is strict, meaning that liability can be imposed regardless of fault or negligence. Liability is joint and several, meaning that one party can be held responsible for the actions of others when the harm is indivisible. Finally, the liability is retroactive, meaning that parties can be held liable for actions that predate CERCLA. The EPA does not have to prove that a particular PRP's waste caused the contamination. EPA only has to prove that there are hazardous substances present at the site that are similar to those associated with a party's hazardous waste treatment and disposal activities.

Enforcement

EPA can force PRPs to conduct and fund cleanup of contaminated sites to which they have been associated in actions termed Private Party Cleanups. Failure to comply with a Private Party Cleanup order may involve fines of up to $25,000 per day and judicial reviews of these cases are not immediately available. Thus, PRPs have little choice but to comply. Failure to report to EPA the release of hazardous substances in quantities greater than the cut-off value for that substance may result in fines amounting to more than $25,000 per day and criminal penalties of three years for a first conviction and five years for a subsequent conviction.

THE EMERGENCY PLANNING AND COMMUNITY RIGHT TO KNOW ACT (EPCRA)

In 1984, the release of methyl isocyanate from a Union Carbide plant in Bhopal, India killed more than 2,500 people and permanently disabled some 50,000 more. This unfortunate incident illustrated the need for communities to develop emergency plans in preparation for releases that might occur from chemical manufacturing facilities. It also highlighted the need for communities to find out what toxic chemicals are being manufactured at facilities and what are the rates and to what media toxic chemicals are being released. Title III of the Superfund Amendments and Reauthorization Act (SARA) contains a separate piece of legislation called the Emergency Planning and Community Right-to-Know Act (EPCRA). There are two main goals of EPCRA: 1) to have states create local emergency units that must develop plans to respond to chemical release emergencies, and 2) to require EPA to compile an inventory of toxic chemical releases to the air, water, and soil from manufacturing facilities, and to disclose this inventory to the public.

Toxic Release Inventory (TRI)

EPCRA requires facilities with more than 10 employees who either use more than 10,000 pounds or manufacture or process more than 25,000 pounds of one of the listed chemicals or categories of chemicals to report annually to EPA. The report must contain data on the maximum amount of the toxic substance on-site in the previous year, the treatment and disposal methods used, and the amounts released to the environment or transferred off-site for treatment and/or disposal. Facilities that are obligated to report must use the Chemical Release Inventory Reporting Form (Form R). Facilities must keep records supporting their TRI submissions for three years from the date of submission of Form R to EPA. The data are compiled by the EPA and entered into a computerized database that is accessible to the public. The TRI is viewed by citizens, environmental groups, states, industry, and others as an environmental scorecard for the chemical manufacturing and allied products industries. As a result of the TRI, many manufacturers have initiated voluntary programs to reduce the releases of toxic chemicals into the environment. In 1990, the EPA implemented the 33/50 Program, a voluntary program for participating facilities to reduce their releases of 17 key chemicals by 33% by 1992 and 50% by 1995 compared to baseline levels. As of October 1999, 1294 companies had committed to the program (EPA, 1999).

Enforcement

Violations of EPCRA's TRI reporting and community emergency planning requirements are subject to civil penalties of up to $25,000 per day. Any person who knowingly and willingly fails to report releases of toxic substances can be fined up to $25,000 and/or be imprisoned for up to 2 years. Second violations may subject persons to fines of up to $50,000 or 5 years imprisonment.

POLLUTION PREVENTION ACT OF 1990

In October 27, 1990, the Pollution Prevention Act was passed by Congress. The act established pollution prevention as the nation's primary pollution management strategy. Pollution prevention is defined as "any practice which: 1) reduces the amount of any hazardous substance, pollutant, or contaminant entering any waste stream or otherwise released into the environment . . . prior to recycling, treatment, and disposal; and 2) reduces the hazards to public health and the environment associated with the release of such substances, pollutants, or contaminants." Thus, pollution prevention not only encourages reductions in waste generation and release from production facilities but also promotes reductions in waste component toxicity or other hazardous characteristics. This strategy is fundamentally different from those of prior environmental statutes, in that pollution prevention encourages steps to reduce pollution generation and toxicity at the source rather than relying on end-of-pipe pollution controls.

The Pollution Prevention Act (PPA) provides for a hierarchy of pollution management approaches. It states that: 1) pollution should be prevented or reduced at the source whenever feasible, 2) pollution that cannot be prevented or reduced should be recycled, 3) pollution that cannot be prevented or reduced or recycled should be treated, and 4) disposal or other releases into the environment should be employed only as a last resort. The Act is not an action-forcing statute, but rather encourages voluntary compliance by industry of the suggested approaches and strategies through education and training. To this end, EPA is required to establish a Pollution Prevention Office independent of the other media-specific pollution control programs. It is also required to set up a Pollution Prevention Information Clearinghouse whose goal is to compile source reduction information and make it available to the public. The only mandatory provisions of the PPA requires owners and operators of facilities that are required to file a form R under the SARA Title III (the TRI) to report to the EPA information regarding the source reduction and recycling efforts that the facility has undertaken during the previous year.

Molecular Connectivity

Correlations for environmentally relevant physical and chemical properties, described in Chapter 5, are primarily based on bulk properties such as boiling point and octanol-water partition coefficient. While these bulk properties are adequate correlating parameters for many properties, they are not adequate for properties that depend on molecular topology, such as soil sorption. In situations where a description of molecular topology is required, a simple alternative is to utilize the molecular connectivity (χ).

The concept of molecular connectivity initially appeared in the pharmaceutical literature and a variety of molecular connectivity indices have been used in predicting drug behavior (Kier and Hall, 1986). This text uses only the most basic of molecular connectivity indices—the simple first order molecular connectivity ($^1\chi$). The goal of this index is to characterize, in a single scalar parameter, the degree of connectedness or the topology of the molecule. A complete description of the rationale behind the molecular connectivity is beyond the scope of this text. The interested reader is referred to Kier and Hall (1986). Instead, the focus here will be on the steps required to calculate $^1\chi$.

The first step in calculating $^1\chi$ is to draw the bond structure of the molecule. For example, isopentane would be drawn as:

$$CH_3$$

$$|$$

$$H_3C\text{-}CH\text{-}CH_2\text{-}CH_3$$

The next step is to count the number of carbon atoms to which each carbon is attached (count any heteroatom as a carbon, but ignore bonds to hydrogen). The assignments of this parameter (δ_i, the connectedness of carbon atom i) for each carbon in isopentane are given below.

$$(\mathbf{1})$$
$$CH_3$$
$$|$$
$$H_3C\text{-}CH\text{-}CH_2\text{-}CH_3$$
$$(\mathbf{1})\ (\mathbf{3})\ (\mathbf{2})\ \ (\mathbf{1})$$

For each bond, identify the connectedness of the carbons connected by the bond (δ_i, δ_j). For isopentane, these pairs are:

$$(1,3), (1,3), (3,2), (2,1)$$

The value of $^1\chi$ is calculated using the equation:

$$^1\chi = \Sigma(\delta_I * \delta_j)^{-0.5}$$

For isopentane,

$$^1\chi = (1/\sqrt{3}) + (1/\sqrt{3}) + (1/\sqrt{6}) + (1/\sqrt{2}) = 2.68$$

Clearly, this calculation yields a simplistic characterization of complex structural features. Note that isopentene would yield exactly the same value as isopentane, as would 1-chloro, 2 methyl propane. Nevertheless, this simple characterization of molecular topology is often used, as described in Chapter 5, in developing property correlations.

Example B-1

Estimate $^1\chi$ for 4-chloro-aniline.

Solution: The molecular structure and the connectedness of each carbon or heteroatom are shown below:

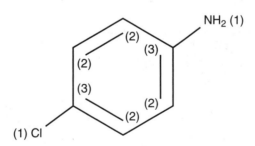

The bond pairs, beginning with the amine and continuing clockwise around the molecule, are (1,3), (3,2), (2,2), (2,3), (3,1), (3,2), (2,2), (2,3)

$$^1\chi = (1/\sqrt{3}) + (1/\sqrt{6}) + (1/\sqrt{4}) + (1/\sqrt{6}) + (1/\sqrt{3}) + (1/\sqrt{6}) + (1/\sqrt{4}) + (1/\sqrt{6}) = 3.787$$

Estimating Emissions from Storage Tanks

Emission rate correlations have been developed by industry and government agencies for storage tanks and secondary emission sources (U.S. EPA 1985, U.S. EPA 1998). Storage tanks are units common to almost every chemical process. They provide a buffer for raw materials availability in continuous processes and allow for storage of finished product before delivery is taken. Tanks have the potential to be major contributors to airborne emissions of volatile organic compounds from chemical facilities because of the dynamic operation of these units. There are two major losses mechanisms from tanks; working losses and standing losses. Working losses originate from the raising and lowering of the liquid level in the tank as a result of raw material utilization and production of product. The gas space above the liquid must expand and contract in response to these level changes. During tank emptying, air from the outside or an inert gas, if provided, enter the tank. Volatile organic vapors from the liquid evaporate in an attempt to achieve an equilibrium condition between the concentrations of each component in the liquid and gas phases. When the tank is filled again, these vapors in the gas exit the unit via the vent to be dispersed into the atmosphere unless pollution control devices are installed. Even if the tank level is static, standing losses from the tank occur as a result of daily temperature and ambient pressure fluctuations which cause a pressure difference between the gas inside the tank and the air outside.

There are four major types of storage tanks; fixed-roof, floating-roof, variable-vapor-space, and pressurized tanks. Equations for estimating emissions from fixed-roof and floating-roof storage tanks will be provided. The total loss (L_T) is the sum of the standing (L_S) and the working (L_W) losses, as shown in the following equations. The standing losses are proportional to the vapor density, the tank vapor space volume, the vapor expansion factor, and an empirical parameter related to the degree of saturation of the gas with the volatile organic chemical. The working losses are proportional to the molecular weight of the liquid, the vapor pressure of the liquid, the throughput of liquid into and out of the tank, and two empirical factors related to the degree of saturation in the gas phase. Tables C-1 and C-2 contain the set of equations needed to calculate storage tank emissions for fixed- and floating-roof tanks, respectively. Equations are defined and numbered sequentially, with references to additional equations for the parameters appearing in each. Also, there are references to tables of data needed to carry out the calculations. Parameters in each equation are defined in subsequent equations in Tables C-1 and C-2 or in the tables referenced. These equations are described in detail in the EPA's emission inventory database, and software is available that performs the emission estimates (the TANKS program) (www.epa.gov/ttn/chief/, under software labeled tanks).

Table C-1 Total Losses from Fixed-Roof Storage Tanks.

Eqn. #	Description	Equation	Parameter	*Reference eqn. # / Notes*
1	Total losses (lb/yr)	$L_T = L_S + L_W$	L_S; L_W	*2; 3*
2	Standing storage losses (lb/yr)	$L_S = 365\ V_V W_V K_E K_S$	V_V; W_V; K_E; K_S	*4; 11; 16; 23*
3	Working losses (lb/yr)	$L_W = 0.0010\ M_V P_{VA} Q K_N K_P$	M_V; P_{VA}; Q; K_N; K_P	Tables C-3 and Table C-4; *12*; Annual net throughput (bbl/yr); *24*; 0.75-crude oil or 1-other organic
4	Vapor space volume (ft^3)	$V_V = (\pi/4)D^2 H_{VO}$	D; H_{VO}	Tank diameter (ft), for horizontal tanks *-5; 6*
5	Effective tank diameter (ft)	$D_E = (L/D / 0.785)^{0.5}$	L; D	Length of tank (ft); Actual tank diameter for horizontal tank (ft)
6	Vapor space outage (ft)	$H_{VO} = H_S - H_L + H_{RO}$	H_S; H_L; H_{RO}	Tank shell height (ft); Liquid height (ft); Cone roof-7 or Dome roof-9
7	Cone roof outage (ft)	$H_{RO} = 1/3\ H_R$	H_R	8
8	Tank cone roof height (ft)	$H_R = S_R R_S$	S_R; R_S	Tank cone roof slope(ft/ft)-0.0625 (ft/ft -if unknown) Tank shell radius (ft)
9	Dome roof outage (ft)	$H_{RO} = H_R \times [0.5 + H_R^2/(6\ R_S^2)]$	H_R; R_S	*10*; Tank shell radius (ft)
10	Tank roof height (ft)	$H_R = R_R - (R_R^2 - R_S^2)^{0.5}$	R_R; R_S	Tank dome roof radius (ft); Tank shell radius (ft)
11	Vapor density (lb/ft^3)	$W_V = M_V P_{VA}/(R\ T_{LA})$	M_V; P_{VA}; R; T_{LA}	Tables C-3 and C-4; *12*; 10.731 (psia*ft^3/lb-mole*R); *13*
12	Vapor pressure (psia)	$P_{VA} = (10 / 51.715)\ *$ $[A - B/\{(T_{LA}-492)/1.8 + C\}]$	A; B; TLA; C	Table C-5; Table C-5; *13*; Table C-5

Table C-1 (*continued*)

Eqn. #	Description	Equation	Parameter	*Reference eqn. # / Notes*
13	Daily average liquid surface temperature (R)	$T_{LA} = 0.44T_{AA} + 0.56T_B + 0.0079\,\alpha I$	T_{AA}; T_B; α; I	*14; 15*; Table C-6; Table A-7
14	Daily average ambient temperature (R)	$T_{AA} = (T_{AX} + T_{AN})/2$	T_{AX}; T_{AN}	Table C-7; Table C-7
15	Liquid bulk temperature (R)	$T_B = T_{AA} + 6\alpha - 1$	T_{AA}; α	*14*; Table C-6
16	Vapor space expansion factor (unitless)	$K_E = \Delta T_V / T_{LA} + \dfrac{(\Delta P_V - \Delta P_B)}{(P_A - P_{VA})}$	ΔT_V; T_{LA}; ΔP_V; ΔP_B; P_A; P_{VA}	*17; 13; 19; 22*; 14.7 (psia); *12*
17	Daily vapor temperature range (R)	$\Delta T_V = 0.72\,\Delta T_A + 0.028\,\alpha I$	ΔT_A; α; I	*18*; Table C-6; Table C-7
18	Daily ambient temperature range (R)	$\Delta T_A = T_{AX} - T_{AN}$	T_{AX}; T_{AN}	Table C-7; Table C-7
19	Daily vapor pressure range (psia)	$\Delta P_V = P_{VX} - P_{VN}$	P_{VX} and P_{VN}	Based on T_{LX}-*20* and T_{LN}-*21* instead of T_{LA} using *12*
20	Daily maximum liquid surface temperature (R)	$T_{LX} = T_{LA} + 0.25\,\Delta T_V$	T_{LA}; ΔT_V	*13; 17*
21	Daily minimum liquid surface temperature (R)	$T_{LN} = T_{LA} - 0.25\,\Delta T_V$	T_{LA}; DT_V	*13; 17*
22	Breather vent pressure setting range (psig)	$\Delta P_B = P_{BP} - P_{BV}$	P_{BP}; P_{BV}	Breather vent pressure setting (psig)- +0.03 psig; Breather vent vacuum setting (psig)- -0.03 psig
23	Vented vapor saturation factor (unitless)	$K_S = (1 + 0.53P_{VA}H_{VO})^{-1}$	P_{VA}; H_{VO}	*12; 6*
24	Turn over factor (unitless)	$K_N = (180 + N)/6N$ $K_N = 1$	N > 36 N ≤ 36	*25*
25	Number of turnovers per year (unitless)	$N = 5.614\,Q\,V_{LX}^{-1}$	Q; V_{LX}	Annual net throughput (bbl/yr); *26*
26	Tank maximum liquid volume, ft^3	$V_{LX} = (\pi/4)D^2 H_{LX}$	D; H_{LX}	Diameter (ft); Maximum liquid height (ft)

Table C-2   Total Losses from External Floating Roof Tanks.

Eqn. #	Description	Equation	Parameter	*Reference eqn. # / Notes*
27	Total loss (lb/yr)	$L_T = L_R + L_{WD} + L_F + L_D$	L_R; L_{WD}; L_F; L_D	*28; 30; 31*
28	Rim seal loss (lb/yr)	$L_R = (K_{Ra} + K_{Rb} v^n) P^*$ $(D\, M_V\, K_C)$	K_R; v; n; P^*; D; M_V; Kc	Table C-8; avg. wind speed (mph); Table C-8; *29*; Tank diameter (ft); Table C-3; 0.4 -crude oils or 1 -all other organic liquids
29	Vapor pressure function (unitless)	$P^* = \dfrac{P_{VA} / P_A}{[1 + (1 - [P_{VA}/P_A])^{.5}]^2}$	P_{VA}; P_A	*12*; 14.7 (psia)
30	Withdrawal loss (lb/yr)	$L_{WD} = 0.943 Q C\, W_L /D$ $\left[1 + \dfrac{N_c F_c}{D} \right]$	Q; C; W_L; D; N_c; F_e	Annual throughput (bbl/yr); Table C-9; Table C-3; Tank diameter (ft); no support columns; column diameter
31	Roof fitting loss (lb/yr)	$L_F = F_F\, P^*\, M_V\, K_C$	F_F; P^*; M_V; K_C	*32*; *29*; Table C-3; 0.4 -crude oils or 1 -all other organic liquids
32	Total roof fitting loss factor (lb-mole/yr)	$F_F = (N_{F1}K_{F1})$ $+ (N_{F2}K_{F2})$ $+ .. + (N_{Fn}K_{Fn})]$	N_{Fi}; K_{Fi}	Table C-10; *33*
33	Loss factor for a particular type fitting (lb-moles/yr)	$K_{Fi} = K_{Fai} + K_{Fbi} (K_v v)^{mi}$	K_{Fai}; K_{Fbi}; K_v; v; m_i	Table C-10; Table C-10; 0.7; avg wind speed (mph); Table C-10

Table C-3   Properties (M_V, P_{VA}, W_L) of Selected Petroleum Liquids.

Petroleum Liquid	Vapor Molecular Weight At 60°F, M_V (lb/lb-mole)	Liquid Density At 60°F, W_L (lb/gal)	True Vapor Pressure, P_{VA} (psi)						
			40°F	*50°F*	*60°F*	*70°F*	*80°F*	*90°F*	*100°F*
Gasoline RVP 13	62	5.6	4.7	5.7	6.9	8.3	9.9	11.7	13.8
Gasoline RVP 10	66	5.6	3.4	4.2	5.2	6.2	7.4	8.8	10.5
Gasoline RVP 7	68	5.6	2.3	2.9	3.5	4.3	5.2	6.2	7.4
Crude oil RVP 5	50	7.1	1.8	2.3	2.8	3.4	4	4.8	5.7
Jet naphtha (JP-4)	80	6.4	0.8	1	1.3	1.6	1.9	2.4	2.7
Jet kerosene	130	7	0	0.01	0.01	0.01	0.02	0.02	0.03
Distillate fuel oil No. 2	130	7.1	0	0	0	0.01	0.01	0.02	0.02
Residual oil No. 6	190	7.9	0	0	0	0	0	0	0

Table C-4 Properties (M_V, P_{VA}, W_L) of Selected Petrochemicals.

Name	Molecular Weight, M_v	Boiling Point At 1 Atmosphere (°F)	Liquid Density At 60°F, W_L (lb/gal)	Vapor Pressure (psia) At						
				40°F	50°F	60°F	70°F	80°F	90°F	100°F
Acetone	58.08	133	6.628	1.682	2.185	2.362	3.713	4.699	5.917	7.251
Acetonitrile	41.05	178.9	6.558	0.638	0.831	1.083	1.412	1.876	2.456	3.133
Acrylonitrile	53.06	173.5	6.758	0.812	0.967	1.373	1.779	2.378	3.133	4.022
Allyl alcohol	58.08	206.6	7.125	0.135	0.193	0.261	0.387	0.522	0.716	1.006
Allyl chloride	76.53	113.2	7.864	2.998	3.772	4.797	6.015	7.447	9.11	11.025
Ammonium hydroxide (28.8% solution)	35.05	83	7.481	5.13	6.63	8.48	10.76	13.52	16.76	20.68
Benzene	78.11	176.2	7.365	0.638	0.87	1.16	1.508	1.972	2.61	3.287
iso-Butyl alcohol	74.12	227.1	6.712	0.058	0.097	0.135	0.193	0.271	0.387	0.541
tert-Butyl alcohol	74.12	180.5	6.595	0.174	0.29	0.425	0.638	0.909	1.238	1.702
n-Butyl chloride	92.57	172	7.43	0.715	1.006	1.32	1.74	2.185	2.684	3.481
Carbon disulfide	76.13	115.3	10.588	3.036	3.867	4.834	6.014	7.387	9.185	11.215
Carbon tetrachloride	153.84	170.2	13.366	0.793	1.064	1.412	1.798	2.301	2.997	3.771
Chloroform	119.39	142.7	12.488	1.528	1.934	2.475	3.191	4.061	5.163	6.342
Chloroprene	88.54	138.9	8.046	1.76	2.32	2.901	3.655	4.563	5.685	6.981
Cyclohexane	84.16	177.3	6.522	0.677	0.928	1.218	1.605	2.069	2.61	3.249
Cyclopentane	70.13	120.7	6.248	2.514	3.287	4.177	5.24	6.517	8.063	9.668
1,1-Dichloroethane	98.97	135.1	9.861	1.682	2.243	2.901	3.771	4.738	5.84	7.193
1,2-Dichloroethane	98.97	182.5	10.5	0.561	0.773	1.025	1.431	1.74	2.243	2.804
cis-1,2-Dichloroethylene	96.95	140.2	10.763	1.45	2.011	2.668	3.461	4.409	5.646	6.807
trans-1,2-Dichloroethylene	96.95	119.1	10.524	2.552	3.384	4.351	5.53	6.807	8.315	10.016
Diethylamine	73.14	131.9	5.906	1.644	1.992	2.862	3.867	4.892	6.13	7.541
Diethyl ether	74.12	94.3	5.988	4.215	5.666	7.019	8.702	10.442	13.342	Boils
Di-iso-propyl ether	102.17	153.5	6.075	1.199	1.586	2.127	2.746	3.481	4.254	5.298

(continued)

Table C-4 Properties (M_V, P_{VA}, W_L) of Selected Petrochemicals. (*continued*)

Name	Molecular Weight, M_V	Boiling Point At 1 Atmosphere (°F)	Liquid Density At 60°F, W_L (lb/gal)	Vapor Pressure (psia) At						
				40°F	50°F	60°F	70°F	80°F	90°F	100°F
1,4-Dioxane	88.1	214.7	8.659	0.232	0.329	0.425	0.619	0.831	1.141	1.508
Ethyl acetate	88.1	170.9	7.551	0.58	0.831	1.102	1.489	1.934	2.514	3.191
Ethyl acrylate	100.11	211.8	7.75	0.213	0.29	0.425	0.599	0.831	1.122	1.47
Ethyl alcohol	46.07	173.1	6.61	0.193	0.406	0.619	0.87	1.218	1.682	2.32
Freon 11	137.38	75.4	12.48	7.032	8.804	10.9	13.4	16.31	19.69	23.6
n-Heptane	100.2	209.2	5.727	0.29	0.406	0.541	0.735	0.967	1.238	1.586
n-Hexane	86.17	155.7	5.527	1.102	1.45	1.876	2.436	3.055	3.906	4.892
Hydrogen cyanide	27.03	78.3	5.772	6.284	7.831	9.514	11.853	15.392	18.563	22.237
Isooctane	114.22	210.6	5.794	0.213	0.387	0.58	0.812	1.093	1.392	1.74
Isopentane	72.15	82.1	5.199	5.878	7.889	10.005	12.53	15.334	18.37	21.657
Isoprene	68.11	93.5	5.707	4.757	6.13	7.677	9.668	11.699	14.503	17.113
Isopropyl alcohol	60.09	180.1	6.573	0.213	0.329	0.483	0.677	0.928	1.296	1.779
Methacrylonitrile	67.09	194.5	6.738	0.483	0.657	0.87	1.16	1.47	1.934	2.456
Methyl acetate	74.08	134.8	7.831	1.489	2.011	2.746	3.693	4.699	5.762	6.961
Methyl acrylate	86.09	176.9	7.996	0.599	0.773	1.025	1.354	1.798	2.398	3.055
Methyl alcohol	32.04	148.4	6.63	0.735	1.006	1.412	1.953	2.61	3.461	4.525
Methylcyclohexane	98.18	213.7	6.441	0.309	0.425	0.541	0.735	0.986	1.315	1.721
Methylcyclopentane	84.16	161.3	6.274	0.909	1.16	1.644	2.224	2.862	3.616	4.544
Methylene chloride	84.94	104.2	11.122	3.094	4.254	5.434	6.787	8.702	10.329	13.342
Methyl ethyl ketone	72.1	175.3	6.747	0.715	0.928	1.199	1.489	2.069	2.668	3.345
Methyl methacrylate	100.11	212	7.909	0.116	0.213	0.348	0.541	0.773	1.064	1.373
Methyl propyl ether	74.12	102.1	6.166	3.674	4.738	6.091	7.058	9.417	11.602	13.729
Nitromethane	61.04	214.2	9.538	0.213	0.251	0.348	0.503	0.715	1.006	1.334
n-Pentane	72.15	96.9	5.253	4.293	5.454	6.828	8.433	10.445	12.959	15.474
n-Propylamine	59.11	119.7	6.03	2.456	3.191	4.157	5.25	6.536	8.044	9.572
1,1,1-Trichloroethane	133.42	165.2	11.216	0.909	1.218	1.586	2.03	2.61	3.307	4.199
Trichloroethylene	131.4	188.6	12.272	0.503	0.677	0.889	1.18	1.508	2.03	2.61
Toluene	92.13	231.1	7.261	0.174	0.213	0.309	0.425	0.58	0.773	1.006
Vinyl acetate	86.09	162.5	7.817	0.735	0.986	1.296	1.721	2.262	3.113	4.022
Vinylidene chloride	96.5	89.1	10.383	4.99	6.344	7.93	9.806	11.799	15.28	23.21

Table C-5   Vapor Pressure Equation Constants for Organic Liquids.

| Name | Vapor Pressure Equation Constants | | |
	A (dimensionless)	B (°C)	C (°C)
Acetaldehyde	8.005	1600.017	291.809
Acetic acid	7.387	1533.313	222.309
Acetic anhydride	7.149	1444.718	199.817
Acetone	7.117	1210.595	229.664
Acetonitrile	7.119	1314.4	230
Acrylamide	11.2932	3939.877	273.16
Acrylic acid	5.652	648.629	154.683
Acrylonitrile	7.038	1232.53	222.47
Aniline	7.32	1731.515	206.049
Benzene	6.905	1211.033	220.79
Butanol (iso)	7.4743	1314.19	186.55
Butanol-(1)	7.4768	1362.39	178.77
Carbon disulfide	6.942	1169.11	241.59
Carbon tetrachloride	6.934	1242.43	230
Chlorobenzene	6.978	1431.05	217.55
Chloroform	6.493	929.44	196.03
Chloroprene	6.161	783.45	179.7
Cresol (-M)	7.508	1856.36	199.07
Cresol (-O)	6.911	1435.5	165.16
Cresol (-P)	7.035	1511.08	161.85
Cumene (isopropylbenzene)	6.963	1460.793	207.78
Cyclohexane	6.841	1201.53	222.65
Cyclohexanol	6.255	912.87	109.13
Cyclohexanone	7.8492	2137.192	273.16
Dichloroethane (1,2)	7.025	1272.3	222.9
Dichloroethylene (1,2)	6.965	1141.9	231.9
Diethyl (N,N) anilin	7.466	1993.57	218.5
Dimethyl formamide	6.928	1400.87	196.43
Dimethyl hydrazine (1,1)	7.408	1305.91	225.53
Dimethyl phthalate	4.522	700.31	51.42
Dinitrobenzene	4.337	229.2	−137
Dioxane (1,4)	7.431	1554.68	240.34
Epichlorohydrin	8.2294	2086.816	273.16
Ethanol	8.321	1718.21	237.52
Ethanolamine (mono-)	7.456	1577.67	173.37
Ethyl acrylate	7.9645	1897.011	273.16
Ethyl chloride	6.986	1030.01	238.61
Ethylacetate	7.101	1244.95	217.88
Ethylbenzene	6.975	1424.255	213.21
Ethylether	6.92	1064.07	228.8
Formic acid	7.581	1699.2	260.7
Furan	6.975	1060.87	227.74
Furfural	6.575	1198.7	162.8
Heptane (iso)	6.8994	1331.53	212.41
Hexane (-N)	6.876	1171.17	224.41
Hydrocyanic acid	7.528	1329.5	260.4

(continued)

Table C-5 Vapor Pressure Equation Constants for Organic Liquids. (*continued*)

Name	Vapor Pressure Equation Constants		
	A *(dimensionless)*	*B* *(°C)*	*C* *(°C)*
Methanol	7.897	1474.08	229.13
Methyl acetate	7.065	1157.63	219.73
Methyl ethyl ketone	6.9742	1209.6	216
Methyl isobutyl ketone	6.672	1168.4	191.9
Methyl metharcrylate	8.409	2050.5	274.4
Methyl styrene (alpha)	6.923	1486.88	202.4
Methylene chloride	7.409	1325.9	252.6
Morpholine	7.7181	1745.8	235
Naphthalene	7.01	1733.71	201.86
Nitrobenzene	7.115	1746.6	201.8
Pentachloroethane	6.74	1378	197
Phenol	7.133	1516.79	174.95
Picoline (-2)	7.032	1415.73	211.63
Propanol (iso)	8.117	1580.92	219.61
Propylene glycol	8.2082	2085.9	203.54
Propylene oxide	8.2768	1656.884	273.16
Pyridine	7.041	1373.8	214.98
Resorcinol	6.9243	1884.547	186.06
Styrene	7.14	1574.51	224.09
Tetrachloroethane (1,1,1,2)	6.898	1365.88	209.74
Tetrachloroethane (1,1,2,2)	6.631	1228.1	179.9
Tetrachloroethylene	6.98	1386.92	217.53
Tetrahydrofuran	6.995	1202.29	226.25
Toluene	6.954	1344.8	219.48
Trichloro (1,1,2) trifluoroethane	6.88	1099.9	227.5
Trichloroethane (1,1,1)	8.643	2136.6	302.8
Trichloroethane (1,1,2)	6.951	1314.41	209.2
Trichloroethylene	6.518	1018.6	192.7
Trichlorofluoromethane	6.884	1043.004	236.88
Trichloropropane (1,2,3)	6.903	788.2	243.23
Vinyl acetate	7.21	1296.13	226.66
Vinylidene chloride	6.972	1099.4	237.2
Xylene (-m)	7.009	1426.266	215.11
Xylene (-o)	6.998	1474.679	213.69

Table C-6   Paint Solar Absorbance for Fixed Roof Tanks.

Paint Color	Paint Shade or Type	Paint Condition, α Good	Paint Condition, α Poor
Aluminum	Specular	0.39	0.49
Aluminum	Diffuse	0.6	0.68
Gray	Light	0.54	0.63
Gray	Medium	0.68	0.74
Red	Primer	0.89	0.91
White	NA	0.17	0.34

Table C-7   Meteorological Data (T_{AX}, T_{AN}, I) for Selected U.S. Locations.

Location	Annual Average T_{AX} (°F)	Annual Average T_{AN} (°F)	Annual Average I (Btu/ft² day)	Wind, V (mph)
Birmingham, AL	73.2	51.1	1345	7.2
Montgomery, AL	75.9	53.9	1388	6.6
Homer, AK	43.6	29.5	838	7.6
Phoenix, AZ	85.1	57.3	1869	6.3
Fort Smith, AR	72.5	49	1404	7.6
Little Rock, AR	72.9	50.8	1404	7.8
Bakersfield, CA	77.7	53.3	1749	6.4
Long Beach, CA	74.2	53.5	1598	6.4
Los Angeles AP, CA	70.1	55	1594	6.2
Sacramento, CA	73.4	47.8	1643	7.9
San Francisco AP, CA	64.9	48.3	1608	8.7
Santa Maria, CA	68.3	45.3	1608	7
Denver, CO	64.3	36.2	1568	8.7
Grand Junction, CO	65.7	39.6	1659	8.1
Wilmington, DE	63.5	44.5	1208	9.1
Atlanta, GA	71.3	51.1	1345	9.1
Savannah, GA	76.7	55.1	1365	7.9
Honolulu, HI	84.2	69.7	1639	11.4
Chicago, IL	58.7	39.7	1215	10.3
Springield, IL	62.6	42.5	1302	11.2
Indianapolis, IN	62	42.2	1165	9.6
Wichita, KS	67.6	45.1	1502	12.3
Louisville, KY	66.1	46.2	1216	8.4
Baton Rouge, LA	78	57	1379	7.6
Lake Charles, LA	77.6	58.3	1365	8.7
New Orleans, LA	77.7	58.7	1437	8.2
Detroit, MI	58.2	38.9	1120	10.2
Grand Rapids, MI	57.2	37.7	1135	9.8
Minneapolis-St. Paul, MN	54.2	35.2	1170	10.6
Jackson, MS	76.3	52.9	1409	7.4
Billings, MT	57.9	35.4	1325	11.2
Las Vegas, NV	79.6	52.8	1864	9.3
Newark, NJ	62.5	45.9	1165	10.2
Roswell, NM	75.3	47.5	1810	8.6

(continued)

Table C-7 Meteorological Data (T_{AX}, T_{AN}, I) for Selected U.S. Locations. (*continued*)

Location	Annual Average			
	T_{AX} (°F)	T_{AN} (°F)	I (Btu/ft² day)	Wind, V (m/s)
Buffalo, NY	55.8	39.3	1034	12
New York, NY	61	47.5	1171	9.4
Cleveland, OH	58.5	40.7	1091	10.6
Columbus, OH	61.5	41.8	1123	8.5
Toledo, OH	58.8	38.3	1133	9.4
Oklahoma City, OK	71.2	48.6	1461	12.4
Tulsa, OK	71.3	49.2	1373	10.3
Astoria, OR	58.1	43.1	1000	12.4
Portland, OR	62	44	1067	7.9
Philadelphia, PA	63.4	45.1	1169	9.5
Pittsburgh, PA	59.9	40.7	1069	9.1
Providence, RI	59.3	41.2	1112	10.6
Columbia, SC	75.3	51.2	1380	6.9
Sioux Falls, SD	56.7	33.9	1290	11.1
Memphis, TN	71.6	51.9	1366	8.9
Amarillo, TX	70.7	43.8	1659	13.6
Corpus Christi, TX	81.6	62.5	1521	12
Houston, TX	79.1	57.4	1351	7.9
Midland-Odessa, TX	77	49.9	1802	11.1
Salt Lake City, UT	64	39.3	1603	8.9
Richmond, VA	68.8	46.5	1248	7.7
Seattle, WA	58.9	43.9	1053	9
Charleston, WV	65.5	44	1123	6.4
Huntington, WV	65.3	45	1176	6.6
Cheyenne, WY	58.3	33.1	1491	13

Table C-8 Rim-Seal Loss Factors, K_{Ra}, K_{Rb}, and n, for Floating Roof Tanks.

Tank Construction and Rim-Seal System		Average-Fitting Seals		
		K_{Ra} (lb-mole/ft-yr.)	K_{Rb} (lb-mole/[mph]ⁿ-ft-yr.)	n (dimensionless)
Welded Tanks				
Mechanical-shoe seal	Primary only	5.8	0.3	2.1
	Shoe-mounted secondary	1.6	0.3	1.6
	Rim-mounted secondary	0.6	0.4	1.0
Liquid-mounted resilient-filled seal	Primary only	1.6	0.3	1.5
	Weather shield	0.7	0.3	1.2
	Rim-mounted secondary	0.3	0.6	0.3
Vapor-mounted resilient-filled seal	Primary only	6.7	0.2	3.0
	Weather shield	3.3	0.1	3.0
	Rim-mounted secondary	2.2	0.003	4.3
Riveted Tanks				
Mechanical-shoe seal	Primary only	10.8	0.4	2.0
	Shoe-mounted secondary	9.2	0.2	1.9
	Rim-mounted secondary	1.1	0.3	1.5

Table C-9   Average Clingage Factors, C.

Product Stored	Shell Condition		
	Light Rust	*Dense Rust*	*Gunite Lining*
Gasoline	0.0015	0.0075	0.15
Single-component stocks	0.0015	0.0075	0.15
Crude oil	0.006	0.03	0.6

Table C-10   Deck-Fitting Loss Factors, K_{Fa}, K_{Fb}, and m, and Typical Number of Fittings, N_F.

Fitting Type and Construction Details	Loss Factors[a]			
	K_{Fa} (lb-mole/yr)	K_{Fb} (lb-mole/ (mph)m-yr)	*m*	*Typical Number of Fittings, N_F*
Access hatch (24-inch diameter well)				1
Bolted cover, gasketed[b]	1.6	0	0	
Unbolted cover, ungasketed	36	5.9	1.2	
Unbolted cover, gasketed	31	5.2	1.3	
Unslotted guidepole well (8-inch diameter unslotted pole, 21-inch diameter well)				1
Ungasketed sliding cover	31	150	1.4	
Gasketed sliding cover	25	13	2.2	
Slotted guide-pole/sample well (8-inch diameter slotted pole, 21-inch diameter well)				c
Ungasketed sliding cover, without float	43	270	1.4	
Ungasketed sliding cover, with float	31	36	2.0	
Gasketed sliding cover, without float	43	270	1.4	
Gasketed sliding cover, with float	31	36	2.0	
Gauge-float well (20-inch diameter)				1
Unbolted cover, ungasketed	14	5.4	1.1[b]	
Unbolted cover, gasketed	4.3	17	0.38	
Bolted cover, gasketed	2.8	0	0	
Gauge-hatch/sample port				1
Weighted mechanical actuation, gasketed	0.47	0.02	0.97	
Weighted mechanical actuation, ungasketed	2.3	0	0	
Vacuum breaker				depends on tank diameter

Table C-10   Deck-Fitting Loss Factors, K_{Fa}, K_{Fb}, and m, and Typical Number of Fittings, N_F. (*continued*)

Fitting Type and Construction Details	Loss Factors			Typical Number of Fittings, N_F
	K_{Fa} *(lb-mole/yr)*	K_{Fb} *(lb-mole/ (mph)m-yr)*	*m*	
Weighted mechanical actuation, gasketed	6.2			
Weighted mechanical actuation, ungasketed	7.8	0.01	4.0	
Deck drain (3-inch diameter)				depends on tank diameter
Open	1.5	0.21	1.7	
90% closed	1.8	0.14	1.1	
Deck leg (3-inch diameter)				depends on tank diameter
Adjustable, pontoon area, gasketed	1.3	0.08	0.65	
Adjustable, center area, gasketed	0.53	0.11	0.13	
Adjustable, internal floating deck	7.9			
Adjustable, pontoon area, ungasketed	2.0	0.37	0.91	
Adjustable, center area, ungasketed	0.82	0.53	0.14	
Adjustable, double-deck roofs	0.82	0.53	0.14	
Fixed	0	0	0	
Rim vent (6-inch diameter)[c]				1
Weighted mechanical actuation, gasketed	0.71	0.1	1.0	
Weighted mechanical actuation, ungasketed	0.68	1.8	1	

[a] The roof-fitting loss factors, K_{Fa}, K_{Fb}, and m, may be used only for wind speeds from 2 to 15 miles per hour.
[b] If no specific information is available, this value can be assumed to represent the most common or typical roof fitting currently in use.
[c] Rim vents are used only with mechanical-shoe primary seals.

Example C-1   Emission Rate from a Fixed-Roof Storage Tank

Determine the annual emission rate of a mixture containing 50/50 wt. % of toluene and ethyl acetate from a vertical coned-roof tank in Chicago, Illinois. The tank is 10.8 ft in diameter, 9.8 ft high, it holds about 80 % of the total volume, and is painted white (good condition). The tank working volume is 5,425 gallons. The number of turnovers per year for the tank is 183 (i. e., the throughput of the tank is 99,278 gal/yr).

Solution:

1. Determine tank type. The tank is a fixed-cone roof, vertical tank.
2. Determine estimating methodology. The product is made up of 2 organic liquids, both of which are well mixed and miscible in each other.
3. Select equations to be used. For a vertical, fixed roof storage tank, the following equations apply:

$$L_T = L_S + L_W \tag{1}$$
$$L_S = 365\ W_V V_V K_E K_S \tag{2}$$
$$L_W = 0.0010\ M_V P_{VA} Q K_N K_P \tag{3}$$

4. Calculate each component of the standing loss and the working loss functions.
   a. Tank vapor space volume, V_V.
      Use Equation 4 from Table C-2. We need to know the vapor space outage, H_{VO}. This can be determined using Equations 6 and 8 from Table C-1. The tank shell height (H_S) is given as 9.8 ft, the stock liquid height (H_L) is stated as 80% of H_S, and the tank roof slope is assumed to be the value shown in notes to Equation 8 of Table C-2. Using these equations, we find

$$V_V = p/4\ (10.8)^2\ (2.113) = 193.65\ ft^3$$

   b. Vapor density, W_V.
      Vapor density is calculated using Equation 11 of Table C-1. We first need to calculate the average liquid surface temperature (T_{LA}) using Equations 13, 14, and 15. Using these equations, we find the value of T_{LA}

$$T_{LA} = (0.44)\ (508.9\ R) + 0.56\ (508.9R) + 0.0079\ (0.17)\ (1215) = 510.5\ R$$

   Next the value of P_{VA}, the average vapor pressure at temperature T_{LA} using Equation 12 of Table C-1. In order to calculate the mixture vapor pressure, the partial pressures need to be calculated for each component. The partial pressure is the product of the pure vapor pressures of each component (calculated above) and the mole fractions of each component in the liquid.
      The mole fractions of each component are calculated as follows:

Component	Amount	Mw	Lb. Moles	Mole fraction	Pure vapor pressure (psia)	Partial vapor pressure (psia)
Toluene	20482 lb	92.13	222.31	0.489	0.247	0.1208
Ethyl Acetate	20482 lb	88.10	232.48	0.511	0.861	0.4400
Total	40964 lb		454.79	1.0000		0.5608

   The vapor pressure of the mixture (P_{VA}) is then 0.56 psia.
      Third, calculate the molecular weight of the vapor, M_V. Molecular weight of the vapor depends upon the mole fractions of the components in the vapor.

$$M_V = S\ M_i y_i$$

where

   M_i = molecular weight of the component
   y_i = vapor mole fraction

The vapor mole fractions, yi, are equal to the partial pressure of the component divided by the total vapor pressure of the mixture. Therefore,

$$y_{toluene} = Ppartial/Ptotal = 0.1208/0.5608 = 0.2154$$
$$y_{ethyl\ acetate} = 0.4400/0.5608 = 0.7846$$

The mole fractions of the vapor components sum to 1.0.

The molecular weight of the vapor can be calculated as follows:

Component	Mw	Partial pressure	Partial Mv
Toluene	92.13	0.2154	19.845
Ethyl Acetate	88.10	0.7846	69.123
Total			88.968

Since all variables have now been solved, the vapor density, W_V, can be calculated:

$$W_V = (88.968 * 0.5608)/(10.731 * 510.5) = 0.00911\ lb/ft^3$$

c. Vapor space expansion factor, K_E.

The vapor space expansion factor is calculated using Equation 16 of Table C-1. We must use equations 17, 18, 19, 20, and 21. The calculation of mixture vapor pressures in equation 19 is similar to the P_{VA} calculation presented immediately above. The resulting vapor space expansion factor is

$$K_E = 19.5/510.5 + (0.168 - 0.06)/(14.7 - 0.56) = 0.0458$$

d. Vented vapor space saturation factor, K_S.

Using previously calculated parameters, Equation 23 provides the value for K_S.

$$K_S = (1 + 0.53 P_{VA} H_{VO})^{-1}$$

5. Calculate the standing storage losses

$$L_S = 365\ W_V V_V K_E K_S$$

Using the values calculated above:

$$W_V = 0.00911\ lb/ft^3$$
$$V_V = 193.65\ ft^3$$
$$K_E = 0.0458$$
$$K_S = 0.941$$
$$L_S = 365\ (0.00911)(193.65)(0.0458)(0.941) = 27.75\ lb/yr$$

6. Calculate working losses.

The amount of VOCs emitted as a result of filling and emptying operations can be calculated from Equation 3 using Equation 24 and using the stated turnover number for the tank, N.

$$L_W = (0.0010)(88.97)(0.56)(2363)(1)(0.3306) = 38.92\ lb/yr$$

7. Calculate total losses, L_T.

$$L_T = L_S + L_W$$

where

$L_S = 27.75 \text{ lb/yr}$
$L_W = 38.92 \text{ lb/yr}$
$L_T = 27.75 + 38.92 = 66.67 \text{ lb/yr}$

Tables of Environmental Impact Potentials

Table D-1   Global Warming Potentials for Greenhouse Gases (CO_2 is the benchmark).

Chemical	Formula	τ (yrs)	BI (atm^{-1} cm^{-2})	GWP[a]
Carbon dioxide	CO_2	120.0		1
Methane	CH_4			21
NOx				40
Nitrous oxide	N_2O			310
Dichloromethane	CH_2Cl_2	0.5	1604	9
Trichloromethane	$CHCl_3$			25
Tetrachloromethane	CCl_4	47.0	1195	1300
1,1,1-trichloroethane	CH_3CCl_3	6.1	1209	100
CFC (hard)				7100
CFC (soft)				1600
CFC-11	CCl_3F	60.0	2389	3400
CFC-12	CCl_2F_2	120.0	3240	7100
CFC-13	$CClF_3$			13000
CFC-113	CCl_2FCClF_2	90.0	3401	4500
CFC-114	$CClF_2CClF_2$	200.0	4141	7000
CFC-115	CF_3CClF_2	400.0	4678	7000
HALON-1211	$CBrClF_2$			4900
HALON-1301	$CBrF_3$			4900
HCFC-22	CF_2HCl	15.0	2554	1600
HCFC-123	$C_2F_3HCl_2$	1.7	2552	90
HCFC-124	C_2F_4HCl	6.9	4043	440
HCFC-141b	$C_2FH_3Cl_2$	10.8	1732	580
HCFC-142b	$C_2F_2H_3Cl$	19.1	2577	1800
HFC-125	C_2HF_5			3400
HFC-134a	CH_2FCF_3			1200
HFC-143a	CF_3CH_3			3800
HFC-152a	$C_2H_4F_2$			150
Perfluoromethane	CF_4			6500
Perfluoroethane	CF_6			9200
Perfluoropropane	C_3F_8			7000
Perfluorobutane	C_4F_{10}			7000
Perfluoropentane	C_5F_{12}			7500
Perfluorohexane	C_6H_{14}			7400
Perfluorocyclobutane	c-C_4F_8			8700
Sulfur hexafluoride	SF_6			23900

Adapted from 1995 IPCC Report (IPCC, 1996 and 1994).
[a] (100 year time horizon).
τ is the tropospheric reaction lifetime (hydroxyl radical reaction dependent) (WMO, 1990a - 1992b)
BI is the infrared absorbence band intensity (Pouchert, 1989)

Table D-2 Ozone-Depletion Potentials for Several Industrially Important Compounds.

Chemical	Formula	τ (yrs)	k (cm^3 molecule^{-1} s^{-1})	X	ODP
Methyl bromide	CH$_3$Br				0.6
Tetrachloromethane	CCl$_4$	47.0	3.1×10^{-10}	4	1.08
1,1,1-trichloroethane	CH$_3$CCl$_3$	6.1	3.2×10^{-10}	3	.12
CFC (hard)					1.0
CFC (soft)					.055
CFC-11	CCl$_3$F	60.0	2.3×10^{-10}	3	1.0
CFC-12	CCl$_2$F$_2$	120.0	1.5×10^{-10}	2	1.0
CFC-13	CClF$_3$				1.0
CFC-113	CCl$_2$FCClF$_2$	90.0	2.0×10^{-10}	3	1.07
CFC-114	CClF$_2$CClF$_2$	200.0	1.6×10^{-10}	2	0.8
CFC-115	CF$_3$CClF$_2$	400.0			0.5
HALON-1201	CHBrF$_2$				1.4
HALON-1202	CBr$_2$F$_2$				1.25
HALON-1211	CBrClF$_2$				4.0
HALON-1301	CBrF$_3$				16.0
HALON-2311	CHClBrCF$_3$				0.14
HALON-2401	CHBrFCF$_3$				0.25
HALON-2402	CBrF$_2$ CBrF$_2$				7.0
HCFC-22	CF$_2$HCl	15.0	1.0×10^{-10}	1	.055
HCFC-123	C$_2$F$_3$HCl$_2$	1.7	2.5×10^{-10}	2	.02
HCFC-124	C$_2$F$_4$HCl	6.9	1.0×10^{-10}	1	.022
HCFC-141b	C$_2$FH$_3$Cl$_2$	10.8	1.5×10^{-10}	2	.11
HCFC-142b	C$_2$F$_2$H$_3$Cl	19.1	1.4×10^{-10}	1	.065
HCFC-225ca	C$_3$HF$_5$Cl$_2$				.025
HCFC-225cb	C$_3$HF$_5$Cl$_2$				.033

τ is the tropospheric reaction lifetime (hydroxyl radical reaction dependent) (WMO, 1990a–1992b).
k is the reaction rate constant with atomic oxygen at 298 K (release of chlorine in the stratosphere).
X is the number of chlorine atoms in the molecule.

Table D-3 Acid Rain Potential for a Number of Acidifying Chemicals.

Compound	Reaction	α	MW$_i$ (mol/kg)	η_i, (mol H$^+$/ kg "i")	ARP$_i$
SO$_2$	SO$_2$ + H$_2$O + O$_3$ → 2H$^+$ + SO$_4^{2-}$ + O$_2$	2	.064	31.25	1.00
NO	NO + O$_3$ +1/2 H$_2$O → H$^+$ + NO$_3^-$ + 3/4 O$_2$	1	.030	33.33	1.07
NO$_2$	NO$_2$ + 1/2 H$_2$O + 1/4 O$_2$ → H$^+$ + NO$_3^-$	1	.046	21.74	0.70
NH$_3$	NH$_3$ + 2 O$_2$ → H$^+$ + NO$_3^-$ + H$_2$O	1	.017	58.82	1.88
HCl	HCl → H$^+$ + Cl$^-$	1	.0365	27.40	0.88
HF	HF → H$^+$ + F$^-$	1	.020	50.00	1.60

Adapted from Heijungs et al., 1992

Table D-4 Maximum Incremental Reactivities (MIR) for Smog Formation (O_3).

Alkanes	normal	MIR	branched	MIR
	methane	0.015	isobutane	1.21
	ethane	0.25	neopentane	0.37
	propane	0.48	iso-pentane	1.38
	n-butane	1.02	2,2-dimethylbutane	0.82
	n-pentane	1.04	2,3-dimethylbutane	1.07
	n-hexane	0.98	2-methylpentane	1.50
	n-heptane	0.81	3-methylpentane	1.50
	n-octane	0.60	2,2,3-trimethylbutane	1.32
	n-nonane	0.54	2,3-dimethylpentane	1.31
	n-decane	0.46	2,4-dimethylpentane	1.50
	n-undecane	0.42	3,3-dimethylpentane	0.71
	n-dodecane	0.38	2-methylhexane	1.08
	n-tridcane	0.35	3-methylhexane	1.40
	n-tetradecane	0.32	2,2,4-trimethylpentane	0.93
	Average	**0.55**	2,3,4-trimethylpentane	1.60
			2,3-dimethylhexane	1.31
	cyclic		2,4-dimethylhexane	1.50
	cyclopentane	2.40	2,5-dimethylhexane	1.60
	methylcyclopentane	2.80	2-methylheptane	0.96
	cyclohexane	1.28	3-methylheptane	0.99
	1,3-dimethylcyclohexane	2.50	4-methylheptane	1.20
	methylcyclohexane	1.80	2,4-dimethylheptane	1.33
	ethylcyclopentane	2.30	2,2,5-trimethylhexane	0.97
	ethylcyclohexane	1.90	4-ethylheptane	1.13
	1-ethyl-4-methylcyclohexane	2.30	3,4-propylheptane	1.01
	1,3-diethylcyclohexane	1.80	3,5-diethylheptane	1.33
	1,3-diethyl-5-methylcyclohexane	1.90	2,6-diethyloctane	1.23
	1,3,5-triethylcyclohexane	1.70	**Average**	**1.20**
	Average	**2.06**		
Alkenes	primary		secondary	
	ethene	7.40	isobutene	5.30
	propene	9.40	2-methyl-1-butene	4.90
	1-butene	8.90	trans-2-butene	10.00
	1-pentene	6.20	cis-2-butene	10.00
	3-methyl-1-butene	6.20	2-pentenes	8.80
	1-hexene	4.40	2-methyl-2-butene	6.40
	1-hepene	3.50	2-hexenes	6.70
	1-octene	2.70	2-heptenes	5.50
	1-nonene	2.20	3-octenes	5.30
	Average	**5.66**	3-nonenes	4.60
			Average	**6.75**
	others			
	1,3-butadiene	10.90		
	isoprene	9.10		
	cyclopentene	7.70		
	cyclohexene	5.70		
	α-pinene	3.30		
	β-pinene	4.40		
	Average	**6.85**		

Alcohols and Ethers

methanol	0.56
ethanol	1.34
n-propyl alcohol	2.30
isopropyl alcohol	0.54
n-butyl alcohol	2.70
isobutyl alcohol	1.90
t-butyl alcohol	0.42
dimethyl ether	0.77
methyl t-butyl ether	0.62
ethyl t-butyl ether	2.00
Average	**1.32**

Acetylenes

acetylene	0.50
methylacetylene	4.10
Average	**2.30**

Aromatics

benzene	0.42
toluene	2.70
ethylbenzene	2.70
n-propylbenzene	2.10
isopropylbenzene	2.20
s-butylbenzene	1.90
o-xylene	6.50
p-xylene	6.60
m-xylene	8.00
1,3,5-trimethylbenzene	10.10
1,2,3-trimethylbenzene	8.90
1,2,4-trimethylbenzene	8.80
tetralin	0.94
naphthalene	1.17
methylnaphthalenes	3.30
2,3-dimethylnaphthalene	5.10
styrene	2.20
Average	**4.34**

Aromatic Oxygenates

benzaldehyde	−0.57
phenol	1.12
alkyl phenols	2.30
Average	**0.95**

Aldehydes

formaldehyde	7.20
acetaldehyde	5.50
C3 aldehydes	6.50
glyoxal	2.20
methyl glyoxal	14.80
Average	**7.24**

Ketones

acetone	0.56
C4 ketones	1.18
Average	**0.87**

Others

Methyl nitrite	9.50

Base Reactive Organic Gas Mixture   **3.10**

Adapted from Carter (1994)

Procedures for Estimating Hidden (Tier II) Costs

Regulatory costs are often charged to overhead, and are therefore "hidden" when project costs are evaluated. Hidden personnel costs are often difficult to account for because environmental reporting and recordkeeping are frequently performed by staff who divide their time between many different processes and facilities. Nevertheless, it is possible to estimate the time required to meet notification, reporting, manifesting and other administrative tasks associated with environmental compliance. Tables E-1 to E-5 give methods for estimating the notification, reporting, recordkeeping, manifesting, and labeling costs required by the Resource Conservation and Recovery Act (RCRA) for hazardous waste generation. Methods for estimating the costs associated with compliance with other statutes are available from the EPA (U.S. EPA, 1989). Tables E-1 to E-5 show that the level of reporting and recordkeeping required depends on whether a facility is a large- or small-quantity-generator of hazardous waste, whether it exports hazardous waste, and on whether they are considered to be a treatment, storage, and disposal site. Costs are estimated by multiplying the frequency of occurrence of reporting and recordkeeping activities by their non-labor and labor costs per occurrence. These tables also show that associated costs depend on the skill and cost of the labor involved and on the frequency of violations. Other RCRA reporting requirements include those for monitoring and testing; planning, studies and modeling; training; inspections; preparedness and protective equipment; closure and post closure assurance; and insurance and special taxes.

The following example illustrates the use of the tables.

Example E-1   Estimating Regulatory Recordkeeping and Reporting Costs for RCRA.

Estimate the costs of RCRA reporting, recordkeeping, manifesting, and labeling for a small-quantity generator of hazardous waste. Assume that the facility generates five drums of hazardous waste per year and that this waste is sent off-site for disposal.

Pick-up of the waste occurs quarterly (4 times per year). To produce your estimate, use the upper value of the wage ranges, the midpoint of the ranges for time required, and the lowest value for frequency when a range of these variables is given in the tables.

Solution: To estimate reporting costs (Table E-2):

$$\frac{0.5 \text{ events}}{\text{yr}} \times \left(\frac{\$5}{\text{event}} + \frac{8 \text{ hr}}{\text{event}} \times \frac{\$100}{\text{hr}} \right) = \frac{\$400.00}{\text{year}}.$$

The facility is a RCRA waste generator and must fill out the RCRA Biennial Report every other year at a cost of $400/year.

The lowest frequency for filling out Small Quantity Generator (SQG) Exception Reports is zero, so the estimated cost of filling them out is zero.

Recordkeeping costs (Table E-3):
The facility is a RCRA waste generator and its recordkeeping costs are

$$\frac{5 \text{ events}}{\text{yr}} \times \left(\frac{\$1}{\text{event}} + \frac{0.25 \text{ hr}}{\text{event}} \times \frac{\$100}{\text{hr}} \right) = \frac{\$130.00}{\text{year}}.$$

Manifesting costs (Table E-4):
The facility sends its waste for disposal and the manifesting costs are

$$\frac{4 \text{ events}}{\text{yr}} \times \left(\frac{\$0.50}{\text{event}} + \frac{\frac{0.25 + 1}{2} \text{ hr}}{\text{event}} \times \frac{\$100}{\text{hr}} \right) = \frac{\$250.00}{\text{year}}.$$

Labeling costs (Table E-5):
An estimate of labeling costs for the facility is

$$\frac{4 \text{ events}}{\text{yr}} \times \left(\frac{\$20}{\text{event}} + \frac{0.75 \text{ hr}}{\text{event}} \times \frac{\$50}{\text{hr}} \right) = \frac{\$230.00}{\text{year}}.$$

The total estimated costs are

$$\frac{\$400}{\text{yr}} + \frac{\$130}{\text{yr}} + \frac{\$250}{\text{yr}} + \frac{\$230}{\text{yr}} = \frac{\$1010.00}{\text{yr}}.$$

Regulatory recordkeeping and reporting requirements are not confined to RCRA. There are other federal requirements, such as reporting in the Toxic Chemical Release Inventory (TRI), and state and local requirements can also be significant.

Table E-1 Estimation Methods for RCRA Notification (US EPA, 1989).

RCRA Category	Notification Provision and Citation for Relevant Legislation	Annual cost ($/yr) = Frequency of occurrence (number per year) Frequency of Notification (Occ/yr)	× Non-labor costs per occurrence Non-Labor Costs per Notification ($/Occ)	+ Labor costs per occurrence Time Required Per Notification (hrs)	Average Wage Rate of Person Completing Notification ($/hr)	Approximate Annual Cost ($/yr)*
Exporter of Hazardous Waste	Exportation of Hazardous Waste Notification 40CFR §262.53	1	2	2-3	25-100	50-300
Treatment Storage or Disposal Facility	RCRA Foreign Source Notification 40CFR §164.12(a), 40CFR §165.12(a)	0-5	1	2	25-100	0-1000
	RCRA Permit Confirmation 40CFR §164.12(b)	1-4	1	2	25-100	50-800
	Local Notification of Operations 40CFR §264.37 40CFR §265.37	1	3	40	25-100	100-4000
	Manifest Discrepancy Notification 40CFR §264.72 40CFR §265.72	0-12	1	2	25-100	0-20,000

*Costs given to one significant figure.

Table E-2 Estimation Methods for RCRA Reporting (US EPA, 1989).

$$\text{Annual cost (\$/yr)} = \text{Frequency of occurrence (number per year)} \times \text{Non-labor costs per occurrence} + \text{Labor costs per occurrence} \qquad \text{Equation 12.2}$$

RCRA Category	Reporting Provision and Citation for Relevant Legislation	Frequency of Reporting (Occ/yr)	Non-Labor Costs per Report ($/Occ)	Time Required Per Report (hrs)	Average Wage Rate of Person Completing Report ($/hr)	Approximate Annual Cost ($/yr)*
RCRA Waste Generator (Large or Small Quantity)	RCRA Biennial Report 40CFR §262.41	0.5	5	8	25-100	100-400
Large Quantity Generator (LQG)	LQG Exception Report 40CFR §262.42(a)	0.1-1.5	1	2	25-100	6-300
Small Quantity Generator (SQG)	SQG Exception Report 40CFR §262.42(b)	0-0.1	1	0.25	25-100	0-4
Waste Exporter	Primary Exporter's Annual and Exception Reports 40CFR §262.55 40CFR §262.56	1-2.5	2	2.5	25-100	60-600
Treatment, Storage, Disposal Facility (TSDF)	TSDF Biennial Report 40CFR §264.35 40CFR §265.75	0.5	5	8-40	25-100	100-2000
Treatment, Storage or Disposal Facility (TSDF)	Unmanifested Waste Report 40CFR §264.76 40CFR §264.75	0-1.25	1	1	25-100	0-1200
TSDR	Release, Fire, Explosion and Closure Reporting	2	2	5	25-100	250-1000

*Costs given to one significant figure.

Table E-3 Estimation Methods for RCRA Recordkeeping (US EPA, 1989).

RCRA Category	Recordkeeping Provision and Citation for Relevant Legislation	Annual cost ($ / yr) = Frequency of occurrence (number per year) Frequency of Recordkeeping (Occ/yr)	× Non-labor costs per occurrence Non-Labor Costs per Record ($/Occ)	+ Labor costs per occurrence Time Required Per Record (hrs)	Average Wage Rate of Person Completing Record ($/hr)	Approximate Annual Cost ($/yr)*
RCRA Waste Generator	Reports, Test Results, Waste Analysis Records 40CFR §262.40	5-100	1	0.25	10-100	10-2500
RCRA Waste Exporter	Exporters Records and Notification Records 40CFR §262.57	5	1	0.25	10-100	10-100
Hazardous Waste Transporter	Manifesting Records 40CFR §263.22	0-200	1	0.25	10-100	0-5000
Treatment, Storage or Disposal Facility	Operating Record 40CFR §264.73, 40CFR §165.73	250	1	0.25	10-100	600-6000

*Costs given to one significant figure.

Table E-4 Estimation Methods for RCRA Manifesting (US EPA, 1989).

RCRA Category	Manifesting Provision and Citation for Relevant Legislation	Frequency of Manifesting (Occ/yr)	Non-Labor Costs per Manifest ($/Occ)	Time Required Per Manifest (hrs)	Average Wage Rate of Person Completing Manifest ($/hr)	Approximate Annual Cost ($/yr)*
RCRA Waste Generator	Off-Site Transport Manifesting 40CFR §262, Subpart B	4-100	0.5	0.25-1	25-100	20-10,000
Treatment, Storage and Disposal Facility (TSDF)	Manifesting 40CFR §264.71 40CFR §265.71	4-500	0.5	0.25-1	25-100	20-50,000

*Costs given to one significant figure.

Table E-5 Estimation Methods for RCRA Labeling (US EPA, 1989).

RCRA Category	Labeling Provision and Citation for Relevant Legislation	Frequency of Labeling (Occ/yr)	Non-Labor Costs per Labeling ($/Occ)	Time Required Per Label (hrs)	Average Wage Rate of Person Completing Label ($/hr)	Approximate Annual Cost ($/yr)*
RCRA Waste Generator	Package Marking and Transportation Labeling 40CFR §262.31 40CFR §262.32 40CFR §262.33	4-500	20	0.75	15-50	100-30,000

*Costs given to one significant figure.

Additional Resources

This appendix is intended to provide a list of supplemental resources to Green Engineering that is readily available on the Internet. It includes resources mentioned in the text as well as support material for more information on related subjects. These resources are categorized by Web Resources, Online Databases, and Software.

WEB RESOURCES

US EPA Office of Pollution Prevention and Toxics
Green Engineering Program website

The goal of the Green Engineering Program is to incorporate risk related concepts into chemical processes and products designed by academia and industry. This program targets 4 major sectors: educators, software, industry, and outreach. This website contains information about the Green Engineering textbook and GE Educator's workshops. It also includes links to software such as Air CHIEF, ChemSTEER, ECOSAR, E-FAST, E-FRAT, EPI Suite™, SMILES, TANKS, and UCSS. More information on specific software can be found in the Software section of Appendix F. This website provides direct links to a majority of the resources listed in this appendix.

    http://www.epa.gov/oppt/greenengineering

US EPA Office of Pollution Prevention and Toxics
Exposure Assessment Tools and Models website

OPPT's Exposure, Assessment Tools and Models website contains several exposure assessment methods, databases, and predictive models to help in evaluating what happens to chemicals when they are used and released to the environment; and how workers, the general public, consumers, and the aquatic ecosystems may be exposed to chemicals. This site includes links to UCSS, ChemSTEER, EPI Suite^{TM}, and other software. EPI Suite^{TM} estimates physical/chemical properties as described in Chapter 5 of this textbook, environmental fate and transport, and includes estimation programs for LogKOW, KOC, Atmospheric Oxidation Potential, Henry's Law Constant, Water Solubility, Melting Point, Boiling Point, Vapor Pressure, Biodegradation, Bioconcentration Factor, Hydrolysis, Sewage Treatment Plant Removal, Fugacity Modeling, and Multimedia Modeling. ECOSAR, Dermwin, and SMILESCAS database are also included. More information specific software can be found in the Software section of Appendix F.

     http://www.epa.gov/oppt/exposure

US EPA Office of Pollution Prevention and Toxics
Green Chemistry Program website

The Green Chemistry Program supports fundamental research in the area of environmentally benign chemistry as well as a variety of educational activities, international activities, conferences and meetings, and tool development. The Green Chemistry Program website includes a link to the Green Chemistry Expert System (GCES). More information on this software can be found in the Software section of Appendix F.

     http://www.epa.gov/oppt/greenchemistry/

US EPA Office of Pollution Prevention and Toxics
Design for the Environment (DfE)

U.S. EPA's Design for the Environment (DfE) Program helps businesses incorporate environmental considerations into the design and redesign of products, processes, and technical management systems. The Design for the Environment website includes a link to Cleaner Technologies Substitutes Assessment (CTSA) Methodology & Resource Guide.

     http://www.epa.gov/dfe

US EPA Office of Pollution Prevention and Toxics
Persistent Bioaccumulative and Toxic (PBT) Chemical Program website
The P2 Assessment Framework

This site includes a link to the P2 Framework manual. This user-friendly manual provides details on how to use some of the models mentioned in this appendix including ECOSAR, OncoLogic, and EPI SUITE^{TM}. The importance of the output of each assessment methodology is discussed, and case studies illustrating how the tools can be used in combination for risk management and pollution prevention strategies are provided.

http://www.epa.gov/oppt/pbt/framwork.htm

US EPA Office of Pollution Prevention and Toxics
Environmental Accounting Project

This web site contains information about the EPA Environmental Accounting Project, including a collection of Environmental Accounting resources. It includes links to overviews, general guidance and software tools, case studies and bench-marking studies, and resources available from other organizations.

http://www.epa.gov/opptintr/acctg

US EPA Office of Air Quality Planning and Standards
Technology Transfer Network
Clearinghouse for Inventories and Emission Factors (CHIEF)

The CHIEF site provides access to the Emission Inventory Improvement Program document series, the emissions modeling clearinghouse, the PM2.5 Inventory Resource Center, and information on the National Emission Inventory (NEI) for criteria and toxic pollutants. This site includes a free downloadable version of the TANKS software, information about the Air CHIEF software, and links to AP-42 emission factors and L & E documents. FIRE, another EFIG (Emission Factor and Inventory Group) software product contains AP-42 emission factors and can be downloaded off the web page. More information on specific software is available in the Software section of Appendix F.

http://www.epa.gov/ttn/chief/index.html

US EPA Office of Air Quality Planning and Standards
Technology Transfer Network
Clearinghouse for Inventories and Emission Factors (CHIEF)
Locating and Estimating (L & E) Documents

This report series, titled Locating and Estimating Air Toxic Emissions from Sources of (source category or substance) characterizes the source categories for which emissions of a toxic substance have been identified. These volumes include general descriptions of the emitting processes, identifying potential release points and emission factors.

http://www.epa.gov/ttn/chief/le/

US EPA Environmental Information Office
TRI (Toxics Release Inventory)

This publicly accessible toxic chemical database known as the Toxics Release Inventory (TRI), contains information concerning waste management activities and the release of toxic chemicals by facilities that manufacture, process, or otherwise use said materials.
     http://www.epa.gov/tri/

US EPA Office of Research and Development (ORD)
Systems Analysis Branch (SAB)

The SAB's research programs in pollution prevention, life cycle assessment, and computer tooling develop and demonstrate cost-effective decision-making tools. Such tools integrate environmental solutions, life cycle concepts, value engineering, environmental engineering, economics, trade-offs and pollution prevention factors. ORD has CRADAs (Cooperative Research and Development Agreements) involving both WAR and PARIS II. This website includes information on WAR. More information on software programs can be found in the Software section of Appendix F.
     http://www.epa.gov/ORD/NRMRL/std/sab/index.html

US EPA Office of Research and Development
National Health and Environmental Effects Laboratory
SMILES Simplified Molecular Input Line Entry System Tutorial

SMILES notation is used in EPI Suite™ models. SMILES (Simplified Molecular Input Line Entry System) is a chemical notation that allows a user to represent a chemical structure in a way that can be used by the computer.
     http://www.epa.gov/medatwrk/databases/smiles.html

US EPA Enviroene

Enviroene provides a single repository for pollution prevention, compliance assurance, and enforcement information and databases. The Enviroene website includes a link to Solvent Substitution Data Systems (SSDS). More information on some of the databases included in Envioene can be found in the Online Database section of Appendix F.
     http://es.epa.gov

US EPA Terminology Reference System (TRS)

The EPA Terminology Reference System (TRS) has been created as a single resource of environmental terminology for the Agency by compiling collections of terms from EPA and other sources.
     http://www.epa.gov/trs/index.htm

American Institute of Chemical Engineers (AIChE)

AIChE is a professional association of more than 50,000 members that provides leadership in advancing the chemical engineering profession. The AIChE website includes links to the AIChE Total Cost Assessment Manual and DIPPR data.

http://www.aiche.org/

AIChE CWRT Total Cost Assessment Manual (TCA)

The Center for Waste Reduction Technologies (CWRT) provides access to technologies and management tools supporting sustainable growth, environmental stewardship, and Responsible Care®. The TCA provides an approach to analyze life cycle costs and benefits related to industrial environmental safety and health issues. These methods are utilized in Chapter 12.

http://www.aiche.org/cwrt/
http://www.aiche.org/cwrt/projects/cost.htm

AIChE Design Institute for Physical Property Data (DIPPR)®

DIPPR develops and evaluates physical and environmental property data. This website includes information on various DIPPR projects including Project 911, Environmental, Safety and Health (ESH) Data Compilation. The goal of DIPPR Project 911 is to develop a comprehensive, critically evaluated database of regulated chemical species that are important to the chemical process industry. Michigan Technological University is coordinating this project.

www.aiche.org/dippr/

Alternative Fluorocarbons Environmental Acceptability Study

The environmentally important properties of HCFCs and HFCs were reviewed by the Alternative Fluorocarbons Environmental Acceptability Study (AFEAS) in 1989. This site provides the results of this study.

http://www.afeas.org

American Conference of Governmental Industrial Hygienists (ACGIH)

This organization publishes workplace chemical exposure concentration limits, Threshold Limit Values (TLVs), which are voluntary, unlike the legally enforceable OSHA PELs. The documentation for the TLVs contains detailed information on the relevant toxicity and exposure concerns related to each chemical with an established TLV.

http://www.acgih.org/home.htm

ChemAlliance Regulatory Tour

ChemAlliance supplies regulatory information to the chemical process industry. This site provides an overview of environmental regulations and what each law requires.
http://www.chemalliance.org/RegTools/regtour/index.asp

MSDS Material Safety Data Sheets

There are many sources for MSDS information on the Internet. The two listed here provide a good starting point. The first site offers general information about MSDS and links to other sites that contain MSDS. The second site contains a searchable online database of MSDS.
http://www.ilpi.com/msds/
http://hazard.com/msds/

NIOSH National Institute for Occupational Safety and Health

This organization performs research for OSHA, the Occupational Safety and Health Administration. The website includes links to databases including the NIOSH Pocket Guide. This resource provides general safety and health information, some chemical properties, OSHA Permissible Exposure Limit concentrations, or PELs, and NIOSH Recommended Exposure Limit concentrations, or RELs.
http://www.cdc.gov/niosh

National Oceanic and Atmospheric Administration (NOAA), Climate Monitoring and Diagnostics Laboratory

The Climate Monitoring and Diagnostics Laboratory (CMDL) of the National Oceanic and Atmospheric Administration conducts research related to atmospheric constituents that are capable of forcing change in the climate of the Earth or that may deplete the ozone layer. CMDL monitors greenhouse gases, aerosols, ozone, ozone-depleting gases and solar and terrestrial radiation at global sites including four Baseline Observatories.
http://www.cmdl.noaa.gov

P2Workshop

The mission of P2Workshop is to facilitate the development and incorporation of pollution prevention instructional material in engineering curricula in higher education. This site serves as a repository for recently developed curricular materials, such as course syllabi, instruction modules, problem sets and solutions, process design and evaluation software, and other curricular components. It has links to other sources of P2 curriculum development information and provides an avenue for open discussion of P2 curriculum topics. Michigan Technological,

Arizona State University, and the National Science Foundation sponsor the P2Workshop.

http://www.p2workshop.org/

US Department of Labor
Occupational Safety and Health Administration (OSHA)

OSHA's mission is to prevent work-related injuries, illnesses, and deaths. This website includes OSHA facts and history, programs and services, information on state plans, and federal regulation citations. Regulatory citations include PELs (Permissible Exposure Limits), which are the legally enforceable workplace chemical exposure limits promulgated by OSHA.

http://www.osha.gov

ONLINE DATABASES

Chemical Health and Safety Data
National Toxicology Program (NTP)

The National Toxicology Program (NTP) has collected health and safety information on thousands of chemicals. This database includes CAS numbers, and primary synonyms for individual chemicals.

http://ntp-server.niehs.nih.gov/Main_Pages/Chem-HS.html

Coating Alternative Guide (CAGE)

Coatings Guide™ is a pollution prevention tool for paints and coatings users. The Coatings Guide™ contains several tools to help users identify low-volatile organic compound/hazardous air pollutant coatings that may serve as drop-in replacements for existing coating operations.

http://cage.rti.org/

The Cumulative Exposure Project—Toxicity Database

This database is a compilation of toxicity data from federal and state agencies for contaminants that were considered in EPA's Cumulative Exposure Project. The database is an extensive compilation of both quantitative data, including peer-reviewed benchmark values and values derived from the primary literature, and qualitative information on health effects (e.g. identification of a HAP as a potential human carcinogen). The database contains a user-friendly interface designed to provide quick and easy access to desired health effects information.

http://www.epa.gov/CumulativeExposure/resource/toxdata.htm

ECOTOX Database System
US EPA Office of Research and Development (ORD)
National Health and Environmental Effects Research Laboratory
(NHEERL)

The ECOTOXicology database is a source for locating single chemical toxicity data for aquatic life, terrestrial plants and wildlife. ECOTOX integrates three toxicology effects databases: AQUIRE (aquatic life), PHYTOTOX (terrestrial plants), and TERRETOX (terrestrial wildlife).

      http://www.epa.gov/ecotox/

Envirofacts
US EPA Office of Environmental Information

This website provides access to several EPA databases that provide you with information about environmental activities that may affect air, water, and land anywhere in the United States.

      http://www.epa.gov/enviro/html/first_time.html

Environmental Fate Database
Syracuse Research Corporation (SRC)

As co developer of EPI Suite™, Syracuse Research Corporation has data on many chemical properties relevant to environmental fate.

      http://esc-plaza.syrres.com/efdb.htm

Enviroene Solvent Substitution Data Systems
US EPA

This web page contains links to several databases, including the Integrated Solvent Substitution Data System (ISSDS), the Solvent Alternatives Guide (SAGE), and the Solvents Database (SOLV-DB).

      http://es.epa.gov/ssds/ssds.html

The Hazardous Substances Data Bank (HSDB®)

The Hazardous Substances Data Bank (HSDB®) from the National Library of Medicine is a toxicology data file that focuses on the toxicology of potentially hazardous chemicals. It is enhanced with information on human exposure, industrial hygiene, emergency handling procedures, environmental fate, regulatory requirements, and related areas.

      http://chem.sis.nlm.nih.gov/hsdb

Health Effects Notebook for Hazardous Air Pollutants
US EPA Office of Air and Radiation
Office of Air Quality Planning and Standards

The Health Effects Notebook for Hazardous Air Pollutants includes data such as LC50s, Threshold Limit Values and Permissible Exposure Levels (see Chapter 8) for approximately 200 chemicals.

http://www.epa.gov/ttn/atw/hapindex.html

IRIS Integrated Risk Information System
US EPA Office of Research and Development
National Center for Environmental Assessment

IRIS is a database of human health effects that may result from exposure to various substances found in the environment. Data, analysis, and uncertainty characterizations are provided for hundreds of common chemicals.

http://www.epa.gov/ngispgm3/iris/index.html

ISSDS Integrated Solvent Substitution Data System

ISSDS facilitates access to solvent alternative information from multiple data systems through a single, easy to use command structure. The data collections available under ISSDS include SAGE, HSSDS, CAGE, and Pollution Prevention Case Studies.

http://es.epa.gov/issds/

NIOSH Pocket Guide to Chemical Hazards.

The Pocket Guide includes safety information, some chemical properties, OSHA Permissible Exposure Limit concentrations, or PELs, and NIOSH Recommended Exposure Limit concentrations, or RELs.

http://www.cdc.gov/niosh/npg/pgdstart.html

SAGE Solvent Alternative Guide

SAGE is a comprehensive guide designed to provide pollution prevention information on solvent and process alternatives for parts cleaning and degreasing. SAGE can be found on Enviro\$en\$e Solvent Substitution Data System

http://es.epa.gov/ssds/ssds.html

Or on

http://clean.rti.org/.

SOLV-DB®
National Center for Manufacturing Sciences (NCMS)

The Solvent Database includes data on physical and chemical properties, environmental fate and regulation of solvents.
http://solvdb.ncms.org/solvdb.htm

TOXNET

TOXNET is a cluster of databases available from the National Library of Medicine on toxicology, hazardous chemicals, and related areas. Both IRIS and the HSDB are available through TOXNET.
http://toxnet.nlm.nih.gov/

TRIAGE Chemical Studies Database
US EPA Office of Pollution Prevention and Toxics

This is a searchable database of scientific studies on the health and environmental effects of chemicals.
http://www.epa.gov/docs/8e_triag/

SOFTWARE

Air CHIEF

The **Air C**learing**H**ouse for **I**nventories and **E**mission **F**actors CD-ROM gives users access to air emission data specific to estimating the types and quantities of pollutants that may be emitted from a wide variety of sources. Updated annually, Air CHIEF contains pages from EPA's most widely used documents as well as the US EPA's Emission Factor and Inventory Group's emission estimation tools. Information on Air CHIEF and how to order it in CD-ROM format is available under the Software and Tools link on the EPA's TTN CHIEF Page.
http://www.epa.gov/ttn/chief/index.html

ChemSTEER: A Software Tool for Screening Level Estimates
of Environmental Release and Worker Exposure

The **C**hemical **S**creening **T**ool for **E**xposures and **E**nvironmental **R**eleases is a personal computer-based software program that uses the U. S. Environmental Protection Agency (EPA) Office of Pollution Prevention and Toxics' most current workplace exposure and release assessment methods. The tool generates screening-level estimates of environmental releases of and worker exposures to a chemical manufactured and used in industrial and commercial workplaces. A draft version of ChemSTEER is available at:

http://www.epa.gov/oppt/exposure/docs/chemsteer.htm

ECOSAR Ecological Structure Activity Relationships

ECOSAR estimates the toxicity of chemicals used in industry and discharged into water. The program predicts the toxicity of industrial chemicals to aquatic organisms such as fish, invertebrates, and algae by using Structure Activity Relationships (SARs). The program estimates a chemical's acute (short-term) toxicity and, when available, chronic (long-term or delayed) toxicity.

http://www.epa.gov/oppt/newchems/21ecosar.htm

E-FAST Exposure & Fate Assessment Screening Tool

E-FAST provides screening-level estimates of the concentrations of chemicals released to air, surface water, landfills, and from consumer products. Estimates provided are potential inhalation, dermal and ingestion dose rates resulting from these releases. Modeled estimates of concentrations and doses are designed to reasonably overestimate exposures, for use in screening level assessment.

http://www.epa.gov/oppt/exposure/docs/efast.htm

EFRAT Environmental Fate and Risk Assessment Tool

EFRAT is a process design software tool to estimate environmental and health impacts of chemical process design options through a combination of screening-level fate and transport calculations and risk assessment indices. EFRAT is intended to be used for the evaluation of a chemical process flowsheet in conjunction with a commercial chemical process simulator (e.g. HYSYS). Information on EFRAT is available from the Green Engineering website and the following link.

http://es.epa.gov/ncerqa_abstracts/centers/cencitt/year3/process/shonn2.html

EPI Suite™

The EPI (estimation program interface) Suite™ is a Windows® based suite of physical/chemical property and environmental fate estimation models developed by the EPA's Office of Pollution Prevention and Toxics and Syracuse Research Corporation (SRC). EPI Suite™ uses a single input for a chemical's structure to run the following estimation models: KOWWIN™, AOPWIN™, HENRYWIN™, MPBPWIN™, BIOWIN™, PCKOCWIN™, WSKOWWIN™, BCFWIN™, HYDROWIN™, STPWIN™, WVOLWIN™, and LEV3EPI™. These models can be divided into two categories, models to estimate physical-chemical properties and models to estimate environmental fate. EPI Suite™ also includes the

SMILESCAS database, which allows the user to input either in SMILES (Simplified Molecular Input Line Entry System) notation, or using a CAS number.

These models within EPI Suite™ can be used to estimate environmental fate:

BCFWIN™   This program calculates the BioConcentration Factor and its logarithm from the log Kow. The methodology is analogous to that for WSKOWWIN. Both are based on log Kow and correction factors.

HENRYWIN™   Calculates the Henry's Law constant (air/water partition coefficient) using both the group contribution and the bond contribution methods.

KOWWIN™   Estimates the log octanol-water partition coefficient, log Kow, of chemicals using an atom/fragment contribution method.

MPBPWIN™   Melting point, boiling point, and vapor pressure of organic chemicals are estimated using a combination of techniques.

PCKOCWIN™   The ability of a chemical to sorb to soil and sediment, its soil adsorption coefficient (Koc), is estimated by this program.

WSKOWWIN™   Estimates an octanol-water partition coefficient using the algorithms in the KOWWIN™ program and estimates a chemical's water solubility from this value. This method uses correction factors to modify the water solubility estimate based on regression against log Kow.

These models can be used to estimate physical/chemical properties:

AOPWIN™   Estimates the gas-phase reaction rate for the reaction between the most prevalent atmospheric oxidant, hydroxyl radicals, and a chemical.

BIOWIN™   Estimates aerobic biodegradability of organic chemicals using 6 different models; two of these are the original Biodegradation Probability Program (BPP™).

HYDROWIN™   Acid- and base-catalyzed hydrolysis constants for specific organic classes are estimated by HYDROWINTM. A chemical's hydrolytic half-life under typical environmental conditions is also determined.

LEV3EPI™   This level III fugacity model predicts partitioning of chemicals between air, soil, sediment, and water under steady state conditions for a default model "environment"; various defaults can be changed by the user.

| STPWIN™ | Using several outputs from EPIWIN, this program predicts the removal of a chemical in a Sewage Treatment Plant; values are given for the total removal and three contributing processes (biodegradation, sorption to sludge, and stripping to air.) for a standard system and set of operating conditions. |
| WVOLWIN™ | Estimates the rate of volatilization of a chemical from rivers and lakes; calculates the half-life for these two processes from their rates. The model makes certain default assumptions—water body depth, wind velocity, etc. |

A downloadable version of EPI Suite™ is available on the Exposure Assessment Tools and Models website.

http://www.epa.gov/oppt/exposure/docs/episuite.htm

FIRE Factor Information Retrieval Data System

The Factor Information Retrieval Data System is a database management system containing EPA's recommended emission estimation factors for criteria and hazardous air pollutants. FIRE includes information about industry emissions and emission factors including AP-42. FIRE is available under the Software and Tools link on the EPA's TTN CHIEF Page.

http://www.epa.gov/ttn/chief/software/fire/index.html

Green Chemistry Expert System (GCES)

The Green Chemistry Expert System (GCES) allows users to assess existing processes, build a green chemical process, design a green chemical, or survey the field of green chemistry. The system is equally useful for new and existing chemicals and their synthetic processes. It includes extensive documentation. A free downloadable version of this software is available on the Green Chemistry website.

http://www.epa.gov/greenchemistry/tools.htm

Mackay Level III Version 2.20

This model gives a description of a chemical's fate in the environment including the important degradation and advection losses and the intermedia transport processes. Physical-chemical properties are used to quantify a chemical's behavior in an evaluative environment. This Level III simulation is based on the publication by Mackay, Donald (1991), "Multimedia Environmental Models: The Fugacity Approach," Lewis Publishers, CRC Press, Boca Raton, FL. Information on this program and a free downloadable version is available at this site.

http://www.trentu.ca/academic/aminss/envmodel/VBL3.html

OncoLogic®

OncoLogic® predicts the potential carcinogenicity of chemicals by applying the rules of structure-activity relationship (SAR) analysis, and incorporating what is known about mechanisms of action and human epidemiological studies. A free downloadable DOS version of the software is available.

http://www.logichem.com/index.html#onco

PARIS II Program for Assisting the Replacement of Industrial Solvents

PARIS II is a program incorporating solvent design using the WINDOWS operating system. The solvent design capability allows the user to match or to enhance desirable solvent properties while simultaneously suppressing undesirable ones such as, for example, toxicity. The composition is manipulated by a solvent search algorithm aided by a library of routines with the latest fluid property prediction techniques, and by another library of routines for calculating solvent performance requirements. The program contains a database of solvents, and lists of solvent properties and solvent performance requirements. PARIS II was developed within EPA/ORD and is being marketed through a CRADA (a Cooperative Research and Development Agreement) between EPA/ORD and Technical Database Services, Inc. A Demo version is available under the link for PARIS II on this website.

http://www.tds-tds.com/

TANKS

TANKS is a program that estimates emissions from the two major loss mechanisms, working and standing losses, for the four major types of storage tanks, fixed-roof, floating roof, variable-vapor-space, and pressurized tanks. TANKS is available under the Software and Tools link on the EPA's TTN CHIEF Page.

http://www.epa.gov/ttn/chief/software/tanks/index.html

UCSS

Use Clusters Scoring System identifies and screens clusters of chemicals ("use clusters") that are used to perform a particular task. A use cluster is a set of chemicals that may be substituted for one another in performing a given task. It also identifies clusters of potential concern and provides an initial ranking of chemicals using human and environmental hazard and exposure data from a number of sources.

http://www.epa.gov/oppt/exposure/docs/ucss.htm

WAR WAste Reduction Algorithm

The method is based on a potential environmental impact balance for chemical processes. Potential environmental impact index is defined as the environmental and human health impact of the waste produced per unit time or mass of a product. This index is used to provide a relative, quantitative measure of the impact of the generated waste within a chemical process; this measure is then compared to indexes from other possible designs for the same chemical process to arrive at an environmentally friendly design for the process. This program was developed within EPA/ORD. WAR software is being developed and marketed under a CRADA (a Cooperative Research and Development Agreement) between EPA/ORD and Horizon Technologies.

More information on this program is available at:
http://www.epa.gov/oppt/greenengineering/software.html and
http://www.epa.gov/ORD/NRMRL/std/sab/sim_war.htm

Index

Octanol-water partition coefficient (K_{ow})	Equilibrium ratio of the concentration of a compound in octanol to the concentration of the compound in water.	Characterizes the partitioning between hydrophilic and hydrophobic phases in the environment and the human body; frequently used as a correlating variable in estimating other properties.
Water solubility (S)	Equilibrium solubility in mol/L.	Characterizes the partitioning between hydrophilic and hydrophobic phases in the environment.
Soil sorption coefficient (K_{oc})	Equilibrium ratio of the mass of a compound adsorbed per unit weight of organic carbon in a soil (in $\mu g/g$ organic carbon) to the concentration of the compound in a liquid phase (in $\mu g/ml$).	Characterizes the partitioning between solid and liquid phases in soil which in turn determines mobility in soils; frequently estimated based on octanol-water partition coefficient, and water solubility.
Bioconcentration factor (BCF)	Ratio of a chemical's concentration in the tissue of an aquatic organism to its concentration in water (reported as L/kg).	Characterizes the magnification of concentrations through the food chain.